STUDENT'S GUIDE

DonnaJean Fredeen
Southern Connecticut State University

FIFTH EDITION

CHEMISTRY

McMURRY · FAY

Upper Saddle River, NJ 07458

Associate Editor: Jennifer Hart
Editor-in-Chief, Science: Nicole Folchetti
Senior Editor: Andrew Gilfillan
Senior Managing Editor, Science: Kathleen Schiaparelli
Assistant Managing Editor, Science: Gina M. Cheselka
Project Manager, Production: Ashley M. Booth
Supplement Cover Manager: Paul Gourhan
Supplement Cover Designer: Victoria Colotta
Senior Operations Supervisor: Alan Fischer
Director of Operations: Barbara Kittle

© 2008 Pearson Education, Inc.
Pearson Prentice Hall
Pearson Education, Inc.
Upper Saddle River, NJ 07458

All rights reserved. No part of this book may be reproduced, in any form or by any means, without permission in writing from the publisher.

Pearson Prentice Hall™ is a trademark of Pearson Education, Inc.

The author and publisher of this book have used their best efforts in preparing this book. These efforts include the development, research, and testing of the theories and programs to determine their effectiveness. The author and publisher make no warranty of any kind, expressed or implied, with regard to these programs or the documentation contained in this book. The author and publisher shall not be liable in any event for incidental or consequential damages in connection with, or arising out of, the furnishing, performance, or use of these programs.

> This work is protected by United States copyright laws and is provided solely for teaching courses and assessing student learning. Dissemination or sale of any part of this work (including on the World Wide Web) will destroy the integrity of the work and is not permitted. The work and materials from it should never be made available except by instructors using the accompanying text in their classes. All recipients of this work are expected to abide by these restrictions and to honor the intended pedagogical purposes and the needs of other instructors who rely on these materials.

Printed in the United States of America

10 9 8 7 6 5 4 3 2 1

ISBN 13: 978-0-13-199348-8

ISBN 10:   0-13-199348-8

Pearson Education Ltd., *London*
Pearson Education Australia Pty. Ltd., *Sydney*
Pearson Education Singapore, Pte. Ltd.
Pearson Education North Asia Ltd., *Hong Kong*
Pearson Education Canada, Inc., *Toronto*
Pearson Educación de Mexico, S.A. de C.V.
Pearson Education—Japan, *Tokyo*
Pearson Education Malaysia, Pte. Ltd.

Student Study Guide

to accompany

Chemistry, 5th Edition — McMurry/Fay

TABLE of CONTENTS

CHAPTER		PAGE
	Preface	v
1	CHEMISTRY: MATTER and MEASUREMENT	1
2	ATOMS, MOLECULES, and IONS	18
3	FORMULAS, EQUATIONS, and MOLES	40
4	REACTIONS IN AQUEOUS SOLUTIONS	69
5	PERIODICITY and ATOMIC STRUCTURE	92
6	IONIC BONDS and SOME MAIN-GROUP CHEMISTRY	108
7	COVALENT BONDS and MOLECULAR STRUCTURE	126
8	THERMOCHEMISTRY: CHEMICAL ENERGY	142
9	GASES: THEIR PROPERTIES and BEHAVIOR	158
10	LIQUIDS, SOLIDS, and CHANGES of STATE	171
11	SOLUTIONS and THEIR PROPERTIES	184
12	CHEMICAL KINETICS	200
13	CHEMICAL EQUILIBRIA	221
14	HYDROGEN, OXYGEN, and WATER	237
15	AQUEOUS EQUILIBRIA: ACIDS and BASES	260
16	APPLICATIONS of AQUEOUS EQUILIBRIA	283
17	THERMODYNAMICS: ENTROPY, FREE ENERGY, and EQUILIBRIUM	296
18	ELECTROCHEMISTRY	317

19	**THE MAIN-GROUP ELEMENTS**	329
20	**TRANSITION ELEMENTS and COORDINATION CHEMISTRY**	346
21	**METALS and SOLID-STATE MATERIALS**	358
22	**NUCLEAR CHEMISTRY**	368
23	**ORGANIC CHEMISTRY**	380
24	**BIOCHEMISTRY**	401

Appendices:

Solutions to Workbook Problems	*A–1*
Putting It Together Solutions	*B–1*
Self–Test Solutions	*C-1*
Inquiry Based Problems Solutions	*D-1*

PREFACE

This study guide was written specifically to assist you, the student, in using CHEMISTRY by McMurry & Fay and presents, in outline form, the major concepts, theories, facts and applications found in the text. The Study Guide also highlights the key concepts found in the *Key Concept Maps* in the textbook and links the key concepts to the chapter learning goals of the Study Guide. The objective of the *Workbook Problems* embedded in each chapter is to lead you toward independence and confidence in your understanding of chemistry. As a consequence, the complete solution to the problem is not provided within the chapter. However, the steps needed to solve a problem are provided along with room for you to work the problem below each step. (Solutions to all *Workbook Problems* can be found in the Appendix.) The *Putting It Together* problems reinforce concepts learned in previous chapters and demonstrate the cumulative nature of the Key Concepts. The *Challenge* problems, found at the end of the Self-Tests, also are multi-conceptual and provide you with the opportunity to simulate the problem-solving strategies utilized by scientists and engineers. *Inquiry Based Problems*, located at the end of some chapters illustrate the nature of applying the concepts and theories covered to the text in the design of laboratory experiments.

Every chapter is keyed to the main text, and is presented in six sections:
- Chapter Learning Goals – alerting you to key concepts that will be covered in each chapter and referenced in the Chapter Outline.
- Chapter in Brief – a brief summary of major material that will be covered in that chapter and placed in its context.
- Chapter Outline – highlighting the major concepts and theories in the text. The Learning Goals are clearly marked and expounded upon within these chapter outlines. The outlines provide you with an initial set of notes, which can then be tailored and annotated with lecture material. Included within the outlines are worked examples and Workbook Problems which are intended to lead you toward confidence and independence in applying the Key Concepts.
- Putting It Together – ties the concepts and topics of each preceding chapter to the next, demonstrating the cumulative nature of the topics.
- Self-Tests and Solutions - modeled on the text problems and linked to the Chapter Learning Goals. This section is intended to test your knowledge of the material covered in the chapters. The student who uses this section as a practice exam will discover which topics have been mastered and which topics need review. The *Challenge* problems are multi-conceptual, providing the opportunity to simulate the problem-solving strategies employed by scientist and engineers.
- Inquiry Based Problems – illustrate the nature of applying the concepts and theories covered to the text in the design of laboratory experiments.

My efforts in writing this Study Guide began nearly fifteen years ago when I first put together a *"Concepts and Definitions"* manual for my students. Since that time, the size and amount of work needed to create this book has dramatically increased. However, I still have only one goal in mind: to help you, the student, understand and appreciate the field of chemistry. It takes hard work. But if you take the time to read and work the problems in your text, use the outline in this study guide and work the problems you find here, you will not only begin to appreciate the role that chemistry plays in your life, you will also do well in your course. (You may even find that you like the subject!)

I would like to take this opportunity to acknowledge the following people: Sandi Hakanson whose belief in me from the very beginning made this all possible; Jennifer Hart, who kindly and patiently pushed me along; Robert Snyder, who accuracy checked the 4th edition; and the entire McMurry and Fay team who worked so diligently to put together the best chemistry package.

Finally, I would like to dedicate this Study Guide to two men who were instrumental in shaping me as a chemist; Mr. Jerry Watts and Mr. Sam Meador, my chemistry teachers at Permian High School in Odessa, Texas. Mr. Watts, it all began in your class watching a candle burn. Mr. Meador, with your encouragement, that flickering flame grew into a fire. If I can inspire one student in the way that you inspired me, I will truly be successful. Thank you!

DonnaJean A. Fredeen
Southern Connecticut State University

CHAPTER 1

CHEMISTRY: MATTER AND MEASUREMENT

Chapter Learning Goals

🔑 A. *Experimentation Involving Observations*
   1. Describe the processes involved in the scientific method.

🔑 B. *Matter: Understanding Composition*
   1. Give the symbol and name of elements mentioned in this chapter.
   2. Identify the group number and period to which an element belongs.
   3. Identify the regions of the periodic table.

🔑 C. *Matter: Understanding Properties*
   1. Identify properties as extensive or intensive and as chemical or physical.
   2. Perform calculations using density.

🔑 D. *Experimentation Involving Measurements*
   1. List the seven basic SI units of measure, and give the numerical equivalent of the common metric prefixes used with these units.
   2. Express numbers in scientific notation. (See Appendix A in your textbook.)
   3. Interconvert Fahrenheit, Celsius, and Kelvin temperatures.
   4. Determine the number of significant digits in a measured quantity.
   5. State the result of a calculation involving measured quantities to the correct number of significant digits.
   6. Use dimensional analysis to solve metric–English conversion problems.

Chapter in Brief

Chemistry is the study of the composition, properties, and transformations of matter and of chemical laws which are responsible for the changes that take place in nature. In this chapter, you will begin your study of chemistry by learning the names and symbols of the elements on the periodic table. You will also begin to see that the periodic table of the elements is the most important organizing principle of chemistry. To understand chemistry, you also must have an understanding of the measurements that we make. You will be introduced to the International System of Units seven base units, which along with other derived units, suffice for all scientific measurements. It is necessary to know the precision of the measurements you make, which requires the use of significant figures. You will learn how to determine the number of significant figures in a measurement and the rules for carrying significant figures through a calculation. Finally, you will be introduced to the method of dimensional analysis, and you will learn how to use this method to convert one unit into an equivalent unit.

1.1 Approaching Chemistry: Experimentation

🔑 *A.1.*

   A. Scientific Method — systematic approach to research.
      1. Questions and experimentation.
      2. Hypothesis — an interpretation that explains the results of many experiments.
      3. Theory — consistent explanation of known observations; logical interpretations of experimental results.
         a. Not laws of nature.
         b. Can never be proven.

Chapter 1 – Chemistry: Matter and Measurement

1.2 Chemistry and the Elements
A. Element — a fundamental substance that can't be chemically changed or broken down into anything simpler.
 1. 114 known elements.
  a. 90 – occur naturally.
  b. 24 – artificially produced.
B. Chemical symbol — used to represent specific elements.
 1. Capitalize the first letter; if second letter is present, use lower case.
 2. Derived from English or Latin name.
C. Periodic Table — a tabular organization of all 114 elements.

EXAMPLE:
Name the following elements: Sr, Sc, In, Mn, Sb

SOLUTION:
Using the table inside the front cover of the textbook, we find that the names for the symbols are:

Sr — strontium   Sc — scandium   In — indium
Mn — manganese   Sb — antimony

HINT: You may find it helpful to make flashcards with the name of the element on one side and the symbol on the other. Learning the names and symbols of the elements takes practice. Remember: Practice makes perfect!

1.3 Elements and the Periodic Table
A. Periodic Table – grid of the elements arranged in 7 horizontal rows and 18 vertical columns.
B. Periods – seven horizontal rows in the periodic table.
C. Groups — 18 vertical columns in the periodic table.
 1. Groups numbered 1A → 8A and 1B → 8B (or 1 →18).
 2. Actually have 32 groups.
  a. lanthanides (14 elements after lanthanum) and actinides (14 elements after actinium) not included in the group numbers.
D. The elements in a given group have similar chemical properties.

E. The Periodic table of the elements is the most important organizing principle of chemistry. (See inside front cover.)
 1. Regular progression in the size of the seven periods.
  a. reflects a similar regularity in atomic structure
 2. Main Group (or Representative) Elements — Groups 1A — 8A; (the two larger groups on the left and the six larger groups on the right of the table).
 3. Transition Metal Elements — Groups 1B — 8B; (the 10 smaller groups in the middle of the table).
 4. Inner Transition Metal (or Rare Earth) Elements (the 14 groups shown separately at the bottom of the table).

1.4 Some Chemical Properties of the Elements
A. Property — any characteristic that can be used to describe or identify matter.
 1. Physical properties — characteristics that can be determined without changing the chemical makeup of the sample.

2. Chemical properties — properties that do change the chemical makeup of the sample.
3. Intensive properties — properties that do not depend on the size of the sample.
4. Extensive properties — properties that depend on the size of the sample.
B. Groups of elements show similarities in chemical properties.
1. Group 7A — Halogens: corrosive, nonmetallic elements; salt formers.
2. Group 8A — Noble gases: gases with low reactivity.
3. Group 1A — Alkali metals: lustrous, silvery metals; react rapidly with water to form highly alkaline products.
4. Group 2A — Alkaline earth metals: lustrous, silvery metals; less reactive than alkali metals.
C. Three major classes of elements in the periodic table.
1. Metals — largest category of elements; found on the left side of the periodic table (left of the heavy zigzag line).
   a. solids (except mercury)
   b. malleable
   c. ductile — can be drawn into thin wires without breaking
   d. conduct heat and electricity
2. Nonmetals — found on the right side of the Periodic table (right of the heavy zigzag line).
   a. gases, liquids, or solids
   b. brightly colored
   c. brittle solids
   d. poor conductors of heat and electricity
3. Semimetals (metalloids) — elements adjacent to the zigzag boundary between metals and nonmetals.
   a. properties fall between metals and nonmetals
   b. brittle
   c. poor conductors of heat and electricity

EXAMPLE:
Determine the group number that corresponds to the following group of elements in the Periodic table: the alkali metals, the alkaline earth metals, and the halogens.

SOLUTION:
From the outline above and the discussion in your book, you learned that the alkali metals are the metals in group 1A, the alkaline earth metals are the metals in group 2A, and the halogens are the nonmetals in group 7A.

EXAMPLE:
Determine the class of the following elements: iridium, antimony, iodine

SOLUTION:
By determining the position of the element in the periodic table, you can determine the class of the element. (It also is necessary to know the symbol for each of the elements.)

Iridium – Ir; iridium is to the left of the zigzag line in the Periodic table and is, therefore, a metal. It belongs to the elements known as the transition metals.

Antimony – Sb; antimony is adjacent to the zigzag line in the Periodic table and is, therefore, a semimetal or metalloid.

Iodine – I; iodine is found to the right of the zigzag line in the Periodic table and is, therefore, a nonmetal. It belongs to the elements known as the halogens.

Chapter 1 – Chemistry: Matter and Measurement

1.5 Experimentation and Measurement
A. International System (SI) of Units — seven base units (and units derived from them) that suffice for all scientific measurements. (See Table 1.3 in textbook, page 10.)
B. Common prefixes used to modify SI units
1. mega (M); factor = 10^6
2. kilo (k); factor = 10^3
3. deci (d); factor = 10^{-1}
4. centi (c); factor = 10^{-2}
5. milli (m); factor = 10^{-3}
6. micro (μ); factor = 10^{-6}
7. nano (n); factor = 10^{-9}

C. Scientific Notation — an exponential format for numbers that are either very large or very small (see Appendix A in your textbook).
D. All measurements of physical quantities contain both a number and a unit label.

EXAMPLE:
Express the following numbers in scientific notation: 1,743,000,000 and 0.000 008 562

SOLUTION:
We will use the rules outlined in Appendix A in your textbook.

For the number 1,743,000,000 we need to move the decimal point nine places to the left so that we obtain the number 1.743, which is between 1 and 10. We now must multiply 1.743 by 10^9. Therefore, the answer is 1.743×10^9.

For the number 0.000 008 562, we need to move the decimal point six places to the right to obtain the number 8.562, which is between 1 and 10. Because we moved the decimal point to the right, we must multiply by 10^{-6}. Therefore, the answer is 8.562×10^{-6}.

EXAMPLE:
State the SI unit and abbreviation used for a) measuring mass, b) measuring temperature, c) measuring the amount of substance and d) measuring time.

SOLUTION:
From Table 1.3 on page 10 in your textbook, we find that the SI unit for a) mass is the kilogram (kg); b) temperature is the Kelvin (K); c) amount of substance is the mole (mol); and d) time is second (s).

1.6 Measuring Mass
A. Mass (SI unit = kg) — the amount of matter in an object.
B. Matter — term used to describe anything physically real.
C. Weight — the pull of gravity on an object by the earth or other celestial body.
D. The mass of an object can be measured on a balance by comparing the weight of the object to the weight of a reference standard of known mass.

1.7 Measuring Length
A. Meter — standard unit of length in the SI system.
1. Distance traveled by light in a vacuum in 1/299,792,458th of a second.

Chapter 1 – Chemistry: Matter and Measurement

1.8 Measuring Temperature   🔑 *D.3.*
- A. Common unit — Celsius degree (°C).
    1. Freezing point of water = 0°C
    2. Boiling point of water = 100°C
- B. Scientific unit — Kelvin.
    1. Coldest possible temperature.
    2. -273.15°C = 0 K
        a. Absolute zero.
- C. Celsius and Kelvin scales have 100 degrees between the freezing point and boiling point of water.

    Temperature in K = temperature in °C + 273.15

    Temperature in °C = temperature in K − 273.15

- D. Fahrenheit scale has 180° between the freezing point and boiling point of water.
    1. Freezing point of water = 32 °F.
    2. Boiling point of water = 212 °F.
- E. Conversion between Fahrenheit and Celsius scale.
    1. Adjust for change in degree size.
        a. 180°F encompasses the same range as 100°C.
            i. $1°C \times \dfrac{180°F}{100°C} = \dfrac{9}{5} °F$ (units can cancel out. This is a method known as *dimensional analysis*, which is discussed on pages 21-22 in your text and page 10 in this book.)
            ii. $1°F \times \dfrac{100°C}{180°F} = \dfrac{5}{9} °C$
    2. Adjust for change in zero point (the numerical difference in the freezing point of water).
        a. Add 32 when converting from °C to °F.
        b. Subtract 32 when converting from °F to °C.
- F. To convert from Celsius to Fahrenheit — do a size adjustment, followed by a zero-point adjustment.
    1. $°F = \left(\dfrac{9}{5} \times °C\right) + 32$
- G. To convert from Fahrenheit to Celsius — do a zero-point adjustment, followed by a size adjustment.
    1. $°C = \dfrac{5}{9} \times \left(°F - 32\right)$

EXAMPLE:
Titanium, an important structural metal used in aircraft, has a melting point of 1725°C. Express this temperature in °F and Kelvin scale.

SOLUTION:
This is a good time to use the "thinking approach." A Fahrenheit degree is smaller than a Celsius degree $\left(1°F = \dfrac{5}{9} °C\right)$. Just think about the freezing point of water. Water freezes at 0° on the Celsius scale and at 32° on the Fahrenheit scale. Therefore, when converting from Celsius to Fahrenheit, the temperature should be higher.

Chapter 1 – Chemistry: Matter and Measurement

Because we are converting from °C to °F, we should first do a size correction (9/5 × °C), followed by a zero-point correction (+ 32). Now, apply the formula given above and see if your answer compares well with your thought process.

$$°F = \left(\frac{9}{5} \times 1725\right) + 32 = 3137°F$$

Your calculated value for the temperature on the Fahrenheit scale agrees well with the thinking approach we used. The same logic can be used in calculating the value of the melting point on the Kelvin scale. The freezing point of water is 273.15° higher on the Kelvin scale, so the melting point of titanium also should be 273.15° higher.

$$K = 1725 + 273.15 = 1998.15 \text{ K}$$

 Workbook Problem 1.1

A baby wakes up in the middle of the night with a temperature of 102.7° F. Calculate the baby's temperature in degrees Celsius and in Kelvin.

Strategy: Consider the differences in the degree size and zero point adjustments for the Fahrenheit and Celsius scales.

Step 1: Based on the discussion of the Fahrenheit scale and the Celsius scale, determine if °C should be higher or lower than °F.

Step 2: Apply the formula for the conversion of °F to °C. (Remember, for this conversion, you do a zero-point correction followed by a size correction.)

Step 3: Does your answer in step 2 agree with your answer in step 1?

Step 4: Apply the formula for the conversion of °C to K.

What key concept did you use?

1.9 Derived Units: Measuring Volume
A. Derived quantities — quantities expressed in terms of one or more of the seven base units. (See Table 1.5 in textbook, page 14.)
B. Volume — the amount of space occupied by an object.
 1. Measured in SI units by the cubic meter (m^3).
 2. Commonly used measurements:
  a. cubic decimeter (dm^3) = metric liter (L)
  b. cubic centimeter (cm^3) = metric milliliter (mL)

Chapter 1 – Chemistry: Matter and Measurement

1.10 Derived Units: Measuring Density C.2.

A. Density — intensive property that relates mass to volume.
  1. Expressed in units of g/cm^3 or g/mL.
  2. Useful property – allows conversion of a liquid's volume to mass.
B. Temperature dependent property.
  1. Substance volume changes when heated or cooled.
     a. Specify temperature when reporting density.
C. Density of water
  1. Density = 1.000 0 g/mL at 3.98 °C.
  2. T < 3.98 °C, water expands and volume is greater.
     a. Density decreases.
  3. If the density of a substance is less than the density of water, the substance floats in water.
  4. If the density of a substance is greater than the density of water, the substance sinks in water.

EXAMPLE:
A student determined that a metal cylinder having a mass of 57.893 g has a volume of 38.32 cm^3. What is the density of this metal cylinder?

SOLUTION:
Knowing both the mass and volume of the metal cylinder, we can determine the density by simply dividing the mass by the volume.

$$\text{Density} = \frac{57.893 \text{ g}}{38.34 \text{ mL}} = 1.511 \text{ g/mL}$$

 *Workbook Problem 1.2*
The density of iron at 25°C is 7.87 g/mL. How many grams of iron are found in a volume of 28.3 mL?

Strategy: Using the information provided about density, set up a mathematical equation that allows you to solve for the mass of iron.

Step 1: Substitute the information given in the problem into the mathematical equation you set up.

Step 2: Solve for the mass of iron.

What key concept did you use?

1.11 Accuracy, Precision, and Significant Figures in Measurement D.4.

A. Accuracy — how close a measurement is to the true value.
B. Precision — how well a number of independent measurements agree with one another.

Chapter 1 – Chemistry: Matter and Measurement

1. To indicate the precision of a measurement use all the digits known with certainty, plus one additional estimated digit.
2. Total number of digits in the measurement equals the number of significant figures.

C. Determining the number of significant figures — see rules, starting on pages 18 and 19 in your textbook.

EXAMPLE:

12.502 g — five significant figures (rule 1)

0.005 25 mL — three significant figures (rule 2); If you write the number as 5.25×10^{-3} mL, it is easier to see the number of significant figures

5.30 m — three significant figures (rule 3)

79,300 s — uncertain (rule 4); If the number is 7.9300×10^4, there are five significant figures. If the number is 7.930×10^4, there are four significant figures. If the number is 7.93×10^4, there are three significant figures.

D. Using scientific notation can be very helpful in determining the number of significant figures (see above example).
E. Exact numbers have an infinite number of significant figures.

1.12 Rounding Numbers

A. Error Analysis – mathematical treatment of data to determine the precision of the measurement.
B. Simplified procedure.
   1. In carrying out a multiplication or division, the answer can't have more significant figures than either of the original numbers.

EXAMPLE:

five significant figures

three significant figures

$$\frac{55.678 \text{ g}}{32.5 \text{ mL}} = 1.71 \text{ g/mL}$$

three significant figures

2. In carrying out an addition or subtraction, the answer can't have more digits to the right of the decimal point than either of the original numbers.

EXAMPLE:

```
  24.7835 g    ends 4 places past decimal point
-  0.45   g    ends 2 places past decimal point
  24.33   g    ends 2 places past decimal point
```

C. Rules for rounding off numbers — see page 20 in text.
   1. In doing calculations, use all figures, significant or not, and then round off the final answer.

Chapter 1 – Chemistry: Matter and Measurement

EXAMPLE:
If it takes 0.75 hours to drive a distance of 26.8 miles, what is the average speed of the car in miles per hour?

SOLUTION:
$$\text{Average speed} = \frac{26.8 \text{ mi}}{0.75 \text{ h}} = 35.73333 \text{ mi/h}$$

Decide how many significant figures should be in your answer. The denominator has only two significant figures; therefore, the answer must also have only two significant figures.
Round off your answer. The first digit to be dropped is greater than 5. 35.733 33 mi/h becomes 36 mi/h.

EXAMPLE:
Round off the numbers to three significant figures: (a) 7.835 g   (b) 0.00208 7 00 g.

SOLUTION:
(a) Because the last digit is a 5 with nothing following, the answer is 7.84 since the digit to the left of the 5 is odd.

(b) Because the last digit is > 5, the answer is 0.00209 g.

 Workbook Problem 1.3
A metal cylinder with a density of 3.678 g/mL has a mass of 98.075 g. Determine the volume of the cylinder and report this volume with the correct number of significant figures.

Strategy: Set up a mathematical equation which allows you to determine the volume of the cylinder. Apply the rules found on page 21 in your text to determine the correct number of significant figures.

Step 1: Solve for the volume of the cylinder.

Step 2: Apply the rules for multiplication and division to determine the number of significant figures in your answer.

Step 3: If necessary, use the rules for rounding off numbers.

What key concept did you use?

Chapter 1 – Chemistry: Matter and Measurement

1.13 Calculations: Converting from One Unit to Another
A. Dimensional analysis method — a quantity described in one unit is converted into an equivalent quantity described in a different unit by using a conversion factor to express the specific relationship between units.

Original quantity × conversion factor = equivalent quantity

B. Conversion factors
   1. Relationship between units can be written as a fraction.
      a. 1 in. = 2.54 cm.
      b. Conversion factor: $\dfrac{1 \text{ in}}{2.54 \text{ cm}}$ or $\dfrac{2.54 \text{ cm}}{1 \text{ in}}$
   2. Quantity in numerator equals the quantity in the denominator.
      a. Multiplying by conversion factor equivalent to multiplying by one.
      b. $\dfrac{1 \text{ in}}{2.54 \text{ cm}} = \dfrac{2.54 \text{ cm}}{1 \text{ in}} = 1$

C. Key to dimensional analysis:
   1. Units are treated like numbers and can be multiplied and divided.
   2. Set up an equation so that all unwanted units cancel, leaving only the desired units.
      a. The right answer is obtained only if the equation is set up so that the unwanted units cancel.
   3. Method:
      a. Start with the information given, including the units.
      b. Identify the information needed, including the units.
      c. Find a relationship between the known information and the unknown answer. (See conversion table, inside back cover.)
      d. Plan a strategy to get from the known information and the unknown answer
      e. Solve.
         i. Use a conversion factor that allows the unit in the information given to be cancelled.
         ii. Continue to use conversion factors to cancel the previous unit until the desired unit is left.
      f. Consider whether the answer obtained is reasonable (referred to the Ballpark Check in the textbook).
   4. The number of significant figures in the answer is determined from the information given.

EXAMPLE:
What would be the weight in kilograms of a person weighing 175.5 lb?

SOLUTION:
Given that 2.205 lb = 1 kg, set up the equation knowing that you want to convert 175.5 lb to kg and that the unit lb should cancel.

$$175.5 \text{ lb} \times \dfrac{1 \text{ kg}}{2.205 \text{ lb}} = 79.59 \text{ kg}$$

Ballpark Check: A kg is a little more than twice a lb. Therefore, your answer in kg should be close to, but less than, half the amount in lb.

Warning: It's easy to get the "right" answer using dimensional analysis without really understanding what you're doing.

Chapter 1 – Chemistry: Matter and Measurement

EXAMPLE:
An international exchange student rents a car while studying in Germany. The speedometer in this car is calibrated in km/hr. Once out on the autobahn, he wants to test the car's performance at 80 mi/hr. How fast must he drive in km/hr?

SOLUTION:
Using the conversion table on the inside back cover, we learn that 1 mi = 1.6093 km. Set up the equation beginning with the information given (80 mi/h), and write the conversion factor so that the unit mi cancels.

$$\frac{80 \text{ mi}}{h} \times \frac{1.6093 \text{ km}}{1 \text{ mi}} = 130 \frac{\text{km}}{h}$$

 *Workbook Problem 1.4*
Calculate the amount of time it would take an athlete to run 3200 m if she normally runs a mile in 4 min, 38 s.

Strategy: Using the conversion table on the inside back cover of your textbook, determine the conversion factor for meters to miles, and convert time to one unit.

Step 1: Calculate the amount of time it will take for the athlete to run that distance.

What key concept did you use?

 **Self-Test**

This section is intended to test your knowledge of the material covered in this chapter. Think through these problems, and make certain you understand what is going on. Ask yourself if your answer makes sense. Many of these questions are linked to the chapter learning goals. Therefore, successful completion of these problems indicates you have mastered the learning goals for this chapter. You will receive the greatest benefit from this section if you use it as a mock exam. You will then discover which topics you have mastered and which topics you need to study in more detail.

True-False

1. The symbol for the element cobalt is CO.

2. The element sodium belongs in the group referred to as the alkaline earth metals.

3. The element bromine is a nonmetal.

Chapter 1 – Chemistry: Matter and Measurement

4. Semimetals have properties somewhere between those of metals and nonmetals, and therefore, are excellent conductors of electricity.

5. The SI base unit for length is the meter.

6. The unit most commonly used in the laboratory for volume is the m^3.

7. The number 0.000 252 contains three significant figures.

8. 1×10^{-6} g is equal to 1 µg.

9. The symbol for iron comes from its Latin root *ferrum*.

10. The number 5,000,000 can be expressed in scientific notation as 5×10^6.

Multiple Choice

1. The symbol for antimony is
    a. At
    b. Ar
    c. As
    d. Sb

2. The symbol Mg represents the element
    a. manganese
    b. magnesium
    c. mercury
    d. molybdenum

3. The symbol for boron is
    a. B
    b. Br
    c. Ba
    d. Bk

4. The definition of mass is
    a. term used to describe anything physically real.
    b. the pull of gravity on an object by the earth or other celestial body.
    c. the amount of matter in an object.

5. The following data were collected by a student during a laboratory exercise:

    25.78 mL, 25.82 mL, 25.65 mL.

    The true volume in this experiment is 30.00 mL. These numbers represent data that is
    a. very precise and accurate
    b. very precise but not accurate
    c. very accurate but not precise
    d. neither precise nor accurate

6. If 5.078 93 were divided by 0.0789, the answer would contain
   a. six significant figures.
   b. four significant figures.
   c. five significant figures.
   d. three significant figures.

7. The element oxygen is found in which region of the periodic table?
   a. metal
   b. semimetal
   c. halogens
   d. nonmetal

8. The transition metals
   a. are the six larger groups on the right of the periodic table.
   b. are the 10 smaller groups in the middle of the periodic table.
   c. are the 14 groups shown separately at the bottom of the table.

9. A physical property is one that
   a. changes the chemical makeup of the sample
   b. depends on the size of the sample
   c. can be determined without changing the chemical makeup of the sample
   d. does not depend on the size of the sample

10. The SI unit that is used for temperature is
    a. °F
    b. °C
    c. °K
    d. K

11. The element bromine is found in the group known as the
    a. alkali metals
    b. alkaline earth metals
    c. chalcogens
    d. halogens

12. The element europium, Eu, is a(an)
    a. alkali metal
    b. lanthanide element
    c. transition metal
    d. actinide element

13. The symbol for potassium is
    a. K
    b. Pt
    c. Po
    d. P

14. The SI prefix, micro, corresponds to the multiplier
    a. 10^{-3}
    b. 10^{-6}
    c. 10^{-9}
    d. 10^{-12}

Chapter 1 – Chemistry: Matter and Measurement

15. Incrementally, the Fahrenheit degree is
    a. smaller than a Celsius degree
    b. smaller than a Kelvin
    c. larger than a Celsius degree
    d. the same as a Celsius degree

Matching

Hypothesis

a. the amount of space occupied by an object

Theory

b. property that does not depend on the size of the sample

Element

c. how close a measurement is to the true value

Main Group Elements

d. how well a number of independent measurements agree with one another

Matter

e. an interpretation that explains the results of many experiments.

Volume

f. a fundamental substance that can't be chemically changed or broken down into anything simpler

Precision

g. a method in which a quantity described in one unit is converted into an equivalent quantity described in a different unit by using a conversion factor to express the specific relationship between units

Accuracy

h. a consistent explanation of known observations; logical interpretations of experimental results

Dimensional analysis

i. term used to describe anything physically real

Intensive property

j. groups 1A — 8A on the periodic table

Fill-in-the-Blank

1. The symbol for the element scandium is _____.

2. The element Ti can be found in the _____ period and the group _____.

3. The element Ge is called _____ and is a _____.

Chapter 1 – Chemistry: Matter and Measurement

4. Collectively, the elements in group 8A are called the _____.

5. The SI unit for pressure is the _____.

6. The definition and SI unit for mass is _____
   _____.

7. All measurements of physical quantities contain both a _____.

8. One nanosecond is equal to _____ seconds.

9. One meter is equal to _____ centimeters.

10. One gram is equal to _____ kilograms.

11. To indicate the precision of a measurement, use all of the digits known with certainty plus _____
    _____.

12. In carrying out a multiplication or division, the answer can't have more significant figures than
    _____.

13. Density is a(an) _____ property that relates _____.

14. The three steps involved in the scientific method are _____.

15. Group 1A metals react _____ with water to form highly _____
    products. These metals are called the _____ metals.

16. Chemistry is defined as _____
    _____
    _____.

17. _____ describe the changes that take place in nature.

18. The system of units used by chemists is the _____ and
    contains _____ fundamental units.

19. The columns in the periodic table are referred to as _____, and the rows are referred
    to as _____.

20. Elements within a group have similar _____.

Problems

1. 3.89 μg = _____ mg = _____ ng

2. Perform the following calculations and report your answer with the correct number of significant figures:

   a. $\dfrac{7.984}{3.2} =$ b. 7.53 × 23.945 62 =   c. 3.268 + 4 =   d. 34.21 − 0.039 =

3. The land area of Australia is 2,941,526 mi^2. Round off this quantity to four significant figures; to two significant figures. Express your answers in scientific notation.

15

Chapter 1 – Chemistry: Matter and Measurement

4. How many inches are there in 35.67 cm?

5. The boiling point of ethanol is 78.5°C. Convert this temperature to °F and K.

6. A student going home for the weekend traveled 700 km at a speed of 103 km/h. How many miles did she travel? What was her average speed in mi/h?

7. A student determines the mass of a metal cylinder to be 53.487 g. She then places the metal cylinder in a graduated cylinder containing 25.34 mL of water. The water level in the cylinder rises to 59.72 mL. What is the density of the metal cylinder?

8. What are the names and symbols of the elements that are vertical and horizontal neighbors of arsenic?

9. Write the names and symbols of all elements that occupy the same column as oxygen in the periodic table. What name is given to this group?

10. How many minutes does it take light from the sun to reach Earth? (The distance from the sun to Earth is 93 million miles; the speed of light = 3.00×10^8 m/s.)

11. The density of titanium is 4.55 g/ml. Calculate the mass of a solid cylinder of titanium 2.75 cm in diameter and 5.05 cm long.

12. A pycnometer is a device used to make precise density determinations. A pycnometer weighs 36.20 g when empty and 50.27 g when filled with water at 20°C. The density of water at 20°C is 0.9982 g/mL. When 12.05 g of iron is placed in the pycnometer, a total mass of 60.79 g is obtained when the pycnometer is again filled with water. What is the density of iron in g/mL?

13. What is the mathematical formula for determining the volume of a cylinder? Determine the density of gallium in g/cm^3 if at 25°C a 0.75 in by 0.75 in by 0.50 in cube has a mass of 0.0599 lbs.

14. Why is density dependent upon temperature?

15. For the following two mixtures, determine which substance will float on top of the other substance:
    a. Olive oil with a density of 0.918 g/mL mixed with water with a density of 0.997 g/mL at 25°;
    b. 1,1,1-trichloroethane with a density of 1.3492 g/mL mixed with water with a density of 0.99820 g/mL at 20° C.

Challenge Problem

Gallium has a melting point of 29.78°C and a boiling point of 2403°C. This very large liquid range coupled with the fact that gallium is a liquid near room temperature makes this metal ideal for the construction of a high temperature thermometer.

You need to measure the melting points of some metals and have decided to construct a thermometer using gallium as the liquid. However, you don't want the thermometer to be too awkward to handle, so you have decided to define the melting point of gallium as 0°C and the boiling point as 500°C. To ensure the accuracy of your thermometer, you measure the melting point of calcium. The melting point of calcium on your thermometer is 170°Ga. The reported melting point for calcium is 839°C. Is your thermometer accurate?

HINT: You need to determine both a size adjustment and a zero-point adjustment in the same manner outlined for the conversion between °F and °C in your textbook. You can then write the equation for conversion between °Ga and °C.

Inquiry Based Problem

On your first day of General Chemistry laboratory, your instructor informs you that you will not be required to buy a textbook for the lab. Instead, you will be assigned problems throughout the semester related to the material you have learned in the lecture. Using that material, you will design your laboratory experiments to solve the problem at hand. Your first problem is stated below.

You are given a metal cylinder and a table of densities for metal cylinders from the *CRC Handbook of Chemistry and Physics*. You are also provided with a caliper (an instrument for measuring thicknesses and external diameters calibrated to 1 µm), a balance, and a graduate cylinder calibrated to 1 mL. You are required to determine the identity of the metal cylinder using two different procedures. Which procedure is more accurate?

The following questions/information may help you determine two procedures to use:

Knowing the definition of density, what two measurements do you need to make?

Are there any mathematical formulas which may be helpful in this experiment? (Please refer to problem 13 above.)

Is there more than one way to determine these measurements given the equipment provided?

How many measurements should you make?

Which method is more accurate?

Once you have written the two procedures, you collected the following data. From this data and the two procedures, determine the identity of the metal cylinders.

Mass	Volume – H$_2$O displaced	Diameter	Length	CRC Handbook Data	
				Alloys	**Density**
28.73 g	3.4 mL	0.6293 cm	11.1742 cm	Stainless Style Type 304	7.9 g/cm^3
28.79 g	3.5 mL	0.6302 cm	11.2437 cm	Yellow Brass (high brass)	8.47 g/cm^3
28.85 g	3.5 mL	0.6298 cm	11.1765 cm	Aluminum bronze	7.8 g/cm^3
				Beryllium copper 25	8.23 g/cm^3
				Red brass, 85%	8.75 g/cm^3

CHAPTER 2

ATOMS, MOLECULES, AND IONS

Chapter Learning Goals

A. Chemical Laws
1. Explain the law of a) definite proportions, b) multiple proportions, c) conservation of mass.
2. For two different compounds comprised of the same two elements, show that the law of multiple proportions is obeyed.

B. Atomic Structure
1. Explain what information about the atom was revealed by the experiments of a) Thomson, b) Millikan, c) Rutherford.
2. Describe the atom in terms of the composition, mass, and volume of the nucleus relative to the mass and volume occupied by the electrons.

C. Elements
1. State the difference between atoms and elements.
2. Given the symbol for an isotope of an atom or ion, determine the number of protons, neutrons, and electrons.
3. Given the mass and natural abundance of all isotopes of a given element, calculate the average atomic mass of that element.
4. For any element, calculate a) the mass in grams of a single atom and b) the number of atoms in a given number of grams.

D. Atoms, Compounds, and Molecules
1. State the difference between a) compounds and mixtures, b) heterogeneous and homogeneous mixtures, and c) atoms and molecules.
2. Identify which substances are ionic and which are molecular.
3. Identify which substances are acids and which are bases.
4. Give the formulas and names of a) common polyatomic ions, b) ionic compounds, c) binary molecular compounds, d) common acids.

Chapter in Brief

To have an understanding of substances that are called atoms, molecules, and ions, you first need to understand the fundamental laws that govern chemistry. You then will begin to look at the structure of the atom and learn about the three subatomic particles: protons, neutrons, and electrons. Knowing the atomic number and atomic mass of an atom allows you to determine the number of protons (equals number of electrons) and neutrons for an element. An atom's mass is measured using the atomic mass unit. You will see how the total mass of an atom in amu, or the atomic mass, can be used to calculate the mass of a single atom. You will also learn the differences between compounds and mixtures, molecules and ions, and acids and bases. Finally, you will begin applying the rules for naming inorganic compounds.

2.1 Conservation of Mass and the Law of Definite Proportions

A. Element — a substance that cannot be further broken down.
B. **Law of Mass Conservation** — Mass is neither created nor destroyed in chemical reactions.
  1. Cornerstone of chemical science.
C. **Law of Definite Proportions** — Different samples of a pure chemical substance always contain the same proportion of elements by mass.
  1. Elements do not combine chemically in random proportion.

EXAMPLE:

2.52 g of barium chloride and 1.72 g of sodium sulfate are each dissolved in water. The two solutions are mixed. An immediate chemical reaction occurs, leading to the formation of barium sulfate, a solid precipitate, and sodium chloride. The reaction mixture is filtered, and the solid barium sulfate is weighed. If the weight of the barium sulfate is 2.82 g, what is the weight of the sodium chloride after the water is evaporated?

SOLUTION:

According to the **Law of Mass Conservation** mass is neither created nor destroyed in chemical reactions. Therefore, the total mass of the reactants (the starting material) must equal the total mass of the products (the substances produced after the reaction has occurred). In this case

g barium chloride + g sodium sulfate = g barium sulfate + g sodium chloride

(2.52 g barium chloride) + (1.72 g sodium chloride) =
  (2.82 g barium sulfate) + (x g sodium chloride)

x g sodium chloride = 4.24 g reactant − 2,82 g barium sulfate

x g sodium chloride = 1.42 g

2.2 Dalton's Atomic Theory and the Law of Multiple Proportions

A. Dalton's Atomic Theory.
  1. Elements are made of tiny particles called atoms.
  2. Each element is characterized by the mass of its atoms.
    a. Atoms of the same element have the same mass.
    b. Atoms of different elements have different masses.
  3. Chemical combination of elements to make different substances occurs when atoms join together in small, whole-number ratios.
  4. Chemical reactions only rearrange the way that atoms are combined; the atoms themselves are not changed.
B. Law of Multiple Proportions — If two elements combine in different ways to form different substances, the mass ratios are small, whole-number multiples of each other.

EXAMPLE:

Carbon monoxide and carbon dioxide are both gases that contain only carbon and oxygen. A sample of carbon monoxide contains 1.00 g of C and 1.33 g of O. A sample of carbon dioxide contains 1.00 g of C and 2.67 g of O. Show that the two substances obey the law of multiple

Chapter 2 – Atoms, Molecules, and Ions

proportions.

SOLUTION:
Find the C:O mass ratio in each compound.

Carbon monoxide: C:O mass ratio $= \dfrac{1.00 \text{ g C}}{1.33 \text{ g O}} = 0.752$

Carbon dioxide: C:O mass ratio $= \dfrac{1.00 \text{ g C}}{2.67 \text{ g O}} = 0.375$

Determine if the two C:O mass ratios are small, whole-number multiples of each other.

$$\dfrac{\text{C:O mass ratio in carbon monoxide}}{\text{C:O mass ratio in carbon dioxide}} = \dfrac{0.752}{0.375} = 2.00$$

 *Workbook Problem 2.1*

Phosphorus forms two compounds with chlorine. In one compound, 0.75 g of phosphorus combines with 2.58 g of chlorine. In another compound, 1.35 g of phosphorus combines with 7.73 g of chlorine. Show that these two substances obey the law of multiple proportions.

Strategy: Determine the chlorine-to-phosphorus mass ratios for both compounds, and compare these ratios to each other.

Step 1: Determine the chlorine to phosphorus ratio for the first compound.

Step 2: Determine the chlorine-to-phosphorus ratio for the second compound.

Step 3: Divide the ratio found for the first compound by the ratio for the second compound.

Step 4: Can your answer be converted to a small, whole-number ratio?

What key concept did you use?

2.3 The Structure of Atoms: Electrons

A. Thomson — found that cathode rays consist of tiny, negatively charged particles called electrons.
   1. Electrons are emitted from electrodes made of two thin pieces of metal.
   2. Many different metals may be used to make electrodes.
      a. Different metals contain electrons.

Chapter 2 – Atoms, Molecules, and Ions

3. Cathode rays can be deflected by bringing either a magnet or an electrically charged plate near the tube. This deflection depends on
   a. the strength of the deflecting magnetic or electric field
   b. the size of the negative charge on the electron
   c. the mass of the electron
4. Charge-to-mass ratio, e/m of the electron = $1.758\,819 \times 10^8$ C/g.

B. Millikan determined that the charge on a drop of oil (see Fig. 2.4 in your textbook, page 38) was always a small, whole-number multiple of e.   *B.2.*
   1. $e = 1.602\,177 \times 10^{-19}$ C.
   2. Knowing the values for e/m and e for an electron, m can be calculated.
      a. $m = 9.109\,390 \times 10^{-28}$ g

2.4 The Structure of Atoms: Protons and Neutrons

A. Alpha (α) particles.
   1. 7000 times more massive than electrons
   2. have a positive charge that is twice the magnitude of, but opposite in sign to, the charge on an electron
B. Rutherford experiment:
   1. Directed a beam of α particles at a thin gold foil.
   2. Most α particles passed through the thin gold foil but a few were deflected at large angles.
   3. Proposed existence of nucleus.
C. Nuclear model of the atom   *B.2.*
   1. Nucleus — a tiny central core in an atom where the mass of the atom is concentrated.
      a. contains the atom's positive charges
   2. Electrons move in space a relatively large distance away from the nucleus.
D. Nucleus is composed of two kinds of particles.
   1. Protons have a mass of $1.672\,623 \times 10^{-24}$ g and are positively charged.
      a. number of protons equals number of electrons in a neutral atom
   2. Neutrons — almost identical in mass to protons but carry no charge.
      a. number of neutrons not directly related to number of protons and electrons.

2.5 Atomic Number

A. Elements differ from one another according to the number of protons in their atoms.   *C.1.*
   1. Atomic number (Z) = the number of protons in an atom = the number of electrons in an atom.
B. Most nuclei also contain neutrons.
   1. Mass number (A) = number of protons + number of neutrons in an atom.
C. Isotopes — atoms with identical atomic numbers but different mass numbers.   *C.2.*
   1. Mass number written as left superscript.
   2. Atomic number written as left subscript. (The atomic number is sometimes left off since all atoms of an element always contain the same number of protons.)
   3. The number of neutrons in an isotope can be calculated from $A - Z$.
   4. Number of neutrons in an atom has little effect on the chemical properties of the atom.

EXAMPLE:
Determine the number of protons, neutrons, and electrons in the following isotopes:

$^{199}_{80}\text{Hg}$ $^{195}_{78}\text{Pt}$ $^{84}_{38}\text{Sr}$

SOLUTION:
The subscript is the atomic number (Z), and the superscript is the mass number(A).
Z = the number of protons = the number of electrons, and $A - Z$ = the number of neutrons.

$^{199}_{80}\text{Hg}$: $Z = 80$; number of protons = number of electrons = 80

$A - Z = 199-80 = 119$; number of neutrons = 119

$^{195}_{78}\text{Pt}$: $Z = 78$; number of protons = number of electrons = 78

$A - Z = 195 - 78 = 117$; number of neutrons = 117

$^{84}_{38}\text{Sr}$: $Z = 38$; number of protons = number of electrons = 38

$A - Z = 84 - 38 = 46$; number of neutrons = 46

Workbook Problem 2.2
Iron, the fourth most abundant element in the earth's crust, plays an important role in both human civilization and in living systems. Two of iron's isotopes have mass numbers of 56 and 58.

Strategy: Use the periodic table to determine the atomic number and the number of neutrons in each isotope. Give the standard symbol for each.

Step 1: First, it's necessary to know the chemical symbol for iron.

Step 2: Use the periodic table to determine the atomic number for iron.

Step 3: The number of neutrons can be determined from the definition for mass number.

Step 4: The standard symbol is written with the mass number as a superscript and the atomic number as a subscript, both to the left of the symbol.

What key concept did you use?

Workbook Problem 2.3
The metal X has a very high electrical conductivity and as a consequence is used to make electrical

wiring and corrosion-resistant water pipes. An atom of the metal contains 29 protons and 34 neutrons. Identify the metal, and write the symbol in standard format.

Strategy: Determine the atomic number from the information given.

Step 1: Knowing the atomic number, identify the element by using the periodic table.

Step 2: The standard symbol is written with the mass number as a superscript and the atomic number as a subscript, both to the left of the symbol.

What key concept did you use?

2.6 Atomic Mass
A. Atomic mass unit (amu) — exactly 1/12th the mass of an atom of $^{12}_{6}C$.
   1. 1 amu = $1.660\,54 \times 10^{-24}$ g
B. Isotopic mass – the mass of an atom in atomic mass units.
   1. Numerically close to the atom's mass number.
C. Atomic mass values are weighted averages for the naturally occurring mixtures of different isotopes.   *C.3., C.4*
   1. Atomic mass of an element = Σ(mass of each isotope $\times$ the abundance of the isotope).
      a. Σ is used for the term "the sum of."
   2. Can use the atomic masses to count the number of atoms by weighing a sample of the element.

EXAMPLE:
Calculate the mass in grams of a single atom of iridium.

SOLUTION:
From the mass given in the periodic table, we know that 1 atom of Ir has a mass of 192.2 amu. We also know that 1 amu = 1.6605×10^{-24} g. The final unit we want to end up with is g/atom of Ir. Therefore, we should set up our problem in the following manner:

$$\frac{192.2 \text{ amu}}{1 \text{ atom Ir}} \times \frac{1.6605 \times 10^{-24} \text{ g}}{1 \text{ amu}} = 3.191 \times 10^{-22} \text{ g}$$

 *Workbook Problem 2.4*
How many atoms of cadmium are present in 25.2 g?

Strategy: Determine the conversion factors needed to convert from grams of sample to number of atoms.

Chapter 2 – Atoms, Molecules, and Ions

Step 1: Set up a mathematical equation such that all of the units except number of atoms cancel.

What key concept did you use?

 *Workbook Problem 2.5*

Magnesium has three naturally occurring isotopes: ^{24}Mg with an abundance of 78.99% and an atomic mass of 23.985 amu; ^{25}Mg with an abundance of 10.00% and an atomic mass of 24.986 amu; and ^{26}Mg with an abundance of 11.01% and an atomic mass of 25.983 amu. Calculate the average atomic mass of Mg.

Strategy: Remember, the atomic mass of an element = Σ(mass of each isotope × the abundance of the isotope).

Step 1: Use the information given to determine the average atomic mass.

What key concept did you use?

2.7 Compounds and Mixtures
A. Different kinds of matter on Earth classified as either pure substances or mixtures (see Fig. 2.7, page 45 in your textbook).
  1. Pure substance —
     a. elements
     b. compounds
B. Chemical compounds - pure substance formed from the combination of atoms of two or more different elements.
  1. Properties completely different from constituent elements.
  2. Have constant composition.
  3. Composition indicated by a chemical formula.
     a. lists the symbols of the individual constituent elements
     b. the number of each atom is given by a subscript.
  4. Formed when atoms undergo chemical combination in a specific manner.
C. Chemical reactions — atoms from two or more different elements combine, creating new materials with properties completely unlike those of their constituent elements.
  1. Chemical equation – format used to write a chemical reaction.
     a. reactants — starting substances in chemical reaction; written on the left side of the arrow

Chapter 2 – Atoms, Molecules, and Ions

   b. products — new materials formed; written on the right side of the arrow
   c. an arrow is placed between the reactants and products to indicate a transformation.
   d. must obey **Law of Mass Conservation**
    i. numbers and kinds of atoms must be equal on both sides of reaction arrow.
  D. Mixtures — formed when 2 or more substances are blended together in some random proportion.
   1. Individual substances do not undergo a chemical change.
   2. Homogeneous mixture (solution) has uniform properties throughout.
    a. liquids are transparent
   3. Heterogeneous mixture has regions with differing compositions.
    a. liquids are cloudy and will separate on standing.

2.8 Molecules, Ions, and Chemical Bonds

  A. Chemical bonds – the connections that join the atoms together in a compound.
   1. Formed by the electrons in the atoms.
   2. Classified as:
    a. Covalent bonds — occur between two nonmetals.
    b. Ionic bonds — occur between a metal and a nonmetal.
  B. Covalent bond — two atoms share electrons.
   1. Molecule — unit of matter that results when two or more atoms are joined by covalent bonds.
    a. Some elements exist as molecules.
     i. H_2, O_2, N_2, F_2, Cl_2, Br_2, I_2
   2. Structural formula shows the specific connections between atoms.
    a. contains more information than the chemical formula
    b. important in organic chemistry – the chemistry of carbon compounds
     i. behavior of molecules governed by structure.

EXAMPLE:
Halothane is an anesthetic which is nonflammable and nonexplosive. It is considered safe even though it does cause depression of cardiovascular and respiratory functions. Its formula is BrC_2ClF_3. The structure has the two carbon atoms bonded to each other. The three fluorine atoms are bonded to one carbon, and the bromine, chlorine, and hydrogen atom are bonded to the other carbon. Draw the structure.

SOLUTION:

$$\begin{array}{c} \quad\;\; Br \quad F \\ \quad\;\; | \quad\;\; | \\ H - C - C - F \\ \quad\;\; | \quad\;\; | \\ \quad\;\; Cl \quad F \end{array}$$

  C. Ionic Bonds — a complete transfer of one or more electrons from one atom to another.
   1. Formed between metals and nonmetals.
    a. Metal tends to give up electrons.
    b. Nonmetal tends to accept electrons.
   2. Ions — charged particles resulting from the loss or gain of electrons.
    a. cation — positively charged particle resulting from the loss of one or more electrons
    b. anion — negatively charged particle resulting from the gain of one or more electrons

Chapter 2 – Atoms, Molecules, and Ions

      c. polyatomic ions — charged, covalently bonded groups of atoms
         i. charged molecules — consist of specific numbers and kinds of atoms joined together by covalent bonds in a definite way
   3. Ionic solids – cations and anions are packed together in a regular manner so that the charges cancel.

EXAMPLE:

Identify the following as molecules or ionic compounds: a) PCl_3, b) MnO, c) PbS, d) SO_3

SOLUTION: The easiest way to differentiate between molecules and ionic compounds is to determine whether the formula contains both metals and nonmetals or only nonmetals. The above discussion points out that covalent bonding occurs with nonmetals (molecule) and ionic bonding occurs with both metals and nonmetals (ionic compounds). PCl_3 and SO_3 contain only nonmetals and, therefore, are molecules. MnO and PbS contain both metals and nonmetals and, therefore, are ionic compounds.

2.9 Acids and Bases
   A. Two important ions: H^+ (a proton) and OH^- (a polyatomic anion).
      1. Fundamental to the concept of acids and bases.
   B. Acid — a substance that provides H^+ ions when dissolved in water.
      1. Produces H^+ along with the corresponding anion.
   C. Base — a substance that provides OH^- ions when dissolved in water.
      1. Produces OH^- along with the corresponding cation.

🗝 *D.3*

EXAMPLE:
Identify the following as an acid or a base: a) HBr, b) KOH, c) $Ba(OH)_2$, d) H_2S

SOLUTION:
To determine if a substance is either an acid or a base, we must ask if the substance is capable of providing either a H^+ or OH^- ion in water. For now, the simplest way to answer that question is to see if the compound contains H^+ or OH^-. HBr and H_2S both contain hydrogen and will produce H^+ when dissolved in water. Therefore, these compounds are acids. KOH and $Ba(OH)_2$ both contain OH^- and will produce OH^- when dissolved in water. Therefore, these compounds are bases.

Workbook Problem 2.6
Identify the following as either a molecule or ionic compound. If applicable, also identify as either an acid or base. a) HI, b) CO_2, c) $Mg(OH)_2$, d) $FeCl_3$

Strategy: Use the periodic table to determine the classification of the elements present, and apply this knowledge to the preceding discussion.

Step 1: Determine whether the formula contains both metals and nonmetals or only nonmetals.

Chapter 2 – Atoms, Molecules, and Ions

Step 2: Identify the compounds as either molecular or ionic, based on your conclusions in step 1.

Step 3: Determine if the substance is capable of providing either an H^+ or an OH^- ion in water.

Step 4: Identify the compounds as either acids or bases (if applicable), based on your conclusions in step 3.

What key concept did you use?

2.10 Naming Chemical Compounds

D.4

A. Binary Ionic Compounds — ionic compounds (contain a cation and an anion) containing only two elements.
  1. Identify the cation first, then the anion.
     a. Cation takes the same name as the element. (Remember metals form cations.)
     b. Anions take the first part of its name from the element and adds the ending *ide*. (Remember nonmetals form anions.)
  2. Common main group and transition metal ions – see Figure 2.11 and 2.12 in textbook (page 53 and 54).
     a. Elements within a group often form similar kinds of ions.
     b. main-group metal cations: charge = group number
     c. main-group nonmetal anions: charge = group number – 8
     d. Some metals form more than one kind of cation.
        i. When naming these ions, the charge is indicated by a Roman numeral in parenthesis.
        ii. Use for transition metal complexes and Sn, Tl, and Pb.
  3. Electrical Neutrality — cations and anions combine in such a manner that the overall charge on a compound is equal to zero.
     a. total positive charge = total negative charge
     b. can determine the number of positive charges on the cation by counting the number of negative charges on the anion (and vice versa).
  4. Formulas for ionic compounds always contain the smallest whole number ratio of cation to anion.

EXAMPLE:
Name the following compounds:
CaS   CuO   CsI   SnCl$_4$

SOLUTION:
  CaS: calcium sulfide   No Roman numeral is needed: calcium is a group 2A metal and forms only Ca^{2+}.

  CuO: copper(II) oxide   The charge on oxygen is a –2. To have electrical neutrality, the charge on copper, which is a transition metal, must be +2. The

Chapter 2 – Atoms, Molecules, and Ions

                                                                     Roman numeral (II) is present because copper is a transition metal.

CsI: cesium iodide   Cesium is a group 1A metal and forms only Cs^+.

$SnCl_4$:   tin(IV) chloride   There are 4 Cl^- ions with a total charge of –4. To have electrical neutrality, the charge on tin must be +4. Tin can form more than one kind of ion; therefore the charge should be indicated with a Roman numeral.

EXAMPLE:
Write formulas for the following compounds:

barium iodide   cobalt(II) phosphide   lead(II) bromide

SOLUTION:
BaI_2 Barium is a group 2A metal and forms only Ba^{2+}. To balance the charge, you need 2 I^- ions.

Co_3P_2 Cobalt(II) has a +2 charge, while phosphorous has a –3 charge. To have electrical neutrality, you need 3 Co^{2+} and 2 P^{3-}. $[(3 \times +2) + (2 \times -3)] = 0$. (HINT: A useful method for writing formulas is to use the number in the charge on the cation as the subscript on the anion and to use the number in the charge on the anion as the subscript on the cation. However, be sure that the formula contains the smallest whole number ratio of cation to anion.)

$PbBr_2$ Lead(II) has a +2 charge, while bromine has a –1 charge. To have electrical neutrality, you need 1 Pb^{2+} and 2 Br^-.

Workbook Problem 2.7
Name or write formulas for the following:

$CaCl_2$ strontium bromide   chromium (II) oxide VO_2 $TiCl_4$

magnesium nitride   aluminum oxide Co_2S_3 MgSe

Strategy: Follow the rules outlined on page 27 of this book.

Step 1: When naming compounds, determine if the metal is a representative metal or a transition metal. If the metal is a transition metal, you must indicate the charge on the metal when naming the compound. The charge on the metal is determined from the number and charge on the anion. (Remember, you must maintain electrical neutrality.)

Chapter 2 – Atoms, Molecules, and Ions

Step 2: When writing formulas, use the periodic table or the name of the compound to determine the charge on the metal. Use the periodic table to determine the charge on the anion. (Remember, you must maintain electrical neutrality.)

Step 3: Make sure the formula contains the smallest whole-number ratio of cation to anion.

What key concept did you use?

   B. Binary molecular compounds — molecular compounds containing only two nonmetal elements.
      1. One of the elements is more cationlike.
         a. takes the name of the element
      2. One of the elements is more anionlike.
         a. takes an *ide* ending
      3. Character depends on the relative positions of the two elements in the periodic table.
         a. more cationlike — farther left and/or toward bottom in the periodic table
         b. more anionlike — farther right and/or toward the top in the periodic table
      4. To specify the numbers of each element present, use numerical prefixes. (See Table 2.2 in the textbook, page 56.)
         a. the *mono* prefix is not used for the atom named first
      5. When naming binary molecular compounds that contain hydrogen, it is necessary to indicate whether the molecule is in the gaseous or aqueous (in water) state.
         a. If the molecule is a gas, use the above rules.
            i. When writing the formula, always indicate the molecule is a gas with (*g*).
         b. If the molecule is in aqueous solution, name the compound as a binary acid (see below).

EXAMPLE:
Name the following compounds:

SF_4 PCl_5 P_4O_{10} SiO_2

SOLUTION:
   SF_4:   sulfur tetrafluoride

   PCl_5:   phosphorus pentachloride

   P_4O_{10}: tetraphosphorus decoxide

   SiO_2:   silicon dioxide

Chapter 2 – Atoms, Molecules, and Ions

EXAMPLE:
Write formulas for the following compounds:

dihydrogen sulfide   xenon tetrafluoride   dinitrogen trioxide

SOLUTION:
dihydrogen sulfide:	H_2S (g)	If naming as a binary molecule, the compound is in a gaseous state that needs to be indicated.
xenon tetrafluoride:	XeF_4	
dinitrogen trioxide:	N_2O_3	

 ***Workbook Problem 2.8***
Name or write formulas for the following:

SF_6 dihyrogen selenide gas   iodine pentafluoride ICl_3 HBr (g)

Strategy: Follow the rules on page 29 of this book.

Step 1: When naming the molecules, remember that the element that is more anionlike uses the *ide* suffix. Also remember to use the numerical prefixes found in Table 2.2 (p. 60) in your textbook.

Step 2: When writing formulas, refer to Table 2.2 in your textbook for the meaning of the numerical prefixes. (Eventually, you'll need to know these numerical prefixes by heart.) Remember to indicate if a binary hydrogen compound is a gas or is in aqueous solution.

What key concept did you use?

    C. Ionic compounds containing polyatomic ions are named by following the rules for naming binary ionic compounds.
       1. Identify the cation.
       2. Identify the anion.
          a. Learn the names, formulas, and charge numbers of the most common polyatomic anions. (See Table 2.3 in the textbook, page 57)
          b. Most names end with the suffix *ite* or *ate*.
          c. Several pairs of ions are related by the presence or absence of a hydrogen atom.

Chapter 2 – Atoms, Molecules, and Ions

3. Oxoanions — an atom of the same element is combined with different numbers of oxygen atoms.
   a. Learn the name and formula of the ion whose name ends with ate (including the charge on the anion).
   b. Add one O; add prefix *per*.
   c. Remove one O; change *ate* to *ite*.
   d. Remove two O's; add prefix *hypo* and change *ate* to *ite*.

EXAMPLE:
Name the following compounds:
$AlPO_3$ $KMnO_4$ $Co_2(CO_3)_3$ $CuBrO_3$

SOLUTION:

$AlPO_3$: Aluminum phosphite.

$KMnO_4$: Potassium permanganate.

$Co_2(CO_3)_3$: Cobalt(III) carbonate. The carbonate anion has a –2 charge. The total charge on the carbonate anion is –6 (3 × –2). Therefore, the total charge on the cobalt cation must be a +6. Each cobalt cation would then have a charge of +3 (+6/2). The charge on the cobalt is indicated with Roman numerals because cobalt is a transition metal.

$CuBrO_3$: Copper(I) bromate. The bromate anion has a –1 charge. Therefore, the charge on copper must be a +1. The charge on the copper is indicated with Roman numerals because copper is a transition metal.

EXAMPLE:
Write formulas for the following compounds:

potassium sulfate   silver(I) phosphate   iron (III) periodate

SOLUTION:
potassium sulfate: K_2SO_4 The sulfate anion has a –2 charge. Potassium has a +1 charge, so 2 K^+ are needed.
silver(I) phosphate: Ag_3PO_4 The phosphate anion has a –3 charge. Silver has a +1 as indicated by the Roman numeral. 3 Ag^+ are needed for every PO_4^{3-} ion.

iron(III) periodate: $Fe(IO_4)_3$ The periodate anion has a –1 charge. Iron has a +3 charge as indicated by the Roman numeral. 3 IO_4^- are needed for every Fe^{3+}.

Workbook Problem 2.9
Name or write formulas for the following:

$Co(NO_3)_3$ sodium sulfite   barium hydroxide $Fe_2(SO_4)_3$

ammonium hypochlorite

Chapter 2 – Atoms, Molecules, and Ions

Strategy: Follow the rules on page 31 in this book.

Step 1: When naming the compounds, make reference to Table 2.3 on page 57 in your textbook. (Eventually, you will need to know the names of these polyatomic ions by heart. Flashcards will certainly come in handy when learning these names.) Also, don't forget that you must indicate the charge on certain metal cations.

Step 2: When writing formulas, use the periodic table or the name of the compound to determine the charge on the metal. To determine the formula for a polyatomic ion, make reference to Table 2.3 on page 57 in your textbook. Remember, you must maintain electrical neutrality.

What key concept did you use?

    D. Oxoacids — contain oxygen in addition to hydrogen and another element.
        1. Yields H^+ and a polyatomic oxoanion when dissolved in H_2O.
           a. See Table 2.4 in your textbook (page 59).
        2. Names are related to the names of the corresponding oxoanions.
           a. Replace the suffix *ite* with *ous acid*.
           b. Replace the suffix *ate* with *ic acid*.
    E. Binary Acids — contain hydrogen and one other element or a polyatomic anion that is not an oxoanion.
        1. Aqueous solutions.
        2. Use the prefix *hydro* followed by the first part of the name of the other element, the suffix *ic* and acid.
        3. When writing the formula, always indicate that it is an aqueous solution by (*aq*).

EXAMPLE:
Name the following acids:
H_3PO_3 $H_2S\ (aq)$ H_2CO_3

SOLUTION:
    H_3PO_3:   This compound is an acid that contains the phosphite anion. It is named phosphorous acid.

    $H_2S\ (aq)$ This compound is a binary compound containing hydrogen in aqueous solution. It is named hydrosulfuric acid.

H₂CO₃ This compound is an acid that contains the carbonate anion. It is named carbonic acid.

Workbook Problem 2.10

Name or write formulas for the following:

MnS   hydrobromic acid   sulfurous acid   N₂O₄

calcium carbonate   strontium nitrate   GaCl₃   aluminum sulfide

Strategy: First, determine the type of compound. Is it a simple binary ionic compound, a molecule, an ionic compound containing a polyatomic ion, or an acid (either binary acid or oxoacid)? Once you have determined the type of compound, apply the appropriate nomenclature rules.

What key concept did you use?

 **Self-Test**

This section is intended to test your knowledge of the material covered in this chapter. Think through these problems, and make certain you understand what is going on. Ask yourself if your answer makes sense. Many of these questions are linked to the chapter learning goals. Therefore, successful completion of these problems indicates you have mastered the learning goals for this chapter. You will receive the greatest benefit from this section if you use it as a mock exam. You will then discover which topics you have mastered and which topics you need to study in more detail.

True-False

1. Different samples of a pure chemical substance always contain the same proportion of elements by mass.

2. In a chemical reaction, the atoms undergo a change.

3. Cathode rays consist of tiny, positively charged particles.

Chapter 2 – Atoms, Molecules, and Ions

4. The nucleus is composed of two kinds of particles: protons and electrons.

5. Isotopes are atoms with identical atomic numbers but different mass numbers.

6. The mass number, A, represents the number of protons and neutrons in an atom.

7. A cation is a positively charged particle resulting from the gain of a proton.

8. An anion is a negatively charged particle resulting from the gain of an electron.

9. The correct name for FeS is iron sulfide.

10. The correct formula for dinitrogen pentoxide is N_2O_5.

Multiple Choice

1. The nucleus of an atom contains
    a. the protons, neutrons, and electrons of the atom
    b. the protons of the atom
    c. the neutrons of the atom
    d. the protons and neutrons of the atom

2. Elements differ from one another according to the number of
    a. protons
    b. neutrons
    c. isotopes
    d. amu

3. The isotope $^{60}_{27}Co$ contains
    a. 27 neutrons
    b. 60 protons
    c. 33 electrons
    d. 33 neutrons

4. Chemical bonds are formed by the interaction of
    a. the protons in an atom
    b. the neutrons in the atom
    c. the electrons in the atom

5. A covalent bond is the result of
    a. a transfer of electrons between atoms
    b. the interaction of a cation and anion
    c. the sharing of electrons between atoms

6. A homogeneous mixture is a mixture
    a. formed when atoms undergo chemical combination in a specific manner.
    b. with uniform properties throughout.
    c. that has regions of differing composition.

7. The correct name for HCl (*aq*) is
   a. hydrogen chloride
   b. hydrochloric acid
   c. chloric acid
   d. hypochloric acid

8. The correct formula for calcium phosphide is
   a. Ca_2P_3
   b. C_3P_2
   c. Ca_3P_2

9. The correct name of $Co(NO_3)_2$ is
   a. cobalt nitrate
   b. cobalt(II) nitrate
   c. cobalt(III) nitrate
   d. cobaltic nitrate

10. The correct formula for bromous acid is
    a. HBr (*aq*)
    b. $HBrO_3$
    c. HBrO
    d. $HBrO_2$

Fill-in-the-Blank

1. If two elements combine in different ways to form different substances, the mass ratios are _____.

2. In comparison with the electron, alpha (α) particles are _____.

3. The atomic number of an atom (Z) is the number of _____ in the atom.

4. The number of electrons in a neutral atom is equal to the number of _____ in the atom.

5. The number of _____ in an atom has little effect on the chemical properties of the atom.

6. The mass of an atom in amu is referred to as the _____.

7. Covalent bonds occur between two _____.

8. Ionic bonds occur between a _____.

9. A substance that provides H^+ ions in water is called _____.

10. In a chemical formula, the number of each atom is given as a _____.

11. Elements are made of tiny particles called _____.

12. Most substances on earth are _____, formed when atoms of two or more elements combine in a _____.

13. The atoms in a compound are held together by one of two fundamental kinds of _____.

Chapter 2 – Atoms, Molecules, and Ions

14. _____ form when one atom completely transfers one or more electrons to another atom, resulting in the formation of _____.

15. According to the nuclear model of an atom, protons and neutrons are clustered into a dense core called the _____.

Matching

Al_2O_3	a. mercury(II) oxide
Cations	b. atoms with identical atomic numbers but different mass numbers.
FeO	c. phosphorus trichloride
Anions	d. fundamental atomic particle that is negatively charged.
NH_4Cl	e. calcium carbonate
Bases	f. fundamental atomic particle that is neutral
$CaCO_3$	g. sulfur tetrafluoride
Isotopes	h. substances that yield OH^- ions when dissolved in water.
SF_4	i. aluminum oxide
Protons	j. positively charged ions
TiF_3	k. chromium(III) hydroxide
Neutrons	l. negatively charged ions.
PCl_3	m. barium sulfite
Electrons	n. fundamental atomic particle that is positively charged
$BaSO_3$	o. iron(II) oxide
HgO	p. ammonium chloride
$Cr(OH)_3$	q. titanium(III) fluoride

Chapter 2 – Atoms, Molecules, and Ions

Problems

1. Nitrogen forms a number of compounds with oxygen. It is found that in one compound, 0.681 g of N is combined with 0.778 g of O; in another compound, 0.560 g of N is combined with 1.28 g of O. Prove that these data support the law of multiple proportions.

2. When ammonia reacts with sodium hypochlorite, hydrazine (a poisonous, colorless gas) is formed. Both ammonia and hydrazine contain only nitrogen and hydrogen. A sample of ammonia contains 8.75 g of nitrogen and 3.75 g of hydrogen. A sample of hydrazine contains 11.48 g of nitrogen and 3.28 grams of hydrogen. Prove that the two substances obey the law of multiple proportions.

3. Europium, Eu, has two naturally occurring isotopes: $^{151}_{63}$Eu with an abundance of 47.82% and an atomic mass of 150.9 amu, and $^{153}_{63}$Eu with an abundance of 52.18% and an atomic mass of 152.9 amu. Calculate the average atomic mass of europium.

4. The four isotopes of lead, along with their abundances and atomic masses are ^{204}Pb, atomic mass = 203.973 amu, abundance = 1.48%; ^{206}Pb, atomic mass = 205.9745 amu, abundance = 23.6%; ^{207}Pb, atomic mass = 206.9759 amu, abundance = 22.6%; and ^{208}Pb, atomic mass = 207.9766 amu, abundance = 52.3%. Calculate the average atomic mass of lead.

5. How many protons, neutrons, and electrons are present in the following elements?

    a. $^{18}_{8}$O   b. $^{57}_{26}$Fe^{2+} c. $^{197}_{79}$Au   d. $^{131}_{53}$I^{-}

6. What is the identity of the element X in the following ions?

    a. X^{3+}, a cation that has 13 protons

    b. X^{2-}, an anion that has 34 protons

7. What anion will the following acids produce when dissolved in water?

    a. HNO_3 b. $HClO_4$ c. H_2SO_4 d. H_2S

8. What cation will the following bases produce when dissolved in water?

    a. KOH   b. $Ca(OH)_2$ c. $Mg(OH)_2$ d. $Li(OH)$

9. What is the mass (in grams) of 150 atoms of iron?

10. If the average mass of a single iodine atom is 2.107×10^{-22} g, what is the mass in grams of 6.02×10^{23} atoms? How does this answer compare with the atomic mass of iodine?

11. A 15.51 g quantity of iron was allowed to react with 25.00 mL of nitric acid that had a density of 1.40 g/mL. The reaction yielded H_2 gas and an iron(II) solution. The density of the gas was determined to be 0.0899 g/L, and the mass of the solution was found to be 49.958 g. How many liters of H_2 were formed? What chemical law did you use to work this problem?

12. Name the following compounds:

37

Chapter 2 – Atoms, Molecules, and Ions

a. LiF b. GaCl₃ c. Ca₃N₂ d. V₂O₅

e. Sc(NO₃)₃ f. Mn₂O₇ g. FeBr₃ h. KIO₃

i. K₂Cr₂O₇ j. Sn(NO₂)₄ k. N₂O₅ l. HBr (g)

m. HClO₂ n. HBr (aq) o. H₂SO₃ p. N₂O

13. Nitroglycerine is most famous for its explosive properties and use in dynamite. It is also used as a medicine to relax the muscles of blood vessels and dilate arteries so that blood flow to the heart is increased. Patients suffering from angina pectoris (reduced blood flow to the heart) which is accompanied by severe chest pain will place nitroglycerin tablets under their tongue when experiencing an episode of angina pectoris. The formula for nitroglycerin is:

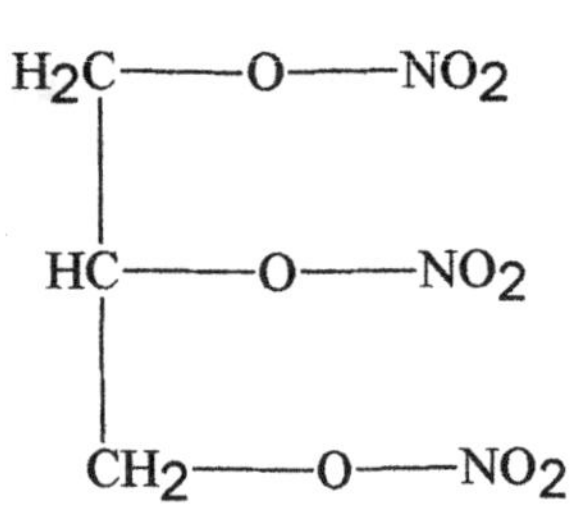

```
H₂C——O——NO₂        NH₃
 |
HC——O——NO₂
 |
CH₂——O——NO₂
```

Write the formula for this compound in alphabetical order.

14. Write formulas for the following compounds:

   a. rubidium iodide
   b. magnesium phosphide
   c. cobalt(II) chloride

   d. copper(I) oxide
   e. ammonium thiosulfate
   f. chromium(III) nitride

   g. hydrogen cyanide gas
   h. sulfur trioxide
   i. hydrofluoric acid

   j. calcium hydride
   k. bromic acid
   l. sodium sulfate

   m. sodium acetate
   n. potassium permanganate
   o. hypochlorous acid

   p. potassium hydrogen carbonate

Challenge Problem

The first anesthetic used in surgery was diethyl ether. However, the use of this compound had many disadvantages and soon other ethers, such as divinyl ether and enflurane replaced diethyl ether in the operating room. The formula for enflurane is C₃H₂ClOF₅. Is this an ionic or covalent compound? The structure of this molecule contains the oxygen bonded to two of the three carbons in a straight chain with the oxygen between the first and second carbon. The first carbon has two fluorines and a hydrogen bonded to it while the second carbon is bonded to two fluorines and the third carbon. The remaining fluorine, hydrogen, and chlorine are bonded to the third carbon. Draw this structure. Using the periodic table, determine the mass of each element in this compound. Determine the molecular mass of enflurane by combining the total mass of each element.

Given that $\% = \dfrac{\text{part}}{\text{whole}} \times 100$, determine the mass percent of each element in ascorbic acid.

CHAPTER 3

FORMULAS, EQUATIONS, AND MOLES

Chapter Learning Goals

A. Balancing Chemical Equations
1. For simple chemical reactions, write and balance chemical equations.

B. Formula Units and Moles
1. Calculate molar mass.
2. Interconvert grams, moles, and numbers of formula units.

C. Stoichiometry
1. Determine the number of moles and grams of one reactant needed to react with a given number of moles and grams of another reactant and the number of moles and grams of product(s) that result from the reaction.
2. Calculate percent yield.
3. Calculate the number of grams of products produced from a given number of grams of reactants when the theoretical yield is less than 100%.
4. Identify the limiting and excess reagents in a reaction mixture.
5. Determine the number of grams of excess reagent remaining at the end of a reaction and the number of grams of product(s) produced.

D. Chemical Reactions Performed in Solution
1. Describe how to prepare a solution of known molarity by a) dissolving a solid in a solvent and b) diluting a more concentrated solution.
2. Interconvert solution molarity, solution volume, solute moles, and solute grams.
3. Determine the volume of one reactant needed to react with a given volume of a second reactant.

E. Elemental Analysis
1. Determine the percent composition and empirical formula of a compound.
2. Understand how to use combustion analysis to obtain the empirical formula of a compound containing carbon, hydrogen, and one other element.
3. From empirical formula and molar mass, determine the molecular formula of a compound.

Chapter in Brief

The central concern of chemistry is the change of one substance into another. As a consequence, chemical reactions are at the heart of the science. In this chapter, you begin your study of chemical reactions by learning how to balance chemical equations. Using balanced chemical equations along with the concepts of the mole and stoichiometry you will learn how to calculate the amounts of reactants needed and the theoretical amount of products produced in a reaction. These calculations are carried out using molar masses for gram ↔ mole conversions or molarities for mole ↔ volume ↔ molarity conversions. You will also learn how to determine an empirical formula from the percent composition of a compound and vice versa. Finally, you will see how combustion analysis, a type of elemental analysis, along with mass spectrometry can be used to determine both the empirical and molecular formula of a compound.

3.1 Balancing Chemical Equations

A. Balanced equations — the numbers and kinds of atoms on both sides of the arrow are the same.
 1. All chemical equations must be balanced.
    a. *obeys law of mass conservation*
B. Formula unit — one unit (an atom, ion, or molecule) corresponding to a given formula.
C. Balancing process for simple equations.
 1. Write the unbalanced equation using the correct chemical formula for all reactants and products.
 2. Use coefficients, the number placed before the formula, to indicate how many formula units are required to balance the equation.
    a. Only the coefficients can be changed: the formulas must stay the same.
 3. Reduce the coefficients to their smallest whole-number values.
 4. Check your answer.

EXAMPLE:
Dinitrogen pentoxide reacts with water to produce nitric acid. Write a balanced equation for this reaction.

SOLUTION:
Step 1: Write the unbalanced equation.

$N_2O_5 + H_2O \rightarrow HNO_3$

Step 2: Use coefficients to balance the equation. Think about one element at a time. (HINT: It usually helps to save oxygen for last.) Notice on the reactant side that there are two N's but only one N on the product side. Begin by placing a 2 in front of HNO_3.

$N_2O_5 + H_2O \rightarrow 2\,HNO_3$

You now have two H's on both the reactant and product side. You also have a total of six O's on both the reactant and product side. (Remember to multiply the coefficient in front of a formula by the subscript of an element to get the total number of atoms of the element; for the oxygen in HNO_3, the total number = 2 (coefficient) × 3 (subscript) = 6.)

Step 3: The coefficients are already reduced to their smallest whole-number ratio.

Step 4: Check your answer.

Reactant side	Product side
2 N	2 N
6 O	6 O
2 H	2 H

EXAMPLE:
Write a balanced equation for the combustion reaction of pentane (C_5H_{12}).

SOLUTION:
The term combustion reaction is used to indicate reaction with oxygen. When hydrocarbons (compounds containing primarily C and H) undergo combustion, carbon dioxide and water are produced.

Step 1: Write the unbalanced equation.

$$C_5H_{12} + O_2 \rightarrow CO_2 + H_2O$$

Step 2: Use coefficients to balance the equation. (Remember, it helps to save oxygen for last.) Begin with carbon. There are five C's on the reactant side, but only one C on the product side. Begin by placing a 5 in front of CO_2.

$$C_5H_{12} + O_2 \rightarrow 5\,CO_2 + H_2O$$

There are 12 H's on the reactant side, but only two H's on the product side. Place a 6 in front of H_2O.

$$C_5H_{12} + O_2 \rightarrow 5\,CO_2 + 6\,H_2O$$

Now balance the oxygens. There are two O's on the reactant side and 16 O's on the product side. Place an 8 in front of O_2.

$$C_5H_{12} + 8\,O_2 \rightarrow 5\,CO_2 + 6\,H_2O$$

Step 3: Reduce the coefficients to their smallest whole-number ratio.

The ratio is 1:8:5:6, which is the smallest whole-number ratio.

Step 4: Check your answer.

Reactant side	Product side
5 C	5 C
12 H	12 H
16 O	16 O

 **Workbook Problem 3.1**
Write a balanced chemical equation for the combustion reaction of octane (C_8H_{18}).

Strategy: Remember, the term *combustion reaction* is used to indicate reaction with oxygen. When hydrocarbons (compounds containing primarily C and H) undergo a combustion reaction, carbon dioxide and water are produced.

Step 1: Write the unbalanced chemical equation.

Step 2: Use coefficients to balance the equation. (Remember, it helps to save oxygen for last.)

Chapter 3 – Formulas, Equations, and Moles

Step 3: Reduce the coefficients to their smallest whole-number ratio.

Step 4: Check your answer.

What key concept did you use to work this problem?

3.2 Chemical Symbols on a Different Level
A. Chemical symbols represent both a microscopic and a macroscopic level.
B. Microscopic level — chemical symbols represent the behavior of individual atoms and molecules.
C. Macroscopic level — formulas and equations represent the large-scale behavior of atoms and molecules that gives rise to observable properties.
   1. Deal with macroscopic behavior in the laboratory.

3.3 Avogadro's Number and the Mole
A. Dealing with macroscopic behavior requires that the number ratio from a balanced equation be converted to a mass ratio.
   1. Number ratio is determined from the coefficients in the balanced equation.
   2. Mass ratio is determined using molecular masses.
B. Molecular mass — sum of the atomic masses of all atoms in a molecule.
   1. Formula mass — sum of atomic masses of all atoms in one formula unit of any substance.
C. The mass ratio of molecules (or formula units) should be equal to the molecular (or formula unit) mass ratio.
   1. Molecular (or formula) mass ratios for different molecules (or formula units) are determined by the coefficients in the balanced chemical equation.
D. One mole of any substance is the amount whose mass is equal to the molecular or formula mass in grams.
   1. Mole (mol) = 6.022×10^{23} particles
      a. Avogadro's number; abbreviated N_A
      b. 1 mol HNO_3 = 6.022×10^{23} molecules of HNO_3; 1 mol electrons = 6.022×10^{23} electrons
   2. Importance of mole — provides a relationship between numbers of molecules and masses of molecules.
E. Molar mass of a substance — one mole of any substance has a mass equal to its molecular or formula mass in grams.   🗝 *B.1.*
   1. Mass of one mole of a substance.
   2. Mass of 6.022×10^{23} molecules, ions, atoms, or compounds.
   3. Molecular mass of substance in grams.
   4. Serves as a conversion factor between numbers of formula units and mass.
   5. Formula of the particles is always specified.
F. The coefficients in a balanced chemical equation indicate the number of moles of each substance needed for the reaction.
   1. Use molar masses to calculate reactant masses.

EXAMPLE:
What is the formula mass of ammonium dihydrogen phosphate? What is its molar mass?

SOLUTION:
The formula for ammonium dihydrogen phosphate is $NH_4H_2PO_4$.

43

Chapter 3 – Formulas, Equations, and Moles

The elements present, along with their atomic masses, (rounded off to one decimal place) are

N (14.0 amu) H (1.0 amu) P (31.0 amu) O (16.0 amu)

Multiply the atomic mass of each element by the number of times that element appears in the formula and add together the results.

$$
\begin{aligned}
&N\,(1 \times 14.0 \text{ amu}) &&= 14.0 \text{ amu} \\
&H\,(6 \times 1.0 \text{ amu}) &&= 6.0 \text{ amu} \\
&P\,(1 \times 31.0 \text{ amu}) &&= 31.0 \text{ amu} \\
&O\,(4 \times 16.0 \text{ amu}) &&= \underline{64.0 \text{ amu}} \\
& &&= 115.0 \text{ amu}
\end{aligned}
$$

The molar mass is the formula mass in grams: 115.0 g/mol.

EXAMPLE:

The balanced equation for the combustion of pentane is

$$C_5H_{12} + 8\,O_2 \rightarrow 5\,CO_2 + 6\,H_2O$$

How many moles of oxygen will react with a) 1 mol C_5H_{12} and b) 0.5 mol C_5H_{12}?

SOLUTION: From the balanced equation, we know that 8 mol of O_2 will react with 1 mol of C_5H_{12} since the coefficients represent the number of moles of each substance. The ratio 8 mol O_2:1 mol C_5H_{12} represents a conversion factor that can be used to calculate the number of moles of O_2 that will react with 0.5 mol C_5H_{12}.

$$0.5 \text{ mol } C_5H_{12} \times \frac{8 \text{ mol } O_2}{1 \text{ mol } C_5H_{12}} = 4 \text{ mol } O_2$$

 **Workbook Problem 3.2**

The Apollo lunar landing module used a fuel composed of hydrazine (N_2H_4) and a derivative of hydrazine, combined with dinitrogen tetroxide (N_2O_4). When hydrazine and nitrogen tetroxide react, they produce nitrogen gas, water and a large amount of energy.
a. Write a balanced chemical equation for the reaction described above.
b. How many moles of hydrazine will react with 0.25 mol of N_2O_4?
c. How many moles of N_2 and H_2O will be produced when 0.25 mol of N_2O_4 are present?

Strategy: To determine mole ratios, you first need a balanced equation. After obtaining the mole ratio from the balanced equation, you can use this knowledge to obtain the conversion factors needed to solve the problem.

Moles N_2O_4 → Moles N_2H_4
↑
*use coefficients
in balanced equation to find
mole ratios*
↙ ↘
Moles H_2O ← Moles N_2O_4 → Moles N_2

Chapter 3 – Formulas, Equations, and Moles

Part a
Step 1: Write an unbalanced chemical equation based on the information given in the problem.

Step 2: Use coefficients to balance the equation. (Remember to save oxygen for last.)

Step 3: Reduce the coefficients to their smallest whole number ratio if necessary.

Step 4: Check your answer.

Part b
Step 1: Determine the mole ratio for hydrazine and dinitrogen tetroxide from your balanced chemical equation.

Step 2: Use the mole ratio from above as a conversion factor, and calculate the number of moles of hydrazine needed to react with 0.25 mol of dinitrogen tetroxide.

Part c
Step 1: Determine the mole ratio for dinitrogen tetroxide and nitrogen from your balanced chemical equation.

Step 2: Use the mole ratio from above as a conversion factor, and calculate the number of moles of nitrogen produced from 0.25 mol of dinitrogen tetroxide.

Step 3: Determine the mole ratio for dinitrogen tetroxide and water from your balanced chemical equation.

Chapter 3 – Formulas, Equations, and Moles

Step 4: Use the mole ratio from above as a conversion factor, and calculate the number of moles of water produced from 0.25 mol of dinitrogen tetroxide.

What key concept did you use to work this problem?

3.4 Stoichiometry: Chemical Arithmetic
A. Stoichiometry — refers to the conversion between moles and grams of reactants and products in a chemical equation.
  1. The comparison of the number of moles of reactant and products are given by the balanced chemical equation.
  2. Grams are used to carry out the chemical reaction in the laboratory.

B. To convert between grams, moles and formula units.
  1. Use the molar mass of the substance as a conversion factor.

EXAMPLE:
How many moles of PCl_3 are present in 8.0 g PCl_3?

SOLUTION: First, calculate the molar mass (in g/mol) of PCl_3.

Molar mass PCl_3 = 31.0 g/mol + (3 × 35.5 g/mol) = 137.5 g/mol

This molar mass also represents a conversion factor of $\dfrac{137.5 \text{ g } PCl_3}{1 \text{ mol } PCl_3}$ or $\dfrac{1 \text{ mol } PCl_3}{137.5 \text{ g } PCl_3}$.

To determine the moles in 8.0 g of PCl_3

$$8.0 \text{ g } PCl_3 \times \dfrac{1 \text{ mol } PCl_3}{137.5 \text{ g } PCl_3} = 0.058 \text{ mol } PCl_3$$

C. To calculate the grams of reactants needed and products produced, if you know the grams and moles of another reactant.
  1. Begin with the known grams of reactant and convert to moles.
  2. Use the balanced chemical equation to convert between moles of substances.
  3. Once you know the moles of reactants and products, you can convert to grams.

Chapter 3 – Formulas, Equations, and Moles

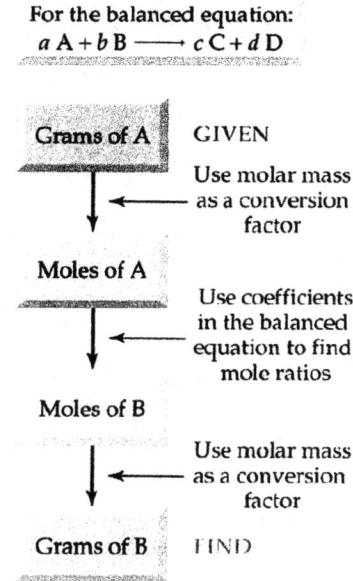

For the balanced equation:
$aA + bB \longrightarrow cC + dD$

Grams of A — GIVEN
↓ Use molar mass as a conversion factor
Moles of A
↓ Use coefficients in the balanced equation to find mole ratios
Moles of B
↓ Use molar mass as a conversion factor
Grams of B — FIND

EXAMPLE:

How many grams of water will react with 8.0 g PCl_3 in the following balanced chemical equation?

$PCl_3 + 3 H_2O \rightarrow H_3PO_3 + 3 HCl$

SOLUTION: From the preceding example, we know that 8.0 g PCl_3 = 0.058 mol PCl_3. From the balanced equation, we know that 3 mol of water will react with 1 mol of PCl_3. We can use this mole ratio to calculate the number of grams of water needed to react with 8.0 g PCl_3.

$$0.058 \text{ mol } PCl_3 \times \frac{3 \text{ mol } H_2O}{1 \text{ mol } PCl_3} \times \frac{18.0 \text{ g } H_2O}{1 \text{ mol } H_2O} = 3.1 \text{ g } H_2O$$

Notice we first converted from moles of PCl_3 to moles of water, then from moles of water to grams of water.

EXAMPLE:

How many grams of phosphorous acid will be produced in the above example?

SOLUTION: Again, we start knowing that 8.0 g PCl_3 = 0.058 mol PCl_3 and that 1 mol PCl_3 will react with 1 mol H_3PO_3. We will first convert from moles of PCl_3 to moles of H_3PO_3, then from moles of H_3PO_3 to grams of H_3PO_3.

$$0.058 \text{ mol } PCl_3 \times \frac{1 \text{ mol } H_3PO_3}{1 \text{ mol } PCl_3} \times \frac{82 \text{ g } H_3PO_3}{1 \text{ mol } H_3PO_3} = 4.8 \text{ g } H_3PO_3$$

Workbook Problem 3.3

The first step in producing $Ca(H_2PO_4)_2$, a water-soluble fertilizer known as triple superphosphate, is to react calcium phosphate with sulfuric acid to produce phosphoric acid and calcium sulfate.

Chapter 3 – Formulas, Equations, and Moles

How many grams of phosphoric acid would be produced if 6.75 g of $Ca_3(PO_4)_2$ were present?

Strategy: You can calculate the grams of H_3PO_4 produced by using the molar mass of $Ca_3(PO_4)_2$ and the mole ratio of $Ca_3(PO_4)_2$ to H_3PO_4 from the balanced chemical equation as conversion factors.

g $Ca_3(PO_4)_2$ → moles $Ca_3(PO_4)_2$ → moles H_3PO_4 → g H_3PO_4

Step 1: Write an unbalanced chemical equation from the information given above.

Step 2: Balance the chemical equation, using the procedure you have learned.

Step 3: From the balanced chemical equation, determine the mole to mole ratio of phosphoric acid to calcium phosphate.

Step 4: Determine the formula mass for calcium phosphate.

Step 5: Determine the molecular mass for phosphoric acid.

Step 6: From the information in steps 3, 4, and 5, create conversion factors. Use these conversion factors, and follow the flow diagram on page 46 in this book, so that you proceed from grams of calcium phosphate to moles of calcium phosphate to moles of phosphoric acid to grams of phosphoric acid.

What key concept did you use to work this problem?

3.5 Yields of Chemical Reactions
A. Actual yield — amount of product actually formed.
   1. Less than the amount predicted by theory.
B. Theoretical yield — amount of product calculated to form if all of the reactants are converted to products.

Chapter 3 – Formulas, Equations, and Moles

C. Percent yield = $\dfrac{\text{Actual yield}}{\text{Theorectical yield}} \times 100\%$

EXAMPLE:

Aspirin is produced from salicylic acid ($C_7H_6O_3$) and acetic anhydride ($C_4H_6O_3$) according to the following balanced equation:

$C_7H_6O_3$ + $C_4H_6O_3$ → $C_9H_8O_4$ + $C_2H_4O_2$
salicylic   acetic   aspirin   acetic
acid   anhydride   acid

How many grams of aspirin would you obtain from 4.5 g salicylic acid if the percent yield of the reaction is 75%?

SOLUTION: We can use the following steps to solve this problem:

1. Calculate the molar mass of both aspirin and salicylic acid.

Molar mass salicylic acid = (7 × 12 g/mol) + (6 × 1.0 g/mol) + (3 × 16.0 g/mol) = 138.0 g/mol

Molar mass aspirin = (9 × 12.0 g/mol) + (8 × 1.0 g/mol) + (4 × 16.0 g/mol) = 180.0 g/mol

2. To determine the actual yield, we must first determine the theoretical yield. To do this, we must first convert grams of salicylic acid to moles of salicylic acid.

$4.5 \text{ g } C_7H_6O_3 \times \dfrac{1 \text{ mol } C_7H_6O_3}{138 \text{ g } C_7H_6O_3} = 0.033 \text{ mol } C_7H_6O_3$

3. To determine the theoretical yield of aspirin, we must convert moles of salicylic acid to moles of aspirin and then convert moles of aspirin to grams of aspirin.

$0.033 \text{ mol } C_7H_6O_3 \times \dfrac{1 \text{ mol } C_9H_8O_4}{1 \text{ mol } C_7H_6O_3} \times 180.0 \text{ g } C_7H_6O_3 = 5.9 \text{ g } C_9H_8O_4$

4. To determine the actual yield of aspirin, we substitute the theoretical yield and percent yield into the equation for percent yield.

$75\% = 100\% \times \dfrac{\text{actual yield}}{5.9 \text{ g aspirin}}$

g aspirin = 0.75 × 5.9 g aspirin = 4.4 g aspirin

Workbook Problem 3.4
Using the reaction in Workbook Problem 3.3, determine the percent yield if 13.5 g of $Ca_3(PO_4)_2$ reacts to produce 6.82 g of H_3PO_4.

Strategy: To determine the percent yield, you first determine the theoretical amount of H_3PO_4 produced when starting with 13.5 g of $Ca_3(PO)_2$.

Chapter 3 – Formulas, Equations, and Moles

Step 1: Follow the same procedure as Workbook Problem 3.3.

Step 2: Calculate the percent yield using the actual yield stated in the problem and the theoretical yield just calculated.

What key concept did you use to work this problem?

3.6 Reactions with Limiting Amounts of Reactants
A. Many reactions are carried out using an excess of one reactant.
B. Excess reactant — reactant that is present in more than the amount that is needed according to the stoichiometry.
   1. Only the amount required by stoichiometry reacts.
   2. Excess reactant acts as a spectator and takes no role in the reaction.
C. Limiting reactant — the reactant present in the least molar amount.
   1. This determines the extent to which a chemical reaction takes place.
D. To determine if a limiting amount of one reactant is present and to calculate the amount of excess reactant consumed:
   1. Find how many moles of each reactant are present.
   2. Look at the coefficients in the balanced equation to see the required mole ratio of the two reactants and compare to the number of moles calculated in the first step.
   3. Compare the mole ratio from the balanced equation to the ratio calculated in step 1.
   4. Subtract the number of moles of excess reactant consumed from the number of moles of excess reactant present. Convert to grams.
      a. This is grams of nonlimiting reactant in excess.
   5. To determine the amount of product formed, use the number of moles of limiting reactant and the balanced equation to first convert to moles of product, then grams of product.

C.4,C

EXAMPLE:
Boron sulfide, B_2S_3, reacts violently with water to form dissolved boric acid, H_3BO_3, and hydrogen sulfide gas. Assume that 5.88 g of B_2S_3 is mixed with 7.85 g of water. Which reactant is limiting, and which reactant is in excess? How many moles of the excess reactant are consumed? How many moles of the excess reactant are left over? How many grams of boric acid are produced?

SOLUTION: First, write a balanced chemical equation. **Always make sure the chemical equation is balanced before beginning any stoichiometry problem!**

$B_2S_3 + 6 H_2O \rightarrow 2 H_3BO_3 + 3 H_2S$

Determine the number of moles of each reactant present.

$$5.88 \text{ g B}_2\text{S}_3 \times \frac{1 \text{ mol B}_2\text{S}_3}{117.6 \text{ g B}_2\text{S}_3} = 0.0500 \text{ mol B}_2\text{S}_3$$

$$7.85 \text{ g H}_2\text{O} \times \frac{1 \text{ mol H}_2\text{O}}{18.0 \text{ g H}_2\text{O}} = 0.436 \text{ mol H}_2\text{O}$$

The required mole ratio is 6 mol of water for every 1 mol of B_2S_3. Therefore, the required number of moles of water to react with 0.0500 mol of B_2S_3 is $6 \times 0.0500 = 0.300$. Since we have more than enough moles of water to react with B_2S_3, water is the excess reactant and B_2S_3 is the limiting reactant.

Moles of B_2S_3 consumed: 0.0500 mol B_2S_3
Moles of H_2O consumed: 0.300 mol H_2O
Moles of H_2O not consumed: 0.436 mol H_2O – 0.300 mol H_2O = 0.136 mol H_2O

To determine the grams of boric acid, use the balanced chemical equation to convert moles of B_2S_3 to moles of boric acid; then convert from moles of boric acid to grams of boric acid.

$$0.0500 \text{ mol B}_2\text{S}_3 \times \frac{2 \text{ mol H}_3\text{BO}_3}{1 \text{ mol B}_2\text{S}_3} \times \frac{61.8 \text{ g H}_3\text{BO}_3}{1 \text{ mol H}_3\text{BO}_3} = 6.18 \text{ g H}_3\text{BO}_3$$

Workbook Problem 3.5

Pure chromium can be obtained by reacting chromium (III) oxide with aluminum. The other product in this reaction is aluminum oxide. Determine the amount of chromium produced when 4.27 g of chromium(III) oxide reacts with 7.48 g of aluminum. Which reactant is limiting, and which is in excess? How many grams of the excess reactant are leftover? What is the percent yield of the reaction if 1.89 g of chromium are produced?

Strategy: WORK SLOWLY AND CAREFULLY! A good approach is to first determine the number of moles of all reactants that are present. Next, compare mole ratios of the actual amounts to the mole ratios required according to the <u>balanced chemical equation</u>. You can then determine which reactant is in excess and which reactant is present in a limiting amount.

Step 1: Using the information given, write an unbalanced chemical equation.

Step 2: Balance the chemical equation.

Step 3: Determine the mole ratio for chromium(III) oxide and aluminum.

Step 4: Determine the number of moles of each reactant present.

Chapter 3 – Formulas, Equations, and Moles

Step 5: Compare the mole ratio in step 3 to the mole ratio in step 4.

Step 6: Compare the number of theoretical moles of aluminum needed to the actual number of moles of aluminum present. Are there enough moles of aluminum present to react? Is aluminum the limiting reactant or the excess reactant?

Step 7: Based on the number of moles of excess reactant consumed and the number of moles of excess reactant present, calculate the number of moles and the number of grams of excess reactant left over.

Step 8: Calculate the number of moles of chromium produced based on the number of moles of limiting reactant and the balanced chemical equation.

Step 9: Determine the percent yield of the reaction.

What key concept did you use to work this problem?

3.7 Concentrations of Reactants in Solution: Molarity
A. Most chemical reactions are carried out in solution.
   1. Reactants must have considerable mobility for a reaction to occur.
B. Solute — a substance dissolved in a solvent to make a solution.
C. Concentration — describes the relative amount of solute to a solvent or solution.
D. Molarity (M) — the number of moles of a solute dissolved in each liter of solution.
   1. Dissolve the solute in enough solution to give a final solution volume of 1.00 L.
   2. These can be used as conversion factors.

$$\text{Molarity} = \frac{\text{Moles of solute}}{\text{Volume of solution (L)}}$$

$$\text{Moles of solute} = \text{Molarity} \times \text{Volume of solution}$$

$$\text{Volume of solution} = \frac{\text{Moles of solute}}{\text{Molarity}}$$

EXAMPLE:
What is the molarity of a solution prepared by dissolving 16.8 g NaOH and diluting to a final volume of 2.5 L?

SOLUTION: First, find the number of moles of NaOH in 16.8 g NaOH (molar mass = 40.0 g/mol).

$$16.8 \text{ g NaOH} \times \frac{1 \text{ mol NaOH}}{40.0 \text{ g NaOH}} = 0.420 \text{ mol NaOH}$$

Next, divide the number of moles of NaOH by the volume (in liters) of solution.

$$\frac{0.420 \text{ mol NaOH}}{2.5 \text{ L}} = 0.17 \text{ M}$$

EXAMPLE: D.2.

How many moles of $K_2Cr_2O_7$ are present in 250 mL of a 0.025 M solution of $K_2Cr_2O_7$?

SOLUTION: You can calculate the number of moles of $K_2Cr_2O_7$ by multiplying the molarity of the solution by the volume (in liters).

$$\frac{0.025 \text{ mol } K_2Cr_2O_7}{1 \text{ L}} \times 0.250 \text{ L} = 0.0063 \text{ mol } K_2Cr_2O_7$$

EXAMPLE: D.2.

A particular reaction requires that you use 36.7 g of H_2SO_4. How many mL of a 5.00 M solution of H_2SO_4 are required to provide 36.7 g of H_2SO_4?

SOLUTION: First, determine the number of moles of H_2SO_4 in 36.7 g H_2SO_4.

$$36.7 \text{ g } H_2SO_4 \times \frac{1 \text{ mol } H_2SO_4}{98.0 \text{ g } H_2SO_4} = 0.374 \text{ mol } H_2SO_4$$

The volume of solution can now be calculated knowing both the moles of solute and the molarity of the solution.

Workbook Problem 3.6

How many grams of $Pb(NO_3)_2$ are needed to prepare 250 mL of a 0.10 M solution?

Strategy: Using the definition of molarity $\frac{\text{mol solute}}{\text{L soln}}$, and the volume of solution, we can determine moles and grams of solute.

Step 1: First, determine the number of moles of $Pb(NO_3)_2$ found in 250 mL of a 0.10 M solution.

Chapter 3 – Formulas, Equations, and Moles

Step 2: Determine the number of grams of $Pb(NO_3)_2$, knowing the number of moles of $Pb(NO_3)_2$.

What key concept did you use?

3.8 Diluting Concentrated Solutions

A. Concentrated solution + Solvent → Diluted solution
B. Dilution — the number of moles of solute remains the same; only the volume is changed by adding more solvent.

$$\text{Moles of solute} = M_i\left(\frac{\text{mol}}{L}\right) \times V_i(L) = M_f\left(\frac{\text{mol}}{L}\right) \times V_f(L)$$

This equation should be used only for problems that deal with the dilution of a solution!

EXAMPLE:
What would the final concentration be if 25 mL of 12.0 M HCl were diluted to a final volume of 500 mL?

SOLUTION: We know that $M_i \times V_i = M_f \times V_f$. Rearranging this equation gives $M_f = \dfrac{M_i \times V_i}{V_f}$.

The final concentration of the above solution would be

$$\frac{12.0 \text{ M} \times 0.025 \text{ L}}{0.500 \text{ L}} = 0.60 \text{ M}$$

 *Workbook Problem 3.7*
Determine the initial volume needed to prepare 500 mL of 0.75 M HCl from a 6.0 M solution.

Strategy: Remember that the number of moles of solute present do not change when a solution is diluted.

Step 1: Rearrange the equation $M_i \times V_i = M_f \times V_f$ to solve for the initial volume.

What key concept did you use?

3.9 Solution Stoichiometry

A. Knowing the molarity of the solutions is critical to carry out stoichiometry calculations on substances in solution.
   1. Can be used to calculate either the number of moles (if volume is known) or volume (if the number of moles is known).

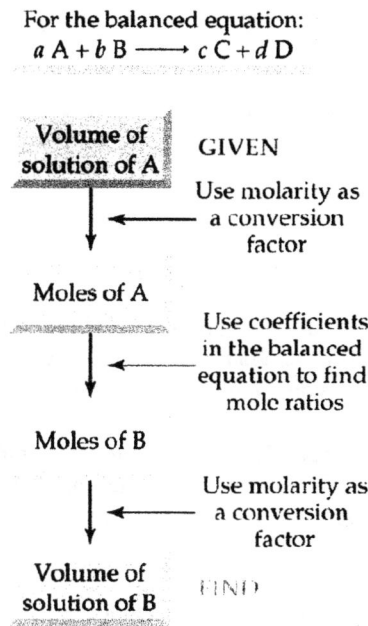

For the balanced equation:
$$a\,A + b\,B \longrightarrow c\,C + d\,D$$

Volume of solution of A — GIVEN
↓ Use molarity as a conversion factor
Moles of A
↓ Use coefficients in the balanced equation to find mole ratios
Moles of B
↓ Use molarity as a conversion factor
Volume of solution of B — FIND

EXAMPLE:

Barium chloride will react with sodium phosphate to produce barium phosphate (a solid) and sodium chloride. How many mL of 0.025 M barium chloride are needed to react with 50 mL of 0.015 M sodium phosphate? How many grams of barium phosphate will be produced?

SOLUTION: First, write the balanced equation for the reaction.

$$3\ BaCl_2 + 2\ Na_3PO_4 \rightarrow Ba_3(PO_4)_2 + 6\ NaCl$$

Next, calculate the moles of Na_3PO_4 present.

$$0.050\ L \times \frac{0.015\ mol\ Na_3PO_4}{1\ L} = 7.5 \times 10^{-4}\ mol\ Na_3PO_4$$

Using the balanced equation, calculate the number of moles of $BaCl_2$ needed to react with 7.5×10^{-4} moles Na_3PO_4.

$$7.5 \times 10^{-4}\ mol\ Na_3PO_4 \times \frac{3\ mol\ BaCl_2}{2\ mol\ Na_3PO_4} = 1.1 \times 10^{-3}\ mol\ BaCl_2$$

The volume of the $BaCl_2$ solution can now be calculated knowing both the molarity and the number of moles.

Chapter 3 – Formulas, Equations, and Moles

$$\frac{1.1 \times 10^{-3} \text{ mol BaCl}_2}{0.025 \text{ mol BaCl}_2 / \text{L}} = 0.044 \text{ L} = 44 \text{ mL}$$

To calculate the grams of barium phosphate produced, we can begin with either moles of sodium phosphate or moles of barium chloride.

$$7.3 \times 10^{-4} \text{ mol Na}_3\text{PO}_4 \times \frac{1 \text{ mol Ba}_3(\text{PO}_4)_2}{2 \text{ mol Na}_3\text{PO}_4} \times \frac{601.9 \text{ g Ba}_3(\text{PO}_4)_2}{1 \text{ mol Ba}_3(\text{PO}_4)_2} = 0.22 \text{ g Ba}_3(\text{PO}_4)_2$$

Workbook Problem 3.8

Hydrochloric acid reacts with zinc metal to produce a solution of zinc chloride and hydrogen gas. What volume of 12 M hydrochloric acid is needed to react with 8.75 g of zinc? If the density of hydrogen gas is 0.0899 g/L, what volume of hydrogen gas will be produced?

Strategy: To solve any stoichiometry problem, we need to know mole ratios of reactants and products as determined by the balanced chemical equation. Knowing the actual number of moles present of one reactant or product, we can then determine the number of moles needed of all other reactants and products.

Step 1: Once again, we begin by writing a balanced chemical equation.

Step 2: Determine the mole ratio of zinc to hydrochloric acid.

Step 3: Determine the number of moles of zinc present.

Step 4: Determine the number of moles of HCl.

Step 5: Calculate the volume of 12 M HCl needed for this reaction.

Step 6: To calculate the volume of H_2 gas produced, we first need to calculate the grams of H_2 gas produced. This is easily accomplished by beginning with the moles of zinc present.

Step 7: Determine the volume of H_2 gas produced, using the grams of H_2 produced and the density of H_2 gas.

What key concept did you use?

3.10 Titration
A. Procedure for determining the concentration of a solution.
   1. Standard solution – a solution of known concentration.
   2. Carefully measured volume of one solution reacts with a known volume of a standard solution.
   3. Indicator – a compound that undergoes a color change during the course of a reaction.

EXAMPLE:
A student finds that 38.4 mL of 0.215 M hydrochloric acid is required to neutralize a 20.0 mL sample of barium hydroxide. What is the molarity of Ba(OH)$_2$?

SOLUTION: First, we need to write a balanced equation for the neutralization reaction.

$$2\ HCl\ (aq)\ +\ Ba(OH)_2\ (aq)\ \rightarrow\ BaCl_2\ (aq)\ +\ H_2O\ (l)$$

From the information given, we can calculate the number of moles of hydrochloric used to reach the equivalence point of the titration.

$$\text{mol HCl} = 0.215 \frac{\text{mol HCl}}{\text{L soln}} \times 0.0384\ \text{L soln} = 8.26 \times 10^{-3}\ \text{mol HCl}$$

From the balanced equation, we know that we have 2 mol HCl for every 1 mol of Ba(OH)$_2$. Therefore, the number of moles of Ba(OH)$_2$ present is 4.13×10^{-3}. Knowing the moles of Ba(OH)$_2$ present allows us to calculate the molarity of Ba(OH)$_2$.

$$\text{Molarity} = \frac{4.13 \times 10^{-3}\ \text{mol Ba(OH)}_2}{0.020\ \text{L soln}} = 0.207\ \text{M}$$

 *Workbook Problem 3.9*
A student determines that 43.35 mL of nitric acid is required to neutralize a 50.0 mL solution containing 0.0316 g of Mg(OH)$_2$. Determine the molarity of the nitric acid solution.

Strategy: Use the same strategy found in Workbook Problem 3.8.

Step 1: Write a balanced chemical equation for the reaction.

Step 2: Determine the mole:mole ratio of nitric acid to Mg(OH)$_2$.

Step 3: Calculate the number of moles of Mg(OH)$_2$ present.

Chapter 3 – Formulas, Equations, and Moles

Step 4: Determine the number of moles of nitric acid that will react with the $Mg(OH)_2$.

Step 5: Calculate the molarity of the nitric acid solution.

What key concept did you use?

3.11 Percent Composition and Empirical Formulas

A. Composition of a compound — the identity and amount of elements present in a compound.
B. Percent composition — the mass percent of each element present in a compound.
  1. Calculate the chemical formula from percent composition.
     a. Find the relative numbers of moles of each element in the compound.
        i. assume 100 g of compound
        ii. use molar masses of the elements as conversion factors
     b. Find the ratio of the numbers of moles by dividing the larger number of moles by the smaller number.
     c. Multiply the subscripts by small integers in a trial-and-error procedure until whole numbers are found.

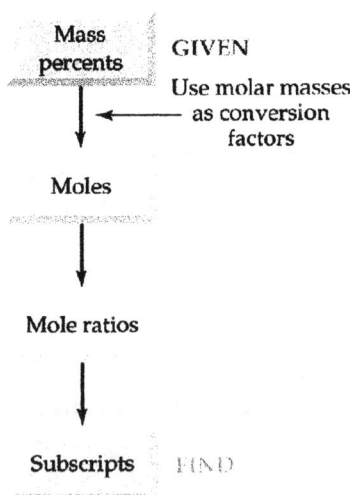

C. Empirical formula — gives only the ratios of atoms in a compound.
D. Molecular formula — gives the actual numbers of atoms in a molecule.
  1. May be the same as the empirical formula.
  2. May be a multiple of the empirical formula.
     a. $\text{Multiple} = \dfrac{\text{Molecular mass}}{\text{Empirical formula mass}}$

Chapter 3 – Formulas, Equations, and Moles

E. Can derive percent composition from a chemical formula.

EXAMPLE:
The percent composition of a solid is known to be 68.4% Ba, 10.3% P, and 21.3% O. What is the empirical formula of the compound?

SOLUTION: Assuming a 100 g sample gives us 68.4 g Ba, 10.3 g P, and 21.3 g O, convert these masses to numbers of moles.

$$68.4 \text{ g Ba} \times \frac{1 \text{ mol Ba}}{137.3 \text{ g Ba}} = 0.498 \text{ mol Ba}$$

$$10.3 \text{ g P} \times \frac{1 \text{ mol P}}{31.0 \text{ g P}} = 0.332 \text{ mol P}$$

$$21.3 \text{ g O} \times \frac{1 \text{ mol O}}{16 \text{ g O}} = 1.33 \text{ mol O}$$

Knowing the relative numbers of moles, find the ratio by dividing the two larger numbers by the smaller number.

$$\frac{1.33}{0.332} = 4 \quad \frac{0.498}{0.332} = 1.5$$

The O:Ba:P ratio of 4:1.5:1 gives an empirical formula of $Ba_{1.5}PO_4$. However, we must multiply the subscripts by a small integer (in this case 2) to find whole numbers for the formula: $Ba_3P_2O_8$, or $Ba_3(PO_4)_2$.

 Workbook Problem 3.10
Determine the empirical formula of a compound that is 32.4% Na, 0.70% H, 21.8% P and 45.1% O.

Strategy: Assume you have a 100 g sample, and convert the percentages of the elements into grams.

Step 1: Convert the grams of each element into moles of each element.

Step 2: Knowing the relative number of moles, find the ratio of moles by dividing the three larger numbers by the smallest number.

Step 3: If necessary, multiply the subscript by a small integer to find whole numbers for the formula, and write the formula of the compound.

What key concept did you use?

EXAMPLE:
What is the percent composition of sodium hydrogen carbonate?

SOLUTION: The formula for sodium hydrogen carbonate is $NaHCO_3$. The Na:H:C:O mole ratio is 1:1:1:3. Convert this mole ratio into a mass ratio by assuming there is a 1 mole sample present.

$$1 \text{ mol } NaHCO_3 \times \frac{1 \text{ mol Na}}{1 \text{ mol } NaHCO_3} \times \frac{23 \text{ g Na}}{1 \text{ mol Na}} = 23 \text{ g Na}$$

$$1 \text{ mol } NaHCO_3 \times \frac{1 \text{ mol H}}{1 \text{ mol } NaHCO_3} \times \frac{1.0 \text{ g H}}{1 \text{ mol H}} = 1.0 \text{ g H}$$

$$1 \text{ mol } NaHCO_3 \times \frac{1 \text{ mol C}}{1 \text{ mol } NaHCO_3} \times \frac{12 \text{ g C}}{1 \text{ mol C}} = 12 \text{ g C}$$

$$1 \text{ mol } NaHCO_3 \times \frac{3 \text{ mol O}}{1 \text{ mol } NaHCO_3} \times \frac{16 \text{ g O}}{1 \text{ g O}} = 48 \text{ g O}$$

To determine the percent composition, divide the mass of each element present by the total mass of the compound and multiply by 100.

Total mass of 1 mole of $NaHCO_3$ = 84 g

$$\% \text{ Na} = \frac{23 \text{ g Na}}{84 \text{ g}} \times 100\% = 27\%$$

$$\% \text{ H} = \frac{1.0 \text{ g H}}{84 \text{ g}} \times 100\% = 1.2\%$$

$$\% \text{ C} = \frac{12 \text{ g C}}{84 \text{ g}} \times 100\% = 14\%$$

$$\% \text{ O} = \frac{48 \text{ g O}}{84 \text{ g}} \times 100\% = 57\%$$

Workbook Problem 3.11
Determine the percent composition of aluminum bromate.

Strategy: Write the formula for aluminum bromate and determine the mole ratio of the elements in the compound.

Chapter 3 – Formulas, Equations, and Moles

Step 1: Assume a 1 mole sample is present and convert the mole ratio into a mass ratio.

Step 2: Determine the percent composition by dividing the mass of each element present by the total mass of the compound and multiplying by 100.

What key concept did you use?

3.12 Determining Empirical Formulas: Elemental Analysis   E.2

A. Combustion analysis — a compound of unknown composition is burned with oxygen to produce the volatile combustion products CO_2 and H_2O.
   1. CO_2 and H_2O are separated by a gas chromatograph and their masses are determined.
   2. The masses of CO_2 and H_2O are used to determine the empirical formula of a compound.
B. To determine the empirical formula of a compound from a combustion reaction:
   1. Carry out gram-to-mole conversions to find the molar amounts of C and H in CO_2 and H_2O.
   2. Carry out mole-to-gram conversions to find the number of grams of C and H in the original sample.
   3. Subtract the masses of C and H from the mass of the starting sample to determine the number of grams of the substance unaccounted for.
   4. Determine the number of moles of the remaining substance.
   5. Find the ratio of the numbers of moles by dividing the larger number of moles by the smaller number of moles.
C. To determine the molecular formula of a substance:
   1. You need to know molecular mass.
   2. Determine ratio of molecular mass to empirical formula mass (see page 58 in this book).

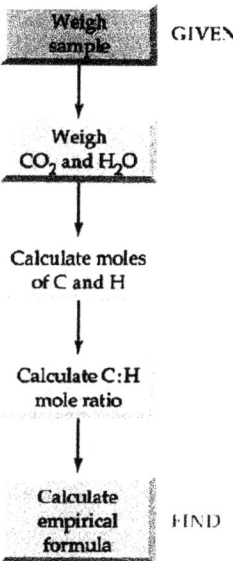

Chapter 3 – Formulas, Equations, and Moles

EXAMPLE:
When 4.50 g of ethyl butyrate, a compound containing C, H, and O, undergoes combustion, 10.24 g of CO_2 and 4.19 g H_2O are produced. Determine the empirical formula of this compound.

SOLUTION: First, find the molar amounts of C and H in CO_2 and H_2O.

$$10.24 \text{ g } CO_2 \times \frac{1 \text{ mol } CO_2}{44.0 \text{ g } CO_2} \times \frac{1 \text{ mol C}}{1 \text{ mol } CO_2} = 0.233 \text{ mol C}$$

$$4.19 \text{ g } H_2O \times \frac{1 \text{ mol } H_2O}{18.0 \text{ g } H_2O} \times \frac{2 \text{ mol H}}{1 \text{ mol } H_2O} = 0.466 \text{ mol H}$$

Next, carry out mole-to-gram conversions to find the number of grams of C and H in the original sample.

$$0.233 \text{ mol C} \times \frac{12.0 \text{ g C}}{1 \text{ mol C}} = 2.79 \text{ g C}$$

$$0.466 \text{ mol H} \times \frac{1.00 \text{ g H}}{1 \text{ mol H}} = 0.466 \text{ g H}$$

Subtract the masses of C and H from the mass of the starting sample to determine the mass of O.

$$4.50 \text{ g} - 2.79 \text{ g} - 0.466 \text{ g} = 1.24 \text{ g}$$

Convert the mass of O to moles of O.

$$1.24 \text{ g O} \times \frac{1 \text{ mol O}}{16.0 \text{ g O}} = 0.0778 \text{ mol O}$$

Find the ratio of the numbers of moles by dividing the larger number of moles by the smaller number of moles.

$$\frac{0.233}{0.0778} = 3 \qquad \frac{0.446}{0.0778} = 6$$

This gives a mole ratio of C:H:O of 3:6:1. Therefore, the empirical formula for ethyl butyrate is C_3H_6O.

EXAMPLE:
Determine the molecular formula of ethyl butyrate knowing that the molecular mass of the compound is 116 g/mol.

SOLUTION: The molecular formula of the compound may be the same as the empirical formula or it may be a multiple of the empirical formula. To determine if the molecular formula is a multiple of the empirical formula the following equation is used.

$$\text{Multiple} = \frac{\text{Molecular mass}}{\text{Empirical formula mass}}$$

For ethyl butyrate:

$$\text{Multiple} = \frac{116 \text{ g/mol}}{58 \text{ g/mol}} = 2$$

The subscripts in the empirical formula are multiplied by 2 giving a molecular formula of $C_6H_{12}O_2$.

 Workbook Problem 3.12

Octyl acetate is an organic compound containing only C, H, and O. This compound is often added to foods to give them a flavor of oranges. When 2.85 g of octyl acetate undergoes combustion, 7.29 g of CO_2 and 2.98 g of H_2O are produced. The molar mass of octyl acetate is 172 g/mol. Determine the empirical and molecular formula of this compound.

Strategy: Follow the steps in the flow chart on page 61 of this book.

Step 1: Find the molar amounts of C and H in CO_2 and H_2O.

Step 2: Carry out mole-to-gram conversions to find the number of grams of C and H in the original sample.

Step 3: Subtract the masses of C and H from the mass of the starting sample to determine the mass of O.

Step 4: Convert the mass of O to moles of O.

Step 5: Find the ratio of the number of moles by dividing the larger numbers of moles by the smallest number of moles.

Step 6: Use the ratio above to write the empirical formula of the compound.

Step 7: Determine the multiple for the empirical formula.

Chapter 3 – Formulas, Equations, and Moles

Step 8: Write the molecular formula.

What key concept did you use?

 **Putting It Together:**

A 1.976 g sample of a metal sulfide in which the metal has a +3 charge reacts with hydrochloric acid to produce an aqueous solution of the metal chloride and hydrogen sulfide gas. 100 mL of 0.75 M HCl was used in the reaction. Once completed, all of the hydrogen sulfide gas was removed and the excess HCl acid was titrated with 36.0 mL of 0.50 M NaOH. Determine the identity of the metal. The density of hydrogen sulfide is 1.539 g/L. Determine the volume of H_2S produced.

 **Self-Test**

This section is intended to test your knowledge of the material covered in this chapter. Think through these problems, and make certain you understand what is going on. Ask yourself if your answer makes sense. Many of these questions are linked to the chapter learning goals. Therefore, successful completion of these problems indicates you have mastered the learning goals for this chapter. You will receive the greatest benefit from this section if you use it as a mock exam. You will then discover which topics you have mastered and which topics you need to study in more detail.

True-False

1. A balanced chemical equation obeys the law of definite proportions.

2. When balancing a chemical equation, the coefficients or the subscripts of the formula units may be changed.

3. When hydrocarbons undergo combustion, carbon dioxide and water are produced.

4. Chemical symbols represent the behavior of individual atoms and molecules only on the microscopic level.

5. Laboratory work involves the behavior of atoms and molecules on a large-scale (macroscopic) level.

6. The coefficients in a balanced chemical equation indicate the number of moles of each substance needed for the reaction.

7. The excess reactant determines the extent to which a chemical reaction takes place.

8. When diluting a solution, the number of moles of solute changes.

9. The theoretical yield of a reaction is always greater than the actual yield.

10. The empirical formula and molecular formula of a compound are always different.

Chapter 3 – Formulas, Equations, and Moles

Matching

Combustion reaction	a. reactant that is present in more than the amount that is actually needed according to the stoichiometry
Microscopic level	b. gives the ratio of atoms in a compound
Macroscopic level	c. reaction with oxygen
Formula mass	d. a procedure for determining the concentration of a solution in which a carefully measured volume of one solution reacts with a second solution of known concentration and volume
Molecular mass	e. the sum of the atomic masses of all atoms in one formula unit of any substance
Molar mass	f. chemical symbols represent the behavior of individual atoms and molecules
Excess reactant	g. the mass percent of each element present in a compound
Limiting reactant	h. give the actual number of atoms in a molecule
Molarity	i. the sum of the atomic masses of all atoms in a molecule
Titration	j. the reactant that is present in the limiting amount and determines the extent to which a chemical reaction takes place
Percent composition	k. the molecular or formula mass in grams of one mole of any substance.
Empirical formula	l. the number of moles of solute dissolved in one liter of solution
Molecular formula	m. the large-scale behavior of atoms and molecules that gives rise to observable properties

Fill-in-the-Blank

1. All chemical equations must be _____.
2. One mole of any object, atom, molecule, or ion contains _____ of formula units.
3. Using the relationship between numbers of moles and numbers of grams to carry out chemical calculations is called _____.

Chapter 3 – Formulas, Equations, and Moles

4. The concentration of a substance in solution is usually expressed as _____.

5. When carrying out a dilution, the _____ of the solution changes, but the _____ stays the same.

6. In carrying out a chemical reaction, the _____ determines the amount of product formed.

7. _____ makes it possible to calculate the amount of volume of one solution needed to react with a given volume of a standard solution.

8. The empirical formula of a substance can be derived from _____.

9. In a combustion analysis, the products _____ and _____ are formed.

10. To do actual laboratory work, it is necessary to convert between _____ and _____ of reactants and products.

Problems

1. Balance the following equations:

    a. $Mg_3N_2 + HCl \rightarrow MgCl_2 + NH_4Cl$

    b. calcium carbonate reacts with nitric acid to produce carbon dioxide gas, water, and calcium nitrate.

    c. Octane (C_8H_{18}) undergoes combustion.

    d. $Al + H_2SO_4 \rightarrow Al_2(SO_4)_3 + H_2$

    e. $AgNO_3 + Na_2SO_4 \rightarrow Ag_2SO_4 + NaNO_3$

2. Phosphine (PH_3) is an extremely poisonous gas with a garlic odor and is used in the process for making flame-resistant cotton cloth. How many grams of phosphine will be produced when 15.4 g of calcium phosphide reacts with water to produce phosphine and calcium hydroxide?

3. Many metal ions will react with hydrogen sulfide to form precipitates with a characteristic color. This reaction has been widely used in the laboratory to identify the presence of metal ions in an unknown solution. The safest way of producing hydrogen sulfide is by the reaction of water with thioacetamide.

$$CH_3C(=S)NH_2 + H_2O \longrightarrow CH_3C(=O)NH_2 + H_2S$$

How many grams of water are needed to react with 5.75 g of thioacetamide? How many grams of hydrogen sulfide will be produced?

4. Sodium sulfite is used in the manufacture of paper pulp and as a food additive to prevent decomposition. It is prepared according to the following equation:

   $NaHSO_3 + Na_2CO_3 \rightarrow Na_2SO_3 + CO_2 + H_2O$

   How many grams of sodium sulfite would you obtain from 75.8 g of sodium hydrogen sulfite if the reaction had a percent yield of 64%?

5. The largest use of nitric acid is in the reaction with ammonia to produce ammonium nitrate for fertilizer. If 156.5 g of nitric acid reacted with 275.0 g of ammonia, which reactant would be in excess? How many moles of the excess reactant would be left over after the reaction? How many grams of ammonium nitrate would be produced?

6. Describe how you would prepare 500 mL of 3.75 M NaOH.

7. Describe how you would prepare 2.50 L of 0.125 M H_3PO_4 beginning with a concentrated solution that has a molarity of 15.0 M.

8. How many milliliters of 0.150 M HCl are needed to react with 2.45 g of zinc in the following reaction?

   $Zn + 2\,HCl \rightarrow ZnCl_2 + H_2$

9. 50 mL of 0.025 M $Pb(NO_3)_2$ is reacted with 50 mL of 0.10 M K_2CrO_4 to produce $PbCrO_4$, a bright yellow solid, and KNO_3. Which reactant is in excess? How many moles of excess reactant will be left over at the end of the reaction? How many grams of $PbCrO_4$ will be produced?

10. What is the percent composition of aspirin, $C_9H_8O_4$?

11. What is the empirical formula for ethylene glycol if the percent composition is 38.7% C, 9.7% H and 51.6% O? What is the molecular formula for this compound if the molar mass is 62 g/mol?

12. For the reaction

    $Ag_2CO_3 + HNO_3 \rightarrow CO_2 + AgNO_3 + H_2O$

    how many grams of silver nitrate will be formed when 9.85 g of silver carbonate reacts with 250 mL of 0.250 M HNO_3?

13. 12.78 g of ethyl acetate, a compound containing C, H, and O, undergoes combustion to produce 25.56 g of carbon dioxide and 10.46 g of water. What is the empirical formula of this compound? What is the molecular formula of this compound if the molar mass is 88.0 g/mol?

14. What is the percent composition of calcium phosphite?

15. 4-*n*-hexylresorcinol is used as an antiseptic in throat lozenges and mouth washes. Its percent composition is 74.2% C, 9.3% H, and 16.5% O. What is its empirical formula? What is its molecular formula if the molar mass is 194 g/mol?

Chapter 3 – Formulas, Equations, and Moles

Challenge Problem

When ascorbic acid, $C_6H_8O_6$, reacts with bromine, dehydroascorbic acid, $C_6H_6O_6$ is produced.

$C_6H_8O_6 + Br_2 \rightarrow C_6H_6O_6 + 2\ HBr$

Any excess bromine can be determined by reacting the bromine with potassium iodide.

$Br_2 + KI \rightarrow I_2 + 2\ KBr$

The iodine produced is titrated with sodium thiosulfate.

$I_2 + 2\ Na_2S_2O_3 \rightarrow 2\ NaI + Na_2S_4O_6$

The bromine used in this reaction is produced by the following reaction:

$KBrO_3 + 5\ KBr + 3\ H_2SO_4 \rightarrow 3\ Br_2 + 3\ H_2O + 3\ K_2SO_4$

Starting with 25 mL of 0.15 M $KBrO_3$, determine the amount of ascorbic acid present if 32.5 mL of 0.25 M $Na_2S_2O_3$ is required for the titration of iodine.

Inquiry Based Problem

You arrive in General Chemistry laboratory to a vial with a solid substance labeled "Unknown Chloride Salt." You have also been provided a solution labeled 0.50 M $AgNO_3$. Your laboratory instructor informs you that reaction of $AgNO_3$ with a chloride salt will cause AgCl to precipitate. Explain to your laboratory instructor a simple experiment using stoichiometry that will allow you to determine the amount of chloride ion in the salt. The chemical equation for this reaction can be written as:

$Cl^- (aq) + Ag^+ (aq) \rightarrow AgCl\ (s)$

> We'll learn about writing chemical equations in this manner in Chapter 4.

Think about the following:

How could you determine the number of moles of AgCl in the laboratory?

If you know the number of moles of AgCl, what data can you obtain about Cl^- using stoichiometric calculations and the chemical equation provided?

From this information, can you write a simple laboratory procedure that will allow you to determine the mass of Cl^- present?

CHAPTER 4

REACTIONS IN AQUEOUS SOLUTIONS

Chapter Learning Goals

⚷ *A. Electrolytes and Nonelectrolytes*
   1. Classify substances as electrolytes or nonelectrolytes.

⚷ *B. Chemical Equations*
   1. Write molecular, ionic, and net ionic equations for precipitation, acid–base, and redox reactions.

⚷ *C. Chemical Reactions in Aqueous Solution: Precipitation and Acid–Base Neutralization*
   1. State solubility rules, and use them to predict whether a precipitate might form when aqueous salt solutions are mixed.
   2. Identify the common strong acids and strong bases.

⚷ *D. Chemical Reactions in Aqueous Solution: Oxidation–Reduction*
   1. Assign oxidation numbers to each atom in a chemical species.
   2. In a redox reaction, identify the species oxidized, the species reduced, the oxidizing agent, and the reducing agent.
   3. Using an activity series, predict whether a redox reaction will occur when a metal is placed in contact with a solution containing an ion of a different metal.
   4. Balance redox reactions by the oxidation-number method and by the half-reaction method.
   5. Determine the concentration of a species, using data from a redox titration.

Chapter in Brief

This chapter discusses three different types of chemical reactions in aqueous solution: precipitation reactions, acid–base reactions, and oxidation–reduction reactions. These reactions are often written as net ionic equations. As a consequence, it is necessary to know the difference between strong and weak electrolytes and nonelectrolytes. You can predict the outcome of each of these reactions if you know the solubility rules, can recognize acids and bases, and know how to assign oxidation numbers to compounds. You are also shown how to balance oxidation–reduction reactions with either the oxidation–number method or the method of half–reactions. Finally, you will apply the concepts you learned in earlier chapters about stoichiometry to the reactions that are discussed in this chapter.

4.1 Some Ways That Chemical Reactions Occur
   A. Three general categories for chemical reactions in aqueous solution.
       1. Precipitaiton reactions
       2. Acid-base neutralization reactions.
       3. Oxidation-reduction reactions.
   B. Precipitation reactions – soluble reactants yield an insoluble solid product.
       1. Solid (precipitate) falls out (precipitates) from solution.
       2. Anions and cations of two ionic compounds change partners.
   C. Acid-base neutralization reaction – an acid reacts with a base to yield water plus a salt (ionic compound).
       1. Removes H^+ and OH^- from solution.

Chapter 4 – Reactions in Aqueous Solution

D. Oxidation-reduction reactions – one or more electrons are transferred between reaction partners (atoms, molecules or ions).
   1. The charges on atoms in the various reactants change

4.2 Electrolytes in Aqueous Solution

A.1.

A. Electrolytes — dissolve in water to produce ionic solutions.
   1. Solution conducts electricity.
B. Nonelectrolytes — do not produce ions in aqueous solution.
C. Dissociation – a dynamic process.
   1. Equilibrium is established between the forward and reverse reaction.
   2. A forward-and-backward double arrow ⇌ indicates the reaction takes place in both directions.
D. Strong Electrolyte — Compound dissociates to a large exten (70-100%) when dissolved in water.
   1. Electrolyte classification of some common substances – see Table 4.1 p. 110 in your textbook.
   2. When writing chemical equations, strong electrolytes are designated by inserting (*aq*) after the formula.
E. Weak Electrolyte — Compounds dissociates only to a small extent.
   1. Establish an equilibrium between the forward and backward reactions.

EXAMPLE:
$BaCl_2$ is a strong electrolyte. Determine the concentration of Cl^- ions in a 1.25 M aqueous solution of $BaCl_2$.

SOLUTION:

From the definition of a strong electrolyte, we know that the $BaCl_2$ will completely dissociate in water. That is, $BaCl_2$ will fall apart to produce 1 mol of Ba^{2+} and 2 mol of Cl^-.

$$1 \text{ mol } BaCl_2 = 1 \text{ mol } Ba^{2+} + 2 \text{ mol } Cl^-$$

We can calculate the molar concentration of Cl^- knowing the mole ratio of 2 mol Cl^- to 1 mol $BaCl_2$.

$$\frac{1.25 \text{ mol } BaCl_2}{L \text{ soln}} \times \frac{2 \text{ mol } Cl^-}{1 \text{ mol } BaCl_2} = \frac{2.50 \text{ mol } Cl^-}{L \text{ soln}} = 2.50 \text{ M } Cl^-$$

 *Workbook Problem 4.1*
Determine the total molar concentration of ions in a $Ca_3(PO_3)_2$ solution with a concentration of 0.275 M.

Strategy: Ca_3P_2 is a strong electrolyte and, therefore, completely dissociates.

Step 1: Determine the total number of moles of ions formed when Ca_3P_2 completely dissociates in water.

Step 2: Create a conversion factor comparing the total number of moles of ions in solution to 1 mol of Ca_3P_2.

Step 3: Use the conversion factor to calculate the molar concentration of ions in solution.

What key concept did you use?

4.3 Aqueous Reactions and Net Ionic Equations
  A. Molecular Equations — reactants and products are written using their complete formulas as if they were molecules.
  B. Ionic Equations — any reactants or products that completely dissociate in water (strong electrolytes) are shown in terms of their free ions.
  C. Spectator Ions — ions that do not undergo a change during the reaction.
   1. The specific identity is not important.
   2. Role is to balance the charge.
  D. Net Ionic Equation — the equation for the net change that takes place during the reaction; the spectator ions are not included.

EXAMPLE:
When solutions of barium chloride and sodium phosphate are mixed, solid barium phosphate and a solution of NaCl are produced. Write molecular, ionic, and net ionic equations for this reaction.

SOLUTION:

Unbalanced Molecular Eqn: $BaCl_2\ (aq)\ +\ Na_3(PO_4) \rightarrow Ba_3(PO_4)_2\ (s)\ +\ NaCl\ (aq)$

Balanced Molecular Eqn: $3\ BaCl_2\ (aq)\ +\ 2\ Na_3(PO_4)\ (aq) \rightarrow Ba_3(PO_4)_2\ (s)\ +\ 6\ NaCl\ (aq)$

Ionic Eqn: $3\ Ba^{2+}\ (aq) + 6\ Cl^-\ (aq) + 6\ Na^+\ (aq) + 2\ PO_4^{3-}\ (aq) \rightarrow Ba_3(PO_4)_2\ (s) + 6\ Na^+\ (aq) + 6\ Cl^-\ (aq)$

Spectator Ions: Na^+ and Cl^-

Net Ionic Equation: $3\ Ba^{2+}\ (aq)\ +\ 2\ PO_4^{3-}\ (aq) \rightarrow Ba_3(PO_4)_2\ (s)$

Workbook Problem 4.2
When solutions of sodium sulfide and hydrochloric acid are mixed together they react to form a solution of sodium chloride and hydrogen sulfide gas. Write the balanced molecular, ionic, and net ionic equations for this reaction.

Strategy: From the information given, write a molecular equation and determine if any of the reactants or products are strong electrolytes.

Chapter 4 – Reactions in Aqueous Solution

Step 1: Write an unbalanced chemical equation.

Step 2: Balance the molecular equation.

Step 3: Write the strong electrolytes in terms of their free ions to obtain the ionic equation.

Step 4: Determine if any spectator ions are present, and write the net ionic equation.

Which key concept did you use?

4.4 Precipitation Reactions and Solubility Rules
A. Solubility — how much of each compound will dissolve in a given amount of solvent at a given temperature.
 1. Dependent upon concentrations of reactant ions.
 2. Low solubility — precipitate forms.
 3. High solubility — no precipitate will form.
 4. Solubilities can be predicted using the guidelines found on page 112 in the textbook.
B. Solubility Guidelines.
 1. Use to predict if a precipitate will form in a reaction.
 2. Use to prepare and isolate a specific compound by carrying out a precipitation reaction.

EXAMPLE:
Write molecular, ionic, and net ionic equations for the reaction between $Pb(NO_3)_2$ and $(NH_4)_2SO_4$; $SnCl_2$ and $NaOH$; $NaNO_3$ and $BaCl_2$.

SOLUTION: First, write an unbalanced chemical equation by exchanging the cations between the two salts. Using the solubility guidelines, determine if any of the products are insoluble.

$Pb(NO_3)_2$ and $(NH_4)_2SO_4$:
 Unbalanced Molecular Eqn:
 $Pb(NO_3)_2 + (NH_4)_2SO_4 \rightarrow PbSO_4 + NH_4NO_3$
 Insoluble Products: $PbSO_4$
 Balanced Molecular Eqn:
 $Pb(NO_3)_2\ (aq) + (NH_4)_2SO_4\ (aq) \rightarrow PbSO_4\ (s) + 2\ NH_4NO_3\ (aq)$
 Ionic Eqn:
 $Pb^{2+}\ (aq) + 2\ NO_3^-\ (aq) + 2\ NH_4^+\ (aq) + SO_4^{2-}\ (aq) \rightarrow PbSO_4\ (s) + 2\ NH_4^+\ (aq) + 2\ NO_3^-\ (aq)$
 Spectator ions: NH_4^+ and NO_3^-
 Net ionic Eqn:
 $Pb^{2+}\ (aq) + SO_4^{2-}\ (aq) \rightarrow PbSO_4\ (s)$

SnCl₂ and NaOH:
   Unbalanced Molecular Eqn:
   $SnCl_2 + NaOH \rightarrow Sn(OH)_2 + NaCl$
   Insoluble Products: $Sn(OH)_2$
   Balanced Molecular Eqn:
   $SnCl_2 (aq) + 2 NaOH(aq) \rightarrow Sn(OH)_2 (s) + 2 NaCl (aq)$
   Ionic Eqn:
   $Sn^{2+} (aq) + 2 Cl^- (aq) + 2 Na^+ (aq) + 2 OH^- (aq) \rightarrow Sn(OH)_2 (s) + 2 Na^+ (aq) + 2 Cl^- (aq)$
   Spectator ions: Na^+ and Cl^-
   Net ionic Eqn:
   $Sn^{2+} (aq) + 2 OH^- (aq) \rightarrow Sn(OH)_2 (s)$

NaNO₃ and BaCl₂:
   Unbalanced Molecular Eqn:
   $NaNO_3 + BaCl_2 \rightarrow Ba(NO_3)_2 + NaCl$
   Insoluble Products: none
   Balanced Molecular Eqn:
   $2 NaNO_3 (aq) + BaCl_2 (aq) \rightarrow Ba(NO_3)_2 (aq) + 2 NaCl (aq)$
   Ionic Eqn:
   $2 Na^+ (aq) + 2 NO_3^- (aq) + Ba^{2+} (aq) + 2 Cl^- (aq) \rightarrow Ba^{2+} (aq) + 2 NO_3^- (aq) + 2 Na^+ (aq) + 2 Cl^- (aq)$
   Spectator ions: all ions are spectator ions
   No net ionic reaction

 **Workbook Problem 4.3**
Give a description for the preparation of 2.50 g $Ba_3(PO_4)_2$ from solutions of two soluble salts.

Strategy: Using the solubility guidelines and solution stoichiometry, determine the reactants to use in the preparation of $Ba_3(PO_4)_3$ and the amount of reactants needed.

Step 1: Determine the reactants that are soluble and will produce the insoluble $Ba_3(PO_4)_2$ and another soluble product.

Step 2: Write a balanced chemical equation.

Step 3: Use solution stoichiometry to determine the molarity of each reactant needed (assume a 1 L solution of each reactant.)

Chapter 4 – Reactions in Aqueous Solution

What key concept did you use?

4.5 Acids, Bases, and Neutralization Reactions
A. Acid — produces H^+ in aqueous solution
  1. Hydronium Ion — H^+ attaches to a water molecule; represented as H_3O^+.
  2. Strong acids — strong electrolytes.
  3. Weak acids — weak electrolytes.
  4. Polyprotic acids — acids with more than one acidic hydrogen; dissociate in steps.
B. Base — produces OH^- in aqueous solution.
  1. Strong bases — strong electrolytes; most metal hydroxides.
  2. Weak bases — weak electrolytes.
    a. $NH_3\ (g) + H_2O\ (l) \rightleftarrows NH_4^+\ (aq) + OH^-\ (aq)$
C. Neutralization Reaction – a reaction between an acid and base that produces a salt and water.
  1. Salt produced from cation of base and anion of acid.
  2. Net reaction for a strong acid + strong base: $H^+\ (aq) + OH^-\ (aq) \rightarrow H_2O$
  3. Net reaction for a weak acid + strong base: $HA\ (aq) + OH^-\ (aq) \rightarrow H_2O + A^-\ (aq)$
D. Common acid and bases – see Table 4.2, page 115 in the textbook.

EXAMPLE:
Determine if the following substances are acids or bases: HBr (*aq*), KOH, $HClO_3$, $Ca(OH)_2$, CH_3COOH, NH_3.

SOLUTION: To determine if a species is either an acid or a base according to the Arrhenius definition, we must look for the ability of the substance to produce either H^+ or OH^- in aqueous solution. (You also can memorize Table 4.2 in your textbook, but we prefer that you understand the concept.)

HBr (*aq*) – produces H^+ in aqueous solution; acid.
KOH – produces OH^- in aqueous solution; base.
$HClO_3$ – produces H^+ in aqueous solution; acid.
$Ca(OH)_2$ – produces OH^- in aqueous solution; base.
CH_3COOH – produces H^+ in aqueous solution; acid.
NH_3 – produces OH^- in aqueous solution; base.

EXAMPLE:
Write the molecular, ionic, and net ionic equations for the reaction between nitric acid and calcium hydroxide.

SOLUTION: Remember that the reaction between an acid and a base produces a salt and water. The cation of the salt comes from the base, and the anion of the salt comes from the acid. In this example, the salt produced is $Ca(NO_3)_2$. Therefore, the equations for this reaction are,

Molecular:
$2\ HNO_3 + Ca(OH)_2\ (aq) \rightarrow Ca(NO_3)_2\ (aq) + 2\ H_2O\ (l)$
Ionic:
$2\ H^+\ (aq) + 2\ NO_3^-\ (aq) + Ca^{2+}\ (aq) + 2\ OH^-\ (aq) \rightarrow Ca^{2+}\ (aq) + 2\ NO_3^-\ (aq) + 2\ H_2O\ (l)$
Spectator ions: Ca^{2+} and NO_3^-
Net ionic:
$2\ H^+\ (aq) + 2\ OH^-\ (aq) \rightarrow 2\ H_2O\ (l)$

Chapter 4 – Reactions in Aqueous Solution

Workbook Problem 4.4
Write balanced molecular, ionic, and net ionic equations for the following reactions:
a. $Fe(NO_3)_3$ and Na_2S
b. NH_4Cl and $Hg_2(NO_3)_2$
c. $Ba(OH)_2$ and H_2SO_4

Strategy: Determine if the reactants are acids and bases, which will lead to a determination of the type of reaction..

Step 1: Write the balanced molecular equation for each reaction.

Step 2: Determine the presence of any strong electrolytes. Write the ionic equation, showing the strong electrolytes in terms of their free ions.

Step 3: Determine the presence of any spectator ions. Write the net ionic equation, dropping out any spectator ions that are present.

What key concept did you use?

4.6 Oxidation–Reduction (Redox) Reactions
A. Oxidation and Reduction
1. Oxidation — the loss of one or more electrons; increase in oxidation number.
2. Reduction — the gain of one or more electrons; decrease in oxidation number.
3. Two processes occur simultaneously.
B. Redox Reaction — electrons are transferred from one substance to another.
1. The number of electrons lost by the substance being oxidized equals the number of electrons gained by the substance being reduced.

Chapter 4 – Reactions in Aqueous Solution

EXAMPLE:
Determine the oxidation number of phosphorus in Na_3PO_4.

SOLUTION:
1. Assign an oxidation number of +1 to Na and –2 to the O.
2. Remember that the sum of all of the oxidation numbers in the compound equals 0.

$Na_3PO_4 = (3\ Na^+)\ (P?)\ (4\ O^{2-})$ $3(+1) + (?) + 4(-2) = 0$ net charge
$? = 0 - 3(+1) - 4(-2) = +5$

 ***Workbook Problem 4.5***
Determine the oxidation number for Mn in $KMnO_4$ and the oxidation number for all elements present in HIO_4.

Strategy: Follow the rules found on pages 118 and 119 of your textbook.

Step 1: Identify the elements that have a fixed oxidation state, and assign that oxidation state to the element.

Step 2: Determine the oxidation number of the remaining element, keeping in mind that the sum of all oxidation numbers must be equal to zero.

What key concept did you use?

4.7 Identifying Redox Reactions
A. Oxidation and reduction occur together.
  1. Whenever an atom loses one or more electrons (is oxidized), another atom must gain those electrons (be reduced).
B. Reducing Agent — the substance that causes reduction to occur.
  1. Loses one or more electrons.
    a. undergoes oxidation
    b. oxidation number of atom increases
  2. Metals act as reducing agents.

Chapter 4 – Reactions in Aqueous Solution

C. Oxidizing Agent — the substance that causes oxidation to occur.
   1. Gains one or more electrons.
      a. undergoes reduction
      b. oxidation number of atom decreases
   2. Reactive nonmetals act as oxidizing agents.

EXAMPLE:
Identify the species oxidized, the species reduced, the oxidizing agent, and reducing agent in the following reaction:

$$16\ H^+\ (aq) + 5\ Sn^{2+}\ (aq) + 2\ MnO_4^-\ (aq) \rightarrow 2\ Mn^{2+}\ (aq) + 5\ Sn^{4+}\ (aq) + 8\ H_2O\ (l)$$

SOLUTION: To solve this problem, we begin by determining the oxidation number of all species present. The oxidation numbers for H and O stay the same (+1 and –2, respectively). The oxidation number for Sn changes from a +2 to a +4. This is an increase in oxidation number; therefore, Sn is oxidized and is the reducing agent. The oxidation of Mn in MnO_4^- is +7. (Remember that the oxidation number of the elements in an ion must add up to the charge on the ion.) If O is –2 and the charge on the ion is –1, then

(oxidation number of Mn) + 4 (–2) = –1
                              └─┬─┘
                              ox. # for
                              all 4 O's

(oxidation number of Mn) = –1 – [4 × (–2)] = +7

oxidation number of Mn = +7

The oxidation number of Mn changes from a +7 to a +2, resulting in a decrease in oxidation number. Mn^{7+} is the species reduced, and MnO_4^- is the oxidizing agent. (MnO_4^- is the oxidizing agent because Mn^{7+} is present in the form of MnO_4^-.)

Workbook Problem 4.6
Identify the species oxidized, the species reduced, the oxidizing agent, and reducing agent in the following reaction:

$$3\ SO_4^{2-}\ (aq) + 12\ H^+\ (aq) + 2\ Al\ (s) \rightarrow 2\ Al^{3+}\ (aq) + 3\ H_2SO_3 + H_2O\ (l)$$

Strategy: Determine the oxidation number of all species present.

Step 1: Identify the species which have a change in oxidation number.

77

Chapter 4 – Reactions in Aqueous Solution

Step 2: Based on the change in oxidation number, identify the species oxidized, the species reduced, and the oxidizing and reducing agent.

What key concept did you use?

4.8 Activity Series
A. Simple redox process — reaction of an aqueous cation with a free element to produce a different ion and a different element.
  1. Process is dependent on the ease of oxidation for the element and the ease of reduction of the cation.
B. Activity Series — ranks the elements in order of their reducing ability in aqueous solution. See Table 4.3, page 123 in textbook.)
  1. Any element higher in the activity series will react with the ion of any element lower in the activity series.
  2. Position of hydrogen — indicates which metals will react with $H^+(aq)$ to produce $H_2 (g)$.
  3. Most reactive metals — top of the activity series.
  4. Least reactive metals — bottom of the activity series.

EXAMPLE:
Using the activity series, write balanced chemical equations for the following reactions:

$Al (s) + Zn^{2+} (aq) \rightarrow$
$Ni (s) + Mn^{2+} (aq) \rightarrow$
$Ni (s) + HCl (aq) \rightarrow$
$Pt (s) + HBr (aq) \rightarrow$

SOLUTION: To predict the outcome of these reactions, remember that any element higher in the activity series will react with the ion of any element lower in the activity series. Also remember that metals above the H^+ ion in the activity series will displace the hydrogen ion from an acid to form H_2 gas.

$2 Al (s) + 3 Zn^{2+} (aq) \rightarrow 2 Al^{3+} (aq) + 3 Zn (s)$
$Ni (s) + Mn^{2+} (aq) \rightarrow$ no reaction; Mn^{2+} lies above Ni in the activity series
$Ni (s) + 2 HCl (aq) \rightarrow NiCl_2 (aq) + H_2 (g)$
$Pt (s) + HBr (aq) \rightarrow$ no reaction; H^+ lies above Pt in the activity series.

Workbook Problem 4.7
Using the activity series, write balanced chemical equations for the following reactions:
a. $Sn (s) + NaCl (aq) \rightarrow$
b. $Ca (s) + H_2S (aq) \rightarrow$
c. $Fe (s) + Pt(NO_3)_2 (aq) \rightarrow$

Strategy: Remember that any element higher in the activity series will react with the ion of any element lower in the activity series. Also remember, metals above the H^+ ion in the activity series will displace the hydrogen ion from an acid to form H_2 gas.

Step 1: Predict the outcome of these reactions.

What key concept did you use?

4.9 Balancing Redox Reactions: The Oxidation–Number Method
A. Oxidation–number method for balancing redox reactions focuses on the chemical changes involved.
B. Key — net change in the total of all oxidation numbers must be zero.
   1. Any increase in oxidation number for the oxidized atoms must be matched by a corresponding decrease in oxidation number for the reduced atoms.
C. Steps — see page 127 of the textbook.

EXAMPLE:
Using the oxidation–number method, balance the following reaction which takes place in acidic solution:

$$MnO_4^- \ (aq) + SO_2 \ (aq) \rightarrow Mn^{2+} \ (aq) + HSO_4^- \ (aq)$$

SOLUTION:
1. Unbalanced ionic equation:

$$MnO_4^- + SO_2 \rightarrow Mn^{2+} + HSO_4^-$$

2. Balance all atoms other than hydrogen and oxygen. Same as unbalanced ionic equation.

3. Assign oxidation numbers to all atoms.

$$\begin{array}{cccc} MnO_4^- + & SO_2 & \rightarrow Mn^{2+} + & HSO_4^- \\ +7 \ -2 & +4 \ -2 & +2 & +1 +6 \ -2 \end{array}$$

4. Atoms that have changed oxidation number:
   Mn: $+7 \rightarrow +2$ (gained 5 e$^-$); S: $+4 \rightarrow +6$ (lost 2 e$^-$)

5. Net increase in oxidation number of oxidized atoms = 2; net decrease in oxidation number of reduced atoms = 5. Multiply the net increase by 5 and the net decrease by 2.

$$2 \ MnO_4^- + 5 \ SO_2 \rightarrow 2 \ Mn^{2+} + 5 HSO_4^-$$

6. Reactant side of the equation has two less oxygens, so add 2 H$_2$O's.

$$2 \ MnO_4^- + 5 \ SO_2 + 2 \ H_2O \rightarrow 2 \ Mn^{2+} + 5 \ HSO_4^-$$

7. Reactant side of the equation has one less hydrogen, so add 1 H$^+$.

$$H^+ \ (aq) + 2 \ MnO_4^- \ (aq) + 5 \ SO_2 \ (aq) + 2 \ H_2O \ (l) \rightarrow 2 \ Mn^{2+} \ (aq) + 5 \ HSO_4^- \ (aq)$$

Chapter 4 – Reactions in Aqueous Solution

8. Check your answer to make sure all atoms and charges are balanced.

 *Workbook Problem 4.8*

Using the oxidation-number method, balance the following reaction which takes place in basic solution:

$$SO_3^{2-} (aq) + CrO_4^{2-} (aq) \rightarrow SO_4^{2-} (aq) + Cr(OH)_3 (s)$$

Strategy: Follow the steps in the worked example above.

Step 1: Write the unbalanced ionic equation.

Step 2: Balance all atoms other than hydrogen and oxygen.

Step 3: Assign oxidation numbers to all atoms.

Step 4: Determine which atoms have changed oxidation number.

Step 5: Determine the net increase and net decrease in oxidation number.

Step 6: Multiply the species oxidized by the net decrease and the species reduced by the net increase.

Step 7: Balance the oxygens by adding the appropriate number of H_2O's.

Step 8: Balance the hydrogens by adding the appropriate number of H^+'s.

Step 9: Make the solution basic by adding 1 OH⁻ for every H⁺.

Step 10: If possible, combine OH⁻ and H⁺ to form water.

Step 11: If necessary, combine water molecules present on the same side of the equation and cancel water molecules on both sides of the equation.

Step 12: Check your answer, making sure atoms and charge are balanced.

What key concept did you use?

4.10 Balancing Redox Reactions by the Half–Reaction Method
A. Half–reaction method of balancing redox reactions focuses on the transfer of electrons.
B. Key — overall reaction can be broken into two parts or half reactions.
   1. Oxidation part of the reaction.
   2. Reduction part of the reaction.
C. Steps — see figure 4.6 on page 131 in your textbook..

EXAMPLE:
Using the half–reaction method balance the following reaction which takes place in acidic medium:

$$Cr_2O_7^{2-} (aq) + HC_2O_4^- (aq) \rightarrow Cr^{3+} (aq) + CO_2 (g)$$

SOLUTION:
1. Unbalanced ionic equation: $Cr_2O_7^{2-} + HC_2O_4^- \longrightarrow Cr^{3+} + CO_2$

2. Two unbalanced half reactions: Chromium is reduced from +6 to +3; C is oxidized from +3 to +4.

   $Cr_2O_7^{2-} \rightarrow Cr^{3+}$

   $HC_2O_4^- \rightarrow CO_2$

3. Balance each half reaction for atoms other than H and O.

   $Cr_2O_7^{2-} \rightarrow 2\,Cr^{3+}$

Chapter 4 – Reactions in Aqueous Solution

$$HC_2O_4^- \rightarrow 2\,CO_2$$

4. Add H_2O for any oxygens that are needed and H^+ for any hydrogens that are needed.

$$14\,H^+ + Cr_2O_7^{2-} \rightarrow 2\,Cr^{3+} + 7\,H_2O$$

$$HC_2O_4^- \rightarrow 2\,CO_2 + H^+$$

5. Balance each reaction for charge.

$$6\,e^- + 14\,H^+ + Cr_2O_7^{2-} \rightarrow 2\,Cr^{3+} + 7\,H_2O$$

$$HC_2O_4^- \rightarrow 2\,CO_2 + H^+ + 2\,e^-$$

6. Make the electron count the same in both reactions.

$$6e^- + 14\,H^+ + Cr_2O_7^{2-} \rightarrow 2\,Cr^{3+} + 7\,H_2O$$

$$3\,HC_2O_4^- \rightarrow 6\,CO_2 + 3\,H^+ + 6\,e^-$$

7. Add the two half–reactions together, canceling anything that appears on both sides of the equation.

$$11\,H^+(aq) + Cr_2O_7^{2-}(aq) + 3\,HC_2O_4^-(aq) \rightarrow 2\,Cr^{3+}(aq) + 7\,H_2O(l) + 6\,CO_2(g)$$

8. Check to make sure all atoms and charges are balanced.

 **Workbook Problem 4.9**

Using the half–reaction method, balance the following reaction which takes place in basic solution:

$$CN^-(aq) + AsO_4^{3-}(aq) \rightarrow AsO_2^-(aq) + CNO^-(aq)$$

Strategy: Follow the steps in the preceding worked example.

Step 1: Write the unbalanced ionic equation.

Step 2: Determine which species is being reduced and which species is being oxidized, and write two unbalanced half–reactions.

Step 3: Balance each half–reaction for atoms other than H and O.

Chapter 4 – Reactions in Aqueous Solution

Step 4: Add H₂O for any oxygens that are needed and H⁺ for any hydrogens that are needed.

Step 5: Balance each reaction for charge.

Step 6: Make the electron count the same in both reactions.

Step 7: Add the two half–reactions together, canceling anything that appears on both sides of the equation.

Step 8: Make the solution basic by adding 1 OH⁻ for every H⁺

Step 9: If possible, combine OH⁻ and H⁺ to form water.

Step 10: If necessary, combine water molecules present on the same side of the equation and cancel water molecules on both sides of the equation.

Step 11: Check your answer to make sure both atoms and charge are balanced.

What key concept did you use?

4.11 Redox Titrations   *D.5.*
    A. Redox titration — a method for determining the concentration of an oxidizing or reducing agent in solution.
        1. Unknown must react in a 100% yield.
        2. Color change should signal the end of the reaction.
    B. Strategy for redox titrations (see Figure 4.7, page 133 in your textbook for the titration of $KMnO_4$ with $H_2C_2O_4$).

Chapter 4 – Reactions in Aqueous Solution

1. Measure a known amount of one substance
2. Using the mole ratio of the balanced equation, determine the number of moles of the second substrate.
3. Determine the volume of solution containing the molar amount of the second substance (from the titration).
4. Determine the molar concentration of the second substance.

 *Workbook Problem 4.10*

When the NaOCl found in a dilute solution of bleach is reacted with I^-, the soluble I_3^- ion is produced. The net ionic equation for this reaction is

$$3\,I^- + OCl^- + 2\,H^+ \rightarrow I_3^- + Cl^- + H_2O$$

A bleach sample with a mass of 1.500 g was reacted with excess I^-. The resulting solution was titrated with 0.0500 M $S_2O_3^{2-}$ (thiosulfate ion) using starch as an indicator. The titration required 42.32 mL of the $S_2O_3^{2-}$ solution. The reaction for the titration is

$$I_3^- + 2\,S_2O_3^{2-} \rightarrow 3\,I^- + S_4O_6^{2-}$$

How many moles of I_3^- reacted? How many moles of OCl^- reacted? What is the weight percent of NaOCl in the bleach?

Strategy: Apply the flow diagram found in Figure 4.7 on page 133 of your textbook.

Step 1: Determine the number of moles of I_3^- that reacted with the thiosulfate ion.

Step 2: Determine the number of moles of OCl^- that reacted.

Step 3: Determine the mass of NaOCl present.

Step 4: Determine the mass percent of NaOCl.

What key concept did you use?

Chapter 4 – Reactions in Aqueous Solution

4.12 Some Applications of Redox Reactions
A. Redox reactions involve almost every element in the periodic table and occur in a vast number of processes.
B. Examples of redox reactions.
1. Combustion
2. Bleaching
3. Batteries
4. Metallurgy
5. Corrosion
6. Respiration

 **Putting It Together**

A classic general chemistry experiment that illustrates the type of reactions discussed in this chapter is the copper cycle experiment. This cycle begins with a redox reaction in which copper reacts with nitric acid to produce copper(II) nitrate and nitrogen dioxide gas. Next, sodium hydroxide is added to the copper(II) nitrate. The product from this reaction is heated and undergoes dehydration (loss of water), producing an oxide. This oxide is then reacted with sulfuric acid, producing a light blue solution. Finally, zinc is added to the light blue solution.

a. Write chemical equations for each of the reactions described above.

b. If you begin with a penny minted before 1982 that weighs 7.087 g, how much copper is present if you use 57 mL of 7.5 M nitric acid (assuming that this is the stoichiometric amount of nitric acid needed to convert all of the copper in the penny to copper(II) nitrate)? What is the percent of copper in the penny?

c. If you recover 5.25 g of precipitate in the last reaction, what comments would you make regarding your laboratory technique?

 **Self–Test**

This section is intended to test your knowledge of the material covered in this chapter. Think through these problems, and make certain you understand what is going on. Ask yourself if your answer makes sense. Many of these questions are linked to the chapter learning goals. Therefore, successful completion of these problems indicates you have mastered the learning goals for this chapter. You will receive the greatest benefit from this section if you use it as a mock exam. You will then discover which topics you have mastered and which topics you need to study in more detail.

True–False

1. $CaCl_2$ is a strong electrolyte.

2. H_3PO_4 is a strong acid.

3. The net ionic equation for the reaction between NaCl and $AgNO_3$ is $Ag^+(aq) + Cl^-(aq) \rightarrow AgCl\,(s)$.

4. In the reaction between NaCl and $Ba(OH)_2$, $BaCl_2$ will precipitate.

Chapter 4 – Reactions in Aqueous Solution

5. The driving force for an acid-base neutralization reaction is the formation of the salt.

6. Reduction is the gain of one or more electrons.

7. The reducing agent is the species being reduced.

8. The oxidation number of Br in $NaBrO_3$ is -1.

9. Ca can react with Zn^{2+} to produce Ca^{2+} and Zn.

10. Hg will react with hydrochloric acid to produce H_2 gas.

Multiple Choice

1. Strong bases are
    a. nonelectrolytes
    b. weak electrolytes
    c. precipitates
    d. strong electrolytes

2. The insoluble chlorides contain
    a. an alkali metal
    b. NH_4^+
    c. Ba^{2+}
    d. Ag^+, Hg_2^{2+}, or Pb^{2+}

3. Insoluble nitrates contain
    a. alkali metals
    b. Ba^{2+}
    c. ammonium
    d. none of the above

4. Soluble hydroxide salts contain
    a. alkali metals
    b. Fe^{3+}
    c. Ag^+
    d. none of the above

5. The weak acid is
    a. HF (*aq*)
    b. HCl (*aq*)
    c. HBr (*aq*)
    d. HI (*aq*)

6. The oxidation number of S in K_2SO_3 is
    a. -2
    b. +4
    c. +6
    d. 0

7. An atom in an uncombined element has an oxidation number of
   a. the charge of its ion
   b. the group number
   c. 0
   d. none of the above

8. For the ionic reaction $Cr_2O_7^{2-}(aq) + Cl^-(aq) \rightarrow Cr^{3+}(aq) + Cl_2(g)$, the species being reduced is
   a. $Cr_2O_7^{2-}$
   b. Cl^-
   c. Cr^{3+}
   d. Cl_2

9. The reducing agent in the reaction $Fe^{2+} + MnO_4^- \rightarrow Fe^{3+} + Mn^{2+}$ is
   a. Fe^{2+}
   b. MnO_4^-
   c. Fe^{3+}
   d. Mn^{2+}

10. Nickel metal will
    a. react with K^+ to produce $K(s)$.
    b. react with cold water to produce H_2 gas.
    c. react with Cr^{3+} to produce chromium.
    d. react with HCl (aq) to produce H_2 gas.

Matching

Electrolyte	a. how much of each compound will dissolve in a given amount of solvent at a given temperature
Strong electrolyte	b. produces H^+ ions in water
Spectator ions	c. causes reduction to occur
Solubility	d. dissociates in water to produce an ionic solution
Oxidation	e. produces OH^- ions in water
Reduction	f. completely dissociates in water
Oxidizing Agent	g. loss of electrons
Reducing Agent	h. ions that do not undergo a change during a reaction
Acid	i. causes oxidation to occur
Base	j. gain of electrons

Chapter 4 – Reactions in Aqueous Solution

Fill-in-the-Blank

1. The _____ ranks the element in order of their reducing ability in aqueous solution.

2. When an acid is mixed with a base the reaction is referred to as a _____, and the products are _____.

3. The concentration of an oxidizing agent or reducing agent in solution can be determined by a _____.

4. _____ are processes in which one or more electrons are transferred between reaction partners.

5. The _____ provides an indication of whether an atom is neutral, electron-rich, or electron-poor.

6. Substances that incompletely dissociate are _____.

7. The _____ method of balancing half-reactions focuses on the chemical change that occurs.

8. In a redox reaction, the oxidation process and the reduction process can be written as two _____.

9. The _____ method of balancing redox reactions focuses on the transfer of electrons.

10. Acids dissociate in aqueous solutions to yield an anion and a _____.

Problems

Write balanced molecular, ionic, and net ionic equations for the reactions between
1. sodium chloride and lead(II) nitrate
2. perchloric acid and potassium hydroxide
3. potassium carbonate and calcium chloride

Determine the oxidation number for each atom in
4. hydrogen carbonate ion
5. magnesium sulfate
6. sulfur hexafluoride

For each unbalanced equation given below, identify the species oxidized and the species reduced; identify the oxidizing agent and the reducing agent.

7. $Mg\,(s) + O_2\,(g) \rightarrow MgO\,(s)$

8. $Cr_2O_7^{2-}\,(aq) + Sn^{2+}\,(aq) \rightarrow Cr^{3+}\,(aq) + Sn^{4+}\,(aq)$

9. $FeS\,(s) + NO_3^-\,(aq) \rightarrow NO\,(g) + SO_4^{2-}\,(aq) + Fe^{2+}\,(aq)$

10. $C_2H_4\,(g) + O_2\,(g) \rightarrow CO_2\,(g) + H_2O\,(g)$

Chapter 4 – Reactions in Aqueous Solution

Using the activity series, predict the outcome of the following reactions.
11. sodium and cold water
12. iron and steam
13. magnesium and cold water
14. copper and zinc sulfate

Balance the following equations using the oxidation number method.
15. $Sn^{2+} + MnO_4^- \rightarrow Sn^{4+} + Mn^{2+}$ (acidic solution)

16. $S_2O_3^{2-} + Cl_2 \rightarrow SO_4^{2-} + Cl^-$ (acidic solution)

17. $MnO_4^- + C_2O_4^{2-} \rightarrow MnO_2 + CO_3^{2-}$ (basic solution)

Balance the following equations using the method of half reactions.
18. $HSO_3^- + Cr_2O_7^{2-} \rightarrow SO_4^{2-} + Cr^{3+}$ (acidic solution)

19. $Pb(OH)_3^- + OCl^- \rightarrow PbO_2 + Cl^-$ (basic solution)

20. $Br_2 \rightarrow Br^- + BrO_3^-$ (basic solution)

21. Using the balanced equation obtained in number 18 above, calculate the molarity of a $K_2Cr_2O_7$ solution if 28.42 mL of the solution reacts completely with a 25.00 mL solution of 0.3143 M Na_2SO_3.

22. A drunk driver is defined as one who drives with a blood alcohol level of 0.1% by mass, or higher. The level of alcohol can be determined by titrating blood plasma with potassium dichromate according to the following unbalanced equation:

$Cr_2O_7^{2-} + C_2H_5OH \rightarrow Cr^{3+} + CO_2$

Balance the above reaction (which takes place in acidic solution). Assuming that the only substance that reacts with dichromate in blood plasma is alcohol, is a person legally drunk if 6.522 mL of 5.000×10^{-3} M potassium dichromate is required to titrate a 0.50 gram sample of blood plasma?

23. A general chemistry student combines 7.5 g of sodium sulfide with 6.8 g of zinc nitrate.
    a. Write the molecular, ionic, and net ionic equations for this reaction.
    b. How much of the solid will be produced?
    c. How much of the excess reactant will be left over?
    d. If the experimental yield is 2.53 g, what will the percent yield be?

24. The next week, the same general chemistry student reacted 4.8 g of Mg with 250 mL of 0.75 M copper(II) sulfate.
    a. Write the molecular, ionic, and net ionic equations for this reaction.
    b. How many grams of the solid will be produced?
    c. How many grams of the excess reactant will be left over?
    d. If the experimental yield is 9.2 g, what will the percent yield be?

25. The copper and zinc found in brass can be dissolved by treating a sample of the alloy with nitric acid. The resulting Cu^{2+} ion will react with a solution of potassium iodide by the following equation:

$2 Cu^{2+} (aq) + 5 I^- (aq) \rightarrow 2 CuI (s) + I_3^- (aq)$

Chapter 4 – Reactions in Aqueous Solution

Determine the weight percent of copper in a 0.500 g brass sample, if 20.80 mL of 0.2500 M $Na_2S_2O_3$ is required to titrate the I_3^- produced in the preceding reaction. (The reaction between I_3^- and $S_2O_3^{2-}$ is found in Workbook Problem 4.10 on page 84 of this book.)

Challenge Problems

1. You have a solution that may contain Ag^+, Pb^{2+}, Ba^{2+}, Cu^{2+}, and Ni^{2+}. You divide this solution into two separate test tubes labeled A and B. To the solution in test tube A, you add hydrochloric acid, and a precipitate forms. You separate out the precipitate by pouring the solution into a test tube labeled C. To test tube C, you add sodium hydroxide (to adjust the basicity of the solution), followed by the addition of sodium carbonate. A precipitate forms. To test tube B, you add hydrogen sulfide gas. No precipitate forms. Determine which ions are present in the original solution.

2. A 0.500 g solid that contains both sodium sulfate and potassium sulfate was dissolved in water. An excess of barium chloride was added. The precipitate that was formed weighed 0.716 g. Write a net ionic equation for the formation of the precipitate, and determine the mass of sodium sulfate and potassium sulfate in the original sample.

Inquiry Based Problem

When a solid mixture containing Na_2SO_4 and $BaCl_2 \cdot 2H_2O$ is added to water, $BaSO_4$ precipitates from the solution. $BaCl_2 \cdot 2H_2O$ is called a hydrate because it contains water molecules. However, the structure of this hydrate is not well known, so we use a dot in writing the formula to indicate the number of water molecules associated with the salt without indicating how the water is bound to the salt. In this problem, the addition of the two water molecules is needed only in determining the molar mass of $BaCl_2 \cdot 2H_2O$. The laboratory assignment for this chapter is to determine the percent composition of the solid mixture. The only instruction given by your laboratory instructor is to make use of the 0.5 M Na_2SO_4 and 0.5 M $BaCl_2$ solutions at your laboratory desk and to use a sample size of 1.25 g of the solid mixture. She also informs you to use 250 mL of water and 1 mL of concentrated HCl (*aq*) to dissolve the solid mixture. You will need to maintain the reaction mixture at a temperature between 80° C and 90° C for 45 minutes. You need to determine any additional experimental procedures. Laboratory equipment available to use include a 400 mL beaker, a Bunsen burner, ring stand, ring, wire guaze, 2 – 50 mL beakers, watch glass, and a vacuum filtration apparatus.

The following may be helpful:

Review the section in Chapter 3 on limiting reactants. Think about the definition of limiting reactant and excess reactant.

Write the molecular, ionic, and net ionic equations for the reaction between Na_2SO_4 and $BaCl_2 \cdot 2H_2O$ using the solubility guidelines.

You know that the solid is $BaSO_4$. Where do you expect to find the excess reactant? Use the solubility guidelines and the net ionic equation.

Chapter 4 – Reactions in Aqueous Solution

Once you dissolve the solid in water and the reaction occurs, what experimental data can you collect?

Using the two solutions provided, how could you determine the limiting or excess reactant?

Write out the steps needed to carry out this experiment. Some helpful experimental techniques may be decanting the supernatant liquid and vacuum filtration.

Using the stoichiometric method outlined in chapter 3 and the following data, which was collected in the laboratory, determine the percent composition of the solid mixture.

Mass of $BaSO_4$ precipitate = 0.478 g
Addition of SO_4^{2-} to reaction mixture – precipitate forms.
Addition of Ba^{2+} to reaction mixture – no precipitate forms.

CHAPTER 5

PERIODICITY AND ATOMIC STRUCTURE

Chapter Learning Goals

A. Electromagnetic Radiation – Characterization
1. Interconvert wavelength, frequency, and energy of electromagnetic radiation.
2. Using the Balmer–Rydberg equation, calculate the wavelength and energy of a photon absorbed or released when an electron changes orbitals.
3. Interconvert the amount of energy associated with a quantum of radiant energy, frequency, and wavelength.
4. Using the de Broglie equation, calculate the mass of an object knowing its wavelength and vice versa.

B. Wave Functions and Quantum Numbers
1. Relate a set of quantum numbers to a particular orbital.
2. Sketch and name each of the s, p, and d orbitals.

C. Electron Configurations
1. State the Pauli Exclusion Principle, Hund's Rule and the Aufbau Principle.
2. Predict ground–state electron configurations for elements; use orbital–filling diagrams to determine the number of unpaired electrons in these species.

D. Valence Electrons and the Periodic Table
1. Explain what is meant by effective nuclear charge, Z_{eff}.
2. Write the general valence–shell electron configuration for each group of the periodic table, and identify the blocks in which the elements are located.
3. Given a set of atoms, determine which atom is expected to have the largest radius.

Chapter in Brief

The periodic table is the most important organizing principle in chemistry. This chapter explains why the elements, when placed in order of increasing atomic weight, have a periodic occurrence of chemical and physical properties. This periodicity can be understood by examining the theory used to describe the electronic structure of atoms. Chapter 5 begins the examination of this theory by introducing electromagnetic radiation and the properties of waves. You will discover how both light and matter can have dual (both wave and particle) properties. With this information, you are introduced to quantum mechanics and quantum numbers, a mathematical theory used to describe the probability of finding an electron in an atom. You are then shown how to use quantum mechanics to determine the electronic configuration of the elements. Finally, you will apply this knowledge to learn how atoms in the same group in the periodic table have similar electronic configurations and how these electronic configurations affects periodic properties such as atomic radii.

5.1 Development of the Periodic Table
A. Creation of the Periodic Table.
  1. Ideal example of how scientific theory comes into being.
     a. random observations
     b. organization of data in ways that make sense
     c. consistent hypothesis emerges
        i. explains known facts and makes predictions about unknown phenomena

Chapter 5 – Periodicity and Atomic Structure

- B. Mendeleev's hypothesis about organizing known chemical information.
    1. Met criteria for a good hypothesis.
        a. listed the known elements by atomic weight
        b. grouped them together according to their chemical reactivity
        c. was able to predict the properties of unknown elements – *eka*-aluminum, *eka*-silicon

5.2 Light and the Electromagnetic Spectrum
- A. Electromagnetic radiation — forms of radiant energy (light in all its varied forms).
    1. Electromagnetic spectrum — a continuous range of wavelengths and frequencies of all forms of electromagnetic radiation. (See figure 5.3, page 150 in your textbook.)
- B. Radiant energy — has wavelike properties.   🔑 A.1.
    1. Frequency (ν) — the number of peaks (maxima) that pass by a fixed point per unit time (s^{-1} or Hz).
    2. Wavelength (λ) — the length from one wave maximum to the next.
    3. Amplitude — the height measured from the middle point between peak and trough (maximum and minimum).
        a. Intensity of radiant energy is proportional to the amplitude^2.
- C. Speed of light (c) — rate of travel of all electromagnetic radiation in a vacuum.
    1. $c = 3.00 \times 10^8$ m/s.
    2. Wavelength × Frequency = Speed.
    λ (m) × ν (s^{-1}) = c (m/s)
    3. Frequency and wavelength are inversely related: $\lambda = \dfrac{c}{\nu}$ or $\nu = \dfrac{c}{\lambda}$.
        a. long λ; low ν
        b. short λ; high ν

EXAMPLE:
Calculate the wavelength, in meters, of radiation with a frequency of 1.18×10^{14} s^{-1}. What region of the electromagnetic spectrum is this?

SOLUTION: We can solve this problem by using the equation $\lambda \times \nu = c$, where $c = 3.00 \times 10^8$ m/s.

$$\lambda = \frac{3.00 \times 10^8 \text{ m/s}}{1.18 \times 10^{14} \text{ s}^{-1}} = 2.54 \times 10^{-6} \text{ m}$$

This particular wavelength is found in the infrared region of the electromagnetic spectrum (see Figure 5.3, page 150 in your textbook.)

5.3 Electromagnetic Radiation and Atomic Spectra
- A. Individual atoms give off light when heated or otherwise excited energetically.
    1. Provides clue to atomic makeup.
    2. Consists of only a few λ.
    3. Line spectrum — series of discrete lines (or wavelengths) separated by blank areas.
    4. Each element has its own unique line spectrum.
- B. Balmer discovered a pattern in atomic line spectra for the hydrogen atom.

Chapter 5 – Periodicity and Atomic Structure

1. All four lines in the hydrogen spectrum are expressed by $\frac{1}{\lambda} = R\left(\frac{1}{2^2} - \frac{1}{n^2}\right)$.
   a. R (*Rydberg constant*) = 1.097×10^{-2} nm^{-1}
2. Rydberg — every line in the entire hydrogen spectrum fits the generalized Balmer–Rydberg equation.

$$\frac{1}{\lambda} = R\left(\frac{1}{m^2} - \frac{1}{n^2}\right); \quad (n > m)$$

A.2.   EXAMPLE:

Calculate the wavelength of light emitted when an electron falls from the $n = 6$ to $n = 4$ levels in the hydrogen atom.

SOLUTION: We solve this problem by using the Rydberg equation.

$$\frac{1}{\lambda} = 1.097 \times 10^{-2} \text{ nm}^{-1}\left(\frac{1}{4^2} - \frac{1}{6^2}\right) = 3.809 \times 10^{-4} \text{ nm}^{-1}$$

$\lambda = 2.625 \times 10^3$ nm

Workbook Problem 5.1

Determine the shortest wavelength and the two longest wavelength (in nm) lines in the Brackett series ($n = 4$) for hydrogen.

Strategy: Determine the values of *m* that will make λ the longest and shortest. Remember that the value of λ is greatest when the value of *m* is smallest and the value λ is smallest when the value of *m* is greatest. Also remember that $m > n$.

Step 1: Determine the value of *m* that will make λ the shortest.

Step 2: Use the Balmer–Rydberg equation with $n = 4$ and solve for λ.

Step 3: Solve for λ using the Balmer–Rydberg equation and the values of *m* that make λ the longest.

What key concept did you use?

Chapter 5 – Periodicity and Atomic Structure

5.4 Particlelike Properties of Electromagnetic Radiation: The Planck Equation
A. Photoelectric effect — irradiating a clean metal surface with light causes electrons to be ejected from the metal.
   1. Einstein — beam of light behaves as if it were composed of photons (stream of small particles).
   2. Energy of photons: $E = h\nu$.
      a. $\nu = \dfrac{c}{\lambda}$
         i. $E = \dfrac{hc}{\lambda}$
      b. $h = 6.626 \times 10^{-34}$ J·s (Planck's constant)
      c. high–energy radiation — higher ν, shorter λ
      d. low–energy radiation — lower ν, longer λ
      e. energy depends only on frequency of photon
      f. intensity of light beam — measure of number of photons, not energy
C. Light energy can behave as both waves and small particles.
D. Both matter and energy occur only in discrete units.
   1. Quantum – discrete unit of energy.
   2. Atoms — emit light quanta (photons) of a few specific energies.
      a. Give rise to a line spectrum.
E. Particle–like nature of electromagnetic energy explains atomic line spectra.
F. Bohr's model of the hydrogen atom.
   1. The electron orbits around the nucleus.
   2. Energy levels of the orbits are quantized.
      a. Only certain specific energies are available to the electron.
   3. Electron falls from a higher-energy outer orbit to a lower-energy inner orbit (see Figure 5.8 p. 156 in your textbook.)
      a. Quantum of energy emitted
         i. Emitted energy is equal to the difference between the higher energy and lower energy orbit.
         ii. Energy is emitted as a photon (electromagnetic energy).
   4. Spectral emission lines due to emission of electromagnetic energy.

A.3.

EXAMPLE:
What energy is emitted in the example on page 94?

SOLUTION: We need to use the equation $E = \dfrac{hc}{\lambda}$ to calculate the energy. However, we first need to convert the wavelength we calculated from nanometers to meters, since the units for the speed of light are in m/s.

$$2.625 \times 10^3 \text{ nm} \times \dfrac{1 \text{ m}}{1 \times 10^9 \text{ nm}} = 2.625 \times 10^{-6} \text{ m}$$

$$E = \dfrac{(6.626 \times 10^{-34} \text{ J·s})(3.00 \times 10^8 \text{ m/s})}{2.625 \times 10^{-6} \text{ m}} = 7.57 \times 10^{-20} \text{ J}$$

Chapter 5 – Periodicity and Atomic Structure

Workbook Problem 5.2

Determine the energy of the wavelengths calculated in Workbook Problem 5.1

Strategy: Use the equation for the energy of a photon.

What key concept did you use?

5.5 Wavelike Properties of Matter: The de Broglie Equation
A. Louis de Broglie – if light can behave like matter, then matter can behave like light.
   1. Matter is wavelike as well as particle-like.
   2. Inverse relationship between energy and wavelength.
      a. $E = \dfrac{hc}{\lambda}$; $\lambda = \dfrac{hc}{E}$
   3. Einstein - $E = mc^2$
      $$\lambda = \dfrac{hc}{mc^2} = \dfrac{h}{mc}$$
   4. de Broglie equation – substitute the speed of the electron, v, for the speed of light, c
      $$\lambda = \dfrac{h}{mv}$$
B. Dual wave/particle description of light and matter is a mathematical model that accounts for atomic properties and behavior.

EXAMPLE:
What velocity would an electron (mass = 9.11×10^{-31} kg) need for its de Broglie wavelength to be 590 nm?

SOLUTION:

Using the de Broglie equation, we can solve for velocity.

$$v = \dfrac{6.626 \times 10^{-34} \text{ J} \cdot \text{sec}}{(9.11 \times 10^{-31} \text{ kg})(5.90 \times 10^{-7} \text{ m})} = 1.23 \times 10^3 \dfrac{\text{J} \cdot \text{sec}}{\text{kg} \cdot \text{m}}$$

Substituting 1 (kg·m^2)/s^2 for the unit J gives

$$1.23 \times 10^3 \dfrac{\left(\dfrac{\text{kg} \cdot \text{m}^2}{\text{s}^2}\right) \cdot \text{s}}{\text{kg} \cdot \text{m}} = 1.23 \times 10^3 \text{ m/s}$$

5.6 Quantum Mechanics and the Heisenberg Uncertainty Principle
A. Bohr – described the structure of the hydrogen atom as containing an electron circling the nucleus.
   1. Specific orbits of the electron correspond to specific energy levels.
B. Schrödinger — quantum mechanical model of the atom.

1. Concentrate on the electron's wavelike properties.
C. Heisenberg Uncertainty Principle
   1. $(\Delta x)(\Delta mv) \geq \dfrac{h}{4\pi}$
   2. Cannot know both the position and the momentum of an electron with a high degree of certainty.
   3. If the momentum is known with a high degree of certainty
      a. Δmv is small.
      b. Δx (uncertainty of the position of the electron) is large.
   4. If the exact position of the electron is known
      a. Δx is small.
      b. Δmv is large.

5.7 Wave Functions and Quantum Numbers

A. Quantum mechanical model of atomic structure.
   1. Mathematical form is a wave equation.
      a. Similar to the equation used to describe wave motions in a fluid.
   2. Wave function (ψ) or orbital — solutions to wave equation.
      a. has a specific energy
      b. contains information about an electron's position in 3-D space
      c. ψ^2 — gives the probability of finding an electron within a given region in space
      d. defines a volume of space around the nucleus where there is a high probability of finding an electron
      e. says nothing about the electron's path or movement
B. Wave function — contains a set of three parameters, quantum numbers.
   1. Describes the energy level of an orbital.
   2. Defines the shape and orientation of the region in space where the electron is most likely to be found.
C. Principal quantum number (n) — describes the size and energy level of the orbital.
   1. A positive integer ($n = 1,2,3,4,..$).
   2. For H and He^+ (one electron atoms or ions), the energy of the orbital only depends on n.
   3. As the value of n increases,
      a. the number of allowed orbitals increases.
      b. size of the orbitals increases.
      c. the energy of the electron in the orbital increases.
   4. Shell — grouping of orbitals according to the principal quantum number.
D. Angular-momentum quantum number (ℓ) — defines the 3-D shape of the orbital.
   1. Integral value from 0 to $n - 1$.
   2. Within each shell, there are n different shapes for orbitals.
   3. Subshells — grouping of orbitals according to the angular-momentum quantum number.
   4. Referred to by letter rather than by number.
      quantum number ℓ: 0 1 2 3 4 ...
      subshell notation: s p d f g ...
E. Magnetic quantum number (m_ℓ) defines the spatial orientation of the orbital along a standard set of coordinate axes.
   1. Integral value from $-\ell$ to $+\ell$.
   2. Within each subshell (same n and same ℓ) there are $2\ell + 1$ different spatial orientations.
F. Energy level of various orbitals (see Figure 5.9, page 161 in your textbook),
   1. Hydrogen — energy levels depend only on n.
   2. Multielectron atoms — energy levels depend on both n and ℓ.
      a. Some crossover of energies from one shell to another.

Chapter 5 – Periodicity and Atomic Structure

   i. 3d greater in energy than 4s.

EXAMPLE:
Give the values of all possible quantum numbers of a 3d subshell.

SOLUTION:
For a 3d subshell, $n = 3$, $\ell = 2$; and $m_\ell = -2, -1, 0, 1, 2$.

 **Workbook Problem 5.3**
Determine the subshell for an electron having $n = 4$, $\ell = 3$.

Strategy: Determine the subshell associated with a value of $\ell = 3$.

Step 1: Identify the subshell with the value of n and the letter designation for $\ell = 3$.

What key concept did you use?

5.8 The Shapes of Orbitals
A. s orbitals (see Figure 5.10, page 163 in your textbook).
  1. Spherical.
  2. Probability of finding an electron depends only on the distance of the electron from the nucleus.
  3. Differences among s orbitals in different shells.
     a. Size increases in successively higher shells.
     b. Electron distribution in outer s orbitals has several different regions of maximum probability separated by a node.
  4. Node — a surface of zero probability.
     a. intrinsic property of a wave — zero amplitude at node (see Figure 5.11, page 176 in textbook).
B. p orbitals (see Figure 5.12, page 164 in your textbook).
  1. Dumbbell shaped.
  2. Electron distribution concentrated in identical lobes on either side of the nucleus.
  3. Nodal plane cuts through the nucleus.
     a. Probability of finding a p electron near the nucleus is zero.
  4. Have different phases (mathematical signs)
     a. crucial for bonding
     b. Only lobes with same phase can interact to form covalent bonds.
  5. Three p orbitals oriented along the x–, y–, and z–axes. (p_x, p_y, p_z).
C. d and f orbitals
  1. d orbitals (see Figure 5.13, page 165 in your textbook).
     a. 4 – d_{xy}, d_{xz}, d_{yz}, $d_{x^2-y^2}$ are cloverleaf shaped.
        i. four lobes of maximum electron probability separated by two nodal planes through the nucleus

Chapter 5 – Periodicity and Atomic Structure

   b. d_{z^2} — similar in shape to a p_z orbital with an additional donut–shaped region of electron probability in the xy plane
   c. Alternating lobes have different phases.
  2. f orbitals — eight lobes of maximum electron probability.
   a. three nodal planes through the nucleus

5.9 Quantum Mechanics and Atomic Spectra
A. Electron in an atom.
  1. Occupies an orbital with a specific energy.
  2. The energies available to electrons are quantized.
   a. have only the specific energy values associated with the orbital
B. Addition of energy to an atom excites the atom
  1. Electron jumps from a lower–energy orbital to a higher–energy orbital.
C. Excited atom is unstable.
  1. Electron returns to a lower–energy level and emits energy.
  2. The energies of the orbitals are quantized.
   a. Emitted energy is equal to the difference between the higher and lower-energy orbitals.
D. Calculate energy differences between orbitals by measuring the frequencies emitted.
C. Balmer–Rydberg equation — variables m and n corresponds to the principal quantum numbers of the two orbitals involved in the transition.
  1. n — principal quantum number of the outer shell (orbital the transition is from)
  2. m — principal quantum number of the inner shell (orbital the transition is to).

5.9 Electron Spin and the Pauli Exclusion Principle
A. Fourth quantum number – m_s.
  1. Related to a property called electron spin.
  2. Electrons behave as if they were spinning around an axis.
   a. gives rise to a tiny magnetic field
  3. Has two values: $+1/2$ (↑) or $-1/2$ (↓).
  4. Independent of other quantum numbers.

B. Pauli Exclusion Principle: No two electrons in an atom can have the same four quantum numbers.   🔑 C.1.
  1. Only two electrons with opposite spins per orbital.

5.11 Orbital Energy Levels in Multielectron Atoms
A. Energy level of an orbital in multielectron atoms depends on both n and ℓ.
  1. Energy difference due to electron—electron repulsions.
   a. Outer–shell electrons are pushed farther away from the nucleus and are held less tightly.
   b. Partially cancels the electron—nucleus attractions
   c. Electrons are shielded from the nucleus by the other electrons.

B. Effective nuclear charge, Z_{eff} — net nuclear charge actually felt by an electron.   🔑 D.1.
  1. $Z_{eff} = Z_{actual}$ – electron shielding.
  2. Lower than the actual nuclear charge.
  3. For the same shell, the lower value of ℓ corresponds to a higher value of Z_{eff}.
   a. corresponds to lower energy for the orbital
  4. Useful for explaining various chemical phenomena.

5.12 Electron Configurations of Multielectron Atoms   🔑 C.2.
A. Electron configuration — describes the orbitals that are occupied by the electrons in an atom.

Chapter 5 – Periodicity and Atomic Structure

      B. Aufbau principle.
          1. Fill the lowest–energy orbitals first.
          2. Only two electrons with opposite spin per orbital.
          3. Follow Hund's rule.
             a. If two or more orbitals with the same energy are available, put one electron with parallel spin in each until all are half full.
      C. Ground–state configuration - lowest–energy electron configuration.
      D. Degenerate orbitals — orbitals with the same energy level.
      E. Orbital–filling diagrams — electrons are represented by arrows.

EXAMPLE:
Give the electron configuration for phosphorus.

SOLUTION: Using the orbital–filling diagram given in Figure 5.9 on page 161 in the textbook, we find that the electron configuration for phosphorus is $1s^2\ 2s^2 2p^6 3s^2 3p^3$.

EXAMPLE:
Give the orbital–filling diagram for phosphorus.

SOLUTION: The orbital–filling diagram for phosphorus would be:

↑↓	↑↓	↑↓ ↑↓ ↑↓	↑↓	↑ ↑ ↑
$1s$	$2s$	$2p_x\ 2p_y\ 2p_z$	$3s$	$3p_x\ 3p_y\ 3p_z$

      F. Shorthand version — give the symbol of the noble gas in the previous row to indicate electrons in filled shells, and then specify only those electrons in unfilled shells.

EXAMPLE:
Give the shorthand electron configuration for phosphorus.

SOLUTION: The shorthand electron configuration for phosphorus would be [Ne] $3s^2\ 3p^3$.

 Workbook Problem 5.4
Give the ground–state electronic configuration (both the complete and shorthand version) for vanadium. Give the orbital–filling, diagram using the shorthand version of the electronic configuration.

Strategy: Use Figure 5.9, page 161 in your textbook, to determine the order of the orbitals.

Step 1: Determine the number of electrons in vanadium.

Step 2: Use the Aufbau principle to determine the ground–state electronic configuration.

Step 3: Determine the noble gas in the previous row. Specify only those electrons in the unfilled subshells.

Chapter 5 – Periodicity and Atomic Structure

Step 4: Draw the orbital–filling diagram.

What key concept did you use?

5.13 Some Anomalous Electron Configurations
A. Half–filled and filled subshells have an unusual stability.
   1. Leads to anomalies in electron configurations.
B. Anomalies occur where the energy differences between subshells are small.
   1. Transfer of an electron from one subshell to another lowers the total energy of the atom.
      a. due to decrease in electron–electron repulsions
      b. $Z > 40$

EXAMPLE:
Give the electron configuration of Ag.

SOLUTION: Silver is one of the elements that will transfer an electron from one subshell to another in order to lower the total energy of the atom. Therefore, the electron configuration of Ag is $[Kr]5s^1 4d^{10}$.

5.14 Electron Configurations and the Periodic Table
A. Valence–shell electrons — outermost shell of electrons.
   1. Elements in each group of the periodic table have similar valence–shell electron configurations.
   2. Most loosely held.
   3. Determine an element's properties.
B. Similar electron configurations explain why the elements in a given group have similar chemical behavior.
C. Blocks of elements in the periodic table — depends upon the valence orbitals being filled.
   1. *s*–block elements — Groups 1A and 2A (filling of an *s* orbital).
   2. *p*–block elements — Groups 3A through 8A (filling of *p* orbitals; *ns* orbitals are filled).
   3. *d*–block elements — transition metals (filling of $(n-1)d$ orbitals).
   4. *f*–block elements — lanthanide and actinide elements (filling of $(n-2)f$ orbitals).

EXAMPLE:
Give the general electron configuration for the elements in Group 5A.

SOLUTION:

The elements in group 5A are *p* block elements. The *p* orbitals begin filling with the 3A elements; therefore, we know that the group 5A elements contain 3 *p* electrons. The electron configuration for these elements is $ns^2 np^3$.

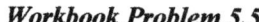 **Workbook Problem 5.5**
Determine the general electronic configuration for elements in groups 7A and 4B.

Chapter 5 – Periodicity and Atomic Structure

Strategy: Determine if the elements are *s*–block, *p*–block, or *d*–block elements.

Step 1: Determine if other outer subshells need to be taken into consideration.

Step 2: Determine the number of electrons in the outer block.

What key concept did you use?

5.15 Electron Configurations and Periodic Properties: Atomic Radii
A. Radius of an atom — half the distance between the nuclei of two identical atoms when they are covalently bonded together.
B. Atomic radius increases down a group.
   1. Successively larger valence–shell orbitals are occupied.
C. Atomic radius decreases across a period.
   1. Due to increase in Z_{eff} for valence–shell electrons across a period.
   2. The value of Z_{eff} is dependent upon the amount of shielding felt by an electron.
      a. Amount of shielding depends on both the shell and subshell of the other electrons.
   3. Valence–shell electron
      a. strongly shielded by electrons in inner shells.
      b. less strongly shielded by electrons in same shell
         i. order: $s > p > d > f$
      c. weakly shielded by electrons in same subshell.

 Putting It Together

The ionization energy is the energy required to remove an electron from a gaseous atom or ion in the ground–state. A particular metal, M, has an ionization energy of 7.0924 eV (1 eV = 1.602×10^{-19} J). A 0.500 g sample of the metal was oxidized to produce MO. This oxide was dissolved in hot concentrated hydrochloric acid producing both the M^{2+} and M^{3+} ions. The M^{3+} ion is converted to M^{2+} according to the reaction

$$M^{3+} + Sn^{2+} \rightarrow M^{2+} + Sn^{4+}$$

Any excess Sn^{2+} that is left over from this reaction is eliminated by the addition of mercury (II) chloride.

$$Sn^{2+} + HgCl_2 \rightarrow Hg_2Cl_2 + Sn^{4+} + 2\ Cl^-$$

The solution of M^{2+} ions is then titrated with potassium permanganate according to the reaction

$$M^{2+} + MnO_4^- \rightarrow M^{3+} + Mn^{2+}.$$

a. Balance all of the chemical equations.

Chapter 5 – Periodicity and Atomic Structure

    b. 35.7 mL of 0.50 M potassium permanganate were required for the titration. How many moles of M are present, and what is the identity of M?
    c. Write the ground–state electron configuration for M.
    d. Determine the wavelength of a photon with an energy equal to the ionization energy.

Self-Test

This section is intended to test your knowledge of the material covered in this chapter. Think through these problems, and make certain you understand what is going on. Ask yourself if your answer makes sense. Many of these questions are linked to the chapter learning goals. Therefore, successful completion of these problems indicates you have mastered the learning goals for this chapter. You will receive the greatest benefit from this section if you use it as a mock exam. You will then discover which topics you have mastered and which topics you need to study in more detail.

True–False

1. Frequency and wavelength are directly related.

2. High–energy radiation consists of higher frequency and shorter wavelength.

3. The energy of photons depends on both the frequency and the number of photons.

4. The line spectra of atoms are due to the emission of light quanta of a few specific energies.

5. The quantum mechanical description of the hydrogen atom is based on the idea of an electron as a small particle moving around the nucleus in a defined path.

6. Wave functions define a volume of space around the nucleus where there is a high probability of finding an electron.

7. The principal quantum number defines the three–dimensional shape of the orbital.

8. The energy level of various orbitals depends only on the value of n.

9. The increase in atomic radius down a group is due to an increase in Z_{eff}.

10. Elements in each period have similar valence–shell electron configurations.

Multiple Choice

1. The quantum number ℓ is referred to as
    a. the principle quantum number
    b. the angular–momentum quantum number
    c. the magnetic quantum number
    d. the spin quantum number

Chapter 5 – Periodicity and Atomic Structure

2. The Heisenberg uncertainty principle states the following:
   a. No two electrons can have the same four quantum numbers.
   b. If two or more orbitals are equal in energy, each is half–filled before any one of them is completely filled.
   c. The lowest energy orbital is filled first.
   d. You can never know both the position and the velocity of an electron beyond a certain level of precision.

3. For the lanthanide elements, the valence electrons are found in a
   a. s subshell
   b. p subshell
   c. d subshell
   d. f subshell

4. The valence–shell of the atom is
   a. the lowest energy shell
   b. the most probable excited state
   c. the outermost shell
   d. represented by a noble gas when writing the electron configuration in shorthand.

5. High energy radiation is associated with
   a. higher ν and longer λ
   b. higher ν and shorter λ
   c. lower ν and longer λ
   d. the speed of light

6. The quantum numbers associated with a $2p$ electron are
   a. $n = 2, \ell = 0, m_\ell = 0$
   b. $n = 2, \ell = 1, m_\ell = 0$
   c. $n = 2, \ell = 1, m_\ell = -1, 0, 1$
   d. $n = 2, \ell = 1, m_\ell = -2, -1, 0, 1, 2$

7. The element Au is a
   a. s–block element
   b. p–block element
   c. d–block element
   d. f–block element

8. The electron configuration for Cl is
   a. $1s^2 2s^2 2p^6 2d^7$
   b. $1s^2 2s^2 2p^6 3s^2 3p^5$
   c. $[Ar]3s^2 3p^5$
   d. $1s^2 2s^2 2p^6 3s^2 3p^7$

9. The atom or ion with the larger radius is
   a. Na > Cs
   b. Na < Mg
   c. Na > Al
   d. Na < Li

Chapter 5 – Periodicity and Atomic Structure

Matching

Photoelectric effect

a. can never know both the position and the velocity of an electron beyond a certain level of precision.

Einstein

b. quantum mechanical model of the atom that concentrates on the electron's wavelike properties.

de Broglie

c. organized the known elements by atomic weight, and grouped them together according to their chemical reactivity.

Schrödinger

d. No two electrons in an atom can have the same four quantum numbers.

Heisenberg

e. Beam of light behaves as if it were composed of photons.

Pauli

f. set of rules that guides the filling of atomic orbitals.

Hund

g. dual wave/particle description of light and matter.

Aufbau principle

h. Irradiating a clean metal surface with light causes electrons to be ejected.

Mendeleev

i. If two or more orbitals with the same energy are available, put one electron in each with parallel spin until all are half full.

Fill–in–the–Blank

1. A continuous range of wavelengths and frequencies of all forms of electromagnetic radiation is referred to as the _____.
2. Each element has its _____ line spectrum.
3. The visible glow that solid objects give off when heated is called _____.
4. Atomic properties and behavior can be accounted for by the _____ description of light and matter.
5. Wave functions contain information about an electron's _____.
6. The grouping of orbitals according to the principal quantum number is referred to as a _____.
7. When an electron returns to a lower–energy level from a higher–energy level energy equal to _____ is emitted.

105

Chapter 5 – Periodicity and Atomic Structure

8. The probability of finding an electron in an *s* orbital depends only upon the _____ _____.

9. The _____ is the net nuclear charge actually felt by an electron.

10. The _____ describes the orbitals that are occupied by the electrons in an atom.

11. The lowest–energy electron configuration is the _____.

12. Elements in each group of the periodic table have similar _____ _____.

13. The similar chemical behavior of elements in a given group is due to _____ _____.

14. The transfer of an electron from one subshell to another can lower the _____.

15. Across a period, the atomic radius _____.

Problems

1. What is the wavelength of infrared light (in nanometers) with a frequency of 1.58×10^{13} Hz? What is the energy of this light?

2. When lithium ions are excited, they emit radiation at $\lambda = 670.8$ nm. What is the frequency of this radiation? What is the color of the radiation emitted?

3. What energy in kJ/mol is associated with a wavelength of 130 nm?

4. How much energy will be released when an electron falls from the $n = 6$ shell to the $n = 2$ shell in the hydrogen atom?

5. What are the allowed quantum numbers for a 4*p* subshell?

6. What is the orbital notation for an electron with $n = 3$ and $\ell = 1$?

7. What quantum numbers are associated with the 5*g* subshell?

8. Lines in the Paschen series of the hydrogen spectrum are caused by emission energy when the electron falls from outer shells to the third shell. Using the Rydberg equation, calculate the wavelength (in nanometers) and the energy (in kilojoules/mol) of the first line in the Paschen series.

9. If a hydrogen atom in the ground–state is excited by light with a wavelength of 97.3 nm, how much energy has been absorbed? Which shell is the electron excited to?

10. Give the ground–state electron configuration for sulfur, strontium, lead, and nickel atoms. Draw the orbital–filling diagrams for the valence–shell electrons of atoms of these elements. How many unpaired electrons does each atom have?

11. What is the general valence–shell electron configuration for the group 5A elements? Which block are these elements located in?

12. Identify the atoms with the following electron configurations:
    $[Ar]4s^2 3d^7$

$[Xe]6s^2 4f^{14} 5d^{10} 6p^4$
$[Kr]5s^2 4d^{10} 5p^2$

13. Group the following atoms by increasing atomic radius: Ir, Ba, and Pb.

14. Practice writing the electron configuration for as many elements as possible.

Challenge Problem

1. An ion having a 3+ charge has 2 electrons in the first shell, 8 electrons in the second shell and 13 electrons in the third shell. Write the ground–state electron configuration for the neutral atom and identify the element. (**Note:** The 4s electrons are lost before the 3d electrons.)

2. The ionization energy is the energy required to remove an electron from a gaseous atom or ion in its ground–state. For ions with multiple charges the electrons are removed in steps. We can depict this process for aluminum in the following manner:

Al (g) → Al$^+$ (g) + e$^-$ E_1 = 580 kJ/mol
Al$^+$ (g) → Al^{2+} (g) + e$^-$ E_2 = 1815 kJ/mol
Al^{2+} (g) → Al^{3+} (g) + e$^-$ E_3 = 2740 kJ/mol
Al^{3+} (g) → Al^{4+} (g) + e$^-$ E_4 = 11,600 kJ/mol

Explain the increase in ionization energy in terms of the electron configuration, as well as an effective nuclear charge.

CHAPTER 6

IONIC BONDS AND SOME MAIN–GROUP CHEMISTRY

Chapter Learning Goals

A. *Formation of Ions*
1. Predict the ground–state electron configuration for ions.
2. Given a set of ions, determine which ion is expected to have the largest radius.

B. *Ionization Energies and Electron Affinities*
1. For any two elements predict which has the higher first ionization energy.
2. For any two elements, predict which has the higher second, third, fourth, etc. ionization energy.
3. For any two elements, predict which has the more negative first electron affinity.

C. *Ionic Solids and Lattice Energies*
1. Identify the energies involved in a Born–Haber calculation of lattice energy. Know whether these energies are positive or negative, large or small, and use the Born–Haber cycle to calculate the lattice energy of an ionic compound.
2. On the basis of ionic charges and ionic radii, predict which of two ionic compounds should have the greater lattice energy.

E. *Octet Rule*
1. Use the octet rule to generalize the chemistry of each family of elements studied in this chapter. Know when to expect the octet rule to be valid and when it can fail.
2. Give the noble gas configuration of cations and anions in ionic compounds.

D. *Main–Group Chemistry: Oxidation–Reduction Reactions*
1. Know which alkali and alkaline earth metals form a) oxides, b) peroxides, and c) superoxides upon reaction with oxygen. Assign oxidation numbers to the oxygen atoms in these compounds.
2. Give the formulas of products formed when alkali metals react with halogens, hydrogen, nitrogen, oxygen, water, or ammonia. Balance the equations.
3. Give the formulas of products formed when alkaline earth metals react with halogens, hydrogen, oxygen, or water. Balance the equations.
4. Give the formulas of products formed when aluminum reacts with halogens, nitrogen, oxygen, acids, or bases. Balance the equations.
5. Give the formulas of products formed when halogens react with metals, hydrogen, or other halogens.
6. Balance redox equations representing reactions studied in this chapter. Identify which species are oxidized and which are reduced and which elements have undergone a change in oxidation state.

Chapter 6 – Ionic Bonds and Some Main–Group Chemistry

Chapter in Brief
Chemical bonds are the forces that hold atoms together. There are two types: ionic bonds and covalent bonds. This chapter concentrates on the reasons ionic bonds are formed and the energies involved in the formation of these bonds. You will learn that the underlying reason for the formation of ions and ionic compounds is the valence–shell configuration of the elements, which determines the amount of energy needed for losing or gaining electrons. Knowing the periodic trends in the energies for ion formation, you will learn how to predict which elements form cations or anions and if the formation of ionic compounds is energetically favorable. You will also learn about the Born–Haber cycle, a series of hypothetical steps used to calculate the overall energy involved in the formation of ionic compounds. Finally, you will learn about the chemistry of the groups 1A – 3A, 7A, and 8A elements and how the chemistry of these elements is governed by the octet rule.

6.1 Ions and Their Electron Configurations   *A.1.*
    A. Metallic elements
       1. Left side of periodic table
       2. Give up electrons in chemical reactions.
       3. Form cations.
    B. Halogens and a few other nonmetals
       1. Right side of periodic table.
       2. Accept electrons in chemical reactions.
       3. Form anions.
    C. Ground state electron configurations for metal and non-metal ions.
       1. Electrons lost by a metal come from the highest–energy occupied orbital.
       2. Electrons gained by a nonmetal go into lowest–energy unoccupied orbital.
    D. Resultant ions have a noble gas configuration.
       1. Group 1A atom [Noble gas]ns^1 → Group 1A cation [Noble gas]
       2. Group 2A atom [Noble gas]ns^2 → Group 2A cation [Noble gas]
       3. Group 6A atom [Noble gas]ns^2np^4 → Group 6A anion [Noble gas]ns^2np^6
       4. Group 7A atom [Noble gas]ns^2np^5 → Group 6A anion [Noble gas]ns^2np^6
    E. Transition metals.
       1. Lose valence–shell s electrons first, then d electrons.
          a. Remaining valence electrons occupy d orbitals.

6.2 Ionic Radii   *A.2.*
    A. Cations — radii are smaller than for neutral atoms.
       1. Electrons are removed from larger valence–shell orbitals.
       2. Increase in Z_{eff} when electrons are removed.
          a. Smaller number of electrons shield to a lesser extent.
    B. Anions — radii are larger than for neutral atoms.
       1. Decrease in Z_{eff} when electrons are added.
       2. Increase in electron–electron repulsions.

EXAMPLE:
Which atom or ion in the following pairs has the larger radius:
a. Fe^{2+} or Fe^{3+}
b. N or P
c. Se or Se^{2-}
d. Ba or Ba^{2+}

SOLUTION:
a. Fe^{2+}: The Z_{eff} felt by Fe^{3+} is greater; therefore, the electrons are more strongly attracted to the nucleus causing a reduction in the radius.

Chapter 6 – Ionic Bonds and Some Main–Group Chemistry

    b. P: Z_{eff} either remains the same or decreases slightly going down the periodic table. Therefore, the outer electrons are not held as tightly (also because of the increase in distance from the nucleus), causing an increase in the radius.
    c. Se^{2-}: The addition of electrons causes a decrease in Z_{eff}. Therefore, the outer electrons are more loosely held, causing an increase in the radius.
    d. Ba: The removal of electrons causes an increase in Z_{eff} creating a decrease in the radius.

6.3 Ionization Energy
  A. Ionization energy (E_i) — the amount of energy required to remove the outermost electron from an isolated neutral atom in the gaseous state.

  B. Periodic trends in ionization energy. (See Figure 6.3 on page 189 in your textbook.)   🔑 B.1.
    1. Minimum E_i — group 1A alkali metals.
    2. Maximum E_i — group 8A noble gases.
    3. E_i increases across a period.
    4. E_i decreases down a group.
  C. Periodicity due to electron configurations.
    1. Single s electron in valence–shell of alkali metals feels a low Z_{eff}.
       a. Single valence electron is well shielded by the core (inner shell) electrons.
       b. Electron is loosely held.
       c. Energy needed to remove electron is low.
    2. Electrons in filled valence–subshell of noble gas elements feel a high Z_{eff}.
       a. Electrons are tightly held.
       b. Radius of atom shrinks.
       c. Energy needed to remove electron is high.
    3. Increase in atomic number down a group.
       a. Value of n increases along with the average distance of the electron from the nucleus.
       b. Valence–shell electrons are less tightly held.
       c. E_i decreases.
  D. Minor irregularities occur from left to right across a row of the periodic table.
    1. Group 2A elements — E_i (Be) > E_i (B) (due to electron configuration).
       a. 2p electron of boron is shielded somewhat by the 2s electrons.
         i. feels a smaller Z_{eff}
         ii. more easily removed
    2. Group 6 A elements — E_i (N) > E_i (O) (due to electron configuration).
       a. 2p electron removed from nitrogen
         i. removed from a half–filled subshell (stable configuration)
       b. 2p electron removed from oxygen
         i. removed from a filled orbital
           – Electrons in filled orbitals are forced together and have a slightly higher energy.
           – easier to remove electrons
         ii. Ion formed has a stable half-filled subshell.

EXAMPLE:
  Which has the higher ionization energy: Na or Al, Al or In?

  SOLUTION: The ionization energy increases across the periodic table and decreases down the table. Therefore, E_i (Al) > E_i(Na) and E_i (Al) > E_i (In)

6.4 Higher Ionization Energies   🔑 B.2.
  A. Remove two, three, or even more electrons sequentially from an atom.
  B. Each ionization step requires successively larger amounts of energy.

Chapter 6 – Ionic Bonds and Some Main–Group Chemistry

1. Harder to remove a negatively charged electron from a positively charged ion.
C. Large jumps in successive ionization energies of the elements (see Table 6.2, page 191 in your textbook).
   1. Due to electron configuration.
      a. high degree of stability associated with filled s and p sublevels
      b. valence–shell electrons are easily lost during ionization
      c. core electrons — relatively difficult to remove
D. Valence–shell electron configuration of an atom controls its chemistry.

EXAMPLE:
Which ion has the higher third ionization energy, Mg^{2+} or Al^{2+}?

SOLUTION: For Mg^{2+} to lose another electron, the electron must come from the $2p$ subshell which is filled. This filled subshell has a high degree of stability and removing an electron will require a large amount of energy. Al^{2+}, on the other hand, can obtain a noble gas configuration by losing another electron. Therefore, the third ionization energy of Mg^{2+} is greater than the third ionization energy of Al^{2+}.

Workbook Problem 6.1
Determine which has the higher ionization energy:
a. Na or Cs
b. P or S
c. B^{2+} or C^{2+}

Strategy: Consider the ground–state electron configuration for each element and compare the Z_{eff} felt by the outer electrons.

Step 1: Determine the highest E_i based on comparisons of Z_{eff} in the strategy step.

What key concept did you use?

6.5 Electron Affinity
A. Electron affinity (E_{ea}) — the energy change that occurs when an electron is added to an isolated atom in the gaseous state.
B. E_{ea} — has a negative value.
   1. Energy is usually released when an atom adds an electron.
      a. Positive change in energy – energy is absorbed.
      b. Negative change in energy – energy is released.
   2. The more negative the value, the greater the tendency of the atom to accept an electron.
      a. the more stable the anion
   3. $E_{ea} > 0$ for atoms that form an unstable anion on addition of an electron.
      a. No experimental measurements can be made.

C. Periodic trends of E_{ea} — related to the electron configurations.
   1. Sign and magnitude of E_{ea} due to offsetting factors.

Chapter 6 – Ionic Bonds and Some Main–Group Chemistry

        a. negative E_{ea} due to attraction between the additional electron and the atomic nucleus
        b. positive E_{ea} due to repulsions between the additional electron and the electrons in the atom
    2. Halogens — large, negative E_{ea}.
        a. high Z_{eff}
        b. room in valence–shell for an additional electron
        c. high attraction between additional electron and atomic nucleus
    3. Noble–gas elements — positive E_{ea}.
        a. filled s and p sublevels
        b. additional electron goes into the next higher shell
            i. feels a low Z_{eff}
        c. small attraction between the additional electron and the atomic nucleus
        d. high electron–electron repulsions
    4. Alkaline earth metals – $E_{ea} \approx 0$.
        a. filled s subshell
        b. added electron goes into a p subshell of higher energy
        c. relatively low Z_{eff}

EXAMPLE:
Determine which has the larger E_{ea}: sulfur or selenium; selenium or bromine
SOLUTION: To determine which element has the larger E_{ea}, we need to determine which element has the larger Z_{eff}. We learned in Chapter 5 that Z_{eff} increases across the periodic table and decreases down the periodic table. Therefore, E_{ea} (S) > E_{ea} (Se) and E_{ea} (Br) > E_{ea} (Se).

6.6 Ionic Bonds and the Formation of Ionic Solids
  A. An element with a low E_i can transfer an electron to an element with a large negative E_{ea}.
    1. Produces a cation and an anion.
    2. Cation and anion are attracted together by electrostatic forces.
    3. Creates an ionic bond.
    4. Form a three–dimensional network of ions (ionic solid).
        a. Each cation is surrounded by and attracted to many anions.
        b. Each anion is surrounded by and attracted to many cations.
  B. Formation of ionic bonds leads to a large gain in stability.
    1. Overcomes unfavorable energy change of electron transfer.

  C. Born–Haber cycle — a series of hypothetical steps each of which contributes to the overall energy change during the formation of an ionic compound.
    1. Experimentally measured energy values available for each step.
    2. Net process is the sum of the individual steps.
    3. Individual steps may involve
        a. sublimation of a metal atom to a gaseous atom
        b. dissociation of nonmetal molecules
        c. ionization of isolated metal atoms
        d. formation of nonmetal anions
        e. formation of solid ionic compound

6.7 Lattice Energies in Ionic Solids
  A. Lattice energy (U) — the measure of the electrostatic interaction energies between ions in a solid.

    1. Measure of the strength of the crystal's ionic bonds.
    2. Refers to the breakup of a crystal into individual ions.
        a. has a positive value

Chemistry

Chapter 6 – Ionic Bonds and Some Main–Group

      b. When using the lattice energy in a Born–Haber cycle, change the sign since the last step is the reverse of the breakup of the crystal.
   3. Coulomb's law – describes the force (F) that results from the interaction of electric charges.
      a. $F = k \times \dfrac{z_1 z_2}{d^2}$
         i. k = constant dependent on the arrangement of ions in the specific compound.
         ii. z_1 and z_2 = charges on the ions
         iii. d = distance between the centers of the ions.
   4. U – largest when d is small and z_1 and z_2 are large.

B. Periodic Trends in lattice energies.
   1. Small d – ions close together.
      a. Small ionic radii
      b. z_1 and z_2 held constant, largest U belong to compounds formed from the smallest ions.
   2. Compounds with same anion but different cations.
      a. trend for increasing lattice energies parallels trend of decreasing cation size
   3. Compounds with same cation but different anion.
      a. Trend for increasing lattice energies parallels trend of decreasing anion size.
   4. Compounds of ions with higher charges have greater lattice energies than compounds of ions with lower charges.

EXAMPLE:
For CsI:
a. Calculate the energy change in kJ/mol if cesium atoms react with iodine atoms to yield isolated Cs^+ and I^- ions. Is the reaction favorable or unfavorable? (E_{ea} (I) = –295.4 kJ/mol; E_i (Cs) = 375.3 kJ/mol.) Is this a favorable energy change?

b. Calculate the net energy change in kJ/mol that takes place on formation of CsI(s) from the elements Cs (s) + 1/2 I_2 (g) → CsI (s), using the following information: heat of sublimation for Cs = 77.6 kJ/mol, bond dissociation energy for I_2 = 106.8 kJ/mol, E_{ea} for I = –295.4 kJ/mol, E_i for Cs = 375.3 kJ/mol; lattice energy for CsI = 602 kJ/mol. Is this a favorable energy change?

SOLUTION:
a. E = –295.4 kJ/mol + 375.3 kJ/mol = 79.9 kJ/mol. The reaction is unfavorable.

  b. sublimation of metal atom
     Cs (s) → Cs (g)   77.6 kJ/mol
     dissociation of nonmetal atom
     1/2 I_2 (g) → I (g)   53.4 kJ/mol
     ionization of the metal
     Cs (g) → Cs^+ + e^- 375.3 kJ/mol
     formation of nonmetal anion
     I (g) + e^- → I^- (g)   –295.4 kJ/mol
     formation of the solid compound
     Cs^+ (g) + I^- (g) → CsI (s)   –602.0 kJ/mol

E = 77.6 kJ/mol + 53.4 kJ/mol + 375.3 kJ/mol + (–295.4 kJ/mol) + (–602.0 kJ/mol) = –391.1 kJ/mol. This reaction is favorable.

EXAMPLE:
Which has the larger lattice energy? $MgCl_2$ or $CaCl_2$, $AlCl_3$ or $AlBr_3$, $FeCl_2$ or $FeCl_3$

Chapter 6 – Ionic Bonds and Some Main–Group Chemistry

SOLUTION: For compounds which have the same anion but different cations, lattice energies increase with decreasing cation size. Since the ionic radius of Ca^{2+} is greater than the ionic radius of Mg^{2+}, then $U(MgCl_2) > U(CaCl_2)$. For compounds that have the same cation but different anions, lattice energies increase with decreasing anion size. The ionic radius of Cl^- is less than the ionic radius of Br^-; therefore, $U(AlCl_3) > U(AlBr_3)$. Compounds of ions with higher charge have greater lattice energies; therefore, $U(FeCl_3) > U(FeCl_2)$.

Workbook Problem 6.2

For $MgCl_2$:

a. Calculate the energy change in kJ/mol if magnesium atoms react with a chlorine molecule to yield isolated Mg^{2+} and Cl^- ions. Is the reaction favorable or unfavorable? $E_{ea}(Cl) = -348.6$ kJ/mol; $E_{i1}(Mg) = 738$ kJ/mol; $E_{i2} = 1451$ kJ/mol)

b. Calculate the net energy change in kJ/mol that takes place on formation of $MgCl_2(s)$ from the elements, $Mg(s) + Cl_2(g) \rightarrow MgCl_2(s)$ using the E_{ea} and E_i given above, along with the additional following information: heat of sublimation for Mg = 149.6 kJ/mol, bond dissociation energy for Cl_2 = 243 kJ/mol, lattice energy for $MgCl_2$ = 2526 kJ/mol.

Strategy: Keeping the stoichiometry in mind, either sum the E_i and the E_{ea} or create a Born–Haber cycle for the reaction.

Step 1: Add the E_{i1} and E_{i2} for Mg to twice the E_{ea} for chlorine. Is the answer positive or negative?

Step 2: Construct a Born–Haber cycle for the formation of $MgCl_2$.

What key concept did you use?

6.8 The Octet Rule

A. Octet rule — main–group elements tend to undergo reactions that leave them with eight valence electrons.
B. Useful for making predictions; provides insights about chemical bonds.
   1. Noble gas electron configuration.
      a. filled octet – s and p subshells are filled
      b. Can't take electrons away – electrons are tightly held by a high Z_{eff}
      c. No low–energy orbitals available to accept electrons.
C. Failure of rule — elements on right side of periodic table in third or lower period.
   1. due to availability of d orbitals
D. Factors that determine the formation of a cation or anion.
   1. Cation — electrons are shielded from nucleus by core electrons.
      a. electrons feel low Z_{eff}

b. electrons easily lost
c. reach next–lower noble gas configuration — more difficult to lose another electron
2. Anion — electrons are poorly shielded.
    a. Electrons feel high Z_{eff}.
    b. can gain electrons because of Z_{eff}
    c. reach noble gas configuration — low–energy orbitals no longer available

6.9 Chemistry of the Alkali Metals (Group 1A)
A. General properties:
  1. Valence–shell electron configuration — ns^1.
     a. Common ion — M^+
     b. Lowest ionization energies of all the elements
     c. Most powerful reducing agents in the periodic table
     d. Ability to donate electrons dominates chemistry.
  2. Low densities and melting points.
  3. Metals.
     a. bright, silvery solids
     b. malleable
     c. good conductors of electricity
     d. soft enough to cut with a knife
  4. Very reactive.
     a. must be stored under oil to prevent their instantaneous reaction with oxygen and water
  5. Occur in nature as salts.
B. Occurrences and Uses of Alkali Metals
  1. Lithium – common occurrence in rocks.
     a. "Lithos" – Greek word meaning stone.
     b. Major industrial use – all-purpose automotive greases.
     c. Other uses:
        1. Li_2CO_3 – pharmaceutical agent for the treatment of bipolar disorder.
  2. Sodium – 6th most abundant element in the earth's crust.
     a. Vast deposits laid down by evaporation of ancient seas.
     b. World's oceans are ~ 3% by mass NaCl.
     c. Use spans nearly the entire range of processes in modern chemical industry.
  3. Potassium
     a. Vast deposits of KNO_3 and KCl.
     b. Use – plant fertilizer.
  4. Rubidium, Cesium
     a. The two most chemically reactive of the common alkali metals.
     b. Cesium – "caesius" (sky blue)
     c. No major commercial importance.
  5. Francium
     a. Highly radioactive
     b. Behavior similar to other alkali metals.
C. Produced by reaction of their chloride salts.
  1. Lithium and sodium – produced by electrolysis.
     a. an electric current is passed through the molten salt.
  2. Potassium, rubidium, and cesium – produced by chemical reduction.

D. Reactions of alkali metals.
  1. Halogens.
     a. general reaction: $2 M (s) + X_2 (g) \rightarrow 2 MX (s)$
     b. products — halides, colorless, crystalline ionic salts
     c. A decrease in ionization energy leads to an increase in reactivity
  2. Hydrogen.

Chapter 6 – Ionic Bonds and Some Main–Group Chemistry

      a. general reaction: $2\,M\,(s) + H_2\,(g) \rightarrow 2\,MH\,(s)$
      b. products — hydrides (oxidation number of H = –1)
         i. white crystalline compounds.
      c. sluggish reaction at room temperature
   3. Nitrogen.
      a. only lithium reacts: $6\,Li\,(s) + N_2\,(g) \rightarrow 2\,Li_3N\,(s)$
   4. Oxygen.   E.1.
      a. All react rapidly.
      b. form different kinds of products
         i. Li → oxide (oxidation number of O = –2)
         ii. Na → peroxide (oxidation number of O = –1)
         iii. K, Rb, Cs → superoxide (oxidation number O = –1/2)
   5. Water.
      a. general reaction: $2\,M\,(s) + 2\,H_2O\,(l) \rightarrow 2\,M^+\,(aq) + 2\,OH^-\,(aq) + H_2\,(g)$
      b. Reactivity increases down the group.
      c. produces an alkaline solution
      d. redox process
   6. Ammonia.
      a. general reaction: $2\,M\,(s) + 2\,NH_3\,(l) \rightarrow 2\,M^+\,(soln) + 2\,NH_2^-\,(soln) + H_2\,(g)$
      b. analogous to reaction between metal and water
      c. Solutions have powerful reducing properties.
         i. MNH_2 — metal amide

6.10 Chemistry of the Alkaline Earth Metals (Group 2A)
  A. General properties.
    1. Similar to alkali metals.
    2. Valence–shell electron configuration — ns^2.
       a. common ion — M^{2+}
       b. powerful reducing agents
       c. Born-Haber cycle – contribution from ionic bonds releases enough energy to drive the entire process.
    3. Soft, silvery metals.
    4. Higher melting points and densities than alkali metals.
    5. Less reactive toward oxygen and water than alkali metals.
    6. Occur in nature as salts.
  B Occurrences and Uses of Alkaline Earth Metals
    1. Beryllium
       a. found in large commercial deposits of beryl.
       b. compounds are extremely toxic
       c. useful in forming allows.
    2. Magnesium
       a. many magnesium-containing minerals
       b. world's oceans provide an infinite supply
       c. used as a structural material.
    3. Calcium
       a. Gypsum and $CaCO_3$ deposits in ancient seabeds.
       b. alloy agent – hardens aluminum.
    4. Strontium, Barium
       a. Strontium – no commercial use for the pure metal
         i. $SrCO_3$ – used in the manufacture of glass for color TX picture tubes.
       b. Barium – no commercial use for the pure metal.
         i. $BaSO_4$ – used as a contrast medium for stomach and intestinal X rays.
    5. Radium – isolated from pitchblende.

a. highly radioactive
b. no commercial uses.
C Produced by reduction of their salts.

D. Reactions of Alkaline Earth Metals.   🗝 *E.3.*
1. Same kinds of redox reactions as alkali metals.
2. Less reactive than alkali metals.
   a. E_i (alkaline earth) > E_i (alkali)
3. Halogens → ionic halide salts, MX_2.
4. Oxygen → oxides, MO.   🗝 *E.1.*
   a. Sr, Ba → peroxides, MO_2
5. Water → metal hydroxides, $M(OH)_2$.

EXAMPLE:
Write general chemical equations for the reactions of the alkaline earth metals described in D above.

SOLUTION:
$M(s) + X_2 \rightarrow MX_2$
$M(s) + O_2(g) \rightarrow 2\,MO$ (oxides, M = Mg, Ca)
$M(s) + O_2(g) \rightarrow MO_2$ (peroxide, M = Ba, Sr)
$M(s) + 2\,H_2O(l) \rightarrow 2\,M^+(aq) + 2\,OH^-(aq) + 2\,H_2(g)$

6.11 Chemistry of the Group 3A Elements: Aluminum
A. General properties.
1. Valence–shell electron configuration — ns^2np^1.
2. B — semimetal.
3. Others — silvery metals.
   a. soft
   b. good conductors of electricity
B. Aluminum — most abundant metal in earth's crust.
1. Name comes from alum, $KAl(SO_4)_2 \cdot 12\,H_2O$.
   a. medicinal salt
2. Occurs in many common minerals and gemstones.
   a. sapphire and ruby impure forms of Al_2O_3
   b. impurity creates color
      i. Cr – impurity in ruby
      ii. Fe and Ti impurity in sapphire
3. Commercially obtained from bauxite, $Al_2O_3 \cdot xH_2O$.
4. Reducing agent.
   a. loses all three valence electrons
   b. common ion: Al^{3+}.

C. Reactions of aluminum.   🗝 *D.4.*
1. Halogens → AlX_3.
   a. vigorous at room temperature
   b. releases large amount of heat
2. Oxygen → Al_2O_3.
   a. vigorous at room temperature only on surface
   b. thin, hard oxide coat protests metal from contact with air.
3. Acids and bases → $Al^{3+} + H_2$.

Chapter 6 – Ionic Bonds and Some Main–Group Chemistry

6.12 The Halogens (Group 7A)
  A. General properties.
   1. Exist as diatomic molecules.
   2. Valence–shell electron configuration — ns^2np^5.
      a. gain electrons
      b. powerful oxidizing agents
      c. large negative E_{ea}
   3. Occur in nature as salts and minerals.
  B. Occurrence and Uses of Halogens
   1. Fluorine – pale, yellow gas
      a. found in several common minerals.
      b. extremely toxic and reactive.
      c. used in the manufacture of polymers, such as Teflon, and the production of UF_6.
         i. UF_6 – used to separate uranium isotopes for nuclear power plants
      d. F^- - added to toothpaste to help prevent tooth decay.
   2. Chlorine – toxic, reactive, greenish-yellow gas.
      a. vast mounts found in the world's oceans
      b. used for the preparation of numerous chlorinated organic chemicals, as bleach during paper manufacture, and as a disinfectant for swimming pools and municipal water supplies.
   3. Bromine – volatile, reddish liquid
      a. toxic fumes
      b. primary use – preparing brominated organic compounds
      c. AgBr used in photographic emulsions
   4. Iodine – volatile purple-black solid
      a. widely used as a skin disinfectant and in the preparation of numerous organic compounds.
   5. Astatine – a radioactive element.
  C. Produced by oxidation of their anions.

  D. Reactions.   🔑 E.5.
   1. Most reactive elements in the periodic table.
   2. Less reactive going down the periodic table.
   3. Metals → metal halides, MX_n.
      a. $2M + nX_2 \rightarrow 2MX_n$
   4. Hydrogen → hydrogen halides, HX.
      a. $H_2(g) + X_2 \rightarrow 2HX(g)$
      b. behave as acids when dissolved in water
   5. Other halogens → XY (X and Y are different halogens).
      a. interhalogen compounds
      b. redox process – lighter, more reactive element is oxidizing agent and the heavier, less reactive element is the reducing agent
      c. properties intermediate between parent elements
      d. act as strong oxidizing agents in redox reactions
      e. number of polyatomic interhalogen compounds

Workbook Problem 6.3
Write balanced chemical equations for the following reactions:
a. Na and NH_3
b. Ca and water
c. Al and HBr (g)

Chapter 6 – Ionic Bonds and Some Main–Group Chemistry

   d. Sr and F_2
   e. Br_2 and F_2

Strategy: Review the chemistry described above and determine the outcome for each reaction.

What key concept did you use?

6.13 Chemistry of the Noble Gases (Group 8A)
   A. General properties.
      1. Colorless, odorless, unreactive gases.
      2. Valence–shell electron configuration — ns^2np^6.
         a. stable configuration
         b. do not form cations or anions
         c. normally do not undergo redox reactions
         d. commercial use – applications that require inert atmospheres
   B. Occurrence and Uses of Noble Gases
      1. Helium – second most abundant element
      2. Main commercial use – applications that require an inert (unreactive) atmosphere.
      3. Radon – radioactive element.
         a. produced by radioactive decay of the radium present in small amounts in many granitic rocks
         b. can slowly seep into basements
         c. can cause radiation damage in the lungs
            i. KEEP BASEMENTS VENTED!
   C. Reactions of noble gases.
      1. Kr and Xe react only with fluorine.
         a. xenon fluorides — powerful oxidizing agents
      2. Lack of reactivity due to valence–shell configuration.
         a. large E_i
         b. small E_{ea}

 **Putting It Together**

An experiment in which students determine the empirical formula of a compound is often included in many general chemistry laboratory manuals. Usually, this experiment involves placing magnesium ribbon in a clean, weighed crucible and heating slowly. During the heating process, the lid is occasionally lifted to allow air to react with the magnesium. This process continues until there is no apparent change in the magnesium ash. The lid is then removed and heating of the crucible continues for several minutes. The crucible is then removed from the heat and water is added to remove any magnesium nitride that may have formed during the process. Once the crucible is cooled it is weighed again. The difference in the mass of the empty crucible and the crucible after heating is the mass of the magnesium oxide.

Chapter 6 – Ionic Bonds and Some Main–Group Chemistry

a. Write the chemical equations for the reaction of magnesium with the oxygen in air, and the reaction of magnesium with the nitrogen in the air. Also write a chemical equation for the reaction of magnesium nitride with water to form magnesium oxide.

b. Reaction of magnesium ribbon in a weighed crucible produced 0.386 g of product. After weighing the product, water was added and the crucible was gently reheated. After cooling, the product in the crucible weighed 0.399 g. How much magnesium oxide was formed? What mass percent of magnesium nitride was formed in the first heating?

 **Self–Test**

This section is intended to test your knowledge of the material covered in this chapter. Think through these problems, and make certain you understand what is going on. Ask yourself if your answer makes sense. Many of these questions are linked to the chapter learning goals. Therefore, successful completion of these problems indicates you have mastered the learning goals for this chapter. You will receive the greatest benefit from this section if you use it as a mock exam. You will then discover which topics you have mastered and which topics you need to study in more detail.

True–False

1. The elements that have the maximum E_i in the periodic table are the noble gases.

2. E_i increases across a period because the value of n increases along with the average distance of the electron from the nucleus.

3. The valence–shell electron configuration of an atom controls its chemistry.

4. The tendency of the atom to accept an electron decreases as the value of E_{ea} becomes more negative.

5. An element with a high E_i can transfer an electron to an atom with a negative E_{ea}.

6. The formation of an ionic bond leads to a large gain in stability.

7. The lattice energy refers to the breakup of a crystal into individual ions and has a positive value.

8. The alkaline earth metals are more reactive than the alkali metals because the E_i (alkaline earth metal) > E_i (alkali metal).

9. The halogens have the valence–shell configuration ns^2np^5 and are powerful reducing agents.

10. In a filled octet, a high Z_{eff} tightly holds the electrons.

Multiple Choice

1. Ionization energies
    a. increase across a period and down a group
    b. decrease across a period and down a group
    c. increase across a period and decrease down a group
    d. decrease across a period and increase down a group

Chemistry

Chapter 6 – Ionic Bonds and Some Main–Group

2. The element with the higher ionization energy is
   a. Rb > Li
   b. Rb > Sr
   c. Rb > Ca
   d. Rb > Cs

3. The element with the more negative electron affinity is
   a. Se > S
   b. Se > Br
   c. Se > Cl
   d. Se > Te

4. The ionic compound with the higher lattice energy is
   a. NaCl > LiCl
   b. CaS > BaS
   c. NaI > NaF
   d. MgS > MgO

5. The alkali metals
   a. are the most powerful oxidizing agents in the periodic table
   b. have high densities and melting points
   c. occur in nature as the free elements
   d. are good conductors of electricity

6. When potassium reacts with oxygen, a(an)
   a. oxide, K_2O, is formed
   b. no reaction occurs
   c. peroxide, KO, is formed
   d. superoxide, KO_2, is formed

7. The oxidation number of oxygen in a peroxide is
   a. −1
   b. −1/2
   c. −2
   d. none of the above

8. The alkaline earth metals
   a. are more reactive toward oxygen and water than the alkali metals
   b. occur in nature as salts
   c. are dull, hard metals
   d. are powerful oxidizing agents

9. The group 3A elements
   a. are all semimetals
   b. are all metals
   c. have the valence–shell configuration ns^2np^1
   d. show an increase in ionization energy going down the group

10. The noble gases
    a. are odorless, colorless, reactive gases
    b. have no commercial use
    c. normally do not undergo redox reactions
    d. have an unstable electron configuration

Chapter 6 – Ionic Bonds and Some Main–Group Chemistry

Matching

    Ionization energy

    Electron affinity

    Born–Haber cycle

    Lattice energy

    Octet rule

a. the energy change that occurs when an electron is added to an isolated atom in the gaseous state.

b. Main–group elements tend to undergo reactions that leave them with eight valence electrons.

c. the amount of energy required to remove the outermost electron from an isolated atom in the gaseous state.

d. the sum of the electrostatic interaction energies between ions in a solid.

e. a series of hypothetical steps, each of which contributes to the overall energy change during the formation of an ionic compound.

Fill–in–the–Blank

1. Alkali metals have the lowest _____ of all the elements.
2. When the alkali metals react with halogens, a decrease in ionization energy leads to an increase in _____.
3. Alkaline earth metals are _____ reactive than the alkali metals because _____ _____.
4. For compounds with the same anion but different cation, the trend for increasing lattice energies parallels the trend of _____.
5. The oxidation number of each oxygen atom in the superoxide ion is _____.
6. When the alkaline earth elements react with water, the product is _____.
7. The halogens act as _____ in redox reactions.
8. When halogens react with each other to form interhalogen compounds, the _____ is the oxidizing agent.
9. Xenon fluorides act as powerful _____.
10. Elements on the right side of the periodic table in the third or lower period can have an expanded octet due to _____.

Chemistry

Chapter 6 – Ionic Bonds and Some Main–Group

Problems

1. Arrange the elements Ge, Si, and P in order of increasing ionization energies.

2. Which has the larger fourth ionization energy, In or Sn?

3. Explain why fluorine has a higher E_{ea} than its neighbors on either side.

4. Calculate the heat of sublimation for aluminum given that the net energy change for formation of $AlBr_3$ (s) from the elements, Al (s) + 3/2 Br_2 (l) → $AlBr_3$ (s) is –527.2 kJ/mol. Use the following information:

   Al: E_{i1} = 578 kJ/mol, E_{i2} = 1,817 kJ/mol, E_{i3} = 2,745 kJ/mol
   Br: E_{ea} = 324.8 kJ/mol
   Bond dissociation energy for Br_2 = 193 kJ/mol
   Heat of vaporization for Br_2 (l) = 30.91 kJ/mol
   Lattice energy for $AlBr_3$ (s) = 5,361 kJ/mol

5. Balance the following redox reactions. Identify which species are oxidized, and which species are reduced, and which elements have undergone a change in oxidation state.
   a. Lithium reacts with water to produce lithium hydroxide.
   b. In the presence of water, chlorine oxidizes iodine and produces hydrochloric acid and iodic acid.

6. Predict the outcome of the following reactions:
   a. Ca (s) + Br_2 (l) →

   b. Al (s) + N_2 (g) →

   c. K (s) + O_2 (g) →

   d. Na (s) + H_2 (g) →

   e. H_2 (g) + I_2 (g) →

7. Which has the higher lattice energy?
   a. $MgCl_2$ or $AlCl_3$
   b. $AlCl_3$ or $GaCl_3$
   c. $SnCl_2$ or $PbBr_2$

8. Determine the oxidation number of the halogens in the following compounds: ClF_3, IF_7, ICl_5, IBr.

9. The reaction used to manufacture beryllium is given below:

   BeF_2 (l) + Mg (l) → Be (l) + MgF_2 (l)

   How much beryllium is produced when 4.50 g of BeF_2 reacts with 8.23 g of Mg?

10. Write and balance the chemical equations for the formation of ClF_3 and BrF_5 from the halogens.

Chapter 6 – Ionic Bonds and Some Main–Group Chemistry

Challenge Problem

A 2.25 g sample of an alkaline earth metal was reacted with a volume of liquid bromine that contains 7.5×10^{22} molecules. The resulting metal bromide was analyzed for bromine by dissolving a 3.75 g sample in water and adding an excess of silver nitrate. The analysis yielded 7.05 g of silver bromide.

a. What is the percent bromide in the alkaline earth bromide?

b. What is the identity of the alkaline earth metal?

c. Write the balanced chemical equations for all reactions.

d. What was the limiting reactant in the reaction between the alkaline earth metal and the bromine? How many grams of the excess reactant did not react?

Inquiry Based Problem

Potassium aluminum sulfate dodecahydrate, $KAl(SO_4)_2 \cdot 12 \, H_2O \, (s)$, is referred to as an alum, a hydrated double salt. This particular alum is used for water purification, sewage treatment and in fire extinguishers. It is easily prepared in a General Chemistry laboratory. The procedure for this preparation is as follows:[1]

1. Potassium aluminate, $KAl(OH)_4$ is produced by reaction of aluminum metal with hot 20% KOH. The $KAl(OH)_4$ remains in solution. During this reaction, H_2 gas is evolved.

   Given the tremendous amount of energy needed to produce pure aluminum metal, what would you suggest be a good resource for the aluminum in this experiment?

   What type of reaction is this?

   The concentration of the KOH solution being used is given as 20% KOH. Given this is an aqueous solution, what else is present and in what concentration (stated as %)?

   With the information provided for this step, write a balanced chemical equation.

2. 6 M H_2SO_4 is added to the solution of $KAl(OH)_4$ produced in the first step. Insoluble $Al(OH)_3$ forms from this reaction

   This reaction produces two salts, one of which is soluble and the other of which is insoluble. Knowing the reactants, can you predict the identity of the soluble salt?

   Write a balanced chemical equation for this step.

3. The mixture containing the $Al(OH)_3$ is reheated until the $Al(OH)_3$ is converted to $Al_2(SO_4)_3 \, (aq)$. The clear solution is then cooled in an ice-bath producing the alum.

---

[1] "Laboratory Manual for *Brady/Holum Fundamentals of Chemistry, 3rd Edition*;" Beran, J.A.; John Wiley & Sons; New York; 1988; pp./223-225.

Chapter 6 – Ionic Bonds and Some Main–Group Chemistry

Keeping in mind that you used the entire reaction solution from both step 1 and step 2 in step 3, what reactants are present in the third step when the solution is reheated?

Write a balanced chemical equation for the reaction that occurs when Al(OH)$_3$ is converted to the soluble aluminum salt.

Write a balanced chemical equation for the reaction that occurs when the alum precipitates from solution. At this point, H$_2$SO$_4$ does not participate in the reaction.

CHAPTER 7

COVALENT BONDS AND MOLECULAR STRUCTURE

Chapter Learning Goals

A. Covalent Bonds
1. From a list of compounds, predict which are ionic and which are molecular.
2. Using only the periodic table, predict which of two elements is more electronegative.
3. Using only the periodic table, predict whether a given bond is ionic, polar covalent, or nonpolar covalent.
4. Using a table of electronegativities, predict which of two bonds is expected to be more polar.

B. Lewis Theory
1. Write Lewis symbols for atoms, and tell how many electrons must be shared to enable the atom to achieve a completed valence shell. Give the symbol of the noble gas with the same number of valence electrons.
2. For each atom in an electron–dot structure, give the number of bonded electron pairs and the number of nonbonded electron pairs.
3. For a given electron–dot structure, give the number of single bonds, double bonds, and triple bonds. Give the bond order of each bond.
4. Draw electron–dot structures of molecules and polyatomic ions, recognizing when multiple bonding and resonance structures are needed.
5. Determine the formal charge on each atom in a resonance structure, and use the formal charges to select the best resonance structure.

C. VSEPR Theory
1. Use the VSEPR model to predict the geometries of molecules and polyatomic ions, including those with more than one central atom.

D. Valence Bond Theory
1. For molecules and polyatomic ions, sketch and identify the orbitals used by each atom to form bonds. Show which orbital overlaps result in σ bonds and which result in π bonds.

E. Molecular Orbital Theory
1. Sketch a molecular orbital diagram for a diatomic molecule. Use the molecular orbital diagram to determine the number of unpaired electrons and to calculate the bond order of the molecule described.

Chapter in Brief
The most important kind of bond in all of chemistry is the covalent bond, which involves the sharing of electrons between atoms. To understand this bonding concept, you need to first learn about the type of interactions that occur between atoms and how to draw electron–dot structures, a useful tool for visualizing the interactions of electrons in a molecule. You will also learn how to determine the type of bonding that occurs in a molecule based on bond polarities, to predict the relative importance of resonance structures based on formal charge calculations, and to determine molecular shapes using VSEPR theory. Once you have mastered molecular shape determinations, you will be able to apply this knowledge to valence bond theory and to ascertain the type of hybridization predicted by valence bond theory. Finally, you will be introduced to molecular orbital theory, a more sophisticated and, at times, more accurate bonding description.

Chapter 7 – Covalent Bonds and Molecular Structure

7.1 The Covalent Bond
A. Formation of a covalent bond.
  1. Two atoms come close together, and electrostatic interactions begin to develop.
     a. Two nuclei repel each other; electrons repel each other.
     b. Each nucleus attracts the electrons; electrons attract both nuclei.
  2. Attractive forces > repulsive forces; then covalent bond is formed.
B. Distance between the two atoms affects the magnitude of the various attractive and repulsive forces.
  1. Bond length — the optimum point where net attractive forces are maximized.
     a. Each covalent bond has a characteristic length that leads to maximum stability.
     b. can predict from atomic radii

7.2 Strengths of Covalent Bonds
A. Formation of covalent bonds leads to lower energy.
B. Bond dissociation energy (D) — the amount of energy necessary to break a chemical bond in an isolated molecule in the gaseous state (positive value).
  1. Equals the amount of energy released when the bond forms (negative value).
C. Bonds between the same pairs of atoms usually have similar bond dissociation energies.

7.3 A Comparison of Ionic and Covalent Compounds ⚷ *A.1.*
A. Ionic compounds.
  1. High–melting solids.
     a. must overcome every ionic attraction in the entire crystal
B. Covalent compounds.
  1. Low–melting solids, liquids, or even gases.
     a. Bonds within individual molecule may be very strong.
     b. Attractive forces between different molecules (intermolecular forces) are relatively weak.

7.4 Polar Covalent Bonds: Electronegativity
A. Ionic and covalent bonds represent extremes of a continuous spectrum of possibilities.
B. Polar covalent bonds — the bonding electrons are attracted somewhat more strongly by one atom in a bond.
  1. Electrons are not completely transferred.
  2. More electronegative atom: δ–. (δ represents the partial charge formed.)
  3. Less electronegative atom: δ+.

C. Electronegativity (EN) — the ability of an atom in a molecule to attract the shared electrons in a bond. ⚷ *A.2, A.3.*
  1. Metallic elements — low electronegativities.
  2. Halogens and other elements in upper right–hand corner of periodic table — high electronegativities.

D. Predicting bond polarity. ⚷ *A.4.*
  1. Atoms with similar electronegativities ($\Delta EN \leq 0.4$) — form nonpolar bonds.
  2. Atoms whose electronegativities differ by more than two ($\Delta EN > 2$) — form ionic bonds.
  3. Atoms whose electronegativities differ by less than two ($\Delta EN < 2$) — form polar covalent bonds.

Chapter 7 – Covalent Bonds and Molecular Structure

EXAMPLE:
Predict whether the following bonds are polar, nonpolar or ionic: Si — Cl, Si — H, and Ca — F.

SOLUTION: Si — Cl: polar (3.0 – 1.8 = 1.2); Si — H: nonpolar (2.1 – 1.8 = 0.3);
Ca — F: ionic (4.0 – 1.0 = 3.0)

7.5 Electron–Dot Structures

    B.1.

A. Electron–dot structure — represents how an atom's valence electrons are distributed in a molecule.
   1. Atom's electrons represented by dots.
B. Filled *s* and *p* subshells for each atom in a molecule leads to stability.
   1. Octet rule.
      a. important guiding principle.
      b. can predict formulas and electron-dot structures.
C. Electron–dot structures for elements.
   1. First, place the electrons (dots) one to each side, until all four sides are occupied.
   2. Next, pair the electrons (dots) until all the valence–shell electrons are used up.
      a. The pairing of dots does not correspond to the pairing of electrons in the electron configuration.

EXAMPLE:
Draw the electron–dot structure for C and N.

SOLUTION:

·Ċ· :Ṅ·

D. Use the octet rule to determine the number of covalent bonds an element will form.
   1. Only unpaired electrons can form bonds.
   2. An atom will share electrons until it is surrounded by four pairs of electrons (an octet) or has no more electrons to share.
   3. Lone pairs (nonbonded pairs) – electron pairs not used in bonding.
   4. Bonding pairs share electrons.

EXAMPLE:
Draw the electron–dot structure for PH$_3$.

SOLUTION:
First draw the electron–dot structure for phosphorus. Phosphorus has five valence–shell electrons.

·Ṗ·
··

Each hydrogen has an electron that it can share with an unpaired electron on phosphorus.

H·

The electron–dot structure is

$$\begin{array}{c} H \\ | \\ H - \underset{..}{P} - H \end{array}$$

E. Multiple covalent bonds — atoms share more than one pair of electrons.
  1. Shorter and stronger than single bonds.
  2. Double bond — atoms share two pairs of electrons.
  3. Triple bond — atoms share three pairs of electrons.
  4. Bond order — refers to the number of electron pairs between atoms.
F. Coordinate covalent bonds — one atom donates both electrons (a lone pair) to another atom that has a vacant valence orbital.
  1. N, O, P, and S frequently form coordinate covalent bonds.

7.6 Electron–Dot Structures of Polyatomic Molecules

A. Compounds containing only hydrogen and second–row elements: C, H, N, O.
  1. Octet rule almost always applies.
  2. Easy to predict the number of bonds formed by each element.
    a. number of bonds formed equals number of unpaired dots in the Lewis formula
  3. Indicate a covalent bond by a line.
    a. double bond – two lines
    b. triple bond – three lines
  4. Small molecules with a few second row atoms and H.
    a. Second row atoms bonded to one another in a central core.
    b. Hydrogen on periphery.

EXAMPLE:
Draw the electron-dot structure for N_2H_4.

SOLUTION:
To draw the electron-dot structure of N_2H_4:

1. First draw the electron-dot structure of N.

   $:\dot{N}\cdot \quad \cdot\dot{N}:$

2. Form a bond between the two nitrogens using one of the unpaired electrons on each nitrogen.

   $:\dot{N} - \dot{N}:$

3. You need to place 4 hydrogens on the periphery. Each nitrogen has two remaining unpaired electrons which can be used to form a bond with a hydrogen. This leads to the structure:

$$\begin{array}{cc} H & H \\ | & | \\ :N - & N: \\ | & | \\ H & H \end{array}$$

Chapter 7 – Covalent Bonds and Molecular Structure

    B. Compounds with elements beyond the second row.
        1. Third period and lower
            a. larger – can accommodate more atoms.
            b. have unfilled *d* orbitals.

    C. Steps for drawing Electron–dot structures (see pages 235 and 236 in your textbook).
        1. Atoms with less than four unpaired dots will not form an octet.
        2. When determining the connection, keep in mind that the central atom is usually written first in the formula.
            a. When unsure, always assign the most symmetrical structure to the atom.

EXAMPLE:
Draw the electron–dot stucture of IF_5.

SOLUTION:
To draw the electron–dot structure of IF_5:

1. Total number of valence–shell electrons = 7 (from I) + 35 (from 5 F) = 42.

2. Determine the connections. In this case, I is the central atom. (This leads to the most symmetrical structure.)

$$\begin{array}{ccc} & F & \\ F & | & F \\ & I & \\ F & & F \end{array}$$

3. Ten of the 42 valence electrons are used in forming the 5 I—F bonds, leaving 32. Thirty of these electrons are used in forming the octet around each fluorine.

$$\begin{array}{ccc} & :\ddot{F}: & \\ :\ddot{F} & | & \ddot{F}: \\ & I & \\ :\ddot{F} & & \ddot{F}: \end{array}$$

4. There are two remaining electrons which are placed on the iodine.

$$\begin{array}{ccc} & :\ddot{F}: & \\ :\ddot{F} & | & \ddot{F}: \\ & \ddot{I} & \\ :\ddot{F} & & \ddot{F}: \end{array}$$

Chapter 7 – Covalent Bonds and Molecular Structure

 *Workbook Problem 7.1*
Draw the electron–dot structure for BrF$_3$.

Strategy: Follow the steps outlined on pages 256 and 257 in your textbook.

Step 1: Determine the total number of valence–shell electrons.

Step 2: Determine the connections.

Step 3: Draw the bonds and subtract the number of electrons used from the total number of valence electrons available.

Step 4: Complete the octets of the outer atoms.

Step 5: Place any remaining electrons on the central atom.

What key concept did you use?

7.7 Electron–Dot Structures and Resonance
  A. More than one electron–dot structure for a molecule.
    1. Resonance hybrid — average of the various possible electron–dot structures for a molecule.
       a. resonance indicated by a straight double–headed arrow (↔)
       b. differ only in placement of valence electrons

EXAMPLE:
Draw the resonance structures for CO$_3^{2-}$.

Chapter 7 – Covalent Bonds and Molecular Structure

SOLUTION:
1. Total number of valence–shell electrons = 4 (from C) + 18 (from 3 O) + 2 (from the 2– charge) = 24

2. Determine the connections. In this anion, C is the central atom.

$$\begin{array}{c} O \\ | \\ O - C - O \end{array}$$

3. Six of the 24 electrons are used for forming the 3 C—O bonds, leaving 18. All 18 of these electrons are used to form the octet around the 3 oxygen atoms.

4. There are no more electrons to distribute; however, carbon does not have a completed octet. To form a completed octet, we can borrow a pair of lone electrons from one of the oxygen atoms. However, since all 3 C—O bonds are equal, we must draw three different resonance structures, each of which has a C=O bond between carbon and a different oxygen.

7.8 Formal Charges
A. Related to ideas of electronegativity and polar covalent bonds.
B. Result of electron bookkeeping.
  1. Compare the number of valence electrons around an atom in a molecule to the number of valence electrons around the isolated atom.
     a. if not equal – atom has gained or lost electrons
        i. atom has a formal charge.
  2. Formal charge = (number of valence electrons in free atom) – 1/2(number of bonding electrons) – (number of nonbonding electrons).
  3. For ions, the sum of the formal charges on all the atoms is equal to the overall charge on the ion.
C. Use to evaluate the relative importance of different resonance structures.
  1. Structures with the smallest formal charges are more stable.

Chapter 7 – Covalent Bonds and Molecular Structure

2. Structures that have the more negative formal charge on the more electronegative atom are more stable.

EXAMPLE:
Determine the formal charge on each atom in:

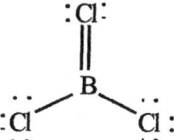

SOLUTION:

Formal charge = number of valence electrons − $\frac{1}{2}$(number of bonded electrons) − number of lone electrons

formal charge (B) = (3) − $\frac{1}{2}$(8) = −1

formal charge (double bonded Cl) = (7) − $\frac{1}{2}$(4) − 4 = +1

formal charge (single bonded Cl) = (7) − $\frac{1}{2}$(2) − 6 = 0

EXAMPLE:
Why is the structure below the better structure for BCl_3?

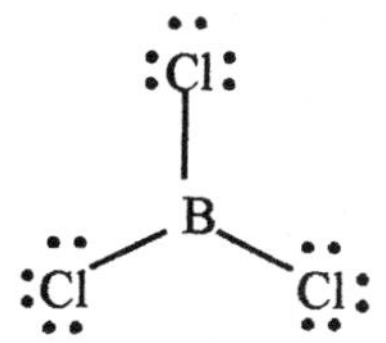

SOLUTION:
To answer this question, we must first calculate the formal charges on all of the atoms.

formal charge (B) = 3 − $\frac{1}{2}$(6) = 0

formal charge (Cl) = 7 − $\frac{1}{2}$(2) − 6 = 0

In the first stucture, boron had a formal charge of −1 and one chlorine had a formal charge of +1. We know that chlorine is more electronegative than boron and therefore, should have the negative charge. In the second structure, boron has a formal charge of zero, making this structure more likely.

 Workbook Problem 7.2
Determine which of the following resonance structures is the best.

133

Chapter 7 – Covalent Bonds and Molecular Structure

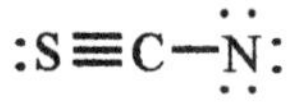

a. 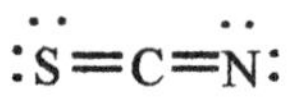 :S̈=C=N̈:

b. :S≡C−N̈:

c. :S̈−C≡N:

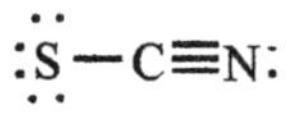

Strategy: The most stable resonance structures are those that have the smallest formal charges and those that have the more negative formal charge on the more electronegative atom.

Step 1: Determine the formal charges for all atoms in each of the compounds.

Step 2: Based upon the formal charges of all of the elements in each compound as well as the electronegativities of the elements, determine the most likely structure.

What key concept did you use?

7.9 Molecular Shapes: The VSEPR Model

A. All molecules have a specific 3-D shape.
   1. Plays an important role in determining the molecule's chemistry.
B. Molecular shape — determined by electronic structure of bonded atoms..
B. Valence–shell electron–pair repulsion model — predicts approximate shape of a molecule.
   1. Bonded and lone electrons occupy charge clouds.
      a. Repel each other
   2. Count the number of charge clouds surrounding the atom.
   3. Predict the geometry.
      a. Charge clouds will orient themselves so that they stay as far away as possible from each other.
      b. See Table 7.4, page 248 in your textbook.

EXAMPLE:
Predict the shape of AsF_5.

SOLUTION:

134

Chapter 7 – Covalent Bonds and Molecular Structure

1. First, draw the electron–dot structure of the molecule.

2. There are five charge clouds around arsenic, all of which are attached to atoms. Therefore, the molecular shape is trigonal bipyramidal.

C. Shapes of larger molecules
   1. Use rules summarized in Table 7.4 on page 248 in your textbook to describe the geometry around each atom.

EXAMPLE:
Butyric acid has the formula $CH_3CH_2CH_2COOH$. It is the organic constituent that gives rancid butter its smell. The structural formula for this compound is

Describe the geometry around each carbon atom.

SOLUTION:
The first three carbons from left to right all have four charge clouds around them. Therefore, these three carbon atoms have a tetrahedral geometry around them. The carbon atom that has a double bonded–oxygen and a hydroxide group bonded to it has only three charge clouds around it. Therefore, the geometry around this carbon atom is trigonal planar.

Workbook Problem 7.3
Predict the molecular shape of BrF_3 and IF_5.

Chapter 7 – Covalent Bonds and Molecular Structure

Strategy: Draw the electron dot structure for each of the molecules and use Table 7.4 on page 248 in your textbook to determine the shape.

Step 1: Determine the number of charge clouds around the central atom. How many are used for bonding and how many are nonbonding?

Step 2: Based upon your answer in step 1, determine the shape.

What key concept did you use?

7.10 Valence Bond Theory
A. Provides a detailed description of electronic nature of covalent bonds.
B. Easily visualized orbital picture of how electron pairs are shared in a covalent bond.
   1. The greater the overlap, the stronger the bond.
   2. Singly occupied valence orbital on one atom overlaps a singly occupied valence orbital on another atom.
      a. atomic orbitals contain one electron of opposite spin.
      b. for orbitals other than s orbitals, gives a direction to the bond
   3. Each of the bonded atoms maintains its own orbital, but shares the electron.
   4. Paired electrons in overlapping orbitals are attracted to nuclei in both atoms.

7.11 Hybridization and sp^3 Hybrid Orbitals
A. Hybrid orbitals — the combination of wave functions for atomic orbitals that form a new set of equivalent wave functions.
B. Combine one s with three p — form four equivalent new orbitals — sp^3.
   1. Each hybrid has two lobes; one larger than the other.
   2. Four large lobes point toward the corners of a tetrahedron.
   3. Bonds formed with sp^3 orbitals are very strong.
C. A tetrahedral arrangement of charge clouds always implies sp^3 hybridization.

7.12 Other Kinds of Hybrid Orbitals
A. Geometries in Table 7.4 of your textbook — accounted for by a specific kind of hybridization.
B. Sigma (σ) bond — a bond that is formed by head–on overlap and has its shared electrons centered about the axis between the two nuclei.
C. Pi (π) bond — a bond that is formed by side–by–side overlap and has its shared electrons occupy a region above and below a line connecting the two nuclei.
D. Atoms with five and six chare clouds.
   1. Combination of five or six atm\omic orbitals.
      a. implies d orbitals are involved.
   2. Transition metals – hybridization involves d orbitals.
   3. Main-group compounds do not use d orbitals in hybridization.
      a. more complex bonding pattern
      b. not explained by valence bond theory.
E. Given charge cloud geometry implies the necessary hybridization (see Table 7.5, page 255 in your textbook).

Chapter 7 – Covalent Bonds and Molecular Structure

EXAMPLE:
Determine the hybridization for the central atom in PH_3 and BCl_3.

SOLUTION: We have determined the electron dot structure for these molecules previously. They are:

PH_3 has four charge clouds around the central atom; therefore, the hybridization is sp^3.
BCl_3 has three charge clouds around the central atom; therefore, the hybridization is sp^2.

7.13 Molecular Orbital: The Hydrogen Molecule
A. Molecular orbital model.
   1. More complex bonding model than valence bond.
   2. Sometimes bonding description offers better agreement with experimental observations.
   3. Considers the molecule as a whole rather than concentrating on individual atoms.
B. Atomic orbitals — a wave function whose square gives the probability of finding the electron within a given region of space in an atom.
   1. Same atom can combine to form hybrids.
   2. Different atoms overlap to form covalent bonds.
      a. Orbitals and the electrons in them remain localized on specific atoms.
   3. Concentrates on individual atoms.
C. Molecular orbital — a wave function whose square gives the probability of finding the electron within a given region of space in a molecule.
   1. Specific energy levels and specific shapes.
   2. Occupied by a maximum of two electrons with opposite spins.
D. Orbital interactions in molecular orbital theory.
   1. Additive interaction.
      a. for H_2 — formation of a molecular orbital that is egg–shaped
      b. denoted σ
      c. lower in energy than the two isolated 1s orbitals
      d. bonding molecular orbital (MO)
         i. Electrons in bonding MO spend most of their time in the region between the two nuclei.
         ii. bond the atoms together
   2. Subtractive interaction.
      a. for H_2 — formation of a molecular orbital that has a node between the two atoms
      b. denoted σ*
      c. higher in energy than the two isolated 1s orbitals
      d. antibonding molecular orbital
         i. Electrons in antibonding MO do not occupy the central region between the nuclei.
         ii. don't contribute to bonding
E. Bond order = 1/2(number of bonding electrons — number of antibonding electrons).
F. Number of molecular orbitals formed = the number of atomic orbitals combined.

Chapter 7 – Covalent Bonds and Molecular Structure

G. Key ideas of molecular orbital theory – see page 259 in your textbook.

7.14 Molecular Orbital Theory of Other Diatomic Molecules
A. Paramagnetic — substances with unpaired electrons are attracted by magnetic fields.
B. Diamagnetic — substances whose electrons are all spin–paired.
C. Molecular orbitals of O_2.
  1. 2s orbitals interact giving σ_{2s} and σ^*_{2s} molecular orbitals.
  2. 2p orbitals on the internuclear axis.
     a. interact head–on
     b. form σ_{2p} and σ^*_{2p} molecular orbitals
  3. Remaining 2p orbitals perpendicular to the internuclear axis.
     a. interact in sideways manner
     b. form π_{2p} and π^*_{2p} molecular orbitals
  4. See Figure 7.17, page 261 in your textbook for energy levels of the molecular orbitals.
D. Molecular orbital diagrams require some experience to generate.

7.15 Combining Valence Bond Theory and Molecular Orbital Theory
A. Molecules with resonance structures.
  1. σ–bond framework — localized; describe with valence bond theory.
  2. π–bond framework — delocalized; describe with molecular orbital theory.

 **Putting It Together**

Pyruvic acid is formed from carbohydrate metabolism in the body and is reduced to lactic acid in the muscles during exercise. Its molar mass is 72.0 g/mol and its percent composition is 40.9% carbon, 54.5% oxygen, and 4.5% hydrogen. The structure for this molecule includes two C—C single bonds, with two carbons sp^2–hybridized. Most of the hydrogen atoms are bonded to a single carbon atom. Determine the empirical formula, molecular formula, and structure.

 **Self–Test**

This section is intended to test your knowledge of the material covered in this chapter. Think through these problems, and make certain you understand what is going on. Ask yourself if your answer makes sense. Many of these questions are linked to the chapter learning goals. Therefore, successful completion of these problems indicates you have mastered the learning goals for this chapter. You will receive the greatest benefit from this section if you use it as a mock exam. You will then discover which topics you have mastered and which topics you need to study in more detail.

True–False

1. A covalent bond is formed when the attractive forces between the two atoms is greater than the repulsive forces between the two atoms.

2. Energy is needed for a covalent bond to form.

3. The bond order refers to the number of electrons between atoms.

Chapter 7 – Covalent Bonds and Molecular Structure

4. The pairing of dots in the Electron–dot structures corresponds to the pairing of electrons in the electron configuration.

5. N, O, and P frequently form coordinate covalent bonds.

6. All elements in the periodic table obey the octet rule when they undergo bonding.

7. Electrons are transferred in polar covalent bonds.

8. Resonance structures that have the more negative formal charge on the more electronegative atom are more stable.

9. The shape of a molecule can be determined from the number of charge clouds surrounding the central atom.

10. The type of hybrid orbitals used by the central atom in a molecule can be determined from the geometry of the molecule.

Fill–in–the–Blank

1. A representation of the distribution of an atom's valence electrons in a molecule is referred to as the _____.

2. An atom will share electrons until _____.

3. In a multiple covalent bond, the atoms share _____.

4. The _____ can expand their valence shell beyond an octet.

5. The _____ represents the average of the various possible Electron–dot structures of a molecule.

6. Nonpolar bonds are formed by atoms with _____.

7. In valence bond theory, a singly occupied valence orbital on one atom _____.

8. A trigonal planar arrangement of charge clouds always implies _____ hybridization.

9. Orbital interactions in molecular orbital theory are both _____.

10. The number of molecular orbitals formed is equal to _____.

Matching

    Bond length   a. the ability of an atom in a molecule to attract the shared electrons in a bond

    Bond dissociation energy   b. predicts approximate shape of a molecule

    Coordinate covalent bond   c. the amount of energy necessary to break a chemical bond in an isolated molecule in the gaseous state

Chapter 7 – Covalent Bonds and Molecular Structure

Polar covalent bond	d. a bond in which one atom donates both electrons to another atom that has a vacant valence orbital
Electronegativity	e. a bond that is formed by head–on overlap and that has its shared electrons centered about the axis between the two nuclei.
VSEPR model	f. a wave function whose square gives the probability of finding the electrons within a given region of space in a molecule
Hybrid orbitals	g. substance whose electrons are all spin paired
σ bond	h. substance with unpaired electrons
π bond	i. the optimum point where net attractive forces are maximized
Molecular orbital	j. bonds in which the bonding electrons are attracted somewhat more strongly by one atom in a bond
Paramagnetic	k. the combinations of wave functions for atomic orbitals which form a new set of equivalent wave functions
Diamagnetic	l. a bond that is formed by side–by–side overlap and that has its shared electrons occupying a region above and below a line connecting the two nuclei

Problems

1. Draw the electron–dot structures for a) the chromate ion, b) IF_6^+, c) ClF_3, d) H_2F^+, e) PF_4^-, f) XeF_4 and g) BF_3.

2. Draw all resonance structures for SO_3.

3. Predict whether the following bonds are nonpolar, polar covalent, or ionic: C—H, Na—Cl, C—N, O—H, F—F.

4. Calculate the formal charges on S and O in SO_3.

5. The best way to depict the structure of nitric acid is to write the formula as $(HO)NO_2$. Draw all resonance structures, and use formal charge calculations to determine the most stable structure(s).

6. NOF, nitrosyl fluoride, has an atom sequence such that the formal charge on all atoms is zero. Draw the Lewis structure for this molecule.

7. Predict the molecular geometries of the molecules in #1.

Chapter 7 – Covalent Bonds and Molecular Structure

8. Predict the shape around the carbon atoms in acetaldehyde, CH_3CHO. The structural formula for acetalaldehyde is

$$H-\underset{\underset{H}{|}}{\overset{\overset{H}{|}}{C}}-C\underset{H}{\overset{O}{\diagup\!\!\!\diagdown}}$$

9. Predict the hybridization of the carbon, nitrogen, and phosphorous in carbamoyl phosphate, a molecule which participates in the catabolism of the nitrogen in amino acids.

$$H_2N-\overset{\overset{O}{\|}}{C}-O-\underset{\underset{O^-}{|}}{\overset{\overset{O}{\|}}{P}}-O^-$$

10. Calculate the bond order for N_2, O_2, and F_2 using the molecular orbital diagram found in Figure 7.16 on page260 in your your textbook.

Challenge Problem

The following mass percent was determined for a compound: 24.3% carbon, 71.6% chlorine, and 4.1% hydrogen. Draw the structure of the empirical formula. From what you have learned in this chapter, does this structural formula look correct. If not, what would be the structure of the simplest molecular formula?

CHAPTER 8

THERMOCHEMISTRY: CHEMICAL ENERGY

Chapter Learning Goals

A. Heat Transfer
1. State the First Law of Thermodynamics, and understand its application to thermochemistry.
2. Differentiate between the concepts of heat and temperature.
3. Identify a state function.

B. Energy Change
1. Define and calculate PV work. Know whether work is being done by the system or on the system.
2. Differentiate between energy and enthalpy, and perform calculations interconverting the two. From ΔH or ΔE, tell whether energy is being lost from or gained by the system.
3. Given a balanced chemical equation and enthalpy change for a chemical reaction, calculate the enthalpy change per mole or per gram of each reactant and product.
4. Perform calculations involving specific heat (or molar heat capacity), heat flow, and temperature change.

C. Enthalpy Change
1. Perform calculations involving Hess's law.
2. Use standard heats of formation to calculate a standard heat of reaction.
3. Use bond dissociation energies to approximate a standard heat of reaction.

D. Spontaneous Reactions
1. Predict whether entropy increases or decreases for a chemical reaction or physical change.
2. Use the equation $\Delta G = \Delta H - T\Delta S$ to determine whether the forward reaction or the reverse reaction is favored.
3. Use ΔH and ΔS to determine the temperature at which a reversible system is at equilibrium.

Chapter in Brief

This chapter introduces you to the concept of thermochemistry: the heat changes that take place during reactions. You begin the study of this topic by learning the difference between heat and energy, and the types of energy changes that can take place. You are then introduced to the Law of Conservation of Energy, the First Law of Thermodynamics, and the concept of state functions. With this background, you will then learn how to calculate the internal energy of the system using $P\Delta V$ work and how the internal energy of the system is related to the enthalpy (ΔH) of the system. You will spend much of the rest of the chapter exploring how to use specific heat calculations in the laboratory and how to use Hess's law, standard heats of formation, and bond dissociation energies to calculate heats of reaction. Finally, you are introduced to the topics of entropy and free energy, topics that will be explored in more detail in later chapters.

Chapter 8 – Thermochemistry: Chemical Energy

8.1 Energy
A. Energy — the capacity to do work or supply heat.
 1. Energy = work + heat
B. Kinetic energy — the energy of motion.
 1. $E_K = 1/2 mv^2$
C. Potential energy — stored energy.
D. Joule — SI unit for energy.
 1. $1\text{ J} = 1(\text{kg} \cdot \text{m}^2)/\text{s}^2$
E. Calorie — the amount of energy necessary to raise the temperature of 1 g of water by 1°C.
 1. 1 cal = 4.184 J
 2. Nutritional calorie (Calorie); 1 Cal = 1000 cal = 1 kcal = 4.184 kJ.

8.2 Energy Changes and Energy Conservation

A. Law of Conservation of Energy: Energy can be neither created nor destroyed
 It can only be converted from one form into another.

B. Many forms of energy.
 1. Thermal energy — kinetic energy of molecular motion.
  a. temperature — a measure of the kinetic energy of molecular motion
 2. Heat — energy transferred from one object to another as the result of a temperature difference between them.
 3. Chemical energy — a type of potential energy in which the chemical bonds of molecules act as the storage medium.

C. First Law of Thermodynamics: The energy of the universe is constant.

8.3 Internal Energy and State Functions
A. System — everything we focus on in an experiment.
B. Surroundings — everything other than the system.
C. Internal energy — energy of the system.
D. System isolated from the surroundings — no energy transfer to the surroundings; $\Delta E = 0$.
 1. First Law of Thermodynamics – The total internal energy of an isolated system is constant.
E. System is not isolated from the surrounding — energy flows to or from the surroundings:
 1. Measure any change in the internal energy of the system.
 2. $\Delta E = E_{\text{final}} - E_{\text{initial}}$.
F. Energy changes are measured from the point of view of the system.
 1. Energy flows out of the system to surroundings — negative value.
 2. Energy flows into the system from the surroundings — positive value.

G. State Functions — a function or property whose value depends only on the present state (condition) of the system, not on the path used to arrive at that condition.
 1. State functions are reversible.
 2. For any state function, the overall change is zero if the system returns to its original condition.

8.4 Expansion Work
A. Work – the force (*f*) that produces the movement of an object times the distance moved (*d*).
 1. $w = F \times d$
B. Expansion work (*PV* work) — work done as the result of a volume change in the system.
 1. $w = -(P \times \Delta V)$

Chapter 8 – Thermochemistry: Chemical Energy

2. Expansion of the system.
   a. System does work on the surroundings.
   b. ΔE is negative
   c. $w = -P\Delta V$
      i. Work is negative.
      ii. ΔV is positive.
3. Contraction of the system.
   a. Surroundings do work on the system.
   b. ΔE is positive.
   c. $w = -P\Delta V$
      i. Work is positive.
      ii. ΔV is negative.
4. No volume change, $\Delta V = 0$.
   a. no work.

EXAMPLE:
Calculate the amount of work done during a reaction in which a volume change from 28.3 L to 35.7 L occurs. What is the direction of work? The external pressure in the reaction is 9.3 atm.

SOLUTION:
$w = -P\Delta V$

$w = -(9.3 \text{ atm}) \times (35.7 \text{ L} - 28.3 \text{ L}) = -69 \text{ atm·L}$

The value for the work is negative; therefore, the surroundings do work on the system.

8.5 Energy and Enthalpy
A. Total energy change of a system.
   1. $\Delta E = q + w$ (q = heat)
   2. $\Delta E = q + (-P\Delta V)$
B. Amount of heat transferred.
   1. $q = \Delta E + P\Delta V$
C. Reactions carried out with constant volume.
   1. $\Delta V = 0$; no PV work is done.
   2. $q_v = \Delta E$
D. Reactions carried out at constant pressure.
   1. $\Delta V \neq 0$; energy change due to both heat transfer and PV work.
   2. $q_p = \Delta E + P\Delta V$.
E. ΔH – heat of reaction or enthalpy change.
   1. Heat change for reaction carried out at constant pressure in an open container.
   2. Enthalpy change of a system.
      a. $\Delta H = \Delta E + P\Delta V$
   3. State function – value depends only on the current state of the system.
   4. $\Delta H = H_{products} - H_{reactants}$
   5. Amount of heat released in a specific reaction depends on the actual amounts of reactants.

8.6 The Thermodynamic Standard State
A. The value of the enthalpy change, ΔH, reported for a reaction represents the amount of heat released when reactants are converted to products in the molar amounts represented by coefficients of the balanced equation.

Chapter 8 – Thermochemistry: Chemical Energy

1. The actual amount of heat involved in a reaction depends on the actual amounts of reactants.
2. The physical states of reactants and products must be specified.
3. Temperature and pressure also must be reported.

B. Thermodynamic Standard State = 298.15 K (25°C) 1 atm pressure of each gas, 1 M concentration (for solutions)
   1. Allows different reactions to be compared.
   2. Indicated by addition of a superscript o, $\Delta H°$.

C. Standard enthalpy of reaction – an enthalpy change measured under standard conditions.

EXAMPLE:
What is the value of $\Delta E°$ if a reaction having $\Delta H° = 135.3$ kJ is carried out at a constant pressure of 18.9 atm and the volume change is 238 L?

SOLUTION:
$\Delta H° = \Delta E° + P\Delta V$

Before solving for $\Delta E°$, we must give serious consideration to our units. Keep in mind that the units for $\Delta H°$ are kJ while the units for $P\Delta V$ are L·atm. Obviously, we would like all of our units to be energy units. A detailed study of conversion factors found on the inside back cover leads us to the following:

$$1 J = \frac{1 kg \cdot m^2}{1 s^2} \qquad 1 atm = 101,325 Pa = \frac{1 kg}{m \cdot s^2} \qquad 1 L = 10^{-3} m^3$$

$$P\Delta V = (18.9 \text{ atm})(238 \text{ L}) = 4500 \text{ atm} \cdot \text{L}$$

$$4500 \text{ atm} \cdot \text{L} \times \frac{101,325 \text{ Pa}}{1 \text{ atm}} \times \frac{1 \frac{kg}{m \cdot s^2}}{1 \text{ Pa}} \times \frac{10^{-3} m^3}{1 L} = 4.56 \times 10^5 \frac{kg \cdot m^2}{s^2} = 4.56 \times 10^5 \text{ J} = 4.56 \times 10^2 \text{ kJ}$$

$$\Delta E° = \Delta H° - P\Delta V = 135.3 \text{ kJ} - 456 \text{ kJ} = -321 \text{ kJ}$$

Workbook Problem 8.1
For the reaction

$\text{Si (s)} + 2 \text{ Cl}_2 \text{ (g)} \rightarrow \text{SiCl}_4 \text{ (l)} \qquad \Delta H = -687.0 \text{ kJ/mol}$

determine the value of ΔE if 1.81 mole of Si react with a stoichiometric amount of Cl_2 (g) when the reaction is carried out at 0.50 atm of pressure and with a volume change of –22.4 L.

Strategy: Use the equation $\Delta H = \Delta E + P\Delta V$.

Step 1: Determine the ΔH for the reaction taking into consideration that 1.81 mol of Si reacted.

Step 2: Determine ΔE from the equation: $\Delta H = \Delta E + P\Delta V$. (Watch your units!)

Chapter 8 – Thermochemistry: Chemical Energy

What key concept did you use?

8.7 Enthalpies of Physical and Chemical Change
A. Enthalpies of physical change.
   1. Heat of fusion — the amount of heat required for melting without changing the temperature.
   2. Heat of vaporization — the amount of heat required for evaporation without changing the temperature.
   3. Sublimation — the direct conversion of a solid to a vapor without going through a liquid state.
      a. Heat of sublimation = Heat of fusion + Heat of vaporization
B. Enthalpies of chemical change.
   1. Heats of reaction — enthalpies of chemical change.
   2. Measure of heat flow into or out of the system at constant pressure.
   3. Endothermic reactions.
      a. $H_{products} > H_{reactants}$
      b. heat flow into the system from the surroundings
      c. ΔH is positive.
   4. Exothermic reactions.
      a. $H_{products} < H_{reactants}$
      b. heat flow to the surroundings from the system
      c. ΔH is negative.
   5. $\Delta H°$ values for a given equation.
      a. The equation is balanced for the number of moles of reactants and products.
      b. All substances are in their standard states.
      c. Physical state of each substance is specified.
      d. Refer to the reaction going in the direction written.
         i. Reverse the direction of the reaction, change the sign of $\Delta H°$.

EXAMPLE:
Sulfuric acid is produced by reacting sulfur trioxide with water according to the equation:

$$SO_3\ (g) + H_2O\ (l) \rightarrow H_2SO_4\ (l) \qquad \Delta H° = -131.8\ kJ/mol$$

How much heat is evolved when 75.0 g of SO_3 reacts with a stoichiometric amount of H_2O?

SOLUTION:
The $\Delta H°$ reported is for the reaction of 1 mol of SO_3. To answer this question, we need to calculate how many moles of SO_3 are found in 75.0 g of SO_3.

$$75.0\ g\ SO_3 \times \frac{1\ mol\ SO_3}{80.0\ g\ SO_3} = 0.938\ mol\ SO_3$$

We can now calculate the amount of heat evolved when 75.0 g of SO_3 reacts.

$$-131.8 \; \frac{kJ}{1 \; mol \; SO_3} \times 0.938 \; mol \; SO_3 = -124 \; kJ$$

8.8 Calorimetry and Heat Capacity

A. Calorimetry – an experimental technique that allows the energy change associated with a chemical or physical process to be determined.
   1. A temperature change is observed when a system gains or loses energy in the form of heat.
   2. Carried out in a calorimeter.
   3. For an exothermic reaction:
      a. amount of heat released by the reaction = amount of heat gained by calorimeter + amount of heat gained by solution.
B. Heat capacity (C) — the amount of heat required to raise the temperature of an object or substance a given amount.
   1. $C = \dfrac{q}{\Delta T}$
   2. Extensive property
C. Specific heat — the amount of heat necessary to raise the temperature of exactly 1 g of a substance by exactly 1°C.
   1. $q = $ (specific heat) $\times$ (mass of substance) $\times$ (ΔT)
D. Molar heat capacity (C_m) — the amount of heat necessary to raise the temperature of 1 mole of a substance by 1°C.
   1. $q_m = (C_m) \times$ (moles of substance) $\times$ (ΔT)

EXAMPLE:
Calculate the amount of heat needed to raise the temperature of 32.8 g of iron 12°C. The specific heat of iron is 25.1 J/g·°C.

SOLUTION:
Use the information given and dimensional analysis to determine the amount of heat (J) needed.

$$25.1 \frac{J}{g \cdot °C} \times 32.8 \; g \times 12.0° \; C = 9.8 \times 10^3 \; J = 9.88 \; kJ$$

Workbook Problem 8.2
A student mixes 50.0 mL of 0.400 M $CuSO_4$ at 23.35°C with 50.0 mL of 0.600 M NaOH also at 23.35°C in a coffee–cup calorimeter with a heat capacity of 25.0 J/°C. The final temperature of the reaction is 26.65°C. Calculate the amount of heat evolved in this reaction. (The density of the solution is 1.02 g/mL. The specific heat of water may be used for the specific heat of the solution.)

Strategy: For an exothermic reaction, the amount of heat released by the reaction equals the amount of heat gained by the calorimeter plus the amount of heat gained by solution. Determine the amount of heat gained by the solution by first calculating the mass of the solution.

Chapter 8 – Thermochemistry: Chemical Energy

Step 1: Determine the amount of heat gained by the calorimeter.

Step 2: Determine the amount of heat released by the reaction.

What key concept did you use?

8.9 Hess's Law

A. Hess's law — the overall enthalpy change for a reaction is equal to the sum of the enthalpy changes for the individual steps in the reaction.
   1. Reactants and products in the individual steps can be added and subtracted like algebraic quantities in determining the overall equation.
B. Combine the individual reactions so that their sums will be the desired reaction (See example 8.6, page 290 in your textbook.)

EXAMPLE:
Calculate $\Delta H°$ for the reaction $NO\ (g) + O\ (g) \rightarrow NO_2\ (g)$ given the following information:

$NO\ (g) + O_3\ (g) \rightarrow NO_2\ (g) + O_2\ (g)$	$\Delta H° = -200$ kJ
$O_3\ (g) \rightarrow \frac{3}{2} O_2\ (g)$	$\Delta H° = -143$ kJ
$O_2\ (g) \rightarrow 2\ O\ (g)$	$\Delta H° = 498$ kJ

SOLUTION:
We will use the first equation as written since we know that the overall equation has 1 mol of NO as a reactant. We need to reverse the third equation and divide by 2 because we also need 1 mol of O as a reactant. Finally, we will need to cancel both O_3 and O_2 from the first and third equations. This requires that we reverse the second equation. Remember, whatever we do to the balanced chemical equation, we have to also do to the value of ΔH. When we reverse the direction of a chemical equation, we change the sign of ΔH. If we divide the coefficients in an equation by 2, we divide the value of ΔH by 2.

$NO\ (g) + O_3\ (g) \rightarrow NO_2\ (g) + O_2\ (g)$	$\Delta H° = -200$ kJ
$O\ (g) \rightarrow \frac{1}{2} O_2\ (g)$	$\Delta H° = -249$ kJ
$\frac{3}{2} O_2\ (g) \rightarrow O_3\ (g)$	$\Delta H° = +143$ kJ
$NO\ (g) + O\ (g) \rightarrow NO_2\ (g)$	$\Delta H° = -306$ kJ

Chapter 8 – Thermochemistry: Chemical Energy

8.10 Standard Heats of Formation

A. Standard heat of formation — The enthalpy change, ΔH_f°, for the *hypothetical* formation of 1 mole of a substance in its standard state from the most stable forms of it constituent elements in their standard states.
   1. The most stable forms of all elements in their standard state have $\Delta H_f^\circ = 0$.
B. The standard enthalpy change for any chemical reaction is found by subtracting the sum of the heats of formation of the reactants from the sum of the heats of formation of the products.
   1. $\Delta H^\circ = \sum \Delta H_{products}^\circ - \sum \Delta H_{reactants}^\circ$

EXAMPLE:

Calculate ΔH° for the reaction $2\ Na_2O_2\ (s) + 2\ H_2O\ (l) \rightarrow 4\ NaOH\ (s) + O_2\ (g)$ using ΔH_f° found in Table 8.2, page 292 and Appendix B of your textbook.

SOLUTION:
Subtract the total heats of formation of the reactants from the total heats of formation of the products.

$\Delta H^\circ = [(4 \times \Delta H_f^\circ\ NaOH\ (s)\ ] - [(2 \times \Delta H_f^\circ\ Na_2O_2\ (s)) + (2 \times \Delta H_f^\circ\ H_2O\ (l))]$
$= [(4 \times -425.6\ kJ)] - [(2 \times -510.9\ kJ) + (2 \times -285.8\ kJ)]$
$= -109\ kJ$

Workbook Problem 8.3

The first step in producing lead from its ore, galena (PbS), is the roasting of lead sulfide in air.

$2\ PbS\ (s) + 3\ O_2(g) \rightarrow 2\ PbO\ (s) + 2\ SO_2\ (g)$

Calculate ΔH° when 2.50 g of PbS reacts with a stoichiometric amount of oxygen.

Strategy: Determine ΔH° for this reaction from the standard heats of formation.

Step 1: Determine the ΔH°, using the ΔH_f° values in Appendix B of your textbook.

Step 2: Knowing the ΔH° for the reaction of 2 mol of PbS, calculate the ΔH° for 2.50 g of PbS. (First convert grams of PbS to moles of PbS.)

What key concept did you use?

Chapter 8 – Thermochemistry: Chemical Energy

Workbook Problem 8.4
Nitric acid is produced industrially by the multi–step Ostwald process given below:

1) $4 NH_3 (g) + 5 O_2 (g) \rightarrow 4 NO (g) + 6 H_2O (g)$
2) $2 NO (g) + O_2 (g) \rightarrow 2 NO_2 (g)$
3) $3 NO_2 (g) + H_2O (l) \rightarrow 2 HNO_3 (aq) + NO (g)$

Determine $\Delta H°$ for the overall process

$4 NH_3 (g) + 2 H_2O (l) + 8 O_2 (g) \rightarrow 4 HNO_3 (aq) + 6 H_2O (g)$

using Hess's Law.

Strategy: Calculate $\Delta H°$ for each step in the reaction from the heats of formation listed in Appendix B of your textbook.

Step 1: For each step in the reaction, $\Delta H°$ can be calculated from the $\Delta H_f°$ values listed in Appendix B of your textbook.

Step 2: Combine the individual reactions so that their sums will be the desired reaction.

What key concept did you use?

8.11 Bond Dissociation Energies
A. Bond dissociation enthalpies — enthalpy changes, $\Delta H°$, for the corresponding bond–breaking reactions.
   1. $\Delta H° = D =$ bond dissociation energy

2. Always positive; always need energy to break a bond.

B. $\Delta H° = \Sigma D$(bonds broken) $- \Sigma D$(bonds formed)

EXAMPLE:
Use the data in Table 7.1 (page 246 in your textbook) to find an approximate $\Delta H°$ for the production of ammonia by the Haber process.
$N_2 (g) + 3 H_2 (g) \rightarrow 2 NH_3 (g)$

SOLUTION:
Three bonds are formed for each NH_3 produced. One bond is broken for each H_2 reacted, and one bond is broken for each N_2 reacted. $\Delta H°$ is calculated in the following manner:

	Bond Energy
H—H	436 kJ/mol
N≡N	945 kJ/mol
N—H	390 kJ/mol

$\Delta H° = [1 D_{N-N} + 3 D_{H-H}] - [6 D_{N-H}]$

$\Delta H° = [(1 \text{ mol} \times 945 \text{ kJ/mol}) + (3 \text{ mol} \times 436 \text{ kJ/mol})] - [(6 \text{ mol} \times 390 \text{ kJ/mol})]$

$\Delta H° = -87 \text{ kJ}$

8.12 Fossil Fuels, Fuel Efficiency, and Heats of Combustion

A. Heat of combustion ($\Delta H_c°$) — amount of energy released on burning a substance.
B. Fuel efficiency — calculate $\Delta H_c°$ in kJ/g or kJ/mL.
   1. Can compare efficiency for different fuels.
C. Common fuels – organic compounds.
   1. Energy derived from the sun through photosynthesis of carbohydrates in green plants.
      a. Net result:
         $6 CO_2 + 6 H_2O \rightarrow C_6H_{12}O_6 + 6 O_2$
         glucose
      b. glucose $\rightarrow$ cellulose + starch
         i. structural materials for plants
         ii. food source for animals
      c. highly endothermic
         i. large input of solar energy
D. Fossil Fuels – decayed remains of organisms from previous geological eras.
   1. Coal, natural gas and petroleum
      a. coal and petroleum – complex mixture of compounds
      b. coal – vegetable origin
         i. compounds structurally similar to graphite
      c. petroleum – viscous liquid mixture of hydrocarbons
         i. primarily marine origin
E. Other sources of energy.
   1. Hydrogen – burns cleanly and relatively nonpolluting.
      a. low availability
      b. low combustion enthalpy per mL.

Chapter 8 – Thermochemistry: Chemical Energy

   2. Ethanol and methanol – produced relatively cheaply and have reasonable combustion enthalpy per mL.
      a. ethanol – produced from wood.
      b. methanol – produced directly from natural gas.

8.13 An Introduction to Entropy
A. Spontaneous process — a process that proceeds on its own without any continuous external influence.
   1. Need either a release of energy or an increase in disorder of the system.
B. Entropy (S) — the amount of molecular disorder or randomness in a system.
   1. The larger the value of S, the greater the molecular randomness.
   2. $\Delta S = S_{final} - S_{initial}$.
   3. $S_{final} > S_{initial}$
      a. ΔS is positive.
      b. system has become more random
   4. $\Delta S_{final} < \Delta S_{initial}$
      a. ΔS is negative.
      b. system has become less random
C. Spontaneous process.
   1. Favored by decrease in H (negative ΔH).
   2. Favored by increase in S (positive ΔS).

EXAMPLE:
Predict whether $\Delta S°$ is likely to be positive or negative for the reaction

$$4\ NH_3\ (g) + 2\ H_2O\ (l) + 8\ O_2\ (g) \rightarrow 4\ HNO_3\ (aq) + 6\ H_2O\ (g)$$

SOLUTION:
On the reactant side of the equation, you have 12 moles of gas and 2 moles of liquid. On the product side of the equation, you have 6 moles of gas and 4 moles of solution. While going from 2 moles of liquid to 4 moles of solution is an increase in entropy, going from 12 moles of gas to 6 moles of gas is a large decrease in entropy. Also, you have a total decrease in the number of moles (14 moles of reactant versus 10 moles of product). $\Delta S°$ should be negative.

8.14 An Introduction to Free Energy
A. Gibbs free–energy change (ΔG); $\Delta G = \Delta H - T\Delta S$.
   1. Sign of ΔG used as a criterion for determining spontaneity of a process.
      a. ΔG negative: spontaneous
      b. ΔG positive: nonspontaneous
      c. $\Delta G = 0$: process is at equilibrium.
B. Temperature dependence ($T\Delta S$) term for ΔG.
   1. Spontaneity of some processes depends on temperature.
   2. Low temperatures — ΔH dominates and controls spontaneity.
   3. High temperatures — $T\Delta S$ dominates and controls spontaneity.
C. $\Delta G = 0$; process is at equilibrium.
   1. Balanced between spontaneous and nonspontaneous.
   2. $T = \dfrac{\Delta H}{\Delta S}$

Chapter 8 – Thermochemistry: Chemical Energy

EXAMPLE:

For the reaction

$$4 NH_3 (g) + 2 H_2O (l) + 8 O_2 (g) \rightarrow 4 HNO_3 (aq) + 6 H_2O (g)$$

$\Delta S° = -832.8$ J/K. From Workbook Problem 8.4, we know that $\Delta H° = -1.39 \times 10^3$ kJ. Calculate $\Delta G°$ for this reaction. Is it spontaneous?

SOLUTION:

$$\Delta G° = \Delta H° - T\Delta S°$$

We are using the symbol to indicate that the reaction is carried out at standard conditions, so we know that the temperature is equal to 25°C (298 K). We can solve for $\Delta G°$. (Watch your units!)

$$\Delta G° = (-1.39 \times 10^3 \text{ kJ}) - (298 \text{ K})(-0.833 \text{ kJ}) = -1.14 \times 10^3 \text{ kJ}.$$

The value for $\Delta G°$ is negative; therefore, the reaction is spontaneous.

Putting It Together

Ammonium nitrate is a powerful oxidizing agent giving rise to many explosive mixtures. When crystals of this substance react with powdered aluminum, nitrogen gas, water vapor and aluminum oxide are produced. How much heat evolves from this reaction, if 0.25 g of ammonium nitrate reacts with aluminum?

Self–Test

This section is intended to test your knowledge of the material covered in this chapter. Think through these problems, and make certain you understand what is going on. Ask yourself if your answer makes sense. Many of these questions are linked to the chapter learning goals. Therefore, successful completion of these problems indicates you have mastered the learning goals for this chapter. You will receive the greatest benefit from this section if you use it as a mock exam. You will then discover which topics you have mastered and which topics you need to study in more detail.

True–False

1. When the system is isolated from the surroundings, $\Delta E > 0$.

2. If energy flows out of the system to the surroundings, $\Delta E < 0$.

3. For any state function, the overall change is zero if the system returns to its original condition.

4. If a system expands doing PV work, the system does work on the surroundings and $w = P\Delta V$.

5. For reactions carried out at a constant pressure, the energy change is due only to PV work.

6. In an endothermic reaction, heat flows into the system from the surroundings and $H_{products} > H_{reactants}$.

Chapter 8 – Thermochemistry: Chemical Energy

7. For a reaction to be spontaneous, there must always be a release of energy.

8. Specific heat is an extensive property.

9. The most stable forms of all elements in their standard state have $\Delta H_f^\circ < 0$.

10. For a spontaneous process, ΔG° is positive.

Matching

Energy	a. the direct conversion of a solid to a vapor without going through a liquid state
Temperature	b. everything we focus on in an experiment
System	c. name given to the quantity $E + PV$.
State function	d. a process that proceeds on its own without any continuous external influence
Work	e. the amount of heat required to raise the temperature of an object or substance a given amount
Enthalpy	f. a function or property whose value depends only on the present state (condition) of the system, not on the path used to arrive at that condition
Heat of fusion	g. the amount of energy released on burning a substance
Sublimation	h. the amount of molecular disorder or randomness in a system
Heat capacity	i. The overall enthalpy change for a reaction is equal to the sum of the enthalpy changes for the individual steps in the reaction.
Hess's Law	j. the capacity to do work or supply heat
Heat of combustion	k. the distance moved times the force that opposes the motion
Spontaneous process	l. a measure of the kinetic energy of molecular motion

Fill–in–the–Blank

1. Chemical energy is a type of _____ in which the chemical bonds of the molecules act as _____.

Chapter 8 – Thermochemistry: Chemical Energy

2. The energy transferred from one object to another as the result of a temperature difference between them is referred to as _____.

3. The statement "the energy of the universe is constant is referred to as _____
_____.

4. The change in the internal energy of the system is positive when energy flows _____
_____.

5. When a system contracts, the _____ do work on the _____, ΔE is _____ and $w =$ _____.

6. The thermodynamic standard state conditions are _____.

7. Enthalpies of chemical change are referred to as _____.

8. During an exothermic process, heat flows _____ from the _____ and ΔH is _____.

9. The enthalpy change ΔH_f° for the hypothetical formation of 1 mole of a substance in its standard state from the most stable forms of its constituent elements in their standard states is referred to as the _____.

10. Bond dissociation enthalpies are the _____
_____.

11. The _____ is used as a criterion for determining the spontaneity of a process.

Problems

1. If 750 mL of a gas is compressed to 350 mL under a constant external pressure of 5.00 atm and if the gas absorbs 15 kJ, what are the values of q, w, and ΔE for the gas? Is work being done on the system or by the system? What is the value of ΔE for the surroundings?

2. What are the values of ΔH and ΔE when a gas at 2.50 atm expands after the addition of 575 J of heat and does 200 J of work on the surroundings?

3. For the reaction

    $10\ N_2O\ (g) + C_3H_8\ (g) \rightarrow 10\ N_2\ (g) + 3\ CO_2\ (g) + 4\ H_2O\ (g)$

    $\Delta H = -2862.7$ kJ. How much heat is evolved from this reaction when 3.98 g of N_2O reacts with a stoichiometric amount of propane?

4. How much heat is needed to raise the temperature of 78.0 g of iron 15.0°C?

5. When 7.75 g of NH_4NO_3 is dissolved in 110 g of water at 25.00°C in a calorimeter, 25.8 kJ/ mol NH_4NO_3 is absorbed. Assuming that the specific heat of the solution is the same as that of pure water, calculate the final temperature of the solution.

Chapter 8 – Thermochemistry: Chemical Energy

6. Calculate the standard enthalpy change for the reaction

   $2\ Al\ (s) + Fe_2O_3\ (s) \rightarrow 2\ Fe\ (s) + Al_2O_3\ (s)$

   from the following:

   $2\ Al\ (s) + \frac{3}{2}\ O_2\ (g) \rightarrow Al_2O_3\ (s)$ $\Delta H° = -1676$ kJ

   $2\ Fe\ (s) + \frac{3}{2}\ O_2\ (g) \rightarrow Fe_2O_3\ (s)$ $\Delta H° = -824.2$ kJ

7. For the following reactions, calculate $\Delta H°$ from the standard heats of formation found in Appendix B in your textbook.

   $SO_3\ (g) + H_2O\ (l) \rightarrow H_2SO_4\ (l)$

   $2\ KClO_3\ (s) \rightarrow 2\ KCl\ (s) + 3\ O_2\ (g)$

8. Calculate the standard heat of reaction for the reaction

   $CH_3CH=CH_2 + HCl \rightarrow CH_3CHClCH_3$

   using the bond dissociation energies found in Table 7.1 in your textbook. The strength of a C=C bond is 635 kJ/mol.

9. Using standard heats of formation, calculate the heat of combustion of propane in kJ/mol and kJ/g.

10. Determine the sign of ΔS for the following reactions:

    $CaO\ (s) + 2\ NH_4Cl\ (s) \rightarrow 2\ NH_3\ (g) + CaCl_2\ (s)$

    $BaCl_2\ (aq) + Na_2SO_4\ (aq) \rightarrow BaSO_4\ (s) + 2\ NaCl\ (aq)$

11. Calculate the enthalpy change for the formation of ethane from graphite and hydrogen, given the following thermochemical equations:

    $C\ (graphite) + O_2\ (g) \rightarrow CO_2\ (g)$ $\Delta H° = -393.5$ kJ

    $H_2\ (g) + \frac{1}{2}\ O_2\ (g) \rightarrow H_2O\ (l)$ $\Delta H° = -285.8$ kJ

    $2\ C_2H_6\ (g) + 7\ O_2\ (g) \rightarrow 4\ CO_2\ (g) + 6\ H_2O\ (l)$ $\Delta H° = -3119.6$ kJ

12. When hydrazine (N_2H_4) reacts with hydrogen peroxide, nitrogen gas and water are produced. Write a balanced reaction and determine $\Delta H°$ for the reaction from the following thermochemical data:

    $N_2H_4\ (l) + O_2\ (g) \rightarrow N_2\ (g) + 2\ H_2O\ (l)$ $\Delta H° = -621.6$ kJ

    $H_2\ (g) + \frac{1}{2}\ O_2\ (g) \rightarrow H_2O\ (l)$ $\Delta H° = -285.8$ kJ

    $H_2\ (g) + O_2\ (g) \rightarrow H_2O_2\ (l)$ $\Delta H° = -187.8$ kJ

Chapter 8 – Thermochemistry: Chemical Energy

13. For the reaction SiO_2 (s) + 2 C (*graphite*) + 2 Cl_2 (g) → $SiCl_4$ (g) + 2 CO (g), $\Delta H° = 32.9$ kJ/mol and $\Delta S° = 226.5$ J/mol K. Calculate $\Delta G°$ for this reaction.

14. For the reaction N_2 (g) + O_2 (g) → 2 NO (g), $\Delta G° = 173.1$. Calculate the temperature at which this reaction becomes spontaneous if $\Delta H° = 180.4$ kJ/mol and $\Delta S° = 421.4$ J/mol K.

15. Determine the ΔH for the reaction between aqueous lithium hydroxide and hydrochloric acid given the following thermochemical equations and using information found in Appendix B of your textbook.

    Li (s) + 1/2 O_2 (g) + 1/2 H_2 (g) → LiOH (s) $\Delta H° = -487.0$ kJ
    2 Li (s) + Cl_2 (g) → 2 LiCl (s) $\Delta H° = -815.0$ kJ
    LiOH (s) → LiOH (aq) $\Delta H° = -19.2$ kJ
    HCl (g) → HCl (aq) $\Delta H° = -77.0$ kJ
    LiCl (s) → LiCl (aq) $\Delta H° = -36.0$ kJ

Challenge Problem

The Haber process is the process by which nitrogen gas reacts with hydrogen gas at a temperature of 400°C, to produce ammonia gas according to the chemical equation:

N_2 (g) + 3 H_2 (g) → 2 NH_3 (g) $\Delta H° = -91.8$ kJ

At 25°C and 1 atm, the density of ammonia is 0.696 g/L, and the density of nitrogen is 1.145 g/L. How much heat is evolved in this reaction if 0.75 L of ammonia is produced? The molar heat capacity of nitrogen is 29.12 J/mol·°C. How much heat is needed so that this reaction can occur?

Inquiry Based Problem

The heat of reaction is measured in the laboratory by assembling a *calorimeter* consisting of two Styrofoam cups nested together and covered by a piece of corrugated cardboard through which is inserted a thermometer. During an exothermic reaction, some heat is absorbed by the calorimeter. We, therefore, need to determine the heat capacity of the calorimeter given that the heat of reaction is equal to:

$\Delta H = \Delta T$(heat capacity of calorimeter + heat capacity of contents)

What simple experiment can you do just using water to determine the heat capacity of the calorimeter?

Consider the following:
1. Heat lost must equal heat gained.
2. What conditions must exist so that one quantity of water loses heat to another quantity of water?
3. What data do you need to collect to determine the heat lost and heat gained?
3. Knowing that $C = \dfrac{q}{\Delta T}$ and therefore, $q = \Delta T \times C$, write a mathematical equation which will allow you to calculate the heat capacity of the calorimeter based on your experimental results.

CHAPTER 9

GASES: THEIR PROPERTIES AND BEHAVIOR

Chapter Learning Goals

A. Gas Laws
1. Explain how the height of a liquid in a barometer depends on the density of the liquid.
2. Interconvert units of pressure.
3. Know how to determine the pressure of a gas, using a manometer.
4. Use the ideal gas law to calculate pressure, volume, moles of gas, or temperature, given the other three variables.
5. Use the ideal gas law to calculate final pressure, volume, moles of gas, or temperature from initial pressure, volume, moles of gas, and temperature.
6. Perform stoichiometric calculations relating the mass of a reactant to the mass, moles, and volume or pressure of a gaseous product.
7. Use the ideal gas law to calculate the molar mass of a gas.
8. Use the ideal gas law to calculate the density of a gas.
9. Use Dalton's law to calculate the partial pressure of a gas in a mixture.

B. Behavior of Gases: Kinetic–Molecular Theory
1. Use the Kinetic–Molecular Theory of gases to explain each of the gas laws.
2. Use Graham's law to calculate the relative rates of effusion of two different gases.
3. State the conditions under which a gas is expected to behave ideally or nonideally.

Chapter in Brief
This chapter looks at the behavior of gases and how that behavior can be explained. You will begin with a general description of gases and how pressures are measured. You then will learn how the behavior of gases can be defined by the four variables: pressure, temperature, volume, and the number of moles; and also how these variables are related through the gas laws and the ideal gas law. You will learn how to apply the ideal gas law to stoichiometric calculations and calculations involving the density and molar mass of a gas. You will also learn how the ideal gas law can be used to describe the behavior of a mixture of gases. Once you are able to describe the behavior of gases through the use of the gas laws, you will be introduced to the Kinetic–Molecular Theory, which explains the reason for that behavior. You will also be introduced to the concepts of effusion and diffusion and the difference between ideal and real gases. Finally, you will take a brief look at some of the chemistry of the atmosphere.

9.1 Gases and Gas Pressure
A. Gases — constituent atoms or molecules have little attraction for one another.
   1. Free to move about in available volume.
B. Some properties of gases.
   1. Mixtures are always homogeneous.
      a. very weak attraction between gas molecules
      b. identity of neighbor is irrelevant
   2. Compressible — volume contracts when pressure is applied.
      a. 0.10% of volume of gas is occupied by molecules
   3. Exert a measurable pressure on the walls of their container.

C. Pressure — force exerted per unit area.

A.2

Chapter 9 – Gases: Their Properties and Behavior

1. $P = \dfrac{\text{Force}}{\text{Unit Area (A)}} = \dfrac{\text{mass} \times \text{acceleration}}{A}$
   a. Force – Newton (N) = $\dfrac{1(\text{kg} \cdot \text{m})}{\text{s}^2}$
   b. Pressure – Pascal (Pa) = $\dfrac{\text{N}}{\text{m}^2}$
2. SI unit equals Pascal (Pa).
3. Alternative units.
   a. millimeters of mercury (mm Hg)
   b. atmosphere (atm)
4. Atmospheric pressure — pressure created from the mass of the atmosphere pressing down on the earth's surface.
   a. standard atmospheric pressure at sea level — 760 mm Hg.

D. Barometer – long thin mercury filled tube sealed at one end and inverted into a dish of mercury. (See figure 9.3, page 317 in your textbook.)   🔑 *A.1.*
   1. downward pressure of Hg in column equals outside atmospheric pressure

E. Measuring pressure.   🔑 *A.3.*
   1. Manometer — U–tube filled with mercury, with one end connected to the gas–filled container and the other end open to the atmosphere. (See figure 9.4, page 318 in your textbook.)
      a. $P_{gas} = P_{atm}$; liquid level in both arms is equal.
      b. $P_{gas} > P_{atm}$; liquid level in the arm connected to the gas–filled cylinder will be lower
         i. $P_{gas} = P_{atm} + P_{Hg}$ (P_{Hg} = the difference in the heights of the two mercury columns)
      c. $P_{gas} < P_{atm}$; liquid level in the arm connected to the gas–filled cylinder will be higher
         i. $P_{gas} + P_{Hg} = P_{atm}$

EXAMPLE:
What is the pressure in mm Hg inside a container of gas connected to a mercury–filled, open–ended manometer when the level in the arm connected to the container is 22.4 mm Hg higher than the level in the arm open to the atmosphere and the atmospheric pressure reading outside the apparatus is 672.2 mm Hg?

SOLUTION:
$P_{gas} < P_{atm}$ since the level of Hg is higher in the arm connected to the gas–filled cylinder; therefore,

$P_{gas} + P_{Hg} = P_{atm}$ and $P_{gas} = 672.2$ mm Hg $- 22.4$ mm Hg $= 649.8$ mm Hg

9.2 The Gas Laws
A. Different gases show similar physical behavior.
   1. Defined by four variables — pressure, temperature, volume, and number of moles.
      a. relationships of variables — gas laws
      b. ideal gas — behavior follows the gas laws exactly
B. Boyle's law — relationship between volume and pressure at constant temperature.
   1. $V \propto 1/P$.
   2. $V = k\left(\dfrac{1}{P}\right)$
C. Charles' law — relationship between volume and temperature at constant pressure.
   1. $V \propto T$ (temperature is expressed in Kelvin).

Chapter 9 – Gases: Their Properties and Behavior

2. $\dfrac{V}{T} = k$

D. Avogadro's law — relationship between volume and amount of gas at constant pressure and temperature.
   1. $V \propto n$ (where n = number of moles of gas)
   2. $\dfrac{V}{n} = k$

   3. 1 mol of gas at 273.15 K and 1.00 atm = 22.4 L of gas.
      a. standard molar volume.

9.3 The Ideal Gas Law
A. Describes how the volume of a gas is affected by changes in pressure, temperature, and amount.
   1. $PV = nRT$; R = gas constant = $0.08206 \ \dfrac{\text{L} \cdot \text{atm}}{\text{K} \cdot \text{mol}}$
   2. Can be rearranged in different ways to take the form of Boyle's law, Charles' law, and Avogadro's law.

B. Standard temperature and pressure (STP): T = 273.15 K; P = 1 atm.
   1. Standard temperature for gas measurements (0° C) is different from standard temperature for thermodynamic measurements (25° C).

EXAMPLE:
What is the volume of 3.57 g of O_2 at a temperature of 18.5°C and a pressure of 0.563 atm?

SOLUTION:

$V = \dfrac{nRT}{P}$; $n = 3.57 \text{ g } O_2 \times \dfrac{1 \text{ mol } O_2}{32.0 \text{ g } O_2} = 0.112 \text{ mol } O_2$; $T = 273.15 + 18.5 = 291.6$ K

$V = \dfrac{0.112 \text{ mol } O_2 \times 0.08206 \dfrac{\text{L} \cdot \text{atm}}{\text{mol} \cdot \text{K}} \times 291.6 \text{ K}}{0.563 \text{ atm}} = 4.76$ L

Workbook Problem 9.1

What would be the volume of a 125 mL sample of gas at 23.5°C and a pressure of 754 mm Hg if the gas is compressed to 725 mm Hg at a temperature of 18.7°C?

Strategy: Rearrange the ideal gas law to isolate the constants (the number of moles of gas and R). Since nR remains constant, that means that the relationship between the final temperature, pressure, and volume is the same as the relationship between the initial temperature, pressure, and volume. Write an equation showing this relationship.

Chapter 9 – Gases: Their Properties and Behavior

Step 1: Rearrange the above equation to solve for the final volume.

What key concept did you use?

9.4 Stoichiometric Relationships with Gases A. 6–8.
A. Many chemical reactions involve gases.
B. Ideal gas law can be used to calculate amounts of gaseous reactants.

 *Workbook Problem 9.2*

How many liters of oxygen are needed to completely react with 15.75 g of propane (C_3H_8) at 400°C and 3.75 atm?

Strategy: Remember that the reaction of a hydrocarbon with oxygen (combustion) produces carbon dioxide and water. With that in mind, write a balance equation for the reaction and calculate the number of moles of propane.

Step 1: Using the balanced chemical equation, determine the number of moles of oxygen needed to react with the propane.

Step 2: Rearrange the ideal gas law, to determine the volume of oxygen needed.

What key concept did you use?

C. Ideal gas law also can be used to determine the density of a gas.

 *Workbook Problem 9.3*

What is the density of 0.275 g of NO at 758 mm Hg and 23.5°C?

Strategy: To begin, you need to think about the information you've been given and the information you're trying to find. You have the mass of the gas. You need the density of the gas. Determine what is needed to solve for the density.

Step 1: Use the ideal gas law to calculate the volume of the gas.

Step 2: Determine the density of the gas.

What key concept did you use?

    D. Ideal gas law also can be used to determine the molar mass of a gas.

 ***Workbook Problem 9.4***

The density of a gas was found to be 3.79 g/L at 45.0°C and 2.25 atm. What is the molar mass of the gas?

Strategy: To solve this problem with the information given, you need to do a little thinking. First, write a mathematical equation for molar mass. Rearrange this equation so that you are solving for n. Now, rearrange the ideal gas law so that you are solving for n. Equate the last two equations.

Step 1: Rearrange the equation so that you are solving for the molar mass of the gas. (Keep in mind that density is grams/volume, $\frac{g}{V}$).

Step 2: Solve for the molar mass of the gas.

What key concept did you use?

9.5 Partial Pressure and Dalton's Law
A. Gas laws apply to mixtures of gases.
B. Dalton's law of partial pressures — $P_{total} = P_1 + P_2 + P_3 + ...$ at constant V, T, where P_1, P_2, ... refer to the pressures of the individual gases in the mixture.
C. Partial pressures refer to the pressure each individual gas would exert if it were alone in the container (P_1, P_2, ...).
   1. Total pressure depends on the total molar amount of gas present.
      a. $P_1 = n_1 \left(\dfrac{RT}{V} \right)$
      b. $P_{Total} = (n_1 + n_2 + n_3 = ...) \left(\dfrac{RT}{V} \right)$
   2. Mole fraction (X) — the number of moles of the component divided by the total number of moles in the mixture.
      a. Mole fraction $(X) = \dfrac{\text{moles of component}}{\text{total moles in mixture}}$
   3. $P_1 = X_1 \cdot P_{total}$.

EXAMPLE:
A 5.0 L flask at 25°C contains N_2 at a partial pressure of 0.28 atm, He at a partial pressure of 0.12 atm, and Ne at a partial pressure of 0.56 atm. What is the total pressure of the mixture? What is the mole fraction of each gas?

SOLUTION: $P_{total} = 0.28$ atm $+ 0.12$ atm $+ 0.56$ atm $= 0.96$ atm

$X_{N_2} = \dfrac{0.28 \text{ atm}}{0.96 \text{ atm}} = 0.29$; $X_{He} = \dfrac{0.12 \text{ atm}}{0.96 \text{ atm}} = 0.12$; $X_{Ne} = \dfrac{0.56 \text{ atm}}{0.96 \text{ atm}} = 0.58$ atm

Workbook Problem 9.5
A particular mixture of gases had the following composition: 13.9 g H_2, 64.8 g N_2 and 78.7 g NH_3. Determine the partial pressure of each gas and the total pressure of the mixture if the gases are in a 25.0 L container at 150° C.

Strategy: Use Dalton's law to solve.

Step 1: Determine the partial pressure for each gas using the ideal gas law.

Chapter 9 – Gases: Their Properties and Behavior

Step 2: Sum the partial pressures to determine the total pressure.

What key concept did you use?

9.6 The Kinetic–Molecular Theory of Gases
A. Model that explains the behavior of gases.
B. Assumptions.
  1. A gas consists of particles in constant random motion.
  2. Most of the volume of a gas is empty space.
  3. The attractive forces between molecules of a gas are negligible.
     a. Gas molecules act independently of one another.
  4. The total kinetic energy of the gas particles is constant at constant T.
  5. Average $E_K \propto T$.
C. Above assumptions can be used to explain the gas laws (see figure 9.11, page 333 in your textbook).

B.1.

9.7 Graham's Law: Diffusion and Effusion of Gases
A. Consequences of constant motion and high velocities of gas particles.
  1. Gases mix rapidly when they come in contact.
     a. diffusion — mixing of different gases by random molecular motion and with frequent collisions
     b. effusion — a process in which gas molecules escape through a tiny hole in a membrane without collisions
B. Graham's law — the rate of effusion of a gas is inversely proportional to the square root of its molar mass.
  1. Rate $\propto \dfrac{1}{\sqrt{M}}$
  2. Two gases at the same temperature and pressure — $\dfrac{\text{Rate}_1}{\text{Rate}_2} = \sqrt{\dfrac{M_2}{M_1}}$.
     a. can use different rates of effusion to separate a gas mixture into separate components

B.2.

EXAMPLE:
Calculate the ratio of effusion rates of NO_2 and SO_3 from the same container at the same temperature and pressure.

SOLUTION: $\dfrac{\text{Rate of effusion of } NO_2}{\text{Rate of effusion of } SO_3} = \sqrt{\dfrac{80.0 \text{ g } SO_3 / \text{mol}}{46.0 \text{ g } NO_2 / \text{mol}}} = 1.32$

9.8 The Behavior of Real Gasesf
A. Ideal gas.
  1. No attractive forces between molecules — true at low pressures.
  2. Molecular volume — most of the volume of a gas is empty space – true at low pressures.
B. Real gas.
  1. Attractive forces are more important at higher pressures
     a. decreases the volume of the gas from that predicted by the ideal gas law

B.3.

2. Molecular volume — actual volume of a gas at high pressure is larger than predicted by the ideal gas law.
C. Two effects — molecular volume and intermolecular forces cancel out at intermediate pressures.
D. van der Waals equation:

$$\left(P + an^2/V^2\right)(V - nb) = nRT$$

1. a — correction for intermolecular attractions.
2. b — correction for molecular volume.

9.9 The Earth's Atmosphere
A. Troposphere — the region nearest the earth's surface.
   1. Has greatest effect on the earth's surface.
   2. Air pollution.
      a. release of unburned hydrocarbon molecules
      b. production of NO from automobiles and production of petroleum products
         i. $NO_2 (g) + h\nu \rightarrow NO (g) + O (g)$
         ii. $O(g) + O_2 (g) \rightarrow O_3 (g)$
      c. produces photochemical smog
   3. Acid rain.
      a. results from the production of SO_2 from burning of sulfur-containing coal.
      b. $2 SO_2 (g) + O_2 (g) \rightarrow 2 SO_3 (g); SO_3 (g) + H_2O (l) \rightarrow H_2SO_4 (aq)$
      c. leads to extinction of fish in acidic lakes, damage of forests, and deterioration of marble.
B. The greenhouse effect and global warming.
   1. Some radiant energy from the sun is radiated back into space.
      a. most passes out through the atmosphere
      b. some absorbed by atmospheric gases
         i. $H_2O (g)$, CO_2, and CH_4
         ii. warms the atmosphere
         iii. maintains stable temperature at the earth's surface.
      c. increase amount of absorbed radiation:
         i. increase atmospheric heating
         ii. global temperature rises.
   2. Increase the concentration of atmospheric CO_2 in last 155 years.
      a. global warming has begun.
      b. last 25 years, temperature of mid-troposphere has increased by 0.4° C.
      c. potential warming of 3° C by 2050.
         i. large increase in melting of glacial ice
C. Ozone layer — an atmospheric band stretching from about 20 to 40 km above the earth's surface.
   1. Absorbs intense ultraviolet radiation from the sun.
      a. shields high-energy solar radiation from reaching the earth's surface
   2. Ozone depletion — due to the presence in the stratosphere of chlorofluorocarbons (CFCs).
      a. reaction sequence — a chain reaction in which the generation of a few chlorine atoms leads to the destruction of a great many ozone molecules.
   3. International ban on industrial production and release of CFC's.

Putting It Together
What is the molecular formula of 2.00 L of a gas that is 64.81% C, 13.60% H and 21.59% O at 25°C and 0.420 atm and weighs 2.57 g?

Chapter 9 – Gases: Their Properties and Behavior

 **Self–Test**

This section is intended to test your knowledge of the material covered in this chapter. Think through these problems, and make certain you understand what is going on. Ask yourself if your answer makes sense. Many of these questions are linked to the chapter learning goals. Therefore, successful completion of these problems indicates you have mastered the learning goals for this chapter. You will receive the greatest benefit from this section if you use it as a mock exam. You will then discover which topics you have mastered and which topics you need to study in more detail.

Multiple Choice

1. The SI unit for pressure is
    a. Newton
    b. mm Hg
    c. pascal
    d. atmosphere

2. When the liquid level in the arm connected to the gas–filled cylinder is lower than the liquid level in the arm open to the atmosphere in an open end manometer
    a. $P_{gas} = P_{atm}$
    b. $P_{gas} > P_{atm}$
    c. $P_{gas} < P_{atm}$
    d. $P_{gas} = P_{Hg}$

3. The gas law that states the relationship between volume and pressure at constant temperature is
    a. Boyle's law
    b. Charles' law
    c. Avogadro''s law
    d. Ideal gas law

4. The gas law that states the relationship between volume and amount of gases at constant pressure and temperature is
    a. Boyle's law
    b. Charles' law
    c. Avogadro's law
    d. Ideal gas law

5. The model used to explain the behavior of gases is
    a. Ideal gas law
    b. Dalton's law of partial pressure
    c. Kinetic–Molecular Theory
    d. Graham's law of effusion

6. The mixing of different gases by random molecular motion and with frequent collisions is
    a. diffusion
    b. effusion
    c. confusion
    d. stirring

Chapter 9 – Gases: Their Properties and Behavior

7. Acid rain is primarily a consequence of
   a. the release of unburned hydrocarbon molecules
   b. the production of SO_2
   c. the production of HNO_3
   d. an increased concentration of CO_2 in the atmosphere

8. Different gases
   a. can be described based on their chemical properties only
   b. show similar physical behavior
   c. produce heterogeneous mixtures when combined
   d. occupy different volumes when one mole is present

9. The region of the atmosphere that has the greatest effect on the earth's surface is
   a. the ozone layer
   b. the mesosphere
   c. the troposphere
   d. the thermosphere

Matching

Gas	a. The rate of effusion of a gas is inversely proportional to the square root of its molar mass.
Pressure	b. describes how the volume of a gas is affected by changes in pressure, temperature, and amount of gas
Atmospheric pressure	c. the pressure each individual gas would exert if it were alone in the container.
Boyle's law	d. a process in which gas molecules escape through a tiny hole in a membrane without collisions
Charles' law	e. a model that explains the behavior of gases
Avogadro's law	f. The total pressure exerted by a mixture of gases in a container at constant V and T is equal to the sum of the pressures exerted by each individual gas in the container
Ideal gas law	g. mixing of different gases by random molecular motion and with frequent collisions
Dalton's law	h. states the relationship between volume and pressure at constant temperature and constant amount of gas
Partial pressures	i. states the relationship between volume and amount at constant pressure and temperature

Chapter 9 – Gases: Their Properties and Behavior

    Kinetic–Molecular Theory   j. a substance whose constituent atoms or molecules have little attraction for one another

    Graham's law   k. force exerted per unit area

    Diffusion   l. pressure created from the mass of the atmosphere pressing down on the earth's surface

    Effusion   m. states the relationship between volume and temperature at constant pressure

Fill–in–the–Blank

1. A gas can be compressed because _____.
2. If $P_{gas} < P_{atm}$ in an open–ended manometer, the liquid level in the arm connected to the gas–filled cylinder will be _____ the liquid in the arm open to the atmosphere.
3. Standard temperature and pressure are defined to be _____.
4. The total pressure of a mixture of gases depends on _____
   _____.
5. Mole fraction is _____
   _____.
6. The average kinetic energy of a gas is proportional to the _____.
7. A gas mixture can be separated into its constituent components by using the different _____ of the gases in the mixture.
8. At higher pressures, the intermolecular forces of gases _____ causing the volume of the gas to _____.
9. The actual volume of a gas at very high pressure is _____ than that predicted by the ideal gas law.
10. The depletion of the ozone layer is due to the presence of _____ in the stratosphere.

Problems

1. An open–ended manometer containing mercury is connected to a container of gas. What is the pressure of the gas (in mm Hg) when a) the level of mercury in the arm connected to the gas is 38 mm lower than in the arm connected to the atmosphere and the atmospheric pressure is 784 mm Hg, and b) the level of mercury in the arm connected to the gas is 23 mm higher than in the arm connected to the atmosphere and the atmospheric pressure is 747 mm Hg?

2. Assume that you are using an open–ended manometer filled with silicon oil rather than mercury. What is the gas pressure in mm Hg if the level of silicon oil in the arm connected to the bulb is a) 140 mm lower and P_{atm} = 760 mm Hg; and b) 175 mm higher and P_{atm} = 760 mm Hg? The density of the oil is 1.30 g/mL and the density of mercury is 13.6 g/mL.

Chapter 9 – Gases: Their Properties and Behavior

3. A 500 mL flask contains 0.40 g of O_2 at a temperature of 23.5°C. What is the pressure of the gas?

4. Calculate the mass in grams of 250 mL of NO_2 at STP.

5. A student carried out a reaction in the lab in which one of the products was a gas. She collected the gas for analysis and found that it contained 82.8% carbon and 17.2% hydrogen. She also observed that 350 mL of the gas at 23° C and 757 mm Hg had a mass of 0.835 g. a) What is the empirical formula of the gas? b) What is the molar mass of the gas? c) What is its molecular formula?

6. What is the density of 3.45 g of H_2S gas at 25°C and 770 mm Hg?

7. What is the molar mass of a gas with a density of 1.49 g/L at 745 mm Hg and 18°C?

8. The oxidation of ammonia is an important reaction in the production of fertilizers,

   $4 NH_3 (g) + 5 O_2 (g) \rightarrow 4 NO (g) + 6 H_2O (g)$

   How many liters of NO at 500°C and 735 mm Hg can be produced from 75 L of O_2 at 100°C and 650 mm Hg?

9. 175 mL of O_2 at 30°C and 723 mm Hg were mixed with 275 mL of NH_3 at 50°C and 613 mm Hg and were transferred to a 500 mL vessel where they underwent reaction according to the above chemical equation. What will be the total pressure (in mm Hg) in the reaction vessel at 200°C after the reaction goes to completion?

10. Calculate the volume of carbon dioxide measured at STP that would be produced from the combustion of 8.92 g of propane (C_3H_8).

11. A gas occupying 575 mL at 23°C is compressed to 350 mL at constant pressure. What is the final temperature?

12. A gas mixture consisting of CH_4, C_3H_8, and C_4H_{10} has a total pressure of 2.0 atm. What are the mole fractions of each gas if the partial pressures are 0.68 atm, 1.05 atm, and 0.27 atm, respectively?

13. Three gases consisting of 4.0 g O_2, 3.0 g CO_2, and an unknown amount of N_2 were added to the same 15.0 L container to give a total pressure of 950 mm Hg at 28.0° C. Calculate a) the total number of moles of gas in the container, b) the mole fraction of each gas, c) the partial pressure of each gas, and d) the number of grams of N_2 in the container.

14. Explain Boyle's law using the Kinetic–Molecular Theory.

15. Calculate the ratio of effusion rates of He and Ar from the same container at the same temperature and pressure.

Chapter 9 – Gases: Their Properties and Behavior

Challenge Problem

The following data was collected for a compound containing C, H, N and O.

All of the nitrogen in a 0.2394 g sample was converted to N_2, which was collected over water at 23.80°C and 746.0 mm Hg. The volume of N_2 collected was 18.90 mL. The vapor pressure of water is 22.110 mm Hg at this temperature.

Combustion of a 6.478 mg sample of the compound produced 17.57 mg of CO_2 and 4.319 mg of H_2O.

The molar mass of the compound is 324 g/mol.

Use this data to determine the molecular formula of the compound.

Inquiry Based Problem

The Dumas method is a laboratory method used to determine the molar mass of a volatile compound as well as the density. Using the gas laws just discussed, provide a brief outline of the steps for this method if provided a volatile liquid.

Consider the following:

1. What information do you need to determine the molar mass of a compound?
2. What information is provided by the ideal gas law to aid in the determination of the molar mass?
3. What data do you need to collect to use the ideal gas law?
4. Do you need to use any simple experimental techniques to collect any (or all) of the data?
5. Given that the density of a gas is mot often recorded at STP, what equation would you use to correct the data collected?

CHAPTER 10

LIQUIDS, SOLIDS, AND CHANGES OF STATE

Chapter Learning Goals

A. *Intermolecular Forces*
1. Using only VSEPR geometries and electronegativity trends, determine whether a molecule is expected to be polar.
2. Identify the major type of intermolecular force present in substances, and determine which of two substances exhibits the stronger intermolecular force.

B. *Relationships Between Phases*
1. For a phase change, determine whether enthalpy is increasing or decreasing and whether entropy is increasing or decreasing.
2. Use the $\Delta G = \Delta H - T\Delta S$ equation to calculate the entropy change for a phase change or the temperature (boiling point, melting point, sublimation point) at which the phase change occurs.
3. Use the Clausius–Clapeyron equation to calculate vapor pressure or heat of vaporization.
4. Sketch a phase diagram, labeling the axes and each of the regions, and locate the triple point, critical point, normal melting point, and the normal boiling point. Use the phase diagram to describe physical changes.

C. *Characterization of Solids*
1. For metals crystallizing in one of the three cubic unit cells, determine the number of atoms, mass, volume, density, atomic radius, and packing efficiency.

Chapter in Brief

Unlike gases, liquids and solids have strong attractive forces between the particles. In this chapter, you will examine the nature of these attractive forces and learn how they arise from the polarity of the particles as well as how these forces affect the properties of particular liquids and solids. You will also learn about the relationship between these attractive forces and the transitions between the three states of matter. You will examine the relationship between the transitions that occur between gases, liquids, and solids, and the ΔG that accompanies these changes, along with the effect that temperature and pressure have on these transitions. You will discover the different types of solids and how you can use information on the structure of the unit cell to calculate both the radius and density of metals. You end this chapter by taking a look at the description of ionic and covalent network solids.

10.1 Polar Covalent Bonds and Dipole Moments

A. Bond dipole — a bond that has a partial positive ($\delta+$) and partial negative ($\delta-$) end due to the difference in electronegativity of the atoms in the bond.

  1. Represented by ⊢→; indicates direction of electron displacement.
     a. point of arrow represents $\delta-$ end of dipole
     b. crossed end represents $\delta+$ end of dipole

B. Polar molecules — due to the net sum of individual bond polarities and lone–pair contributions in the molecule.

  1. Molecular dipoles — center of mass of positive charge (nuclei) doesn't coincide with the center of mass of all negative charges (electrons).
     a. leads to a net polarity in molecule

Chapter 10 – Liquids, Solids, and Changes of State

       2. Lone pairs of electrons make substantial contributions to the net molecular polarity.
       3. Individual bond polarities will cancel in symmetrical molecules.
   C. Dipole moment (μ) — the magnitude of the charge Q at either end of the molecular dipole times the distance r between the charges.

EXAMPLE:
Determine which molecule is expected to be polar: ClF_3 or BF_3.

SOLUTION:
A rough estimate of molecular polarity can be achieved by drawing the electron–dot structure of the molecule.

For ClF_3:

For BF_3:

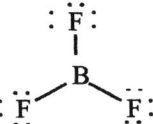

BF_3 has bond dipoles pointing outwards along the B—F bond. Because individual bond polarities will cancel in symmetrical molecules, this molecule is nonpolar. ClF_3 has two lone pairs of electrons in the equatorial plane. These lone pairs of electrons make a substantial contribution to the net polarity of the molecule. Furthermore, they are oriented such that they will not cancel out. As a consequence, ClF_3 is polar.

10.2 Intermolecular Forces

   A. Intermolecular forces — attractive forces between molecules that hold them together at certain temperatures.
       1. Often called van der Waals forces.
       2. Divided into categories.
          a. ion–dipole
          b. dipole–dipole
          c. London dispersion forces
          d. hydrogen bonding
       3. Electrical in nature.
   B. Ion–dipole forces — result of electrical interactions between an ion and the partial charges on a polar molecule.
       1. Favored orientation
          a. $\delta+$ end of molecule near the anion
          b. $\delta-$ end of molecule near the cation
       2. Important in aqueous solutions of ionic substances, where dipolar water molecules surround the ions.
   C. Dipole–dipole forces — result from electrical interactions among dipoles on neighboring molecules.
       1. Generally weak.
          a. strength depends on the sizes of the dipole moments involved
       2. Significant only when molecules are in close contact.
       3. Correlation between dipole moment and boiling point.
          a. high dipole moment corresponds to strong intermolecular forces

b. substance must overcome intermolecular forces to boil
c. stronger intermolecular forces require higher boiling points
D. London Dispersion Forces — result from the motion of electrons around atoms.
1. Any given instance, electron distribution may be unsymmetrical.
   a. creates a short–lived dipole moment (instantaneous dipole)
2. Instantaneous dipole induces a temporary dipole on neighboring atoms.
3. Weak attractive forces.
4. Polarizability — ease with which a molecule's electron cloud can be distorted by a nearby electric field.
   a. smaller molecules and lighter atoms with fewer electrons
      i. relatively nonpolarizable
      ii. smaller dispersion forces
   b. larger molecules and heavier atoms with more electrons
      i. more polarizable
      ii. larger dispersion forces
   c. molecules with more spread–out shapes
      i. maximize molecular surface area
      ii. greater contact between molecules
      iii. give rise to higher dispersion forces
E. Hydrogen Bonds — an attractive interaction between a hydrogen atom bonded to an electronegative O, N, or F atom and an unshared electron pair on another nearby electronegative atom.
1. Quite strong.
2. Responsible for water's remarkable properties.
3. Give rise to higher boiling points than might be expected.
4. Combination of forces.
   a. dipole–dipole interactions
      i. H–F, H–O, H–N bonds are highly polar
   b. Hydrogen can be approached very closely.
      i. no core electrons and small size
F. Comparison of different intermolecular forces – see Table 10.5, page 363 in your textbook.

EXAMPLE:

Order the following molecules by increasing strength of intermolecular forces: C_2H_5OH, PH_3, SF_6.

SOLUTION: The order is $SF_6 < PH_3 < C_2H_5OH$. SF_6 has an octahedral geometry and no lone pairs; therefore, it is a nonpolar molecule. The only intermolecular forces present are London forces. PH_3 has trigonal pyramidal geometry and a lone pair of electrons. Therefore, PH_3 is a polar compound. The strongest type of intermolecular forces present are dipole–dipole forces. C_2H_5OH has an hydrogen atom bonded to oxygen. The strongest type of intermolecular force present is hydrogen bonding.

Workbook Problem 10.1

Determine the strongest intermolecular forces that are present in the following molecules and order by increasing strength: PCl_3, PCl_5, $(CH_3)_2NH$.

Step 1: First, determine if any of the molecules listed are capable of hydrogen bonding.

Chapter 10 – Liquids, Solids, and Changes of State

Step 2: Determine if any of the molecules listed are capable of ion–dipole or dipole–dipole interactions.

Step 3: If any of the molecules are incapable of hydrogen bonding, ion–dipole interactions or dipole–dipole interactions, which type of intermolecular force will be present?

Step 4: Order the molecules from the weakest intermolecular force to the strongest intermolecular force.

What key concept did you use?

10.3 Some Properties of Liquids
A. Viscosity — the measure of a liquid's resistance to flow.
   1. SI unit — N·s/m^2.
   2. Ease with which molecules move around in the liquid.
   3. Related to intermolecular forces.
      a. The stronger the forces, the higher the viscosity.
B. Surface tension — the resistance of a liquid to spreading out and increasing its surface area.
   1. Due to the difference in intermolecular forces felt by the molecules on the surface of the liquid and the molecules in the interior of the liquid.
   2. Related to intermolecular forces.
      a. The stronger the forces, the greater the surface tension.
C. Properties are temperature dependent.
   1. An increase in temperature corresponds to an increase in kinetic energy.
      a. Molecules with high kinetic energies can more easily overcome intermolecular forces.

10.4 Phase Changes
A. Phase changes (changes of state) — the physical form but not the chemical identity of a substance changes.
   1. Matter in any one state can change into either of the other two.
      a. fusion (melting): solid → liquid
      b. freezing: liquid → solid
      c. evaporation: liquid → gas
      d. condensation: gas → liquid
      e. sublimation: solid → gas
      f. deposition: gas → solid
B. Phase change is associated with a free–energy change, ΔG.
   1. $\Delta G = \Delta H - T\Delta S$.
   2. Enthalpy part — energy change associated with making or breaking the intermolecular attractions that hold liquids and solids together.
   3. Entropy part — associated with the change in disorder between various states.
   4. Solid → liquid, solid → gas, liquid → gas: ΔH and ΔS are both positive.

Chapter 10 – Liquids, Solids, and Changes of State

5. Gas → liquid, gas → solid, liquid → solid: ΔH and ΔS are both negative.
6. Can calculate temperature at which two phases are in equilibrium, knowing ΔH and ΔS for a phase transition.
   a. at equilibrium, $\Delta G = 0$; $T = \Delta H / \Delta S$

C. Heating curve — graphically displays the results of adding heat to a sample (see Figure 10.10, page 367 in your textbook).
   1. Melting point — temperature at which solid and liquid coexist in equilibrium as molecules break free from their position in the crystal and enter the liquid phase.
      a. heat of fusion (ΔH_{fusion}) — the amount of energy required for overcoming enough intermolecular forces to convert a solid into a liquid
   2. Boiling point — temperature at which liquid and vapor coexist in equilibrium as molecules break free from the surface of the liquid and enter the gas phase.
      a. heat of vaporization (ΔH_{vap}) the amount of energy necessary to convert a liquid into a gas
   3. $\Delta H_{vap} \gg \Delta H_{fusion}$
      a. for vaporization, must overcome all intermolecular forces in compound
      b. for fusion, must overcome fewer intermolecular forces

EXAMPLE:
The boiling point for NH_3 is $-33.4°C$, and the $\Delta H°_{vap} = 23.4$ kJ/mol. Calculate $\Delta S°_{vap}$ for NH_3.

SOLUTION:
When substances boil, $\Delta G°_{vap} = 0$. Therefore, $T_b = \Delta H°_{vap} / \Delta S°_{vap}$. Rearranging this equation and solving for $\Delta S°_{vap}$ gives

$$\Delta S°_{vap} = \frac{23.4 \text{ kJ/mol}}{240 \text{ K}} = 0.0976 \frac{\text{kJ}}{\text{mol} \cdot \text{K}} = 97.6 \frac{\text{J}}{\text{mol} \cdot \text{K}}$$

Workbook Problem 10.2
Determine the freezing point for dimethylamine, $(CH_3)_2NH$, given that $\Delta H_{fusion} = 5.94$ kJ/mol and $\Delta S_{fusion} = 64.4$ J/mol·K.

Strategy: Rearrange the equation $\Delta G = \Delta H - T\Delta S$ to solve for the temperature.

What key concept did you use?

10.5 Evaporation, Vapor Pressure, and Boiling Point
A. Evaporation — the escape of molecules from the surface of a liquid.
B. Vapor pressure — the pressure exerted by the molecules in a vapor over a liquid in a closed container.
   1. Reaches a dynamic equilibrium.
      a. Number of molecules escaping the liquid equals number of molecules returning to the liquid.
      b. Total number of molecules in both liquid and vapor phases are steady.
      c. Individual molecules are constantly passing back and forth from one phase to another.
C. Processes explained by kinetic molecular theory.

Chapter 10 – Liquids, Solids, and Changes of State

          1. An increase in temperature results in a higher fraction of molecules with sufficient kinetic energy to overcome the surface tension and escape into the vapor.
    D. Value of vapor pressure related to intermolecular forces and temperature.
        1. The smaller the intermolecular forces, the higher the vapor pressure.
            a. Molecules are loosely held in liquid and can easily escape.
        2. The higher the temperature, the higher the kinetic energy.
            a. Molecules will have sufficient kinetic energy to escape the liquid.
    E. Clausius–Clapeyron Equation — relates the vapor pressure of a liquid to the inverse of its temperature.
        1. $\ln(P_{vap}) = \left(\dfrac{-\Delta H_{vap}}{RT}\right) + C$.
            a. C – constant characteristic of each substance.
        2. Plot $\ln(P_{vap})$ versus $1/T$ — straight line
            a. slope $= -\Delta H_{vap}/R$
        3. Can calculate ΔH_{vap} of a liquid knowing the vapor pressure at several temperatures.
            a. $\Delta H_{vap} = \dfrac{(\ln P_2 - \ln P_1)R}{\left(\dfrac{1}{T_1} - \dfrac{1}{T_2}\right)}$

    F. Boiling point — the temperature at which the vapor pressure of a liquid is equal to the external pressure pushing on the surface and all of the liquid is able to change into the vapor phase.
        1. Normal boiling point — external pressure = 1 atm.
        2. External pressure < 1 atm; liquid boils at a lower temperature.
        3. External pressure > 1 atm; liquid boils at a higher temperature.

EXAMPLE:
Propanoic acid has a $\Delta H_{vap} = 32.14$ kJ/mol at a temperature of 25°C. Calculate the vapor pressure for propanoic acid at this temperature. (C = 7.76)

SOLUTION:
Using the Clausius–Clapeyron equation, we can calculate the vapor pressure for propanoic acid.

$$\ln P = \dfrac{-\Delta H}{RT} + C = \dfrac{-32{,}140 \, \dfrac{J}{mol}}{(298 \text{ K})\left(8.314 \, \dfrac{J}{mol \cdot K}\right)} + 7.76 = -5.21$$

Taking the antilog of both sides gives $P = 5.46 \times 10^{-3}$ atm.

10.6 Kinds of Solids
    A. Crystalline solids — solids whose atoms, ions, or molecules have an ordered arrangement extending over a long range.
        1. Seen on visible level — have flat faces and sharp angles.
    B. Amorphous solid – constituent particles are randomly arranged and have no long range ordered structure.
    C. Crystalline solids – categorized as ionic, molecular, covalent network, or metallic. (See Table 10.9, p.374 in your textbook).
        1. Ionic solids — constituent particles are ions (example, NaCl).
            a. ordered into a 3–D arrangement and held together by ionic bonds

Chapter 10 – Liquids, Solids, and Changes of State

2. Molecular solids — constituent particles are molecules held together by intermolecular forces (example, sucrose or ice).
3. Covalent network solids — atoms are linked together by covalent bonds into a giant 3-D array (example, diamond or quartz).
4. Metallic solids — comparable to network solids but they consist of metal atoms (example, Ag or Fe).
   a. have metallic properties

10.7 Probing the Structure of Solids: X–Ray Crystallography
A. Diffraction — a beam of electromagnetic radiation is scattered by an object containing regularly spaced lines or points (*i.e.*, atoms in a crystal).
   1. Spacing must be comparable to the λ of the radiation.
   2. Due to interference between two λ's passing through the same region of space at the same time.
      a. constructive interference — waves are in–phase (peak–to–peak and trough–to–trough)
         i. increases intensity of wave
      b. destructive interference — waves are out–of–phase (waves cancel)
B. Bragg analysis — X–rays are diffracted by different layers of atoms in the crystal, leading to constructive and destructive interference.
   1. Bragg equation: $n\lambda = 2d \times \sin\theta$, $d = \dfrac{n\lambda}{2\sin\theta}$.
      a. λ = known; $\sin\theta$ = angle at which incoming rays are reflected (can be measured); n = an integer (usually 1)
C. X-ray diffraction analysis – measures the interatomic distance between any two atoms in a crystal.
   1. determines the structures of molecules.

10.8 Unit Cells and the Packing of Spheres in Crystalline Solids   ⌕ C.1.
A. Particles pack together in crystals so that they can be as close together as possible to maximize intermolecular attractions.
B. Simple cubic packing — the spheres in one layer sit directly on top of those in the previous layer.
   1. All layers are identical.
   2. Coordination number = 6: Sphere touches four neighbors in the same layer, one above and one below.
   4. Uses only 52% of available volume.
C. Body–centered cubic packing — spheres are in alternate layers in an *a–b–a–b* arrangements where the spheres in the *b* layers fit into the small depressions between spheres in the neighboring *a* layers.
   1. Coordination number = 8; four neighbors above and four neighbors below.
   2. Occupies 68% of the available volume.
D. Hexagonal closest–packed — noncubic unit cell with two alternating layers (*a–b–a–b*).
   1. Hexagonal arrangement of touching spheres.
   2. Spheres in a *b* layer fit into the small triangular depressions between spheres in an *a* layer.
   3. Coordination number = 12.
      a. six neighbors in the same layer, three above and three below
E. Cubic closest–packed — face–centered cubic unit cell with three alternating layers, *a–b–c–a–b–c*.
   1. *a–b* layers identical to hexagonal closest–packed.
   2. Third layer is offset from both *a* and *b*.
   3. Coordination number = 12.
F. Unit cells — small repeating units found in crystals.

Chapter 10 – Liquids, Solids, and Changes of State

1. Symmetrical geometries.
2. Stack to minimize space.
3. Fourteen different geometries.
   a. parallelepipeds (six–sided geometric solids whose faces are parallelograms)
   b. differ in lengths of cell edges and the angles between the edges

G. Cubic cells — all edges are equal in length, and all angles are 90°.
   1. Primitive–cubic unit cell for metals — an atom at each of the eight corners.
      a. Each atom is shared with seven other neighboring cubes that come together at the same point.
      b. One–eighth of each corner atom is in any one cube.
      c. found in cubic, closest-packing.
   2. Body–centered cubic unit cell — additional atom in the center of the cube.
      a. found in body-centered cubic packing.
   3. Face–centered cubic unit cell — additional atom on each of its faces.
      a. shared with one other neighboring cube
      b. one–half of each face atom belongs to any one cube

EXAMPLE:

Determine the atomic radius (in picometers) of an argon atom if solid argon has a density of 1.623 g/cm^3 and crystallizes at low temperatures in a face–centered cubic cell.

SOLUTION:

A face–centered cubic unit cell has a total of 4 atoms.

8 corner atoms are shared by eight unit cells: 8 corner atoms × $\frac{1}{8}$ = 1 atom.

6 face atoms are shared by 2 unit cells: 6 face atoms × $\frac{1}{2}$ = 3 atoms.

We now can calculate the mass of argon in the face–centered cubic unit cell from the molar mass of argon and Avogadro's number.

$$\text{Mass of Ar} = 4 \text{ atoms} \times \left(\frac{39.95 \frac{\text{g Ar}}{\text{mol}}}{6.023 \times 10^{23} \frac{\text{atoms}}{\text{mol}}} \right) = 2.654 \times 10^{-22} \text{ g Ar}$$

Knowing mass and density, we can calculate the volume of the unit cell.

$$d = \frac{\text{Mass}}{\text{Volume}} ; \qquad \text{Volume} = \frac{\text{Mass}}{d} ;$$

$$\text{Volume} = \frac{2.654 \times 10^{-22} \text{ g Ar}}{1.623 \frac{\text{g Ar}}{\text{cm}^3}} = 1.635 \times 10^{-22} \text{ cm}^3$$

Knowing volume, we can calculate the length of the edge of the unit cell, d.

$$d^3 = \text{volume} = 1.635 \times 10^{-22} \text{ cm}^3 ;$$

$$d = \sqrt[3]{1.635 \times 10^{-22} \text{ cm}^3} = 5.468 \times 10^{-8} \text{ cm}$$

Looking at the face of the unit cell, we see that the face atom touches the corner atoms but the

Chapter 10 – Liquids, Solids, and Changes of State

corner atoms do not touch each other along the edge. Each diagonal of the face equals four atomic radii, $4r$. (See Example 10.8, p. 407 in your textbook.) The diagonal ($4r$) and two edges of the cube (d) form a right triangle. Therefore, we can use the Pythagorean theorem to solve for the radius of Ar.

Pythagorean theorem: $d^2 + d^2 = (4r)^2$

$(5.468 \times 10^{-8} \text{ cm})^2 + (5.468 \times 10^{-8} \text{ cm})^2 = 16r^2$

$5.980 \times 10^{-15} \text{ cm}^2 = 16r^2$

$r = \sqrt{\dfrac{5.980 \times 10^{-15} \text{ cm}^2}{16}} = 1.933 \times 10^{-8} \text{ cm}$

Converting our units to picometers gives

$1.933 \times 10^{-8} \text{ cm} \times \dfrac{1 \text{ m}}{100 \text{ cm}} \times \dfrac{1 \times 10^{12} \text{ pm}}{1 \text{ m}} = 193.3 \text{ pm}$

10.9 Structure of Some Ionic Solids
  A. Spheres are not all the same size.
   1. Anions are larger than cations.
  B. Adopt a variety of different unit cells depending on the size and charge of the particles.
   1. Face–centered cubic — NaCl, KCl.
   2. Other common ionic unit cells — see Figure 10.25, page 383 in the your textbook.

10.10 Structure of Some Covalent Network Solids
  A. Carbon.
   1. Allotropes — different structural forms of the same element which differ in physical and chemical properties.
   2. More than 40 amorphous forms of carbon.
   3. Diamond — each carbon atom is sp^3 hybridized and covalently bonded with tetrahedral geometries.
   4. Graphite — 2–dimensional sheets of fused six–membered rings; each carbon is sp^2–hybridized.
   5. Fullerene — a spherical C_{60} molecule with the shape of a soccer ball.
  B. Silica (SiO_2) — four single bonds between silicon and four oxygens in a covalent network structure.
   1. Quartz glass — result of heating silica above 1600°C and then cooling the viscous liquid.
      a. Si–O bonds reform in a random arrangement
      b. amorphous solid
      c. mix in additives — prepare a wide variety of glass
         i. window glass — add $CaCO_3$ and Na_2CO_3
         ii. colored glass — add transition metal ions
         iii. borosilicate glass (Pyrex) — add B_2O_3; resistant to thermal shock because it doesn't expand much on heating

10.11 Phase Diagrams
B.4.
  A. Change any one state of matter spontaneously into either of the other two, depending on the temperature and pressure.

Chapter 10 – Liquids, Solids, and Changes of State

B. Phase diagram — a graphical method of illustrating the pressure and temperature dependencies of a pure substance in a closed system.
1. Boundary line — points on this line represent pressure/temperature combinations at which the two phases are in equilibrium.
2. Triple point — a unique combination of pressure and temperature at which all three phases coexist in equilibrium.
3. Critical temperature — the temperature beyond which a gas cannot be liquefied.
4. Critical pressure — the pressure beyond which a liquid cannot be vaporized.
5. Critical point — a point defined by the critical temperature and critical pressure.
6. Supercritical fluid — a substance that is neither a liquid nor a gas.
   a. Pressure of a gas at the critical point is so high and the molecules are so close together that it is hard to distinguish between the gas and a liquid.
   b. Temperature of a liquid at the critical point is so high that it is hard to distinguish between the liquid and the gas.
7. Effect of pressure on the slope of solid/liquid boundary line depends on the relative densities of the solid and liquid phases.

Putting It Together

Sulfur dioxide is one of the compounds that causes acid rain. Acid rain is produced first by the oxidation of sulfur dioxide to sulfur trioxide in air. The sulfur trioxide then reacts with water, producing sulfuric acid. What is the molecular shape of sulfur dioxide? What intermolecular forces are present? Calculate the heat of reaction for the production of sulfur trioxide and sulfuric acid.

Self–Test

This section is intended to test your knowledge of the material covered in this chapter. Think through these problems, and make certain you understand what is going on. Ask yourself if your answer makes sense. Many of these questions are linked to the chapter learning goals. Therefore, successful completion of these problems indicates you have mastered the learning goals for this chapter. You will receive the greatest benefit from this section if you use it as a mock exam. You will then discover which topics you have mastered and which topics you need to study in more detail.

True/False

1. If a molecule has polar bonds, then the molecule is polar.

2. The different types of intermolecular forces are all electrical in nature.

3. Smaller molecules and lighter atoms with fewer electrons are easily polarizable.

4. Hydrogen bonding is one of the strongest types of intermolecular forces.

5. Strong intermolecular forces result in a large surface area of a liquid.

6. For the phase change from a solid to liquid, there is a decrease in enthalpy and an increase in entropy.

7. The value of the vapor pressure of a given liquid is related to the intermolecular forces and temperature.

8. Constructive interference of two waves leads to cancelation of the waves.

9. The structure of silica (SiO_2) is quite similar to the structure of CO_2.

10. A supercritical fluid is a liquid at very high pressures.

Multiple Choice

1. Interactions that are the result of the attraction between the δ– end of one molecule with the δ+ end of another molecule are
   a. ion–dipole forces
   b. dipole–dipole forces
   c. London dispersion forces
   d. hydrogen bonding

2. The ease with which a molecule's electron cloud can be distorted by a nearby electric field is referred to as
   a. van der Waals forces
   b. London dispersion forces
   c. polarizability
   d. viscosity

3. The strongest type of intermolecular force present in C_2H_5OH is
   a. ion–dipole forces
   b. dipole–dipole forces
   c. London dispersion forces
   d. hydrogen bonding

4. Which of the following liquids will have the highest boiling point?
   a. He
   b. H_2Se
   c. CH_3OH
   d. $HgCl_2$

5. Which of the following liquids has the highest vapor pressure?

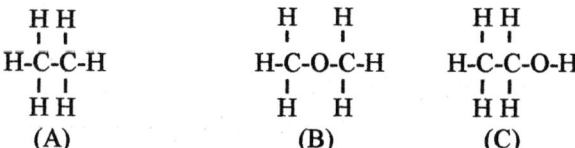

   a. A
   b. B
   c. C
   d. both B and C will have comparable vapor pressures

6. If the external pressure is >1 atm, then a liquid boils at
   a. the normal boiling point
   b. a temperature below the normal boiling point
   c. a temperature above the normal boiling point

Chapter 10 – Liquids, Solids, and Changes of State

7. A solid whose constituent particles are molecules held together by intermolecular forces is a(an)
   a. ionic solid
   b. molecular solid
   c. covalent network solid
   d. metallic solid

8. The type of packing present when the spheres in a cubic unit cell are in alternate layers in an *a–b–a–b* arrangement is
   a. simple cubic packing
   b. body–centered cubic packing
   c. hexagonal closest packed
   d. cubic closest packed

9. Fullerene is an allotrope of
   a. carbon
   b. diamond
   c. silica
   d. phosphorus

10. The slope of the solid/liquid boundary line in a phase diagram depends on
    a. the triple point
    b. the critical pressure
    c. the critical temperature
    d. the relative densities of the solid and liquid phases

Fill–in–the–Blank

1. When the electron distribution in a molecule is temporarily uneven, an _____ is created, which _____ on a neighboring atom.

2. The type of intermolecular force described in question number one is a _____ _____.

3. The _____ is the resistance of a liquid to spreading out and increasing its surface area.

4. Both ΔH and ΔS are negative during the following phase changes: _____ _____.

5. $\Delta H_{vap} \gg \Delta H_{fusion}$ because _____.

6. The value of the vapor pressure is related to the temperature because _____ _____.

7. A solid whose constituent particles are randomly arranged and have no ordered long–range structure is _____.

8. The type of packing present in a noncubic unit cell with two alternating layers is referred to as _____.

9. In diamond, each carbon is _____ hybridized and covalently bonded with _____ geometry.

10. A supercritical fluid exists at the critical point because _____
    _____
    _____.

Problems

1. Determine whether the following molecules are polar or nonpolar:
   a. PCl_5 b. XeF_4 c. SF_4

2. Which compound has the stronger intermolecular forces, C_2H_4 or N_2H_4?

3. For sodium, ΔH_{vap} = 98.0 kJ/mol and ΔS_{vap} = 84.8 J/K·mol. What is the boiling point of sodium?

4. The normal boiling point of propylene glycol is 188.2°C and ΔH_{vap} = 56.8 kJ/mol. What is the vapor pressure of propylene glycol at a temperature of 150°C?

5. Calculate the reflection angle of X–rays with a wavelength equal to 172 pm when they strike a crystal with planes spaced a) 350 pm apart; and b) 975 pm apart. Assume that n = 1.

6. Calculate the spacing between planes (in picometers) that correspond to reflections of θ = 15.0°, 25°, and 35° by X–rays with a wavelength of 175 pm.

7. The edge of the unit cell of palladium is 389 pm. The density of palladium is 12.02 g/cm³. If palladium has a cubic crystal structure, how many palladium atoms are in a unit cell? Identify the cubic unit cell.

8. Nickel crystallizes in a face–centered arrangement with the edge of the unit cell being 352 pm long. What is the radius of a nickel atom?

Challenge Problem

Liquid Q has a heat of vaporization of 35.0 kJ/mol. At 40°C, liquid Q has a vapor pressure of 150 mm Hg. Liquid M has a heat of vaporization of 22.0 kJ/mol. At 40°C, liquid M has a vapor pressure of 300 mm Hg. At what temperature will liquids Q and M have the same vapor pressure?

CHAPTER 11

SOLUTIONS AND THEIR PROPERTIES

Chapter Learning Goals

A. Formation of Solutions
1. Explain the rule of thumb, "like dissolves like" by analyzing the solution process in terms of forces overcome in the solute and solvent and forces formed between solute and solvent particles.
2. Use the equation $\Delta G = \Delta H - T\Delta S$ to calculate normal boiling point, heat of vaporization, or entropy of vaporization, given the other two.

B. Solutions and Concentration Units
1. Define solution density, molarity, mole fraction, mass percent, parts per million, parts per billion, and molality, and perform calculations using these quantities.

C. Solutions and Solubilities
1. Perform calculations using Henry's law.

D. Solutions and Colligative Properties
1. Use Raoult's law to calculate the vapor pressure over a solution containing a nonvolatile solute and a solution containing two volatile liquids.
2. Perform calculations involving freezing–point depression and boiling–point elevation, and determine the molar mass of the solute.
3. Perform calculations involving the osmotic pressure equation, and determine the molar mass of the solute.
4. Describe fractional distillation with the aid of a liquid/vapor phase diagram.

Chapter in Brief

This chapter concentrates on the topic of homogeneous mixtures, with particular emphasis given to solutions. You begin this study by examining the solution process and the energy changes that occur. This is followed by an in–depth description of the concentration units used to indicate the amounts of solute and solvent in a solution. You will learn how to perform calculations to determine the concentration of a solution and how to interconvert between the different units. Next, you will examine the meaning of solubility and how it is affected by both temperature and pressure. Finally, you will explore four types of colligative properties, the free–energy change that is associated with these properties, and some useful applications of these properties.

11.1 Solutions

A. Heterogeneous mixtures — mixtures in which the mixing of components is visually nonuniform.
B. Homogeneous mixtures (solutions) — mixtures in which the mixing of components is visually uniform.
 1. Classified according to the size of the constituent particles.
  a. solutions — contain particles the size of a typical ion or covalent molecule
  b. colloids — contain particles with diameters in the range of 2–1000 nm
   i. od not separate on standing.
  c. suspensions — contain particles that are greater than about 1000 nm in diameter and are visible with a low–power microscope.
C. Any one state of matter can form a solution with any other state.

Chapter 11 – Solutions and Their Properties

1. Seven different kinds of solutions (Table 11.1, page 400 in your textbook).
2. Gas or solid dissolved in a liquid.
    a. solute — the dissolved substance
    b. solvent — the liquid
3. Liquid dissolved in a liquid.
    a. solute — the minor component
    b. solvent — the major component

11.2 Energy Changes and the Solution Process
A. Most solutions involve condensed phases.
   1. Intermolecular forces are important for explaining the properties of solutions.
B. Three types of interactions among particles to take into account.
   1. Solvent–solvent interactions.
   2. Solute–solute interactions.
   3. Solvent–solute interactions.
C. "Like dissolves like."   ⚷ A.1.
   1. Solutions can form when the three types of interactions are similar in kind and in magnitude.
       a. Ionic solids dissolve in polar solvents.
       b. Nonpolar organic substances dissolve in nonpolar organic solvents.
D. Solvated – ions are surrounded and stabilized by a shell of solvent molecules.
   1. Hydrated — water is the solvent.
E. Free–energy change when a solution is formed.
   1. ΔG is negative — spontaneous process and the substance dissolves.
   2. ΔG is positive — nonspontaneous process and the substance does not dissolve.
   3. ΔH_{soln} — the enthalpy of solution.
       a. measures heat flow into or out of the system
       b. difficult to predict — can be either negative (exothermic) or positive (endothermic)
   4. ΔS_{soln} — the entropy of solution.
       a. measures the amount of molecular randomness in the system.
       b. usually positive — increase in molecular randomness during dissolution
F. Variations in heats of solution due to the interplay of the three kinds of interactions.
   1. Solvent–solvent interactions — positive ΔH.
       a. need to overcome intermolecular forces between solvent molecules
       b. separate molecules to make room for solute molecules
   2. Solute–solute interactions — positive ΔH.
       a. need to overcome intermolecular forces between solute molecules
       b. lattice energy for ionic solids
   3. Solvent–solute interactions — negative ΔH.
       a. solvent molecules cluster around solute particles and solvate them
       b. ionic substances in water
           i. increase in hydration energy with a decrease in cation size
           ii. increase in hydration energy with an increase in the charge on the ion
   4. Sum of three interactions determines whether ΔH_{soln} is endothermic or exothermic (see Figure 11.4, page 404 in your textbook).

11.3 Units of Concentration   ⚷ B.1
A. Concentration of a solution — the exact amount of solute dissolved in a given amount of solvent.
B. Molarity $= \dfrac{\text{Moles of solute}}{\text{Liter of solution}}$

Chapter 11 – Solutions and Their Properties

1. Advantages.
   a. simplifies stoichiometry calculations — uses moles instead of mass
   b. Amounts of solution are measured by volume rather than by mass.
      i. simplifies volumetric titrations
2. Disadvantages.
   a. Exact concentration depends on the temperature.
      i. volume changes as temperature changes
   b. need to know the density of the solution to determine the amount of solvent present

C. Mole fraction $(X) = \dfrac{\text{Moles of component}}{\text{Total moles in the solution}}$

1. Advantages.
   a. independent of temperature
   b. useful for calculations involving gas mixtures
2. Disadvantages.
   a. not convenient for liquid solutions

D. Mass percent (mass %) $= \dfrac{\text{Mass of component}}{\text{Total mass of solution}} \times 100$

1. For very dilute solutions:
   a. Parts per million (ppm) $= \dfrac{\text{Mass of component}}{\text{Total mass of solution}} \times 10^6$

   b. Parts per billion (ppb) $= \dfrac{\text{Mass of component}}{\text{Total mass of solution}} \times 10^9$

2. For dilute aqueous solutions at room temperature, 1 kg solution = 1 L solution.
   a. $1\text{ ppm} = \dfrac{1\text{ mg solute}}{1\text{ L soln}}$
   b. $1\text{ ppb} = \dfrac{1\text{ μg solute}}{1\text{ L soln}}$

3. Advantage — values are independent of temperature.
   a. Masses don't change when temperature changes.
4. Disadvantages.
   a. difficult to measure mass of a liquid solution
   b. need to know the density of a solution to convert to molarity

E. Molality $(m) = \dfrac{\text{Moles of solute}}{\text{Mass of solvent (kg)}}$

1. Advantages.
   a. temperature–independent
   b. well suited for calculating certain properties of solutions
2. Disadvantages.
   a. difficult to measure mass of a liquid solution
   b. need to know the density of a solution to convert to molarity
3. When dissolving ionic compounds, moles of solute = total moles of ions.

F. Comparison of four concentration units – see Table 11.3, page 407 in our textbook.

EXAMPLE:
A 1 liter sample of water is analyzed and found to contain 5.68 ppb of lead. How many grams of lead are present? What is the molar concentration of this sample?

SOLUTION:

A concentration of 1 ppb means that each liter of an aqueous solution contains 1 μg (0.001 mg) of solute. Therefore, the number of grams of lead in this solution is 5.68 μg. To calculate the molarity of this solution, we need to convert milligrams to moles.

$$5.68 \, \mu g \, Pb \times \frac{1 \, g}{1 \times 10^6 \, \mu g} \times \frac{1 \, mol}{207.19 \, g \, Pb} = 2.74 \times 10^{-8} \, mol$$

Since we have 1 liter of solution, the molarity is equal to 2.74×10^{-8} M. You can see why the unit of ppb is more useful for solutions with trace impurities.

 **Workbook Problem 11.1**

A solution is prepared by dissolving 17.84 grams of glucose ($C_6H_{12}O_6$) in 250 g of water. The density of this solution is 1.16 g/mL. Calculate the molality, mass percent, and molarity of the solution.

Strategy: From the different concentration units, you need to calculate moles of glucose and grams of solution.

Step 1: You first need to calculate the moles of glucose.

Step 2: Calculate the molality of the solution.

Step 3: Determine the mass percent.

Step 4: Determine the molarity of the solution. (You need to use the density of the solution in this calculation.)

What key concept did you use?

Chapter 11 – Solutions and Their Properties

Workbook Problem 11.2

Which has the greater molality, a 0.75% mass $CaCl_2$ solution or a 0.75% mass citric acid $(C_6H_8O_7)$ solution?

Strategy: Assume 100 g of solution. Remember to calculate the total moles of ions present in $CaCl_2$.

Step 1: If you have 100 g of solution, then you have 0.75 g of $CaCl_2$ and 99.25 g of H_2O. You first need to calculate the number of moles of $CaCl_2$ present.

Step 2: You can calculate the molality of $CaCl_2$ from the moles of $CaCl_2$ and the mass of water. (Remember to use the total number of moles of ions.)

Step 3: Repeat steps 1 and 2 for citric acid.

Step 4: Compare the molality values for $CaCl_2$ and citric acid.

What key concept did you use?

Chapter 11 – Solutions and Their Properties

11.4 Some Factors Affecting Solubility
A. Saturated solution — a solution in which the number of ions leaving a crystal to go into solution is equal to the number of ions returning from solution to the crystal.
 1. Solute + Solvent ⇌ Solution
 2. At equilibrium with undissolved solid.
B. Supersaturated solution — contain a greater than equilibrium amount of solute.
 1. Formed by substances which are more soluble at high temperatures than low temperatures.
C. Solubility — the amount of solute per unit of solvent needed to form a saturated solution.
 1. Physical property characteristic of a particular substance.
 2. Temperature–dependent.
 3. Miscible – solute and solvent are soluble in all proportons.
D. Effect of temperature on solubility.
 1. Gases become less soluble as the temperature is increased.
 2. More difficult to predict effect on the solubility of solids.
E. Effect of pressure on solubility.
 1. None on solids and liquids.
 2. Profound effect on solubility of gases.
 3. Henry's law: solubility = $k \cdot P$ (k = Henry's law constant and P = partial pressure of gas).   ☛ *C.1.*
 4. Increase in pressure leads to an increase in solubility.
  a. due to change in position of the equilibrium between dissolved and undissolved gas
  b. More gas particles are forced into solution.

EXAMPLE:
Determine the solubility of H_2S (g) at 0° C and a partial pressure of 18.9 mm Hg.
$$k = 0.195 \frac{mol}{L \cdot atm}$$

SOLUTION:
Using Henry's law, we can calculate the solubility. CAUTION! Check units on both the Henry's law constant and the partial pressure of H_2S (g).

$$\text{Solubility} = 0.195 \frac{mol}{L \cdot atm} \times 18.9 \text{ mm Hg} \times \frac{1 \text{ atm}}{760 \text{ mm Hg}} = 4.85 \times 10^{-3} \frac{mol}{L}$$

11.5 Physical Behavior of Solutions: Colligative Properties
A. Colligative properties — properties that depend on the amount of dissolved solute but not on the chemical identity of the solute.
 1. Boiling–point elevation.
 2. Freezing–point depression.
 3. Vapor pressure of a solution.
 4. Osmosis — the migration of solvent and other small molecules through a semipermeable membrane.
B. Allow for the comparison of properties of a pure solvent with properties of a solution.
 1. Vapor pressure of the solution is lower.
 2. Freezing point of solution is lower.
 3. Boiling point of solution is igher.
 4. Solution gives rise to osmosis.

Chapter 11 – Solutions and Their Properties

11.6 Vapor–Pressure Lowering of Solutions: Raoult's Law
A. Solutions with a nonvolatile solute — lower vapor pressure than pure solvent.
   1. Raoult's law: $P_{soln} = P_{solv} \times X_{solv}$.
   2. For ionic substances, calculate mole fractions based on the total number of solute particles rather than on the number of formula units.
B. Raoult's law applies only to ideal solutions.
   1. Solute concentrations are low.
   2. Solute and solvent particles have similar intermolecular forces.
C. Solute–solvent intermolecular forces < solvent intermolecular forces
   1. vapor pressure is higher than predicted by Raoult's law.
D. Solute–solvent intermolecular forces > solvent intermolecular forces; vapor pressure is less than predicted by Raoult's law.

E. Free–energy change accompanies vapor pressure lowering.
   1. More negative ΔG, the easier the vaporization process.

F. Liquid → gas:
   1. ΔH is positive (need energy to overcome intermolecular forces).
      a. similar value for both solvent and solution
   2. ΔS is positive (molecular disorder increases).
      a. ΔS for liquid → vapor is smaller for solution than for pure solvent
   3. Larger ΔG for solution (subtracting a smaller $T\Delta S$ from ΔH).
      a. vaporization is more difficult
      b. lower vapor pressure
G. Solutions with a volatile solute — vapor pressure of the solution is always intermediate between the vapor pressures of the two pure substances.
   1. $P_{total} = P_A + P_B$.
   2. P_A and P_B are calculated from Raoult's law.
      a. $P_A = X_A(P_A^\circ)$; $P_B = X_B P_B^\circ$
   3. $P_{total} = X_A(P_A^\circ) + X_B(P_B^\circ)$

EXAMPLE:
The vapor pressure of water at 35°C is 42.175 mm Hg. The vapor pressure of ethyl alcohol (C_2H_5OH) at 35°C is 100.5 mm Hg. What is the vapor pressure of a solution prepared by dissolving 250 g of C_2H_5OH in 375 g of H_2O?

SOLUTION:
$$P_{total} = X_{C_2H_5OH} P_{C_2H_5OH} + X_{H_2O} P_{H_2O}$$

To begin, we need to calculate the number of moles of both ethanol and water so that we can calculate the mole fraction of the two.

$$\text{mol } C_2H_5OH = 250 \text{ g } C_2H_5OH \times \frac{1 \text{ mol } C_2H_5OH}{46.0 \text{ g } C_2H_5OH} = 5.43 \text{ mol } C_2H_5OH$$

$$\text{mol } H_2O = 375 \text{ g } H_2O \times \frac{1 \text{ mol } H_2O}{18.0 \text{ g } H_2O} = 20.8 \text{ mol } H_2O$$

$$X_{C_2H_5OH} = \frac{5.43 \text{ mol } C_2H_5OH}{5.43 \text{ mol } C_2H_5OH + 20.8 \text{ mol } H_2O} = 0.206$$

Chapter 11 – Solutions and Their Properties

$$X_{H_2O} = \frac{20.8 \text{ mol } H_2O}{5.43 \text{ mol } C_2H_5OH + 20.8 \text{ mol } H_2O} = 0.793$$

$$P_{total} = (0.206 \times 100.5 \text{ mm Hg}) + (0.793 \times 42.2 \text{ mm Hg}) = 54.2 \text{ mm Hg}$$

 **Workbook Problem 11.3**

A student needs to prepare an aqueous solution of sucrose at a temperature of 20°C with a vapor pressure of 15.0 mm Hg. How many grams of sucrose does she need if she uses 375 g H_2O? (The vapor pressure of water at 20°C is 17.5 mm Hg.)

Strategy: Substitute the data provided into Raoult's law, and solve for X_{solv}.

Step 1: Solve for the number of moles of sucrose.

Step 2: Calculate the mass of sucrose.

What key concept did you use?

11.7 Boiling–Point Elevation and Freezing–Point Depression of Solutions
A. Solution of a nonvolatile solute
   1. Lower vapor pressure than a pure solvent.
      a. heat to a higher temperature to boil.
      b. phase diagram – liquid/vapor pressure line lower for solution.
         i. lower triple point temperature
         ii. solid/liquid line shifted to lower temperature.
         iii. lower temperature for freezing
B. Boiling–point elevation and freezing–point depression of a solution relative to that of a pure solvent depends on the number of solute particles.
   1. $\Delta T_b = K_b \cdot m$
   2. $\Delta T_f = K_f \cdot m$
   3. Use molality as the concentration unit because the numbers of solute and solvent particles are independent of temperature.

C. Due to an entropy difference between pure solvent and solvent in a solution.

1. $T_b = \dfrac{\Delta H_{vap}}{\Delta S_{vap}}$ at equilibrium.

2. ΔS_{vap} is smaller for a solution, therefore T_b is larger.

3. $T_f = \dfrac{\Delta H_{fusion}}{\Delta S_{fusion}}$ at equilibrium.

4. ΔS_{fusion} is larger for a solution, therefore T_f is smaller.

EXAMPLE:
What will be the freezing point and boiling point of an aqueous solution containing 55.0 g of glycerol, $C_3H_5(OH)_3$, and 250 g of water? $K_b(H_2O) = 0.51°\ C/m$ and $K_f = 1.86°C/m$.

SOLUTION: To determine the freezing point and boiling point of this solution, we need to first calculate the molality of the solution. This requires that we determine the number of moles of glycerol present.

$$55.0 \text{ g glycerol} \times \dfrac{1 \text{ mol glycerol}}{92.0 \text{ g glycerol}} = 0.598 \text{ mol glycerol}; \qquad \dfrac{0.598 \text{ mol glycerol}}{0.250 \text{ kg } H_2O} = 2.39\ m$$

$$\Delta T_f = \left(1.86 \dfrac{°C}{m}\right) \times 2.39\ m = 4.45°C; \qquad \text{f.p. soln.} = 0° - 4.45° = -4.45°C$$

$$\Delta T_b = \left(0.51 \dfrac{°C}{m}\right) \times 2.39\ m = 1.22°C; \qquad \text{b.p. soln.} = 100° + 1.22° = 101.22°C$$

Workbook Problem 11.4
How many grams of $(NH_4)_3PO_4$ need to be added to 500 g of H_2O so that the freezing-point of the solution is lowered to $-8.3°C$? $(K_f = 1.86°C/m)$

Strategy: Using the equation for freezing-point depression, solve for molality.

Step 1: Determine the number of moles of $(NH_4)_3PO_4$, taking into consideration that the molality is determined by the number of moles of solute particles in solution.

Step 2: Determine the mass of $(NH_4)_3PO_4$.

What key concept did you use?

11.8 Osmosis and Osmotic Pressure
A. Semipermeable membrane — membranes that allow water or other small molecules to pass through, but block the passage of large solute molecules or ions.
B. Osmosis — the migration of solvent and other small molecules through a semipermeable membrane.
  1. Passage from pure solvent side to solution side.
     a. more favored
     b. decrease in amount of liquid on solvent side
     c. increase in amount of liquid on solution side
     d. decrease in concentration of solution
C. Osmotic pressure of a solution (Π) — the pressure the pressure needed to create an equilibrium in which the rate of the forward and reverse passage becomes equal.
  1. $\Pi = MRT$
  2. Can use molarity since measurements are made at the temperature specified in the equation.
  3. Due to an increase in entropy when pure solvent passes through the membrane and mixes with the solution.

EXAMPLE:
Determine the osmotic pressure of a 0.075 M solution of aspartic acid at 18.5° C.

SOLUTION:

$$0.075 \frac{\text{mol}}{\text{L}} \times 0.0821 \frac{\text{L} \cdot \text{am}}{\text{mol} \cdot \text{K}} \times 291.7 \text{ K} = 1.80 \text{ atm}$$

11.9 Some Uses of Colligative Properties
A. Common uses.
  1. Freezing–point depression.
     a. sprinkling of salt to melt snow
     b. antifreeze in automobiles
     c. de–icing of airplane wings
B. Desalination of seawater.
  1. Reverse osmosis — increase osmotic pressure so that solvent molecules are forced from solution side to the solvent side.
C. Molar mass determinations.
  1. Can use any four colligative properties.
  2. Most accurate is osmotic pressure — magnitude of osmosis effect is so great.

 *Workbook Problem 11.5*
A solution is prepared from 2.50 g of a compound with an empirical formula of C_6H_5P and 25.0 g of benzene, C_6H_6. The freezing point of this solution is 4.3°C. Determine the molar mass and

Chapter 11 – Solutions and Their Properties

molecular formula of the compound. The freezing point of benzene is 5.5°C and K_f (benzene) = 5.12°C/m.

Strategy: Using the equation for freezing point depression, determine the molality of the solution.

Step 1: Calculate the actual number of moles in this solution.

Step 2: Calculate the molar mass of the sample.

Step 3: Determine the molecular formula.

What key concept did you use?

11.10 Fractional Distillation of Liquid Mixtures
  A. Fractional distillation — separation of a mixture of volatile liquids into fractions by boiling and condensing the vapors.
   1. Vapor is enriched in the more volatile component.
   2. Condensed vapor is also enriched in more volatile component.
   3. Repeat boil/condense cycle with condensed vapor phase many times — can have complete purification of the more volatile liquid component.
      a. occurs naturally in a distillation column (see Figure 11.19, page 459 in your textbook).
  B. Represent fractional distillation with a liquid/vapor phase diagram.
   1. Plot T vs. composition (see Figure 11.18, page 428 in textbook).
   2. Lower region represents the liquid phase.
   3. Upper region represents the vapor phase.
   4. Equilibrium region — between lower and upper region.
      a. liquid and vapor coexist
      b. lower curve — represents composition of liquid phase at a particular temperature
      c. upper curve — represents composition of vapor phase at a particular temperature
      d. tie line — connects points on upper and lower curve at a particular temperature
   5. Fractional distillation — walk across tie lines to achieve desired purity.

Chapter 11 – Solutions and Their Properties

Putting It Together

Esters, organic compounds containing C, H, and O, have very pleasant odors. They are present in natural flavors and are used to make artificial flavors. The ester ethyl butyrate has the odor of pineapples. A 1.50 g sample of ethyl butyrate underwent combustion and produced 3.41 g of CO_2 and 1.40 g of H_2O. Another 1.50 g sample was dissolved in enough solvent to make 250 mL of solution. The osmotic pressure of this solution at 25°C was 1.26 atm. Determine the molecular formula of ethyl butyrate.

Self–Test

This section is intended to test your knowledge of the material covered in this chapter. Think through these problems, and make certain you understand what is going on. Ask yourself if your answer makes sense. Many of these questions are linked to the chapter learning goals. Therefore, successful completion of these problems indicates you have mastered the learning goals for this chapter. You will receive the greatest benefit from this section if you use it as a mock exam. You will then discover which topics you have mastered and which topics you need to study in more detail.

True–False

1. A mixture in which the mixing of components is visually uniform is referred to as a solution.

2. Polar solutes can be dissolved in either nonpolar or polar solvents.

3. The entropy of solution is usually negative.

4. The molality of a solution is determined by dividing the number of moles of solute by the mass (in kg) of the solvent.

5. A solution in which the number of ions leaving a crystal to go into solution is equal to the number of ions returning from solution to the crystal is said to be supersaturated.

6. According to Henry's law, an increase in pressure leads to an increase in solubility.

7. Solutions with a nonvolatile solute will have a higher vapor pressure than the pure solvent.

8. If the solute–solvent intermolecular forces are greater than the intermolecular forces of the pure solvent, the vapor pressure of the solution will be higher than the vapor pressure that is predicted by Raoult's law.

9. ΔS_{vap} is smaller for a solution than for a pure solvent; therefore the T_b will be higher.

10. The colligative property that is considered to be the most accurate for determining the molar mass of a compound is freezing–point depression.

Multiple Choice

1. When a gas or solid is dissolved in a liquid, the solute is
   a. the major component
   b. the liquid

Chapter 11 – Solutions and Their Properties

   c. the dissolved substance
   d. the minor component

2. ΔH_{soln} will be exothermic if
   a. ΔH for solvent–solvent interactions is negative
   b. ΔH for solute–solute interactions is negative
   c. ΔH for solute–solvent interactions is negative
   d. the sum of the three types of interactions leads to a negative ΔH

3. Benzene, a nonpolar organic compound, is most likely to dissolve in
   a. CCl_4
   b. CH_3CH_2OH
   c. water
   d. NH_3

4. The disadvantage of using mole fraction to express the concentration of a solution is
   a. the exact concentration depends on the temperature
   b. you need to know the density of the solution to determine the amount of solvent present
   c. it is not convenient for liquid solutions
   d. it is difficult to measure the mass of a liquid solution

5. The vapor pressure of a solution will be higher than that predicted by Raoult's law when
   a. solute–solvent interactions > solute interactions
   b. solute–solvent interactions < solute interactions
   c. solute–solvent interactions > solvent interactions
   d. solute–solvent interactions < solvent interactions

6. The ionic substance with the highest hydration energy is
   a. NaCl
   b. $BaCl_2$
   c. $AlCl_3$
   d. $PbCl_4$

7. Which of the following is more likely to form a solution?
   a. an ionic solid is mixed with a nonpolar solvent
   b. a nonpolar solute is mixed with water
   c. a nonpolar solute is mixed with NH_3
   d. a nonpolar solute is mixed with a nonpolar solvent

8. The ionic substance with the lowest hydration energy is
   a. $BaSO_4$
   b. $SrSO_4$
   c. $CaSO_4$
   d. $MgSO_4$

9. If you are performing a volumetric titration in the laboratory, the most likely concentration unit you would use is
   a. molality
   b. molarity
   c. ppb
   d. mole fraction

Chapter 11 – Solutions and Their Properties

10. A mixture that contains particles large enough to be visible with a low-power microscope is a
    a. heterogeneous mixture
    b. solution
    c. suspension
    d. colloid

Fill-in-the-Blank

1. When a liquid is dissolved in a liquid, the solute is _____ and the solvent is _____.

2. ΔS_{soln} is usually _____ due to _____.

3. One disadvantage of using molarity as a concentration unit is that it is dependent on _____. This dependence is due to the change in _____ as the _____ changes.

4. Gases become _____ soluble as the temperature increases.

5. Colligative properties are properties that depend on _____ but not on the _____ of the substance.

6. The value of ΔS_{vap} is _____ for a solution than for a pure solvent.

7. Dissolved ions that are surrounded and stabilized by a shell of solvent molecules are said to be _____. If water is the solvent, the term used is _____.

8. The freezing point of solution is lower because _____.

9. Osmosis is _____.

10. Molarity can be used to calculate osmotic pressure because _____.

Matching

Colloids

a. the amount of solute per unit of solvent needed to form a saturated solution

Supersaturated solution

b. the pressure needed to prevent the osmotic flow of solvent through a semipermeable membrane

Solubility

c. mixtures which contain particles with diameters in the rage of 2 to 1000 nm

Semipermeable membrane

d. a process in which a mixture of volatile liquids is boiled and the vapors are condensed

Osmotic pressure

e. a solution which contains a greater than equilibrium amount of solute

Chapter 11 – Solutions and Their Properties

        Fractional distillation   f. membrane that allows water or other small molecules to pass through, but block the passage of large solute molecules or ions

Problems

1. Identify the intermolecular forces present in both the solute and solvent, and predict whether a solution will form between the two.

    a. CCl_4 and Br_2

    b. CH_3OH and Br_2

    c. KCl and NH_3

    d. NH_3 and H_2O

2. The density of a NaOH solution is 1.109 g/mL and has a mass percent of 9.99 %. Calculate the molality and molarity of the solution.

3. A 0.838 M acetic acid, CH_3COOH, solution has a density of 1.0055 g/mL. Calculate the molality and mass percent of this solution.

4. An 8.0 *m* solution of NH_3 has a density equal to 0.950 g/mL. Calculate both the molarity and mass percent of this solution.

5. A 50 mL sample of water was found to have 32.5 ppb of Hg. Calculate the number of grams and molarity of Hg in this sample.

6. The Henry's law constant for methane is 3.34×10^{-3} mol/L·atm. Calculate the solubility of methane at 852 mm Hg.

7. The solubility of N_2 gas at 25°C and 650 mm Hg is 5.85×10^{-4} mol/L. What is the solubility of N_2 at 725 mm Hg?

8. The vapor pressure of water at 35°C is 42.175 mm Hg. Calculate the vapor pressure of an aqueous solution that contains 15.8 g NaCl and 72.1 g of water at this temperature.

9. The vapor pressure of benzene, C_6H_6, at 25°C is 93.4 mm Hg. The vapor pressure of toluene, $C_6H_5CH_3$, is 26.9 mm Hg at 25°C. Calculate the vapor pressure of a solution prepared from 8.0 g of benzene and 15.0 g of toluene.

10. Heptane, C_7H_{16}, has a vapor pressure of 791 mm Hg at 100°C. 250 g of heptane were mixed with 150 g of an unknown substance whose vapor pressure is 352 mm Hg at 100°C. The vapor pressure of the resulting solution is 639 mm Hg. Calculate the molar mass of the unknown substance.

11. Calculate the boiling point and freezing point of a solution prepared by mixing 10.0 g $CaCl_2$ with 90.0 g of water.

12. A solution was prepared by mixing 50.0 g of benzene (freezing point = 5.5°C) and 2.50 g of an unknown substance. The freezing point of the solution was 3.5°C. Calculate the molar mass of the unknown substance.

13. 2.5 g of ethanol (CH_3CH_2OH) were mixed with 10.0 g of water. Calculate the freezing point of the solution and ΔS_{fusion} of the solution. ΔH_{fusion} for water = 6.01 kJ/mol.

14. Calculate the osmotic pressure of a solution at 15°C containing 8.0 g of glucose ($C_6H_{12}O_6$) in 1 L of solution.

15. Calculate the molar mass of 0.250 g of a tripeptide in 1 L of water at 15°C that has an osmotic pressure of 16.4 mm Hg.

Challenge Problem

1.455 g of a metal, M, reacts with hydrochloric acid to form MCl_x and H_2 gas. At 23.0°C and 757 mm Hg, 258.6 mL of H_2 is produced. Dissolution of the MCl_x produced in 25.0 g of water creates a solution with a freezing point of –2.37°C. Determine the identity of M.

Inquiry Based Problem

Your laboratory instructor gave you an unknown liquid that is soluble in cyclohexane. You are also provided with cyclohexane. An example of the experimental set-up – a test tube with a two-hole rubber stopper clamped to a ring stand and inserted in an ice-water bath – is on displayed. A thermometer is inserted through one hole in the stopper and a stirrer is inserted through the other hole. With this experimental set-up, outline the procedure you will use to determine the molar mass of your unknown liquid.

Consider the following:
1. What colligative property are you directed to use from the experimental set-up on display?
2. How should the formula for this colligative property be rearranged to determine the molar mass?
3. From the re-arranged formula for this colligative property, what data to you need to collect?
4. What physical property is involved? How does the phase change curve for this property influence the manner in which you should collect data?

CHAPTER 12

CHEMICAL KINETICS

Chapter Learning Goals

A. Concentration and Reaction Rates
1. Use a table of concentration versus time data to calculate an average rate of reaction over a period of time.
2. From the coefficients of a balanced chemical equation, express the relative rates of consumption of reactants and formation of products.
3. From a table of initial concentrations of reactants and initial rates, determine the order of reaction with respect to each reactant, the overall order of reaction, the rate law, the rate constant, and the initial rate for any other set of initial concentrations.
4. Use integrated first–, second–, and zeroth–order rate laws to find the value of one variable, given values of the other variables.
5. From plots of log concentrations versus time and 1/concentration versus time, determine the order of reaction.
6. Use the expression for half–life of a first– or second–order reaction to determine $t_{1/2}$ from k, or vice versa.
7. From a plot of concentration versus time, estimate the half–life of a first–order reaction.

B. Reaction Mechanisms
1. Given a reaction mechanism and an experimental rate law, identify the reaction intermediates, determine the molecularity of each elementary reaction, and determine if the mechanism is consistent with the experimental rate law.

C. Temperature and Reaction Rates
1. Prepare an Arrhenius plot, and determine the activation energy from the slope of the line.
2. Solve the Arrhenius equation for any variable, given the others.
3. Sketch a potential energy profile, showing the activation energies for the forward and reverse reactions and how they are affected by the addition of a catalyst.

Chapter in Brief

Chemical kinetics is the area of chemistry concerned with reaction rates (how fast a reaction occurs) and the sequence of steps by which reactions occur. In this chapter, you will discover how to describe reaction rates and examine how they are affected by variables such as reactant concentrations and temperatures. You will learn how to determine reaction rates from plots of concentration versus time and how to relate the rates of disappearance or appearance of the individual reactants and products. You will examine experimental rate laws and the order of reaction and learn how to obtain the experimental rate law from initial rate data. You will study integrated rate laws and explore how these rate laws can be used to calculate the concentration of a reactant at any time t, the fraction of reactant that remains at any time, or the time required for the initial concentration of a reactant to drop to any particular value or fraction of its initial concentration. You will also learn how to use plots of concentration versus time to determine the reaction order. You will examine how kinetics allows us to postulate a reaction mechanism, how rate constants depend on temperature, and how collision theory leads to the Arrhenius equation. Finally, you will gain an understanding of the effect of a catalyst on the rate of a reaction and the difference between a homogeneous and a heterogeneous catalyst.

Chapter 12 – Chemical Kinetics

12.1 Reaction Rates
A. Rate of a reaction — how fast the concentration of a reactant or a product changes per unit time.
   1. Rate = $\dfrac{\Delta \text{concentration}}{\Delta \text{time}}$
      a. increase in the concentration of a product per unit time
      b. decrease in the concentration of a reactant per unit time
   2. Units — M/s or mol/(L· s).
      a. allows rate to be independent of the scale of the reaction
      b. use a minus sign in calculating the rate of disappearance of a reactant
      c. indicate reactant or product on which rate is based.
B. Relative rates of product formation and reactant consumption depend on the coefficients in the balanced equation.   ☛ A.2.
   1. Specify the reactant or product when quoting a rate.
C. Rate changes as the reaction proceeds.
   1. Specify the time.
   2. Reaction rates decrease as the reaction mixture runs out of reactants.   ☛ A.1.
D. Plot concentration (y–axis) versus time (x–axis).
   1. Δconcentration and Δtime represent vertical and horizontal sides of a right triangle.
   2. Slope of hypotenuse of triangle is the average rate during that time period.
E. Instantaneous rate at time t — the slope of the tangent to a concentration–time curve at time t.
F. Initial rate — the instantaneous rate at the beginning of a reaction ($t = 0$).

EXAMPLE:
It was found that the rate of formation of N_2 (g) in the following reaction

$$4\,NH_3\,(g) + 3\,O_2\,(g) \rightarrow 2\,N_2\,(g) + 6\,H_2O\,(g)$$

is 0.52 M·s^{-1} at a particular point in time. Determine the rate of disappearance of NH_3.

SOLUTION: Knowing that the rate of appearance is 0.52 $\dfrac{\text{mol } N_2}{L \cdot s}$, we can use the stoichiometry of the balanced equation to determine the rate of disappearance of NH_3.

$$0.52\,\dfrac{\text{mol } N_2}{L \cdot s} \times \dfrac{4 \text{ mol } NH_3}{2 \text{ mol } N_2} = 1.04\,\dfrac{\text{mol } NH_3}{L \cdot s}$$

This rate should be reported as –1.04 M·s^{-1} because we are reporting the rate of *disappearance* of NH_3.

12.2 Rate Laws and Reaction Order
A. Rate law — states the dependence of the reaction rate on concentration.
   1. Equation that tells how the rate depends on the concentration of each reactant.
   2. For the reaction $a\,A + b\,B \rightarrow$ products, the rate law is
      $$\text{Rate} = -\dfrac{\Delta[A]}{\Delta t} = k[A]^m[B]^n$$
      a. k = proportionality constant called the rate constant
B. Reaction order — determined by the values of the exponents.
   1. ***The values of the exponents in the rate law must be determined by experiment; they cannot be deduced from the stoichiometry of the reaction.***

Chapter 12 – Chemical Kinetics

2. Values of m and n indicate the reaction order with respect to A and B.
   a. exponent = 1; first order
   b. exponent = 2; second order
   c. exponent = 3; third order
3. Overall reaction order = $m + n$.
4. Indicates how the change in concentration can affect the rate.
5. **Unrelated to the coefficients in the balanced equation.**
6. Usually small positive integers, but can be negative, zero, or even fractions.
   a. exponent = 1; rate depends linearly on the concentration of the corresponding reactant
   b. exponent = 0; the rate is independent of the concentration of the corresponding reactant
   c. exponent < 1; the rate decreases as the concentration of the corresponding reactant increases

12.3 Experimental Determination of a Rate Law

A. To determine the reaction order (values of the exponents in a rate law) — measure the initial rate of a reaction as a function of different sets of initial concentrations.
   1. Design pairs of experiments to investigate the effect of the initial concentration of a single reactant on the initial rate of change.
      a. hold concentration of other reactants constant
   2. If, by doubling the concentration of a reactant, the rate also doubles, then the reaction is first order with respect to that reactant.
   3. If, by doubling the concentration of a reactant, the rate increases by a factor of $2^2 = 4$, the reaction is second order with respect to that reactant.
   4. If, by doubling the concentration of a reactant, the rate of the reaction increases by a factor of $2^3 = 8$, the reaction is third order with respect to that reactant.
   5. Use initial rates to avoid complications from the reverse reaction.
      a. initial rates measure only the rate of the forward reaction
      b. only reactants and catalysts appear in the rate law
B. Can determine the value of k from the rate law.
   1. Value is characteristic of a reaction.
   2. Depends on temperature.
   3. Does not depend on concentration.
   4. Units depend on the number of concentration terms in the rate law and on the values of the exponents.

EXAMPLE:
The reaction

$$2 \text{ NO } (g) + 2 \text{ H}_2 (g) \rightarrow \text{N}_2 (g) + 2 \text{ H}_2\text{O } (g)$$

is first order in H_2 and second order in NO.

a. Write the rate law.
b. What is the overall order of the reaction?
c. How does the reaction rate change if the concentration of H_2 is doubled and the concentration of NO is held constant?
d. How does the reaction rate change if the concentration of NO is cut in half and the concentration of H_2 is held constant?

SOLUTION:
a. Rate = $k\,[H_2][NO]^2$
b. Overall order = 1 + 2 = 3

c. If the concentration of H_2 is doubled while the concentration of NO is held constant, the rate will double, $(2)^1 = 2$.

d. If the concentration of NO is cut in half while the concentration of H_2 is held constant, the rate will be cut by $\frac{1}{4}$, $(\frac{1}{2})^2 = \frac{1}{4}$

Workbook Problem 12.1

The following data was collected for the reaction

$$2\ NO\ (g) + H_2\ (g) \rightarrow N_2O\ (g) + H_2O\ (g)$$

Exp	$[NO]_I$	$[H_2]_I$	Rate (M·s^{-1})
1	0.15	0.15	8.54×10^{-6}
2	0.30	0.15	3.42×10^{-5}
3	0.45	0.15	7.68×10^{-5}
4	0.15	0.30	1.71×10^{-5}
5	0.15	0.45	2.56×10^{-5}

Determine the rate law from this data. What is the order of the reaction with respect to each reactant? What is the overall order of the reaction? Calculate the value of k.

Strategy: The rate law for the reaction is:

Rate = $k[NO]^m[H_2]^n$.

To find m, compare the change in concentration for NO in experiments 1 and 2 with the change in rate. Verify your answer by comparing the change in concentration for NO in experiments 1 and 3 with the change in rate. ($[H_2]$ is held constant in experiments 1, 2 and 3.) Repeat the process to find n, looking at experiments 1 and 4 and experiments 1 and 5. ([NO] is held constant in experiments 1, 4, and 5.)

Step 1: Determine the order of the reaction with respect to NO.

Step 2: Determine the order of the reaction with respect to H_2.

Step 3: Write the equation for the rate law.

Chapter 12 – Chemical Kinetics

Step 4: Calculate the value of k.

What key concept did you use?

12.4 Integrated Rate Law for a First–Order Reaction
A. Integrated rate law — a concentration–time equation that allows us to calculate the concentration of a reactant at any time t or the fraction of a reactant that remains at any time t.
  1. Can be used to calculate the time required for the initial concentration of a reactant to drop to any particular value or to any particular fraction of its initial concentration.
B. First-order reaction-rate depends on the concentration of a single reactant raised to the first power.
C. For the reaction $a\,A \to$ products, the integrated rate law is

$$\ln \frac{[A]_t}{[A]_0} = -kt$$

  1. Can rearrange the equation to give
  $$\ln[A]_t = -kt + \ln[A]_0$$
  2. Plot of ln [A] versus time gives a straight line if the reaction is first order in A.
  3. $-k =$ (slope)

EXAMPLE:
When sucrose reacts with water, glucose is formed according to the reaction

$$C_{12}H_{22}O_{11} + H_2O \to 2\,C_6H_{12}O_6$$

This reaction follows first order–kinetics with respect to the sucrose. Calculate the value of k if it takes 9.70 hours for the concentration of sucrose to decrease from 0.00375 M to 0.00252 M. Determine the amount of time required for the reaction to go to 80% completion.

SOLUTION: To calculate the value of k, we simply substitute the data given into the first order integrated rate equation.

$$\ln \frac{0.00252}{0.00375} = -k(9.70\text{ h}) \qquad k = 4.10 \times 10^{-2}\text{ h}^{-1}$$

We can now calculate the amount of time required for the reaction to be 80% complete. To determine the concentration at this time, we multiply the initial concentration by 0.80.

$0.80 \times 0.00375 = 0.00300$

This value represents the amount of sucrose that has reacted. The amount of sucrose remaining after the reaction is 80% complete is $0.00375 - 0.00300 = 0.00075$. We now have the initial concentration and the concentration at time t, which can be substituted into the first–order integrated rate law, along with the value of k we calculated in the first part of the problem.

$$\ln \frac{0.00075}{0.00375} = -(4.10 \times 10^{-2}\text{ h}^{-1})t \qquad t = 39.3\text{ h}$$

Chapter 12 – Chemical Kinetics

Half–Life of a First–Order Reaction

A. Half–life ($t_{1/2}$) — the time required for the reactant concentration to drop to one–half of its initial value.

1. $t_{1/2} = \dfrac{0.693}{k}$
2. For first–order reaction, half–life is a constant.
   a. depends only on the rate constant

EXAMPLE:
Determine the half–life for the reaction of sucrose with water.

SOLUTION: We can calculate the half–life by substituting the rate constant determined in the previous example into the equation for the first–order half–life.

$$t_{1/2} = \frac{0.693}{4.10 \times 10^{-2} \text{ h}^{-1}} = 16.9 \text{ h}$$

12.6 Second–Order Reactions

A. Second-order reaction – rate depends on:
   1. The concentration of one reactant raised to the second power
   2. The concentration of two reactants, each raised to the first power.

B. Integrated rate law for a second–order reaction of the type: aA → products
   1. Rate = $k[A]^2$
   2. Rate law:

   $$\frac{1}{[A]_t} = kt + \frac{1}{[A]_0}$$

   3. Plot of $\dfrac{1}{[A]_t}$ versus time gives a straight line if the reaction is second order.
      a. slope of line = k; intercept = $\dfrac{1}{[A]_0}$

C. Half–life for a second–order reaction
   1. Depends on both the rate constant and the initial concentration.

$$t_{1/2} = \frac{1}{k[A]_0}$$

EXAMPLE:
The reaction 2 NOBr (g) → 2 NO (g) + Br_2 (g) is a second order reaction with respect to NOBr. The rate constant for this reaction is $k = 0.810$ $M^{-1} \cdot s^{-1}$ when the reaction is carried out at a temperature of 10°C. If the initial concentration of NOBr = 7.5×10^{-3} M, how much NOBr will be left after a reaction time of 10 minutes? Determine the half–life of this reaction.

SOLUTION: We can solve for the amount of NOBr after 10 minutes by substituting the given data into the integrated rate law for a second–order reaction.

$$\frac{1}{[NOBr]_t} = (0.810\,M^{-1}\cdot s^{-1})\times(600\,s) + \frac{1}{7.5\times10^{-3}\,M}$$

$$\frac{1}{[NOBr]_t} = 6.19\times10^2\,M^{-1} \qquad [NOBr]_t = 1.6\times10^{-3}\,M$$

To determine the half-life for this reaction, we substitute the initial concentration of NOBr and the rate constant for the reaction into the equation for the half-life of a second-order reaction.

$$t_{1/2} = \frac{1}{0.810\,M^{-1}\cdot s^{-1}(7.5\times10^{-3}\,M)} = 160\,s$$

Workbook Problem 12.2

The following data were collected for the general reaction

$$D\,(g) + K\,(s) \rightarrow E\,(l) + 2\,A\,(g)$$

Time (s)	0	100	200	300	400
[D]	0.175	0.151	0.132	0.118	0.106

Determine the order of the reaction and the rate constant. Determine the amount of time required for the reaction to reach 50% completion and the amount of time required for the reaction to reach 95% completion.

Strategy: The order of the reaction can be determined by creating plots of ln [A] versus time and 1/[A] versus time. Once the order has been determined, the correct equation is used to determine the other requested information.

Step 1: Plot ln[A] versus time and 1/[A] versus time to determine the order of the reaction.

Step 2: Use the integrated rate law of the appropriate order to determine k.

Chapter 12 – Chemical Kinetics

Step 3: Use the equation for half–life to determine the time required for 50% completion. (At 50% completion, half of the reactant has reacted.)

Step 4: Determine the concentration of D at 95% completion (5% of D remaining). Substitute the concentrations into the integrated rate law, and solve for *t*.

What key concept did you use?

12.7 Zeroth–Order Reactions

A. Integrated rate law for a zeroth–order reaction for the reaction aA → products.   *A.4.*
 $[A] = -kt + [A]_0$
 1. Rate is independent of the concentration of the reactant.
 2. Plot [A] versus time gives a straight line if zeroth–order.
  a. slope = $-k$; intercept = $[A]_0$
 3. Relatively uncommon; occur under special circumstances.

12.8 Reaction Mechanisms   *C.1.*

A. Reaction mechanism — the sequence of molecular events, or reaction steps, that defines the pathway from reactants to products.
 1. Reaction steps — involve the breaking of chemical bonds and/or the making of new bonds.
 2. Knowing reaction mechanisms allows for better control of known reactions and predictions of new reactions.
B. Elementary step — a single step in a reaction mechanism.
 1. Description of an individual molecular event (collisions of individual molecules).
 2. Describe the reaction mechanism.
 3. Classified on the basis of their molecularity.
  a. molecularity — the number of molecules on the reactant side of the chemical equation
  b. unimolecular reaction — elementary reaction that involves a single reactant molecule
  c. bimolecular reaction — elementary reaction that results from energetic collisions between two reactant molecules
  d. termolecular reaction — involve three atoms or molecules; rare
 4. Elementary steps must sum to give the overall reaction.
 5. Reaction intermediate — a species that is formed in one step of a reaction mechanism and consumed in a subsequent step.
  a. do not appear in the net equation for the overall reaction
  b. Presence is noticed only in the elementary steps.
C. Balanced equation for an overall reaction.
 1. Provides no information about how the reaction occurs.

Chapter 12 – Chemical Kinetics

   2. Describes reaction stoichiometry.

12.9 Rate Laws for Elementary Reactions
A. Rate law for overall reaction — determined by experimentation.
B. Rate law for an elementary reaction — determined from its molecularity.
   1. Elementary reaction is an individual molecular event.
   2. Contains the concentration of each reactant raised to an exponent equal to its coefficient in the chemical equation for the elementary reaction.
   3. *Applies only to elementary reactions, not overall reactions.*
   4. Rate of a unimolecular reaction is first order in the concentration of the reactant molecule.
   5. Overall reaction order for an elementary reaction is equal to its molecularity (see Table 12.5, page 466 in your textbook).

12.10 Rate Laws for Overall Reactions
A. Experimentally observed rate law for an overall reaction depends on the reaction mechanism.
B. Overall reaction occurs in two or more steps.
   1. One step is slower than the others.
      a. rate–determining step — the slowest step in a reaction mechanism
         i. limits the rate at which reactants can be converted to products
      b. overall reaction can occur no faster than the speed of the rate–determining step
C. Two criteria for an acceptable reaction mechanism.
   1. The elementary steps must sum to give the overall reaction.
   2. The mechanism must be consistent with the observed rate law for the overall reaction.
D. Procedure used for establishing a reaction mechanism.
   1. Determine the overall rate law experimentally.
   2. Devise a series of elementary steps.
   3. Predict the rate law based on the reaction mechanism.
   4. If observed and predicted rate laws agree, the proposed mechanism is a plausible pathway for the reaction.
   5. Easy to disprove a mechanism; impossible to "prove" a mechanism.

EXAMPLE:
Given the following reaction mechanism:

$Z_2 \rightarrow 2\,Z$
$2\,Z + 3\,H_2O \rightarrow 2\,ZH_3 + 3/2\,O_2$
$2\,ZH_3 + 4\,O_2 \rightarrow 2\,HZO_3 + 2\,H_2O$

a. Determine the overall reaction.
b. Identify the reaction intermediates, and determine the molecularity of each step.
c. Determine the rate law if the first step is the rate–determining step.

SOLUTION:
To determine the overall reaction, we simply cross out the species common on both the reactant and product sides of the reactions.

$Z_2 \rightarrow \cancel{2\,Z}$
     $1\,H_2O$
$\cancel{2\,Z} + \cancel{3\,H_2O} \rightarrow \cancel{2\,ZH_3} + \cancel{3/2\,O_2}$
$\cancel{2\,ZH_3} + \cancel{8/2\,O_2} \rightarrow 2\,HZO_3 + \cancel{2\,H_2O}$
          $5/2\,O_2$

Overall reaction: $Z_2 + H_2O + 5/2\, O_2 \rightarrow 2\, HZO_3$

The reaction intermediates are those species in the reaction mechanism that are not included in the overall reaction: Z, ZH_3.

The molecularity of the first step = 1 (unimolecular).
The molecularity of the second step = 5.
The molecularity of the third step = 6.

Rate law = $k[Z_2]$

12.11 Reaction Rates and Temperature: The Arrhenius Equation
A. Reaction rates tend to double when the temperature is increased by 10°C.
B. Collision theory model — a bimolecular reaction occurs when two properly oriented reactant molecules come together in a sufficiently energetic collision.
C. For the reaction: $A + BC \rightarrow AB + C$.
 1. For a single–step reaction, a new bond, A–B, develops at the same time as the old bond, B–C, breaks.
 2. Nuclei pass through a configuration in which all three atoms are weakly linked together.
  a. $A + B–C \rightarrow A\text{- -}B\text{- -}C \rightarrow A–B + C$
 3. Need energy to overcome repulsions.
  a. comes from kinetic energy of the colliding particles
  b. stored as potential energy in A- -B- -C.
   i. A- -B- -C has more potential energy than either the reactants or products
 4. Potential energy barrier that must be surmounted before reactants can be converted to products.
  a. potential energy profile — plot of potential energy *versus* reaction progress

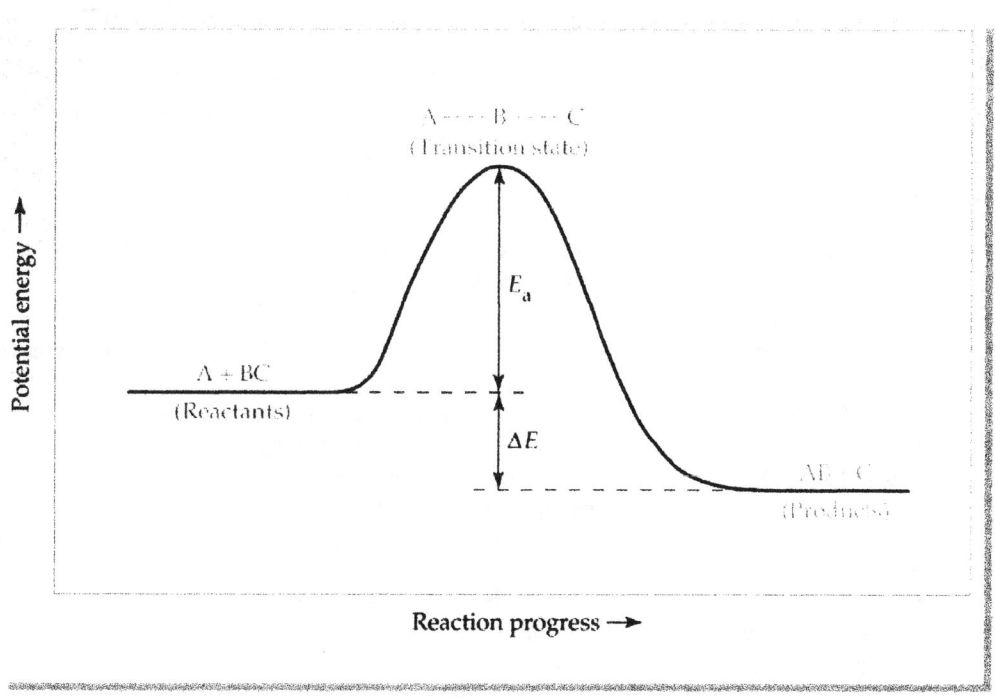

b. activation energy, E_a, — the height of the barrier
c. transition state (or activated complex) —configuration of atoms at the maximum in the potential energy profile
5. All the energy needed to climb the potential energy barrier must come from the kinetic energy of the colliding molecules.

D. Comparison of collision rates and reaction rates leads to experimental evidence for the idea of an activation energy barrier.
1. Only a small fraction of collisions lead to reaction.
a. Very few collisions occur with a kinetic energy as large as the activation energy.

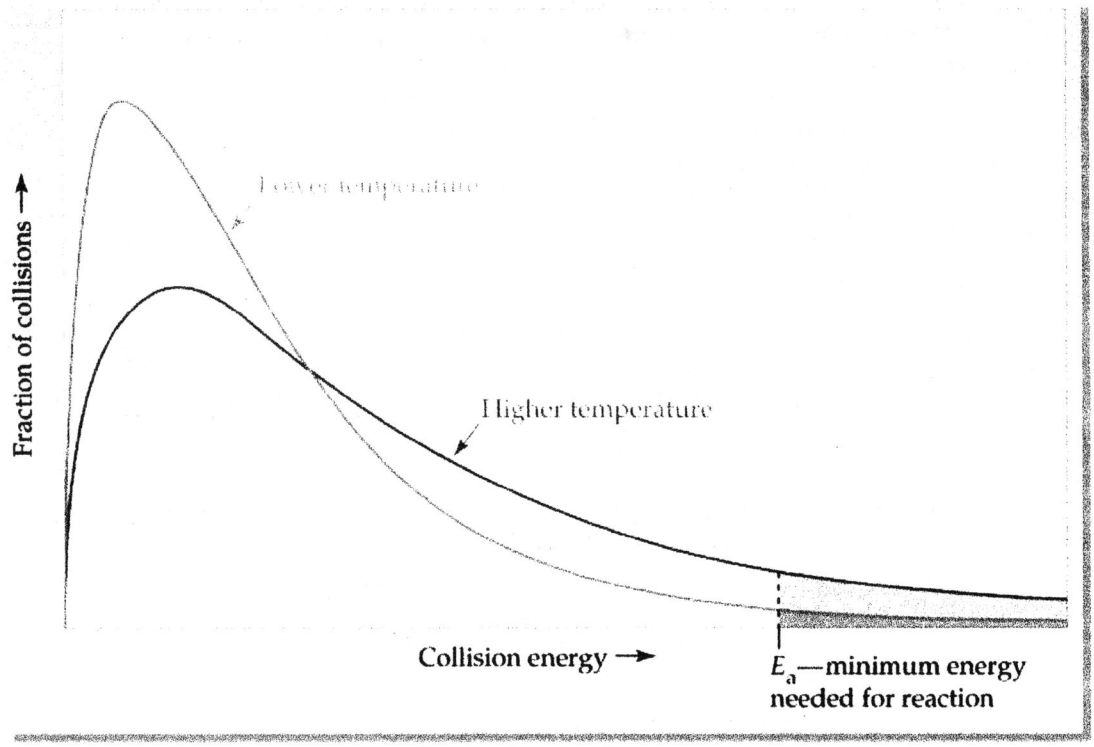

b. Area under the curve to the right of E_a represents the fraction of the collision with an energy equal to or greater than the activation energy.
i. fraction, $f = e^{-E_a/RT}$
2. Distribution of collision energies broadens and shifts to higher energies with an increase in temperature.
a. exponential increase in the fraction of collisions that lead to products
3. Accounts for exponential dependence of reaction rates on temperature.
4. Explains why reaction rates are so much lower than collision rates.

E. Orientation requirement also reduces the fraction of collisions that lead to products.
1. Reactants won't react unless the orientation of the reaction partners is correct for formation of the transition state.
2. Steric factor, p, — fraction of collisions having proper orientation.

F. For a bimolecular collision between A and B.
1. Collision rate = $Z[A][B]$.
a. Z is a constant related to collision frequency

Chapter 12 – Chemical Kinetics

   2. Reaction rate < collision rate by a factor of $p \times f$.
      a. Reaction rate = $pfZ[A][B]$
   3. Reaction rate = $k[A][B]$.
      a. $k = pfZ = pZe^{-E_a/RT}$
G. Arrhenius equation.
   1. $k = Ae^{-E_a/RT}$
   2. A = frequency factor = pZ

12.12 Using the Arrhenius Equation
A. Can determine the activation energy if values of the rate constant are known at different temperatures.
B. Rearrange the equation to get

$$\ln k = \ln A - \frac{E_a}{RT}$$

   1. Plot of $\ln k$ versus $1/T$ gives a straight line.
      a. slope = $-E_a/R$

C. Can estimate the activation energy from rate constants at just two temperatures, using another form of the equation.

$$\ln\left(\frac{k_2}{k_1}\right) = \left(\frac{-E_a}{R}\right)\left(\frac{1}{T_2} - \frac{1}{T_1}\right)$$

EXAMPLE:
The activation energy for the reaction $ClO_2F\ (g) \rightarrow ClOF\ (g) + O\ (g)$ is 186 kJ/mol. If the value of k is $6.76 \times 10^{-4}\ s^{-1}$ at 322°C, what is the value of k at 50°C?

SOLUTION: To solve this problem, we simply substitute the given information into the Arrhenius equation.

$$\ln\left(\frac{k_2}{6.76 \times 10^{-4}\ s^{-1}}\right) = \left(\frac{-1.86 \times 10^5\ \text{J/mol}}{8.314\ \text{J/mol}\cdot\text{K}}\right)\left(\frac{1}{323\ \text{K}} - \frac{1}{595\ \text{K}}\right)$$

$$\ln\left(\frac{k_2}{6.76 \times 10^{-4}\ s^{-1}}\right) = -31.7$$

To determine the value of k_2, we need to take the antilog of both sides of the equation.

$$\frac{k_2}{6.76 \times 10^{-4}\ s^{-1}} = 1 \times 10^{-14} \qquad k_2 = 1 \times 10^{-17}\ s^{-1}$$

Chapter 12 – Chemical Kinetics

12.13 Catalysis
A. Catalyst — a substance that increases the rate of a reaction without being consumed in the reaction.
   1. Important in chemical industry and in living organisms.
   2. Chemical industry — favor formation of specific products and lower reaction temperatures.
   3. Living organisms — enzymes (catalysts) facilitate specific reactions of crucial biological importance.
B. Accelerates the rate of a reaction by making a new and more efficient pathway available for the conversion of reactants to products.   🗝 C.3.
   1. Speeds up reaction in two ways.
      a. increases the frequency factor A
      b. decreases the activation energy
         i. Rate constant is more sensitive to the E_a and a catalyst usually functions by lowering the activation energy.

12.14 Homogeneous and Heterogeneous Catalysts
A. Homogeneous catalyst — one that exists in the same phase as the reactants.
B. Heterogeneous catalyst — exists in a different phase from the reactants.
   1. Mechanism is complex and not well understood.
   2. Important steps.
      a. adsorption of reactants onto the surface of the catalyst
      b. conversion of reactants to products on the surface
      c. desorption of products from the surface
   3. Adsorption steps.
      a. chemical bonding of reactants to the highly reactive metal atoms on the surface
      b. breaking or weakening of bonds in the reactants
   4. Industrial chemical processes use mostly heterogeneous catalysts due to the ease of separation of the catalyst from the reaction products.
   5. Used in automobile catalytic converters.

 Putting It Together

The reaction mechanism for the reaction of triphenylphosphine, $P(C_6H_5)_3$, with $Ni(CO)_4$ is

$Ni(CO)_4 \rightarrow Ni(CO)_3 + CO$

$Ni(CO)_3 + P(C_6H_5)_3 \rightarrow Ni(CO)_3[P(C_6H_5)_3]$

a. What is the molecularity of each step?

b. Doubling the concentration of $Ni(CO)_4$ doubles the rate. However, doubling the concentration of $P(C_6H_5)_3$ does not effect the rate. What is the rate law for the reaction? Which step is the slow step in the reaction mechanism?

c. $k = 9.9 \times 10^{-3}\ s^{-1}$ at 20°C. If the initial concentration of $Ni(CO)_4$ is 0.25 M, how long will it take the reaction to be 80% complete?

d. $Ni(CO)_4$ is formed by the reaction of nickel metal with carbon monoxide. Calculate the $\Delta H°$ if $\Delta H°_f$ [$Ni(CO)_4$] = –602.91 kJ/mol.

e. How many grams of Ni(CO)$_4$ will be formed if you start with 0.148 g Ni and 730 mL of CO at 0.95 at and 28°C?

 **Self–Test**

This section is intended to test your knowledge of the material covered in this chapter. Think through these problems, and make certain you understand what is going on. Ask yourself if your answer makes sense. Many of these questions are linked to the chapter learning goals. Therefore, successful completion of these problems indicates you have mastered the learning goals for this chapter. You will receive the greatest benefit from this section if you use it as a mock exam. You will then discover which topics you have mastered and which topics you need to study in more detail.

True/False

1. The relative rates of product formation and reactant consumption depend on the coefficients in the balanced equation.

2. The rate of a reaction stays the same as the reaction proceeds.

3. If the rate law for the reaction a A + b B → products is rate = $k[A]^m[B]^n$, the values of m and n are equal to the coefficients a and b, respectively, in the chemical equation.

4. For the second–order reaction, A → products, doubling the concentration of A will cause the rate to also double.

5. The units for the rate constant, k, depend on the number of concentration terms in the rate law and on the values of the exponents.

6. For a first–order reaction, the half–life depends only on the rate constant.

7. The molecularity of a reaction is the number of molecules that participate in the reaction.

8. The balanced equation for an overall reaction describes the reaction mechanism.

9. The experimentally observed rate law for an overall reaction can be determined by the slowest step in the reaction mechanism.

10. Industrial chemical processes use mostly homogeneous catalysts due to the ease of recovery.

Multiple Choice

1. The overall reaction order for the rate law, rate = $k[A]^2[B]^3$ is
   a. two
   b. three
   c. five
   d. six

Chapter 12 – Chemical Kinetics

2. A reaction has the rate law, rate = $k[A]^3$. If the initial rate for this reaction is 0.15 mol/L·sec, and the concentration of A is doubled, the new initial rate for this reaction will be
   a. 0.45 mol/L·sec
   b. 1.20 mol/L·sec
   c. 0.30 mol/L·sec
   d. 1.35 mol/L·sec

3. If a plot of $\dfrac{1}{[A]_t}$ versus time gives a straight line for a particular reaction, that reaction is
   a. first order
   b. 1/2 order
   c. second order
   d. zeroth order

4. The reaction order for the elementary reaction step, 2 A + B → products, is
   a. second order
   b. third order
   c. cannot be determined from the balanced equation
   d. first order

5. For every collision that occurs in a reaction
   a. the products are formed
   b. the particles have enough kinetic energy to overcome the potential energy barrier
   c. the fraction of molecules with enough kinetic energy to overcome the potential energy barrier is
      $f = e^{-E_a/RT}$
   d. none of the above

6. The exponents in a rate law indicate the
   a. dependence of the rate on concentration
   b. sum of the coefficients in a balanced equation
   c. time required for a certain amount of reactant to disappear
   d. none of the above

7. The half–life of a first–order reaction with $k = 1.65 \times 10^{-3}$ s^{-1}
   a. cannot be determined without knowing the initial concentration of the reactant
   b. equals 1.14×10^{-3} s
   c. equals 4.06×10^{-2} s
   d. 4.2×10^2 s

8. The half–life of a second–order reaction depends on
   a. both the rate constant and the initial concentration
   b. only the rate constant
   c. only the initial concentration
   d. none of the above

9. Reaction rates tend to double when the temperature is
   a. doubled
   b. increased by 100°C
   c. multiplied by 2
   d. increased by 10°C

10. The activated complex has
   a. more energy than either the reactants or products
   b. is the configuration of the atoms at the minimum of the potential energy barrier
   c. less energy than the reactants but more energy than the products
   d. the same configuration as an intermediate in an elementary step in a reaction mechanism

Matching

Kinetics	a. a single step in a reaction mechanism
Integrated rate law	b. a plot of potential energy versus reaction progress
Half–life	c. configuration of atoms at the maximum in the potential energy profile
Reaction mechanism	d. a substance that increases the rate of a reaction without being consumed in the reaction
Elementary step	e. the area of chemistry concerned with reaction rates and the sequence of steps by which reactions occur
Molecularity	f. the time required for the reactant concentration to drop to one–half of its initial value
Bimolecular reaction	g. the sequence of molecular events, or reaction steps, that defines the pathway from reactants to products
Potential energy profile	h. a concentration–time equation that allows us to calculate the concentration of a reactant at any time t or the fraction of a reactant that remains at any time t
Transition state	i. one that exists in the same phase as the reactants
Catalyst	j. the number of molecules on the reactant side of the chemical equation for an elementary step
Homogeneous catalyst	k. elementary reaction that results from energetic collisions between two reactant molecules

Fill–in–the–Blank

1. The relative rates of product formation and reactant consumption depend on _____ _____.

Chapter 12 – Chemical Kinetics

2. The reaction order indicates _____
   _____.
3. The values of the exponents in a rate law are determined _____.
4. A reaction is third order if, by _____ the concentration of a reactant, the rate
   increases by a factor of _____.
5. If a plot of log [A] versus time gives a straight line, the reaction is _____ with
   respect to A.
6. An elementary step in a reaction mechanism gives a description of an _____
   _____.
7. The rate law for an elementary reaction is determined from _____.
8. If an overall reaction occurs in two or more steps, the rate–determining step is the _____
   _____.
9. Reaction rates tend to double when the temperature is increased by _____.
10. Reactants won't react unless the _____ of the reaction partners is correct for
    formation of the _____.

Problems

1. The following data for the reaction A + B → C was collected. Calculate the average rate for the formation of C between 45 and 75 seconds.

[C] (mol/L)	Time (s)
0	0
0.005	15
0.010	30
0.015	45
0.017	60
0.019	75
0.020	90
0.021	105
0.022	120

2. The rate of disappearance of PH_3 for the reaction

   $4\ PH_3(g) \rightarrow P_4(g)\ +\ 6\ H_2\ (g)$

   is 1.47×10^{-3} mol/L·sec. What is the rate of appearance of H_2?

3. The following data were collected for the reaction

   $2\ NO\ (g)\ +\ Cl_2\ (g) \rightarrow 2\ NOCl\ (g)$

216

Chapter 12 – Chemical Kinetics

[NO] (mol/L)	[Cl$_2$] (mol/L)	Initial Rate mol/(L·sec)
0.025	0.025	39.1
0.050	0.025	156.3
0.075	0.025	351.6
0.025	0.050	78.1
0.025	0.075	117.2

Determine the rate law from this data. What is the order of the reaction with respect to each reactant? What is the overall order of the reaction? Calculate the value of k.

4. The following data were collected for the reaction

$$3\,A + 2\,B \rightarrow 2\,D$$

[A] (mol/L)	[B] (mol/L)	Initial Rate mol/(L·sec)
0.150	0.150	1.56×10^{-2}
0.300	0.150	0.125
0.450	0.150	0.421
0.150	0.300	1.56×10^{-2}
0.150	0.450	1.56×10^{-2}

Determine the rate law from this data. What is the order of the reaction with respect to each reactant? What is the overall order of the reaction? Calculate the value of k. If the initial concentration of A = 0.639 M and the initial concentration of B = 0.75 M, what would the initial rate be?

5. The conversion of cyclopropane to propene obeys first order kinetics. If the initial concentration of cyclopropane is 0.500 M, how long will it take for the reaction to reach 35% completion? 70% completion? ($k = 1.16 \times 10^{-6}$ s^{-1})

6. The following data were collected for the first order reaction

$$4\,PH_3\,(g) \rightarrow P_4\,(g) + 6\,H_2\,(g)$$

[PH$_3$] (mol/L)	t (sec)
0.450	0.0
0.248	30.0
0.137	60.0
0.0757	90.0
0.0418	120.0

Determine the half–life for this reaction from the plotted data. Calculate the value of k. From the calculated value of k, calculate the half–life of this reaction. Do your two answers for half–life agree?

7. The rate law for the decomposition of O$_3$ to O$_2$ is second order in ozone. The rate constant for this reaction equals 1.40×10^{-2} L/(mol·sec). If the initial concentration of ozone is 2.75 M, how much O$_3$ will be present after 24 hours?

217

Chapter 12 – Chemical Kinetics

8. The following data were collected for the reaction $2\,HI\,(g) \rightarrow H_2\,(g) + I_2\,(g)$

[HI] (mol/L)	Time (min)
2.50	0
1.45	3
1.02	6
0.788	9
0.641	12
0.541	15
0.468	18
0.412	21
0.368	24
0.332	27
0.303	30

Determine the order of this reaction. Calculate the value of k and the half life at the beginning of the reaction.

9. The proposed reaction mechanism for the overall reaction

$$2\,NO_2\,(g) + F_2\,(g) \rightarrow 2\,NO_2F\,(g)$$

is

$$NO_2\,(g) + F_2\,(g) \rightarrow NO_2F\,(g) + F\,(g)$$

$$F\,(g) + NO_2\,(g) \rightarrow NO_2F\,(g)$$

What is the molecularity of each step? If the first step is the slow step in the mechanism, what would the rate law be?

10. The reaction mechanism for the decomposition of hydrogen peroxide is

$$H_2O_2 \rightarrow 2\,OH$$

$$H_2O_2 + OH \rightarrow H_2O + HO_2$$

$$HO_2 + OH \rightarrow H_2O + O_2$$

What is the overall reaction? What are the reaction intermediates? What is the molecularity of each step? Which step is the slowest step if the experimental rate law is rate = $k[H_2O_2]$.

11. The following data were collected for the reaction

$$A + B \rightarrow Product$$

$k\,(s^{-1})$	T (°C)
5.39×10^{-4}	100
6.53×10^{-4}	110
7.82×10^{-4}	120
9.29×10^{-4}	130

k (s^{-1})	T (°C)
1.09×10^{-3}	140
1.28×10^{-3}	150

Prepare an Arrhenius plot and determine th4e activation energy from the slope of the line.

12. If the rate constant for a certain reaction is $k = 1.67 \times 10^{-4}$ L/mol·sec at a temperature of 68°C and the activation energy for the reaction is 111.8 kJ/mol, what is the value of A?

13. The activation energy is $E_a = 100.25$ kJ/mol for the reaction

    $2 \text{ NOCl} \rightarrow 2 \text{ NO} + \text{Cl}_2$

    If $k = 9.3 \times 10^{-6}$ s^{-1} at 77°C, at what temperature will $k = 2.8 \times 10^{-3}$ s^{-1}?

14. Draw a potential energy profile for the endothermic reaction A + B → products. Indicate the change in activation energy that will occur when a catalyst is added to this reaction.

Challenge Problem

For the reaction

$C_2H_5I + OH^- \rightarrow C_2H_5OH + I^-$

$E_a = 86.8$ kJ/mol and $A = 2.10 \times 10^{11}$ M^{-1}s^{-1} at 35°C. If the concentration of C_2H_5I is doubled while [OH$^-$] remains constant, the rate doubles. If the [OH$^-$] is doubled while [C_2H_5I] remains constant, the rate doubles. If a 250 mL solution of 0.475 g KOH in ethanol is mixed with a 250 mL solution of 1.378 g C_2H_5I in ethanol, what is the initial rate at 35°C?

Inquiry Based Problem

You arrive in lab to learn that you will be investigating the rate law for the reaction between iodite and sulfite ions. Your laboratory instructor has written the steps for this reaction on the board:

$IO_3^- (aq) + 3 SO_3^{2-} (aq) \rightarrow I^- (aq) + 3 SO_4^{2-} (aq)$

$5 I^- (aq) + 6 H^+ (aq) + IO_3^- (aq) \rightarrow 3 H_2O + 3 I_2 (s)$

$3 I_2 (s) + 3 SO_3^{2-} (aq) + 3 H_2O \rightarrow 6 I^- (aq) + 3 SO_4^{2-} (aq) + 6 H^+ (aq)$

$2 IO_3^- (aq) + 6 SO_3^{2-} (aq) \rightarrow 2 I^- (aq) + 6 SO_4^{2-} (aq)$

S/he informs you that I_2, which is produced in the second step and consumed in the third step, reacts with starch to produce a deep blue color due to the formation of the I_2•starch complex. You are provided a 0.10 M solution of HIO_3, a 0.05 M solution of H_2SO_3, a solution of starch, and a

Chapter 12 – Chemical Kinetics

graduated pipet. Armed with this information, you must develop an experimental procedure to determine the rate law for the overall reaction, and the activation energy for the reaction.

Given this information, design an experiment in which you determine the rate constant and activation energy for the overall reaction for the overall reaction.

Consider the following:
1. Determine the form of the rate law based on the overall reaction.
2. How can you determine the impact of each individual reactant on the overall rate?
3. How are you going to monitor the progress of the reaction? What does this tell you about the consumption of reactants?
4. How are you going to define rate? What unit will you use?
5. How are you going to determine the activation energy for this reaction once you determine the rate law?

CHAPTER 13

CHEMICAL EQUILIBRIUM

Chapter Learning Goals

A. *Extent of Chemical Reactions*
1. Given any balanced chemical equation representing a homogenous or heterogeneous equilibrium, write the equilibrium equation.
2. From the equilibrium concentrations of products and reactants, calculate the equilibrium constant K_c.
3. Given the equilibrium partial pressures of reactants and products, calculate the equilibrium constant K_p.
4. From a value of K_c and a balanced equation, calculate K_p. From a value of K_p and a balanced equation, calculate K_c.

B. *Equilibrium Mixture Composition*
1. From the value of K_c or K_p, determine whether mainly products or mainly reactants exist at equilibrium.
2. For a given mixture of reactants and products, determine whether a system is at equilibrium. If it is not, determine the direction in which the reaction must go to achieve equilibrium.
3. Given K_c and initial concentrations of reactants and/or products, calculate the final concentrations of reactants and/or products.

C. *Systems Under Stress*
1. Determine the reaction direction when a system at equilibrium reacts to a stress applied to the system, including changes in concentrations, pressure and volume, or temperature.
2. Describe the effect of adding a catalyst to a system at equilibrium.

D. *Chemical Kinetics and Chemical Equilibrium*
1. Describe the relationship between the equilibrium constant and the ratio of the rate constants for the forward and reverse reactions. Solve problems involving this relationship.

Chapter in Brief

In this chapter, you begin the study of chemical equilibrium, the state reached when the concentration of reactants and products remain constant over time. The concepts found in this chapter will be applied to various chemical systems in Chapters 15 and 16. A state of chemical equilibrium is achieved when, with time, the concentrations of both the reactants and products in a reversible reaction level off at constant equilibrium values. Your study of chemical equilibria will explore several aspects of equilibrium mixtures. You will study the relationship between the concentration of the reactants and products, as well as how to determine the equilibrium concentrations knowing the initial concentrations. You will also study which factors can be employed to change the composition of an equilibrium mixture. Finally, you will examine the link between chemical equilibrium and kinetics.

Chapter 13 – Chemical Equilibrium

13.1 The Equilibrium State
A. Reversible reaction — reaction in which the product molecules recombine to form reactant molecules.
   1. Reactants — substances on the left side of the equation.
   2. Products — substances on the right side of the equation.
   3. Use a double arrow to indicate a reaction that is proceeds in both the forward and reverse direction.
      a. Reactants $\rightleftarrows$ Products
B. With time, the concentration of reactant decreases and the concentration of product increases until both concentrations level off at constant equilibrium values.
   1. Rates of the forward and reverse reactions are equal.
   2. State of chemical equilibrium.
   3. Equilibrium mixture — a mixture of reactants and products in the equilibrium state.
   4. Dynamic state — forward and reverse reactions continue at equal rates.
      a. no net conversion of reactants to products

13.2 The Equilibrium Constant K_c
A. Equilibrium mixture obeys an equilibrium equation (or law of mass action).
   1. For the reaction: $a\,A + b\,B \rightleftarrows c\,C + d\,D$.
      a. $K_c = \dfrac{[C]^c [D]^d}{[A]^a [B]^b}$
B. K_c = equilibrium constant
   1. K_c for a reaction at a particular temperature always has the same value.
   2. Temperature dependent – state T when giving a value for K_c.
   3. Units depend on the particular equilibrium expression.
      a. Units usually omitted when quoting values of equilibrium constants.
C. Equilibrium constant expression — expression on right side of equilibrium equation
D. For the reverse reaction.
   1. equilibrium expression = reciprocal of the original expression
   2. equilibrium constant $K_c' = \dfrac{1}{K_c}$

EXAMPLE:
For the reaction $CO\ (g) + 2\,H_2\ (g) \rightleftarrows CH_3OH\ (g)$, the equilibrium concentrations are $[CO]_e = 2.58$ M; $[H_2]_e = 0.280$ M; $[CH_3OH]_e = 2.93$ M at a temperature of 210°C. Calculate the equilibrium constant, K_c, for this reaction.

SOLUTION: Before we calculate K_c for this reaction, we need to write an equilibrium equation for the balanced chemical reaction.

$$K_c = \frac{[CH_3OH]}{[CO][H_2]^2}$$

Once we have the equilibrium equation, we can substitute the equilibrium concentrations and solve for K_c.

$$K_c = \frac{(2.93)}{(2.58)(0.280)^2} = 14.5$$

13.3 The Equilibrium Constant K_p

A. For gas–phase reactions, use partial pressures rather than molar concentrations in equilibrium equations.

1. K_p — equilibrium constant is defined using partial pressures.

B. Relationship between K_c and K_p.

1. $K_p = K_c(RT)^{\Delta n}$.

   a. Δn = the sum of the coefficients of the gaseous products minus the sum of the coefficients of the gaseous reactants

EXAMPLE:
Calculate K_p for the reaction in the previous example.

SOLUTION: We first need to determine Δn for the reaction. The product side of the equation has 1 mol of CH_3OH while the reactant side of the equation has 1 mol CO and 2 mol H_2. Therefore, $\Delta n = -2$. We can now calculate the value of K_p from the value of Δn and the value of K_c determined in the previous example.

$$K_p = (14.5)\left[\left(0.0821\frac{L \cdot atm}{mol \cdot K}\right)(483K)\right]^{-2} = 9.22 \times 10^{-3}$$

Workbook Problem 13.1

Determine the value of K_c and K_p for the reaction

$$PCl_5\,(g) \rightleftarrows PCl_3\,(g) + Cl_2\,(g)$$

if the concentrations are $[PCl_5]_e = 2.75$ M and $[PCl_3]_e = [Cl]_e = 1.26$ M at 500 K.

Strategy: Write the equilibrium equation for K_c.

Step 1: Solve for K_c using the molar concentrations given.

Step 2: Determine Δn.

Step 3: Solve for K_p, using the equation that expresses the relationship between K_c and K_p.

What key concept did you use?

13.4 Heterogeneous Equilibria
A. Homogeneous equilibria — reactants and products are in a single phase.
B. Heterogeneous equilibria — reactants and products are present in more than one phase.
  1. Solids — molar concentrations are constants.
     a. can be calculated from the densities and molar masses
     b. independent of its amount

  2. Concentrations of pure solids or pure liquids are not included when writing the equilibrium equation for any heterogeneous equilibrium.   ⌨ *A.1.*
     a. The concentrations are constant and included in value of equilibrium constant.

13.5 Using the Equilibrium Constant

A. Judging the extent of reaction (how much reactant is converted to product).   ⌨ *B.1.*
  1. Large value of K_c ($> 10^3$)
     a. Products predominate over reactants.
     b. Reaction proceeds almost to completion.
     c. Position of reaction is to the right.
  2. Small value of K_c ($< 10^{-3}$)
     a. Reactants predominate over products.
     b. Reaction hardly proceeds.
     c. Position of the reaction is to the left.
  3. $10^{-3} < K_c < 10^3$
     a. appreciable amounts of both reactants and products present

B. Predicting the direction of reaction.   ⌨ *B.2.*
  1. Reaction quotient, Q_c, same as the equilibrium constant expression except that the concentrations are not necessarily equilibrium values.
  2. Predict direction of reaction by comparing the value of Q_c to K_c.
     a. $Q_c < K_c$; reaction goes from left to right
     b. $Q_c > K_c$; reaction goes from right to left
     c. $Q_c = K_c$; reaction is at equilibrium

EXAMPLE:
$K_c = 4.18 \times 10^{-9}$ at 425°C for the following reaction:

$$2\ HBr\ (g) \rightleftarrows H_2\ (g) + Br_2\ (g)$$

What is the position of equilibrium? If the concentrations of all species present are [HBr] = 0.75 M; [H_2] = [Br_2] = 2.5×10^{-4} M is the reaction mixture at equilibrium? In which direction will the reaction proceed?

SOLUTION: We can determine the position of equilibrium by looking at the value of K_c. Since $K_c < 10^{-3}$, we know that reactants predominate over products and that the position of equilibrium is to the left. To determine if the reaction mixture is at equilibrium, we need to calculate Q_c and compare that value to K_c.

$$Q_c = \frac{[H_2][Br_2]}{[HBr]^2} = \frac{(2.5 \times 10^{-4})(2.5 \times 10^{-4})}{(0.75)^2} = 1.1 \times 10^{-7}$$

Since $Q_c > K_c$, the reaction is not at equilibrium and will proceed from right to left.

C. Calculating equilibrium concentrations.

B.3.

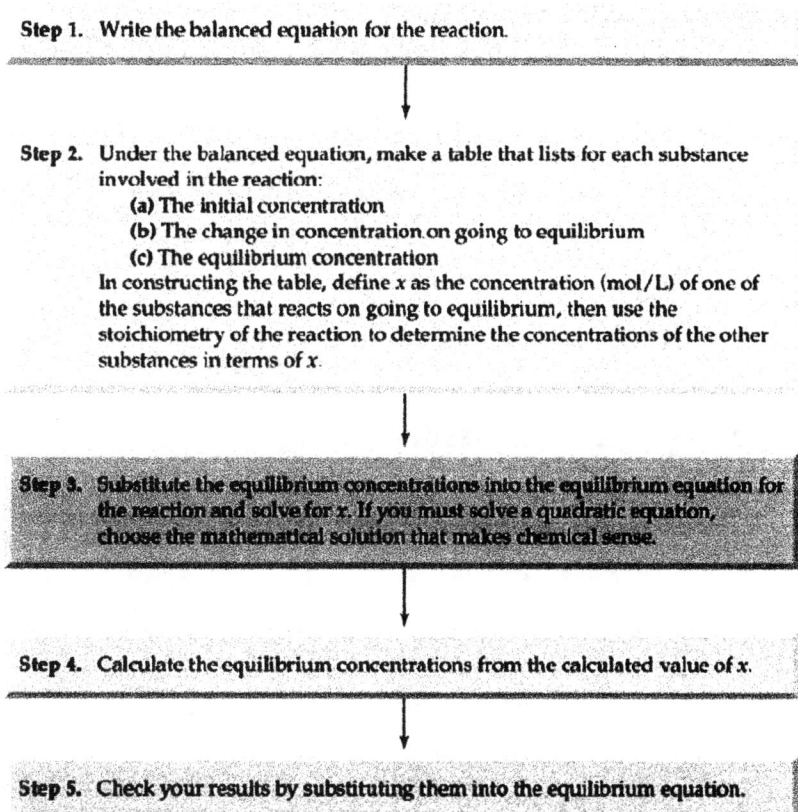

EXAMPLE:

The reaction $PCl_5 (g) \rightleftarrows PCl_3 (g) + Cl_2 (g)$ has $K_c = 85.0$ at a temperature of 760°C. Calculate the equilibrium concentrations of PCl_5, PCl_3, and Cl_2 if the initial concentration of PCl_5 is 5.00 M. Assume a volume of 1.00 L

To solve this problem, follow the steps which are outlined above (see Figure 13.6, page 543 in your textbook).

1. The balanced equation is given.

2. The initial concentration of PCl_5 is 5.00 M. Let x be the concentration of PCl_5 that reacts on going to the equilibrium state. Since the mole to mole ratios are 1:1:1, if x amount of

Chapter 13 – Chemical Equilibrium

PCl$_5$ reacts, then x mol/L of PCl$_3$ and Cl$_2$ will be present at equilibrium. This can be summarized in the following table.

	PCl$_5$ (g) ⇌	PCl$_3$ (g) +	Cl$_2$ (g)
Initial Concentration (M)	5.00	0	0
Change (M)	$-x$	$+x$	$+x$
Equilibrium Concentration (M)	$5.00 - x$	x	x

3. The equilibrium expression for the reaction is:

$$K_c = \frac{[PCl_3][Cl_2]}{[PCl_5]}$$

The equilibrium concentrations from the above table can be substituted into this expression, and we can solve for x.

$$85.0 = \frac{(x)(x)}{(5.00 - x)}$$

We need to rearrange the equation and use the quadratic equation to solve for x.

$$425 - 85.0x = x^2 \quad x^2 + 85.0x - 425 = 0$$

$$x = \frac{-85.0 \pm \sqrt{(7.23 \times 10^3) + (1.7 \times 10^3)}}{2}$$

$x = -89.7$ or 4.74.

The mathematical solution that makes chemical sense is 4.74.

4. The equilibrium concentrations can now be calculated.
   $[PCl_5] = 5.00 - 4.74 = 0.26$ M $[PCl_3] = [Cl_2] = 4.74$ M

Workbook Problem 13.2
The value of K_c for the reaction

C (s) + H$_2$O (g) ⇌ CO (g) + H$_2$ (g)

is 3.0×10^{-2}. Determine the equilibrium concentration if the initial concentration of water is 8.75 M.

Strategy: Follow the steps outlined on previous page (Figure 13.6, page 512 in your textbook).

Step 1: Write the balanced equation for the reaction.

Chapter 13 – Chemical Equilibrium

Step 2: Make a table listing the initial concentration, the change in concentration, and the equilibrium concentration. (Let x = the amount of substance that reacts.)

Step 3: From the balanced equation, write the equilibrium equation. Substitute the equilibrium concentrations in the equilibrium equation, and solve for x. If necessary, use the quadratic equation.

What key concept did you use?

13.6 Factors That Alter the Composition of an Equilibrium Mixture: Le Châtelier's Principle
A. Factors that alter the composition of an equilibrium mixture:
1. Concentration of reactants or products.
2. Pressure and volume.
3. Temperature.

B. Le Châtelier's Principle: If a stress is applied to a reaction mixture at equilibrium, the net reaction occurs in the direction that relieves the stress.
1. Predicts changes in the composition of an equilibrium mixture if one of the above changes occurs.
2. Stress – change in concentration, pressure, volume an temperature.
3. Reaction occurs to change composition of mixture until a new equilibrium is reached.
4. Reaction will move in a direction that relieves the stress.

13.6 Altering an Equilibrium Mixture: Changes in Concentration
A. Disturb an equilibrium by adding or removing a reactant or product.
1. Stress of an added reactant or product is relieved by reaction in the direction that consumes the added substance.
   a. add reactant — reaction shifts right (toward product)
   b. add product — reaction shifts left (toward reactant)
2. Stress of removing reactant or product is relieved by reaction in the direction that replenishes the removed substance.
   a. remove reactant — reaction shifts left

Chapter 13 – Chemical Equilibrium

    b. remove product — reaction shifts right
  B. Changes occur due to change in value of Q_c.
    1. Add reactant — denominator in Q_c expression becomes larger.
      a. $Q_c < K_c$
      b. To return to equilibrium, Q_c must increase.
        i. Numerator of Q_c expression must increase and the denominator must decrease.
        ii. Implies net conversion of reactants to products.
        iii. Reaction shifts right.
    2. Remove reactant — denominator in Q_c expression becomes smaller.
      a. $Q_c > K_c$
      b. To return to equilibrium, Q_c must decrease.
        i. Numerator of Q_c expression must decrease and the denominator must increase.
        ii. Implies net conversion of products to reactants.
        iii. Reaction shifts left.

13.8 Altering an Equilibrium Mixture: Changes in Pressure and Volume

  A. Number of moles of gaseous reactants in the balanced equation is different from the number of moles of gaseous products.
    1. Change in pressure (due to changing volume) changes composition of equilibrium mixture.
  B. Increase in pressure (due to decrease in volume) results in reaction in the direction that decreases the number of moles of gas.
  C. Decrease in pressure (due to increase in volume) results in reaction in the direction that increases the number of moles of gas.
  D. Changes occur due to change in value of Q_c.
    1. Decrease volume — molarity ($= n/V$) increases.
    2. If reactant side has more moles of gas
      a. increase in denominator is greater than increase in numerator
      b. $Q_c < K_c$
      c. To return to equilibrium, Q_c must increase.
        i. Numerator of Q_c expression must increase and the denominator must decrease.
        ii. Implies net conversion of reactants to products (shifts toward fewer moles of gas).
    3. If product side has more moles of gas
      a. Increase in numerator is greater than increase in denominator.
      b. $Q_c > K_c$
      c. To return to equilibrium, Q_c must decrease.
        i. Denominator of Q_c expression must increase and the numerator must decrease.
        ii. Implies net conversion of products to reactants (shifts toward fewer moles of gas).
  E. Reaction involves no change in the number of moles of gas.
    1. No effect on composition of equilibrium mixture.
  F. For heterogeneous equilibrium.
    1. Effect of pressure changes on solids and liquids can be ignored.
      a. Volume is nearly independent of pressure.
  G. Change in pressure due to addition of an inert gas.
    1. No change in the molar concentrations of reactants or products.
    2. No effect on composition of equilibrium mixture.

13.9 Altering an Equilibrium Mixture: Changes in Temperature

  A. Disturb an equilibrium by a change in concentration, pressure, or volume.
    1. $Q_c \neq K_c$.
    2. Value of K_c remains constant with constant T.
  B. Change in temperature always changes equilibrium constant.

Chapter 13 – Chemical Equilibrium

1. Depends on sign of $\Delta H°$ for the reaction.
2. Exothermic reaction ($-\Delta H°$) — equilibrium constant decreases as the temperature increases.
3. Endothermic reaction ($+\Delta H°$) — equilibrium constant increases as the temperature increases.

C. Le Châtelier's principle — add heat to an equilibrium mixture, net reaction occurs in the direction that relieves the stress of the added heat.
1. Endothermic reaction — heat is absorbed by reaction in the forward direction (ΔH is written on the reactant side of the reaction).
   a. Equilibrium mixture contains more product than reactant.
   b. K_c increases with increasing temperature.
2. Exothermic reaction — heat is absorbed by reaction in the reverse direction (ΔH is written on the product side of the reaction).
   a. equilibrium mixture contains more reactant than product
   b. K_c decreases with increasing temperature.

EXAMPLE:
How will the following changes alter the equilibrium for the reaction

$$3\ Fe\ (s)\ +\ 4\ H_2O\ (g) \rightleftarrows Fe_3O_4\ (s)\ +\ 4\ H_2\ (g) \qquad \Delta H° = -150\ kJ$$

a. H_2O is removed from the system.
b. H_2 is removed from the system.
c. The volume of the container is increased.
d. Fe_3O_4 is added to the system.
e. The temperature is raised.

SOLUTION:
a. When H_2O is removed from the system, the equilibrium shifts left.
b. When H_2 is removed from the system, the equilibrium shifts right.
c. Increasing the volume of the container decreases the pressure in the container. However, the number of moles of gas is equal on both sides of the reaction. Therefore, a change in volume and pressure does not affect the equilibrium.
d. Adding a solid to the system does not affect the equilibrium since solids are not included in the equilibrium expression.
e. Raising the temperature causes the equilibrium to shift left.

13.10 The Effect of a Catalyst

C.2.

A. Catalyst increases the rate of a chemical reaction.
1. Provides a new, lower energy pathway.
2. Forward and reverse reactions pass through the same transition state.
   a. Rate for forward and reverse reactions increases by the same factor.
3. Does not affect the composition of the equilibrium mixture.
   a. does not alter the equilibrium constant or equilibrium concentrations
   b. does not appear in the balanced chemical equation
4. Can influence choice of optimum conditions for a reaction.

13.11 The Link Between Chemical Equilibrium and Chemical Kinetics

D.1.

A. For the reaction $A\ +\ B \rightleftarrows C\ +\ D$.
1. Rate of forward reaction $= k_f[A][B]$

Chapter 13 – Chemical Equilibrium

    2. Rate of reverse reaction = $k_r[C][D]$

    3. At equilibrium: $k_f[A][B] = k_r[C][D]$ or $\dfrac{k_f}{k_r} = \dfrac{[C][D]}{[A][B]} = K_c$

  B. Relative values of k_f and k_r determine the composition of the equilibrium mixture.

    1. $k_f \gg k_r$; K_c is very large; reaction goes to completion.
       a. irreversible reaction
       b. Reverse reaction is too slow to be detected.
    2. $k_f \approx k_r$; $K_c \approx 1$; reactants and products are present at equilibrium.
       a. reversible reaction

  C. $\dfrac{k_f}{k_r} = K_c$

    1. Explains why K_c depends on temperature.
    2. Rate increases with an increase in temperature.
       a. Arrhenius equation:
          $k = Ae^{-E_a/RT}$

 **Putting It Together** Calculate the pressure of all species at equilibrium for the reaction

$$2\,NO\,(g) + Br_2\,(g) \rightleftarrows 2\,NOBr\,(g)$$

given that the initial pressure of NO = 98.4 mm Hg and the initial pressure of Br_2 = 41.3 mm Hg. The total pressure at equilibrium = 110.5 mm Hg. Determine the value of K_p

 **Self–Test**

This section is intended to test your knowledge of the material covered in this chapter. Think through these problems, and make certain you understand what is going on. Ask yourself if your answer makes sense. Many of these questions are linked to the chapter learning goals. Therefore, successful completion of these problems indicates you have mastered the learning goals for this chapter. You will receive the greatest benefit from this section if you use it as a mock exam. You will then discover which topics you have mastered and which topics you need to study in more detail.

True–False

1. The reaction quotient Q_c is the same as the equilibrium constant expression.

2. Le Châtelier's principle predicts the changes in the composition of an equilibrium mixture if a stress is applied to a system at equilibrium.

3. If Q_c is greater than K_c the reaction proceeds from left to right.

4. If a product is added to a system at equilibrium, the reaction will shift to the right.

5. If the pressure of a system at equilibrium is increased, the reaction will shift in the direction of the fewer number of moles of gas.

Chapter 13 – Chemical Equilibrium

6. If an inert gas is added to a system at equilibrium, the reaction will shift in the direction of the fewer number of moles of gas.

7. A change in the temperature of a reaction will change the value of the equilibrium constant.

8. Adding a catalyst to a system at equilibrium will change the composition of an equilibrium mixture.

9. The relative values of k_f and k_r determine the composition of the equilibrium mixture.

10. For a large value of K_c, the position of the reaction is to the right.

Multiple Choice

1. A state of chemical equilibrium is reached when
    a. the rate of the forward reaction is greater than the rate of the reverse reaction
    b. the concentration of the products and reactants are equal
    c. more product is present than reactant
    d. the concentrations of the products and reactants have reached constant value

2. If $Q_c > K_c$, the reaction
    a. proceeds from left to right
    b. proceeds from right to left
    c. is at equilibrium

3. When a reactant is added to a system at equilibrium
    a. the reaction shifts towards the left
    b. Q_c must increase because the denominator in the Q_c expression becomes larger
    c. there is a net conversion of products to reactants
    d. the value of Q_c does not change

4. When a system at equilibrium that has more moles of gas on the reactant side experiences a decrease in pressure
    a. the molarity of the species present increases
    b. $Q_c < K_c$
    c. Q_c must decrease for the system to return to equilibrium
    d. the numerator of Q_c expression must increase and the denominator must decrease

5. When the temperature is increased for a system at equilibrium:
    a. the value of K_c remains the same
    b. the effect is independent of the value of ΔH
    c. the value of K_c decreases as the temperature increases for an exothermic reaction
    d. the value of K_c decreases as the temperature decreases for an exothermic reaction

6. When a catalyst is added to a system at equilibrium:
    a. the value of the equilibrium constant changes
    b. the equilibrium concentrations change
    c. a new, lower energy pathway is established
    d. the rate of the forward and reverse reactions are no longer equal

Chapter 13 – Chemical Equilibrium

7. When $K_c > 10^3$:
   a. the reaction hardly proceeds
   b. appreciable amounts of both reactants and products are present
   c. the reactants predominate over the products
   d. the position of the reaction is to the right

8. K_p
   a. is the same as K_c
   b. is the reciprocal of K_c
   c. is defined as the equilibrium constant using partial pressures
   d. includes the concentration of pure liquids in the equilibrium expression

9. The reaction quotient, Q_c, is
   a. calculated in the same manner as K_c
   b. less than K_c at equilibrium
   c. greater than K_c at equilibrium
   d. equal to K_p

10. When a reactant is removed from a system at equilibrium
    a. $Q_c < K_c$
    b. the reaction shifts left
    c. the reactants are converted to products
    d. the denominator in the K_c expression becomes larger

Matching

Chemical equilibrium	a. state in which the forward and reverse reactions continue at equal rates
Equilibrium constant, K_c	b. If a stress is applied to a reaction mixture at equilibrium, reaction occurs in the direction that relieves the stress.
Homogeneous equilibria	c. reactants and products are present in more than one phase
Heterogeneous equilibria	d. reaction in which the product molecules can recombine to form reactant molecules
Reaction quotient, Q_c	e. Reactants and products are present in the same phase.
Reversible reaction	f. a dynamic state in which the concentrations of reactants and products remain constant because the rates of the forward and reverse reactions are equal
Dynamic state	g. the number obtained when equilibrium concentrations (in mol/L) are substituted into the equilibrium constant expression

Chapter 13 – Chemical Equilibrium

Le Châtelier's principle

h. the number obtained when the concentration of all species (not necessarily the equilibrium concentration) are substituted into the equilibrium constant expression

Fill–in–the–Blank

1. When approaching equilibrium from the reactant side of a reaction, the concentration of reactant _____ and the concentration of product _____ until both concentrations _____.

2. The equilibrium constant, K'_c, for the reverse reaction is equal to the _____.

3. The concentrations of pure solids or pure liquids are _____ when writing the equilibrium equation for any heterogeneous equilibrium.

4. If $K_c < 10^{-3}$, the reaction _____.

5. The stress of an added reactant or product to a system at equilibrium is relieved by reaction _____.

6. A decrease in pressure to a system at equilibrium results in reaction _____.

7. For heterogeneous equilibrium, the effect of pressure changes on solids and liquids can be _____.

8. There is _____ on the composition of the equilibrium mixture when an inert gas is added to a system at equilibrium because _____.

9. For an exothermic reaction, K_c _____ with decreasing temperature.

10. The composition of the equilibrium mixture is determined by the relative values of _____.

Problems

1. Write equilibrium expressions for the following balanced equations.

    a. $7 H_2 (g) + 2 NO_2 (g) \rightleftharpoons 2 NH_3 (g) + 4 H_2O (g)$

    b. $3 Cl_2 (g) + CS_2 (l) \rightleftharpoons CCl_4 (l) + S_2Cl_2 (l)$

    c. $Br_2 (g) + 2 HI (g) \rightleftharpoons I_2 (g) + 2 HBr (g)$

    d. $2 NaHCO_3 (s) \rightleftharpoons Na_2CO_3 (s) + CO_2 (g) + H_2O (g)$

2. The partial equilibrium pressures for N_2, O_2, and NO in the reaction

Chapter 13 – Chemical Equilibrium

$$N_2(g) + O_2(g) \rightleftharpoons 2\,NO(g)$$

are $p(N_2) = 0.25$ atm, $p(O_2) = 0.198$ atm, and $p(NO) = 0.050$ atm. Calculate K_p for this reaction.

3. Calculate K_c for the reaction in Question 2.

4. From the value of K_c, determine whether mainly products or mainly reactants exist at equilibrium for the following balanced equations:

   a. $N_2(g) + O_2(g) \rightleftharpoons 2\,NO(g)$ $\quad K = 1 \times 10^{238}$

   b. $SO_2(g) + NO_2(g) \rightleftharpoons NO(g) + SO_3(g)$ $\quad K = 85.0$

   c. $2\,CO_2(g) \rightleftharpoons 2\,CO(g) + O_2(g)$ $\quad K = 6.4 \times 10^{-7}$

5. For the reaction

$$PCl_5(g) \rightleftharpoons PCl_3(g) + Cl_2(g)$$

$K_c = 33.3$ at 760°C. Is this system at equilibrium if $[PCl_5] = 1.25 \times 10^{-3}$ M, $[PCl_3] = [Cl_2] = 0.75$ M? If not, determine the direction the reaction must go to reach equilibrium.

6. Nitrogen dioxide is produced from the reaction of dinitrogen oxide and oxygen according to the equation

$$2\,N_2O(g) + 3\,O_2(g) \rightleftharpoons 4\,NO_2(g)$$

At 25°C, 0.0292 moles of N_2O and 0.0713 moles of O_2 are placed in a 1.00 L container and allowed to react. The $[NO_2]$ at equilibrium is 0.0284 moles. What are the equilibrium concentrations of N_2O and O_2? What is the value of K_c?

7. The equilibrium constant for the reaction

$$SO_2(g) + NO_2(g) \rightleftharpoons NO(g) + SO_3(g)$$

is $K_c = 85.0$ at 460°C. Calculate the concentrations of all species at equilibrium when the initial reaction mixture contains 0.0650 M SO_2 and 0.0650 M NO_2.

8. Consider the reaction

$$2\,SO_3(g) \rightleftharpoons 2\,SO_2(g) + O_2(g) \qquad \Delta H° = 197\text{ kJ}$$

What will be the direction of the reaction when the following stress is applied?

   a. addition of SO_3
   b. decrease in temperature
   c. increase in volume
   d. addition of N_2 gas
   e. addition of SO_2 gas

Chapter 13 – Chemical Equilibrium

9. For the reaction $A + B \rightleftarrows 2C$, $k_f = 7.5 \times 10^{-7}$ s^{-1} and $k_r = 3.2 \times 10^{-2}$ s^{-1}. Calculate K.

10. Calculate the equilibrium concentration for all species present in the reaction

$$HCONH_2 (g) \rightleftarrows NH_3 (g) + CO (g)$$

if the initial concentration of formamide ($HCONH_2$) is 0.250 M. $K_c = 4.84$ at 127°C.

11. For the reaction

$$2 H_2S (g) + CH_4 (g) \rightleftarrows 4 H_2 (g) + CS_2 (g)$$

$K_c = 5.27 \times 10^{-8}$ at 700°C. What is the position of equilibrium for this reaction? Is the reaction mixture at equilibrium when the concentrations of the reactants and products are $[CH_4] = 1.50$ M, $[H_2S] = 3.00$ M, $[H_2] = 0.300$ M, and $[CS_2] = 0.075$ M. If not, in which direction will the reaction proceed?

Challenge Problem

The following reaction was carried out in a 4.00 L vessel:

$$NO_2 (g) + NO (g) \rightleftarrows N_2O (g) + O_2 (g)$$

Determine the number of moles of reactants and products present at equilibrium if the initial number of moles are 0.200 mol NO_2, 0.300 mol NO, 0.150 mol N_2O and 0.250 mol O_2. Use a value of $K_c = 0.914$.

Inquiry Based Problem

Your laboratory assignment for the week is to determine the equilibrium constant for the hydrolysis reaction between ethyl acetate and water to produce ethyl alcohol and acetic acid.[1]

$$CH_3CH_2O-\overset{\overset{O}{\|}}{C}-CH_3 + H_2O \rightleftarrows CH_3CH_2-OH + CH_3-\overset{\overset{O}{\|}}{C}-OH$$

ethyl acetate   water   ethyl alcohol   acetic acid

This reaction is acid catalyzed. Therefore, it should be carried out in a 6 M HCl. You are told to prepare the following solutions.

---

[1] Weiss, Gerald S., Wismer, Robert K., Greco, Thomas G.; "Determination of an Equilibrium Constant" in Experiments in General Chemistry: A laboratory Program to accompany Petrucci's General Chemistry, 5th Ed.; Macmillan Publishing Company, New York; 1985; p. 147.

Chapter 13 – Chemical Equilibrium

1. 5.00 mL 6 M HCl and 5.00 mL distilled water
2. 5.00 mL 6 M HCl and 2.00 mL ethyl acetate and 3.00 mL distilled water

After the solutions reach equilibrium, each one is titrated with 1.00 M NaOH. You collected the following data from these titrations:

Solution	Volume of NaOH
1	28.90 mL
2	38.70 mL

You are instructed to use the above data and the densities of 6 M HCl and the ethyl acetate solution (1.11 g/mL and 0.893 g/mL, respectively) to determine the number of moles of reactants and products at equilibrium. How would you carry out these calculations?

Consider the following:
1. How would you calculate the mass of a 5 mL, 6 M HCl solution?
2. How would you calculate the mass of water in a 5 mL, 6 M HCl solution?
3. How would you calculate the mass of ethyl acetate and the mass of water in solution 2? (Consider all sources for the presence of water.) Why do you need to know these masses?
4. How much NaOH is used to titrate the acetic acid produced in solution 2? (Consider how much NaOH is needed to titrate the 6 M HCl solution.) Why do you need to know this volume?
5. What can you surmise from the stoichiometry of the reaction regarding the number of moles of ethyl alcohol produced?
6. Calculate the equilibrium constant, K_c from this data.

CHAPTER 14

AQUEOUS EQUILIBRIA: ACIDS AND BASES

Learning Goals

A. Arrhenius Acids and Bases
1. Define an acid and a base according to the Arrhenius, Brønsted–Lowry, and Lewis theories.

B. Dissociation of Water
1. Calculate H_3O^+ concentration from OH^- concentration and vice versa. From these concentrations, determine whether the solution is acidic, neutral, or basic.
2. Interconvert pH and $[H_3O^+]$. Classify the solution as acidic, neutral, or basic.
3. Given the molar concentration of a strong acid or a strong base, determine the pH of the solution.

C. Brønsted–Lowry Acids and Bases: Proton–Transfer Reactions
1. From Lewis structures, determine which chemical species can act as a Brønsted–Lowry acid, a Brønsted–Lowry base, or both.
2. From a chemical equation for a proton–transfer reaction, identify the conjugate acid–base pairs.
3. Given the extent of dissociation of an acid in water, determine whether the acid is a stronger or weaker acid than water and whether the conjugate base of the acid is a stronger or weaker base than water.
4. Given a chemical equation representing a proton–transfer reaction and the relative strengths of each acid and base involved in the reaction, determine whether the reaction is favored to the right or to the left.
5. Interconvert K_a and K_b.
6. Classify salt solutions as acidic, neutral, or basic. Calculate the pH of these solutions.

D. Brønsted–Lowry Acids and Bases: Dissociation in Water
1. Given the pH of a weak acid solution, determine the K_a of the acid.
2. Given the K_a value and the initial concentration of a weak monoprotic acid, calculate the concentrations of all species at equilibrium, the pH of the solution, and the percent dissociation of the acid.
3. Given the K_a value and the initial concentration of a weak diprotic acid, calculate the concentrations of all species at equilibrium and the pH of the solution.
4. Given the K_b value and the initial concentration of a weak base, calculate the concentrations of all species at equilibrium and the pH of the solution.
5. Identify which of two substances is more acidic.

E. Lewis Acids and Bases
1. Identify the Lewis acid and the Lewis base in a chemical reaction.

Chapter in Brief

In this chapter, you continue the study of chemical equilibrium by applying the concepts you learned in Chapter 13 to acid–base chemistry. This chapter begins with the definition of acids, bases, and conjugate acid–base pairs using the Brønsted–Lowry concept. You will learn the pH scale and how to determine the pH from the dissociation of a strong acid or strong base. You also will learn how to

Chapter 14 – Aqueous Equilibria: Acids and Bases

determine the strength of an acid or a base relative to water, given the extent of dissociation in water, and how to determine the pH of a weak acid or weak base, using either K_a or K_b. Finally, you will determine the pH of salt solutions and examine the Lewis acid–base concept.

14.1 Acid–Base Concepts: The Brønsted–Lowry Theory

A. Arrhenius theory of acids and bases.
  1. Acids — substances that dissociate in water to produce hydrogen ions.
     a. $HA\,(aq) \rightleftarrows H^+\,(aq) + A^-\,(aq)$
  2. Bases — substances that dissociate in water to produce hydroxide ions.
     a. $MOH\,(aq) \rightleftarrows M^+\,(aq) + OH^-\,(aq)$
  3. Limitations to theory.
     a. restricted to aqueous solutions
     b. doesn't account for the basicity of substances that don't contain OH groups

B. Brønsted–Lowry theory of acids and bases
  1. Acid — any substance that can transfer a proton to another substance (proton donor).
  2. Base — any substance that can accept a proton (proton acceptor).
  3. Acid–base reaction — proton–transfer reactions

C. Conjugate acid–base pair — chemical species whose formulas differ only by one proton.
  1. Conjugate base has one less proton than its acid.
     a. A^- is the conjugate base of HA.
  2. Conjugate acid has one more proton than its base.
     a. BH^+ is the conjugate acid of B.

D. Acid–dissociation equilibrium.
  1. Acid transfers a proton to the solvent, which acts as a base.
  2. $HA\,(aq) + H_2O\,(l) \rightleftarrows H_3O^+\,(aq) + A^-\,(aq)$

E. Base–dissociation equilibrium.
  1. Base accepts a proton from the solvent, which acts as an acid.
  2. $NH_3\,(aq) + H_2O\,(l) \rightleftarrows OH^-\,(aq) + NH_4^+\,(aq)$

F. All Brønsted–Lowry bases have one or more lone pairs of electrons.
  1. Unshared pair of electrons are used for bonding to the proton.

EXAMPLE:
Write the proton–transfer equilibria for the following acids or bases in aqueous solution, and identify the conjugate acid–base pairs in each one: $C_6H_5NH_3^+$ (anilinium ion), $H_2PO_4^-$ (acid reaction), NH_2NH_2 (hydrazine, a base), and PO_4^{3-}.

SOLUTION:
$C_6H_5NH_3^+\,(aq) + H_2O \rightleftarrows C_6H_5NH_2 + H_3O^+\,(aq)$
  Conjugate acid–base pairs: $C_6H_5NH_3^+$ (acid)/$C_6H_5NH_2$ (base); H_2O (base)/H_3O^+ (acid)

$H_2PO_4^-\,(aq) + H_2O \rightleftarrows HPO_4^{2-}\,(aq) + H_3O^+\,(aq)$
  Conjugate acid–base pairs: $H_2PO_4^-$ (acid)/ HPO_4^{2-} (base); H_2O (base)/H_3O^+ (acid)

$NH_2NH_2 + H_2O \rightleftarrows NH_2NH_3^+\,(aq) + OH^-\,(aq)$
  Conjugate acid–base pairs: NH_2NH_2 (base)/$NH_2NH_3^+$ (acid); H_2O (acid)/OH^- (base)

$PO_4^{3-}\,(aq) + H_2O \rightleftarrows HPO_4^{2-}\,(aq) + OH^-\,(aq)$
  Conjugate acid–base pairs: PO_4^{3-} (base)/HPO_4^{2-} (acid); H_2O (acid)/OH^- (base)

Chapter 14 – Aqueous Equilibria: Acids and Bases

14.2 Acid Strength and Base Strength

A. In an acid–dissociation equilibrium, the two bases, H_2O and A^-, are competing for protons.

    1. Direction of reaction to reach equilibrium is proton–transfer from the stronger acid to the stronger base.

B. Strong acid — completely dissociated in aqueous solution.

    1. Has weak conjugate base.
       a. H_2O is a stronger base than A^-

C. Weak acid — partially dissociates in aqueous solution.

    1. Has strong conjugate base.
       a. A^- is a stronger base than H_2O

D. Inverse relationship between the strength of an acid and the strength of its conjugate base (see Table 14.1, page 548 in your textbook).

EXAMPLE:

Determine the direction of reactions involving acetic acid, acetate ion, hydrosulfuric acid, and HS^-; acetic acid, acetate ion, hydrogen sulfate, and SO_4^{2-}.

SOLUTION: To determine the direction of these acid–base reactions, we need to use Table 15.1 on page 616 of your textbook. Keep in mind that the proton is transferred to the stronger base. Therefore, the direction of the reaction to reach equilibrium is proton–transfer from the stronger acid to the stronger base to give the weaker acid and the weaker base. For acetic acid and HS^-, the stronger acid is CH_3COOH and the stronger base is HS^-. The reaction in the forward direction is

$$CH_3COOH + HS^- (aq) \rightleftarrows CH_3CO_2^- (aq) + H_2S$$

For the reaction between acetic acid and SO_4^{2-}, the stronger acid is HSO_4^- and the stronger base is $CH_3CO_2^-$. Therefore, the reaction in the forward direction is

$$HSO_4^- (aq) + CH_3CO_2^- (aq) \rightleftarrows SO_4^{2-} (aq) + CH_3COOH$$

Workbook Problem 14.1

Write the proton–transfer reaction between NH_4^+ and CO_3^{2-} and the proton–transfer reaction between HCN and $H_2PO_4^-$.

Strategy: Use Table 14.1 (page 458 in your textbook) to determine which species is the stronger acid and which is the stronger base.

Step 1: Write the reaction in the forward direction. (Remember that the stronger acid will react with the stronger base.)

Chapter 14 – Aqueous Equilibria: Acids and Bases

What key concept did you use?

14.3 Hydrated Protons and Hydronium Ions
A. Hydronium ion, H_3O^+ — ion in which H^+ is bonded to the oxygen atom of a solvent water molecule.
   1. Simplest hydrate of the proton — $[H(H_2O)]^+$.
      a. produces higher hydrates through hydrogen bonding with other water molecules
         i. general formula — $[H(H_2O)_n]^+$
   2. Symbols H^+ (*aq*) and H_3O^+ (*aq*) used to represent a proton hydrated by an unspecified number of water molecules.

14.4 Dissociation of Water
A. Water can act both as an acid and as a base.
B. Dissociation of water — one water molecule donates a proton to another water molecule.
   1. $H_2O\ (l) + H_2O\ (l) \rightleftarrows H_3O^+\ (aq) + OH^-\ (aq)$.
   2. Ion–product constant for water, K_w.
      a. $K_w = [H_3O^+][OH^-]$
         i. omit $[H_2O]$ – water is a pure substance
   3. Forward and reverse reactions are rapid.
   4. Position of the equilibrium lies far to the left.
   5. In an aqueous solution, $[H_3O^+][OH^-] = 1.0 \times 10^{-14}$.
      a. $[H_3O^+] = [OH^-] = 1.0 \times 10^{-7}$.

B.1. C. Relative values of H_3O^+ and OH^- concentrations can be used to determine if an aqueous solution is neutral, acidic or basic.

   1. $[H_3O^+] = \dfrac{1.0 \times 10^{-14}}{[OH^-]}$

   2. $[OH^-] = \dfrac{1.0 \times 10^{-14}}{[H_3O^+]}$

   3. $[H_3O^+] = [OH^-]$, neutral solution and $[H_3O^+] = 1.0 \times 10^{-7}$ M.
   4. $[H_3O^+] > [OH^-]$, acid solution and $[H_3O^+] > 1.0 \times 10^{-7}$ M.
   5. $[H_3O^+] < [OH^-]$, basic solution and $[H_3O^+] < 1.0 \times 10^{-7}$ M.

EXAMPLE:
Calculate the molarity of OH^- in a solution with an H_3O^+ concentration of 0.35 M. Is this solution acidic, basic or neutral? Is a solution with an OH^- concentration of 8.5×10^{-5} M acidic, basic, or neutral?

SOLUTION: If we know either the $[H_3O^+]$ or $[OH^-]$, we can calculate the other from the relationship $K_w = [H_3O^+][OH^-]$.

$$[OH^-] = \dfrac{1 \times 10^{-14}}{0.35} = 2.9 \times 10^{-14};\ \text{since } [H_3O^+] > 1.0 \times 10^{-7}\ \text{this solution is acidic.}$$

$$\left[H_3O^+ \right] = \frac{1.0 \times 10^{-14}}{8.5 \times 10^{-5}} = 1.2 \times 10^{-10} \text{ ; in this case, } [H_3O^+] < 1.0 \times 10^{-7}; \text{ therefore, the solution is basic.}$$

14.5 The pH Scale

A. pH scale — logarithmic scale used to express the hydronium ion concentration.
   1. pH = –log[H_3O^+].
   2. [H_3O^+] = antilog(–pH) = 10^{-pH}
   3. Significant figures in a logarithm are the digits to the right of the decimal point.
      a. number to the left of the decimal point is an exact number related to the integral power of 10 in the exponential expression for [H_3O^+]
B. pH decreases as [H_3O^+] increases.
   1. Acidic solutions, pH < 7.
   2. Basic solutions, pH > 7.
   3. Neutral solutions, pH = 7.

EXAMPLE:
Calculate the [H_3O^+] for orange juice, which has a pH of 3.80.

SOLUTION:
   [H_3O^+] = antilog (–pH) = antilog (–3.80) = 1.6×10^{-4}

Workbook Problem 14.2
Calculate the pH of a solution whose [OH^-] = 2.35×10^{-8} M. Is the solution acidic, basic, or neutral?

Strategy: Calculate the [H_3O^+], using the expression for the dissociation of water.

Step 1: Calculate the pH.

Step 2: Determine if the solution is acidic, basic, or neutral.

What key concept did you use?

14.5 Measuring pH

A. Acid–base indicators (HIn) — substances that change color in a specific pH range. (See figure 14.4 on page 554 in your textbook.)
   1. Weak acids.
   2. Have different colors in their acid (HIn) and conjugate base (In^-) forms.
   3. Change color over a range of ≈ 2 pH units.
   4. Can determine the pH of a solution to within ≈ ±1 pH unit.

Chapter 14 – Aqueous Equilibria: Acids and Bases

  B. pH meter.
   1. An electronic instrument that measures a pH–dependent electrical potential of the test solution.
   2. Determines pH values more accurately than acid–base indicators.

14.7 The pH of Strong Acids and Strong Bases

  A. Strong monoprotic acids — 100% dissociated in aqueous solution ($HClO_4$, HCl, HNO_3).
   1. Contains a single dissociable proton.
   2. $pH = -\log[H_3O^+]$
   3. $[H_3O^+] = [A^-]$ = initial concentration of the acid
   4. [undissociated HA] = 0
  B. Strong bases.
   1. Alkali metal hydroxides, MOH.
    a. water–soluble ionic solids
    b. exist in aqueous solution as alkali metal cations and hydroxide anions
    c. calculate pH from $[OH^-]$
   2. Alkaline earth metal hydroxides, $M(OH)_2$ (M = Mg, Ca, Sr, Ba).
    a. less soluble than alkali hydroxides, therefore lower $[OH^-]$
    b. most important strong base — lime, CaO
     i. produces $Ca(OH)_2$ when mixed with water
     ii. used in steelmaking, water purification, and chemical manufacture

EXAMPLE:
Calculate the pH of a 1.25×10^{-3} M HNO_3 solution.

SOLUTION: HNO_3 is a strong acid that completely dissociates in water. Therefore, $[H_3O^+]$ = initial concentration of the undissociated acid.

$[H_3O^+] = 1.25 \times 10^{-3}$ M

$pH = -\log [H_3O^+] = -\log (1.25 \times 10^{-3}) = 2.903$

 Workbook Problem 14.3
Calculate the pH of a 5.0×10^{-4} M KOH solution.

Strategy: Determine the $[H^+]$ from the K_w expression and then calculate pH.

Step 1: Determine the concentration of $[OH^-]$.

Step 2: Calculate the $[H_3O^+]$.

Step 3: Calculate the pH.

What key concept did you use?

14.8 Equilibria in Solutions of Weak Acids
A. Weak acids — partially dissociated.
   1. Can write an acid–dissociation equilibrium equation.
      a. HA (aq) + H_2O (l) ⇌ H_3O^+ (aq) + A^- (aq)
B. Position of acid–dissociation equilibrium — characterized by the acid–dissociation constant, K_a.
   1. $K_a = \dfrac{[H_3O^+][A^-]}{[HA]}$
   2. $pK_a = -\log K_a$
   3. pK_a decreases as the K_a increases.
C. The larger the value of K_a, the stronger the acid.
D. K_a values can be determined from pH measurements.

🔑 *D.1.*

EXAMPLE:
The pH of a 0.200 M solution of nicotinic acid ($HC_6H_4NO_2$) is 2.78. Determine the value of K_a for this acid.

SOLUTION: To determine K_a, first write the balanced equation for the dissociation equilibrium and the equilibrium equation that defines K_a.

$HC_6H_4NO_2$ (aq) + H_2O (l) ⇌ $H_3O^+(aq)$ + $C_6H_4NO_2^-$ (aq)

$$K_a = \frac{[H_3O^+][C_6H_4NO_2^-]}{[HC_6H_4NO_2]}$$

To determine K_a, we need to know the concentrations of the species in the equilibrium mixture. We can determine the concentration of H_3O^+ from the pH.

$[H_3O^+]$ = antilog (–pH) = antilog (–2.780) = 1.66×10^{-3} M.

Dissociation of one $HC_6H_4NO_2$ molecule produces one H_3O^+ and $C_6H_4NO_2^-$ ion; the H_3O^+ and $C_6H_4NO_2^-$ concentrations are equal.

$[H_3O^+] = [C_6H_4NO_2^-] = 1.66 \times 10^{-3}$ M

The $[HC_6H_4NO_2]$ concentration at equilibrium is equal to the initial concentration minus the amount of $HC_6H_4NO_2$ that dissociates.

$[HC_6H_4NO_2] = 0.200 - (1.66 \times 10^{-3}) = 0.198$ M

Chapter 14 – Aqueous Equilibria: Acids and Bases

$$K_a = \frac{[H_3O^+][C_6H_4NO_2^-]}{[HC_6H_4NO]} = \frac{(1.66 \times 10^{-3})(1.66 \times 10^{-3})}{(0.198)} = 1.39 \times 10^{-5}$$

14.9 Calculating Equilibrium Concentrations in Solutions of Weak Acid
A. Equilibrium concentrations and the pH of a solution of a weak acid can be calculated from the value of K_a.
B. To solve acid–base equilibrium problems, think about the chemistry, and use the following steps:

Step 1. List the species present before dissociation and identify them as Brønsted–Lowry acids or bases.

▼

Step 2. Write balanced equations for all possible proton-transfer reactions.

▼

Step 3. Identify the principal reaction—the reaction that has the largest equilibrium constant.

▼

Step 4. Make a table that lists the following values for each of the species involved in the principal reaction:
  (a) The initial concentration
  (b) The change in concentration on proceeding to equilibrium
  (c) The equilibrium concentration
In constructing this table, define x as the concentration (mol/L) of the acid that dissociates.

▼

Step 5. Substitute the equilibrium concentrations into the equilibrium equation for the principal reaction, and solve for x.

▼

Step 6. Calculate the "big" concentrations—the concentrations of the species involved in the principal reaction.

▼

Step 7. Use the big concentrations and the equilibrium equations for the subsidiary reactions to calculate the small concentrations—the concentrations of the species involved in the subsidiary equilibria.

▼

Step 8. Calculate the pH = $-\log[H_3O^+]$.

Chapter 14 – Aqueous Equilibria: Acids and Bases

EXAMPLE:
Calculate the concentration of all species present (H_3O^+, $C_4H_7O_2^-$, $HC_4H_7O_2$, and OH^-) and the pH in 0.250 M butyric acid ($HC_4H_7O_2$).

SOLUTION: To solve this problem, use the outline above.
1. The species present initially are $HC_4H_7O_2$ (acid) and H_2O (acid or base)

2. The possible proton–transfer reactions are

    $HC_4H_7O_2\,(aq) + H_2O\,(l) \rightleftarrows H_3O^+\,(aq) + C_4H_7O_2^-\,(aq)$ $\qquad K_a = 1.5 \times 10^{-5}$

    $H_2O\,(l) + H_2O\,(l) \rightleftarrows H_3O^+\,(aq) + OH^-\,(aq)$ $\qquad K_w = 1.0 \times 10^{-14}$

3. Since $K_a \gg K_w$, the principal reaction is dissociation of $HC_4H_7O_2$.

4. Make a table.

Principal reaction	$HC_4H_7O_2\,(aq)$	$\rightleftarrows$	H_3O^+	+	$C_4H_7O_2^-$
Initial concentration (M)	0.250		0		0
Change (M)	$-x$		$+x$		$+x$
Equilibrium concentration (M)	$0.250 - x$		$+x$		$+x$

5. Substitute the equilibrium concentrations into the equilibrium expression.

$$K_a = 1.5 \times 10^{-5} = \frac{[H_3O^+][C_4H_7O_2^-]}{[HC_4H_7O_2]} = \frac{(x)(x)}{(0.250-x)}$$

Assume that x is negligible compared with the initial concentration of the acid; therefore $0.250 - x \approx 0.250$. Using this value in the denominator, solve for x.

$$x^2 \approx (1.5 \times 10^{-5})(0.250)$$

$$x \approx 1.9 \times 10^{-3} \,{}^*$$

6. The big equilibrium concentrations are $[HC_4H_7O_2] = 0.250 - 0.0019 = 0.248$ M; $[H_3O^+] = [C_4H_7O_2^-] - 1.9 \times 10^{-3}$ M

7. The concentration of OH^- is obtained from the dissociation of water.

$$[OH^-] = \frac{K_w}{[H_3O^+]} = \frac{1.0 \times 10^{-14}}{1.9 \times 10^{-3}} = 5.3 \times 10^{-12}$$

8. $pH = -\log\,(1.9 \times 10^{-3}) = 2.72$

---

* The initial concentration of $HC_4H_7O_2$ is known to the third decimal place (0.250). Usually x is considered negligible compared to the value of the initial concentration only if $x < 0.001$. In this case, $x > 0.001$. However, solving for x using the quadratic equation gives $x = 1.9 \times 10^{-3}$. Both values for x give an equilibrium concentration of 0.248 M for butyric acid.

Chapter 14 – Aqueous Equilibria: Acids and Bases

Workbook Problem 14.4
Determine the pH of a 2.75 M lactic acid ($HC_3H_5O_3$) solution.

Strategy: Follow the steps outlined previously.

Step 1: Determine the species present initially.

Step 2: Write the possible proton–transfer reactions.

Step 3: Determine the principal reaction and write the equilibrium equation for that reaction.

Step 4: Make a table showing the principal reaction and the initial and equilibrium concentrations.

Step 5: Substitute the equilibrium concentrations into the equilibrium equation, and solve for x. (Since $K_a < 1.0 \times 10^{-3}$, assume that x is negligible and that $2.75 - x \cong 2.75$.)

Step 6: Calculate the big equilibrium concentrations.

Step 7: Calculate the pH.

What key concept did you use?

14.10 Percent Dissociation in Weak acid Solutions

A. Percent dissociation $= \dfrac{[HA] \text{ dissociated}}{[HA] \text{ undissociated}} \times 100$

1. Useful measure of acid strength.

2. Value depends on the acid.
3. Increases with increasing value of K_a.
4. Increases with increasing dilution.

EXAMPLE:
Calculate the percent dissociation of the lactic acid solution in WP 15.4.

SOLUTION:

$$\% \text{ Dissociation} = \frac{[HC_3H_5O_3]_{diss}}{[HC_3H_5O_3]_{undiss}} \times 100$$

$[HC_3H_5O_3]_{diss} = [HC_3H_5O_3]_{undiss} - [HC_3H_5O_3]_{eq} = 2.75 \text{ M} - 2.73 \text{ M} = 0.02 \text{ M}$

(Note, the concentration of the dissociated lactic acid equals the change in concentration calculated in Workbook Problem 11.4.)

$$\% \text{ Dissociation} = \frac{0.02 \text{ M}}{2.75 \text{ M}} \times 100 = 0.73\%$$

Workbook Problem 14.5
Calculate the percent dissociation of a 0.25 M benzoic acid (C_6H_5COOH) solution. ($K_a = 6.5 \times 10^{-5}$)

Strategy: To determine the percent dissociation we need to first determine the concentration of the dissociated HA. This requires that we determine the equilibrium concentration of H_3O^+ and $C_6H_5COO^-$. We will use the procedure found on page 239 of the Study Guide (Figure 15.7, page 630 in your textbook.)

Step 1: Determine the species present initially, write the possible proton–transfer reactions, and determine the principal reaction.

Step 2: Make a table showing the principal reaction, initial concentration, change in concentration, and the equilibrium concentration.

Step 3: Substitute the equilibrium concentrations into the equilibrium expression, and solve for x.

Step 4: Calculate the concentrations of the major species present.

Step 5: Calculate the percent dissociation.

What key concept did you use?

14.11 Polyprotic Acids
  A. Polyprotic acid — acids that contain more than one dissociable proton.
  B. Dissociate in a stepwise manner.
    1. Each dissociation step has its own K_a.
  C. Stepwise dissociation constants decrease in the order $K_{a1} > K_{a2} > K_{a3}$.
    1. More difficult to remove a positively charged proton from a negative ion.
  D. Diprotic acid solution contains a mixture of acids: H_2A, HA^-, and H_2O.
    1. Strongest acid — H_2A.
        a. principal reaction — dissociation of H_2A
        b. All of H_3O^+ comes from first dissociation step.

 Workbook Problem 14.6
Write the stepwise dissociation for arsenic acid (H_3AsO_4) and determine the pH of a 2.500×10^{-4} M solution. ($K_{a1} = 5.62 \times 10^{-3}$; $K_{a2} = 1.70 \times 10^{-7}$; $K_{a3} = 3.95 \times 10^{-12}$)

Step 1: Write the stepwise dissociation for arsenic acid.

Chapter 14 – Aqueous Equilibria: Acids and Bases

Step 2: Determine the principal reaction.

Step 3: Make a table showing the principal reaction, initial concentration, change in concentration and equilibrium concentration of the reactant and products.

Step 4: Substitute the equilibrium concentrations into the equilibrium expression.

Step 5: Calculate the pH of the solution.

What key concept did you use?

14.12 Equilibria in Solutions of Weak Bases   *D.4.*
A. Weak bases — partially dissociated.
 1. Can write a base–dissociation equilibrium equation.
 2. B (aq) + H_2O (l) $\rightleftarrows$ BH^+ (aq) + OH^- (aq)
B. Position of base–dissociation equilibrium — characterized by the base–dissociation constant, K_b.
 1. $K_b = \dfrac{[BH^+][OH^-]}{[B]}$.
C. Amines — organic compounds that are weak bases.
 1. Derivatives of ammonia.
 2. One or more hydrogen atoms are replaced by another group.
 3. Basicity due to lone pair of electrons on nitrogen atom.   *C.1.*
  a. can be used for bonding to a proton
D. To solve for equilibrium concentrations and solution pH, use the same procedure as for solving weak acid problems.

249

Chapter 14 – Aqueous Equilibria: Acids and Bases

Workbook Problem 14.7
Determine the pH of a 0.075 M trimethylamine, $(CH_3)_3N$, solution. $K_b = 6.5 \times 10^{-5}$.

Strategy: Follow the procedure found on page 630 of your textbook.

Step 1: Determine the principal reaction.

Step 2: Construct a table with concentrations of the reactant and products.

Step 3: Substitute the equilibrium concentrations into the equilibrium expression, and solve for x.

Step 4: Determine the equilibrium concentrations of the species present.

Step 5: Calculate the pH of the solution.

What key concept did you use?

Chapter 14 – Aqueous Equilibria: Acids and Bases

14.13 Relation Between K_a and K_b

A. For a conjugate acid–base pair, can calculate either K_a or K_b from the other.
B. Sum of an acid–dissociation reaction and a base–dissociation reaction is the dissociation of water.
   1. Equilibrium constant for the net reaction (dissociation of water) = the product of the equilibrium constants for the individual reactions.
   2. $K_a \times K_b = K_w$
   3. For $K_a \times K_b$ to remain constant, the strength of a conjugate base must decrease as the strength of the acid increases.

EXAMPLE:
Calculate K_b for lactic acid ($HC_3H_5O_3$), given $K_a = 1.4 \times 10^{-4}$. Write the equilibrium reaction for K_b.

SOLUTION:

$$K_w = K_a \times K_b \qquad K_b = \frac{K_w}{K_a}$$

$$K_b = \frac{1 \times 10^{-14}}{1.4 \times 10^{-4}} = 7.14 \times 10^{-11}$$

$C_3H_5O_3^- \ (aq) + H_2O \rightleftarrows HC_3H_5O_3 + OH^-$

14.14 Acid–Base Properties of Salts

A. pH of a salt solution is determined by the acid–base properties of the constituent cations and anions.
   1. In an acid–base reaction, the influence of the stronger partner is dominant.
      a. Strong acid + Strong base → Neutral solution
      b. Strong acid + Weak base → Acidic solution
      c. Weak acid + Strong base → Basic solution
B. Salts that yield neutral solutions.
   1. Derived from a strong base and a strong acid.
   2. Neither the cation nor the anion reacts with water to produce H_3O^+ or OH^- ions.
   3. Cations that do not react with H_2O,
      a. group 1A
      b. group 2A
   4. Anions that do not react with H_2O,
      a. anions from strong, monoprotic acids
C. Salts that yield acidic solutions.
   1. Derived from a weak base and a strong acid.
   2. Anion is neither an acid nor a base.
   3. Cation is a weak acid.
      a. reacts with water (hydrolysis reaction) to produce H_3O^+
      b. Small, highly charged cations are also acidic.
D. Salts that yield basic solutions.
   1. Derived from a strong base and a weak acid.
   2. Cation is neither an acid nor a base.
   3. Anion is a weak base.

Chapter 14 – Aqueous Equilibria: Acids and Bases

   a. reacts with water to produce OH^-
E. Salts that contain acidic cations and basic anions.
   1. Derived from a weak acid and a weak base.
   2. Both the cation and the anion can undergo proton–transfer reactions.
   3. pH depends on the relative acid strength of the cation and the base strength of the anion.
      a. K_a (for the cation) > K_b (for the anion); acidic solution
      b. K_b (for the anion) > K_a (for the cation); basic solution
      c. K_a (for the cation) ≈ K_b (for the anion); neutral solution

EXAMPLE:
Determine whether aqueous solutions of the following salts are acidic, neutral, or basic. Write the hydrolysis reaction for those solutions that are either acidic or basic. NH_4Br, K_3PO_4, CsI, $[(CH_3)_3NH]F$.

SOLUTION: To determine the acidity of an aqueous solution of a salt, we must determine if the salt is derived from 1) a strong acid/strong base reaction, 2) weak acid/strong base reaction, 3) strong acid/weak base reaction, or 4) weak acid/weak base reaction.

NH_4Br: derived from the weak base NH_3 (*aq*) and the strong acid HBr (*aq*). Acidic (NH_4^+ is the weak conjugate acid of NH_3).

$$NH_4^+ (aq) + H_2O \rightleftarrows NH_3 (aq) + H_3O^+ (aq)$$

K_3PO_4: derived from the strong base KOH (*aq*) and the weak acid HPO_4^{2-} (*aq*). Basic (PO_4^{3-} is the conjugate base of HPO_4^{2-}.)

$$PO_4^{3-} (aq) + H_2O \rightleftarrows HPO_4^{2-} (aq) + OH^-$$

CsI: derived from the strong base CsOH (*aq*) and the strong acid HI (*aq*). Neutral.

$[(CH_3)_3NH]F$: derived from the weak base $(CH_3)_3N$ and the weak acid HF (*aq*). To determine the acidity of this solution we must calculate the K_a of the cation and the K_b of the anion of the salt and compare the two.

$$K_a = \frac{1.0 \times 10^{-14}}{K_b[(CH_3)_3N]} = \frac{1.0 \times 10^{-14}}{6.5 \times 10^{-5}} = 1.5 \times 10^{-10}$$

$$K_b = \frac{1.0 \times 10^{-14}}{K_a(HF)} = \frac{1.0 \times 10^{-14}}{3.5 \times 10^{-4}} = 2.9 \times 10^{-11}$$

$K_a > K_b$; therefore, the solution is acidic. The hydrolysis reaction is:

$$(CH_3)_3NH^+ (aq) + H_2O (l) \rightleftarrows (CH_3)_3N (aq) + H_3O^+ (aq)$$

 **Workbook Problem 14.8**
Calculate the pH of a 0.250 M KOCl solution. (K_a for HOCl = 3.5 × 10^{-8})

Strategy: From the preceding discussion, determine if KOCl produces an acidic, basic, or neutral aqueous solution, and write the hydrolysis reaction for this salt.

Chapter 14 – Aqueous Equilibria: Acids and Bases

Step 1: Write the hydrolysis reaction for this salt.

Step 2: Write an equilibrium expression for this reaction.

Step 3: Determine the principal reaction.

Step 4: Construct a table.

Step 5: Substitute the equilibrium concentrations into the equilibrium expression.

Step 6: Calculate the pH of the solution.

What key concept did you use?

14.15 Factors that Affect Acid Strength

A. The extent of dissociation of an acid, HA, is often determined by the strength and polarity of the H–A bond.
   1. The weaker the H–A bond, the stronger the acid.
   2. The more polar the H–A bond, the stronger the acid.
B. For binary acids.
   1. H–A bond strength decreases down a group in the periodic table.
      a. Increase in the size of A down a group produces poorer orbital overlap and a weaker bond.
      b. H–A bond strength is the most important factor for determining acid strength for binary acids of elements in the same column.
      c. Acidity increases down a group.
   2. H–A bond polarity is the most important factor for determining acid strength for binary acids of elements in the same row.
      a. increase in H–A bond polarity as the electronegativity of A increases
      b. In general, electronegativities increase from left to right across the periodic table
      c. acidity increases across a period.
C. Oxoacids, H_nYO_m (Y = a nonmetallic atom, n and m are integers).
   1. Y is always bonded to one or more hydroxyl (OH) groups.
      a. can also be bonded to one or more oxygen atoms
   2. Dissociation of oxoacid involves breaking an O–H bond.
   3. Strength of the acid increases as the O–H bond is weakened or becomes more polar.
      a. For oxoacids that contain the same number of OH groups and the same number of O atoms, acid strength increases with increasing electronegativity of Y.
      b. For oxoacids that contain the same atom Y but different numbers of oxygen atoms, acid strength increases with increasing oxidation number of Y, which increases, in turn, with an increasing number of oxygen atoms.

EXAMPLE:
Order the following by increasing acid strength: a) HF, NH_3, H_2O; b) NH_3, AsH_3, PH_3; c) HIO_3, HIO, HIO_2, HIO_4; d) H_2SeO_4, H_2TeO_4, H_2SO_4.

SOLUTION: a) Across a period, acid strength is determined from the electronegativity of the nonmetal. An increase in electronegativity gives rise to an increase in acid strength. Therefore, the order of increasing acidity is $NH_3 < H_2O < HF$. B) Down a group, acid strength is determined from the size of the nonmetal atom. An increase in size gives rise to an increase in acid strength. Therefore, the order of increasing acidity is $NH_3 < PH_3 < AsH_3$. c) For oxoacids that contain the same atom Y but different numbers of oxygen atoms, acid strength increases with an increasing number of oxygen atoms (increasing oxidation number of Y). Therefore, the order of increasing acidity is $HIO < HIO_2 < HIO_3 < HIO_4$. d) For oxoacids that contain the same number of OH groups and the same number of O atoms, acid strength increases with increasing electronegativity of Y. Therefore, the order of increasing acidity is $H_2TeO_4 < H_2SeO_4 < H_2SO_4$.

14.16 Lewis Acids and Bases

A. Lewis base — electron–pair donor.
   1. All Lewis bases are Brønsted–Lowry bases.
   2. All Brønsted–Lowry bases are Lewis bases.
B. Lewis acid — electron–pair acceptor.
   1. More general definition than Brønsted–Lowry.
   2. Include cations and neutral molecules having vacant valence orbitals that can accept a share in a pair of electrons from a Lewis base.

Chapter 14 – Aqueous Equilibria: Acids and Bases

 **Putting It Together**

Sulfanilic acid, a compound that is used in making dyes, is produced from the reaction of aniline and sulfuric acid in aqueous solution.

$C_6H_5NH_2$ (aq) + H_2SO_4 (aq) → $C_6H_4NH_2SO_3H$ (aq)

If you prepare 200 g of sulfanilic acid with an experimental yield of 90%, how much aniline will you use? ($d_{aniline}$ = 1.02 g/mL) The K_a of sulfanilic acid is 5.9×10^{-4}. What is the pH of an aqueous solution prepared by dissolving 3.75 g of $Na(C_6H_4NH_2SO_3H)$ in 500 mL of water?

 **Self–Test**

This section is intended to test your knowledge of the material covered in this chapter. Think through these problems, and make certain you understand what is going on. Ask yourself if your answer makes sense. Many of these questions are linked to the chapter learning goals. Therefore, successful completion of these problems indicates you have mastered the learning goals for this chapter. You will receive the greatest benefit from this section if you use it as a mock exam. You will then discover which topics you have mastered and which topics you need to study in more detail.

True–False

1. The Arrhenius theory of a base accounts for the basicity of substances that do not contain OH^- groups.

2. An acid–base reaction is a proton–transfer reaction according to the Bronsted–Lowry theory of acids and bases.

3. A conjugate base has one more proton than its acid.

4. In every acid–base reaction, the proton is transferred to the weaker base.

5. The strength of an acid and the strength of its conjugate base are directly related (the stronger the acid, the stronger its conjugate base).

6. The value for the percent dissociation of an acid depends on the acid.

7. For a diprotic acid, H_2A, the principal reaction is the dissociation of HA^-.

8. The pH of a salt solution prepared by reacting a strong acid with a weak base is greater than 7.

9. The strength of a binary acid decreases down a group in the periodic table.

10. For oxoacids with the same number of OH groups and the same number of O atoms, acid strength decreases with increasing electronegativity of Y.

Chapter 14 – Aqueous Equilibria: Acids and Bases

Multiple Choice

1. For polyprotic acids
   a. the pH is determined from the last dissociation step
   b. more than one dissociable proton is present
   c. all of the acidic hydrogens are lost in one step
   d. the stepwise dissociation constants increase as each H^+ is lost

2. The pH of a salt solution
   a. = 7 for all salts
   b. > 7 for salts derived from a strong acid and a strong base
   c. < 7 for salts derived from a strong acid and a weak base
   d. = 7 for salts derived from a weak acid and a weak base

3. The strength of a binary acid
   a. decreases across a period
   b. increases as the strength of the H–A bond increases
   c. increases as the electronegativity of A decreases
   d. increases down a group

4. For oxoacids, H_nYO_m,
   a. the H is bonded to Y
   b. the strength increases as the H–O bond becomes stronger or more polar
   c. the strength increases as the H–O bond becomes weaker or less polar
   d. increases as the H–O bond becomes weaker or more polar

5. Lewis acids
   a. donate electron pairs.
   b. have a more limited definition than Brønsted acids
   c. include cations and neutral molecules having filled valence orbitals.
   d. include metal ions which can act as cationic Lewis acids.

6. For the reaction HF (*aq*) + NH_2^- (*aq*) ⇌ F^- (*aq*) + NH_3 (*aq*), the stronger base is
   a. HF
   b. NH_2^-
   c. F^-
   d. NH_3

7. Acid–base indicators
   a. are strong acids
   b. are substances that change color in a specific pH range
   c. have the same colors in their acid (HIn) and conjugate base (In^-) forms
   d. can be used to determine the exact pH of a solution

8. The conjugate base
   a. has one less proton than its acid
   b. transfers a proton to the solvent in an acid–dissociation equilibrium
   c. is comparable in strength to its conjugate acid
   d. is always weaker than its conjugate acid

Chapter 14 – Aqueous Equilibria: Acids and Bases

9. Weak acids
   a. completely dissociate in water
   b. have large values of K_a
   c. have dissociation reactions in which the position of equilibrium lies far to the right
   d. have dissociation reactions in which the position of equilibrium is characterized by K_a

10. Alkaline earth hydroxides, $M(OH)_2$ (aq)
    a. are stronger bases than the alkali metal hydroxides
    b. have a higher pH than the alkali metal hydroxides
    c. are more soluble than the alkali metal hydroxides
    d. have a lower pH than the alkali metal hydroxides

Matching

Ion–product for water　　　　　　　　a. contains more than one dissociable proton and dissociate in a stepwise manner

Conjugate acid–base pair　　　　　　　b. H^+ acceptor

Base–dissociation equilibrium　　　　　c. ion in which H^+ is bonded to the oxygen atom of a solvent water molecule

Polyprotic acids　　　　　　　　　　　d. given by $K_w = [H_3O^+][OH^-]$

Hydronium ion　　　　　　　　　　　e. chemical species whose formulas differ by only one proton

pH scale　　　　　　　　　　　　　　f. a reaction in which the base accepts a proton from the solvent, which acts as an acid

Acid–base indicator　　　　　　　　　g. logarithmic scale used to express the hydronium ion concentration

Lewis acid　　　　　　　　　　　　　h. a substance that changes color in a specific pH range

Brønsted–Lowry base　　　　　　　　　i. substance that accepts a pair of electrons

Fill–in–the–Blank

1. The position of equilibrium for the dissociation of water lies _____.

2. The position of an acid–dissociation equilibrium is characterized by the _____ _____.

3. The sum of an acid–dissociation reaction and a base–dissociation reaction is the _____ _____.

4. The pH of a salt solution is determined by the _____ of the constituent cations and anions.

5. For salts that contain acidic cations and basic anions, the pH depends on the relative _____ of the cation and the relative _____ of the anion.

257

Chapter 14 – Aqueous Equilibria: Acids and Bases

6. The extent of dissociation of an acid is often determined by the _____ of the H–A bond.

7. The H–A bond _____ is the most important factor for determining acid strength for binary acids of elements in the same row.

8. For oxoacids that contain the same atom Y but different number of oxygen atoms, acid strength increases with _____ of Y.

9. Lewis acids include cations and neutral molecules having _____ that can _____ from a Lewis base.

10. _____ Lewis acids include the halides of group 3A elements and oxides of nonmetals.

Problems

1. What is the [OH$^-$] for a solution of Ca(OH)$_2$ whose pH = 9.87?

2. Calculate the pH of a solution prepared by dissolving 0.583 g of Ba(OH)$_2$ in 125 mL of water.

3. Write chemical equations so that they proceed from left to right for reactions involving HSO$_4^-$, SO$_4^{2-}$, CH$_3$COOH and CH$_3$CO$_2^-$; NH$_4^+$, NH$_3$, HNO$_2$, and NO$_2^-$.

4. The percent dissociation for a 1.50×10^{-3} M diethylbarbituric acid (veronal) solution is 0.51%. Calculate the pH of this solution and determine the K_a for the acid.

5. Determine the K_a of histidine, a weak organic acid, if the pH of a 2.50×10^{-3} M solution is 5.89.

6. What is the pH of a 0.150 M solution of pyridine, a weak base whose formula is C$_6$H$_5$N? $K_b = 1.8 \times 10^{-9}$

7. Calculate the concentration of all species present and the pH of a 0.025 M H$_2$C$_2$O$_4$ solution.

8. Calculate the pH of a 0.750 M solution of saccharin. ($K_a = 2.10 \times 10^{-12}$)

9. Calculate the pH of a 0.150 M solution of benzylamine. ($K_b = 2.14 \times 10^{-5}$)

10. Determine a) K_b for the conjugate base of formic, ascorbic, and hypochlorous acids and b) the K_a for the conjugate acid of hydrazine, aniline, and methylamine.

11. Determine if a solution containing the OCl$^-$ and NH$_4^+$ ions is acidic, neutral, or basic.

12. 2.75 g of KCN is dissolved in 250 mL of water. What is the concentration of HCN (*aq*) at equilibrium? What is the pH of the solution?

13. Determine which acid is stronger, and explain why.
    a. HBr (*aq*) or HI (*aq*)
    b. H$_2$Se (*aq*) or HBr (*aq*)
    c. H$_3$AsO$_4$ or H$_3$PO$_4$
    d. H$_3$PO$_4$ or H$_3$PO$_3$

14. Identify the Lewis acid and the Lewis base in the following reaction:

$$Co^{3+} (aq) + 6\,CN^- (aq) \rightarrow Co(CN)_6^{3-} (aq)$$

Challenge Problem

A 50.0% by mass solution of H_3PO_4 has a density of 1.3334 g/mL. Calculate the pH and concentrations of all phosphate–containing species present.

Inquiry Based Problem

You have to determine the identity of the anion in an unknown salt. Based on the information in this chapter, how would you do this?
Consider the following:

1. What do you know about the hydrolysis reactions of anions? Write a generic equation for this reaction. Write the equilibrium equation for this reaction.
2. From this reaction, can you predict the pH of the solution of the salt?
3. What simple experiment can you do to determine the pH of the solution?
4. Knowing the pH, how can you determine the identity of the anion?

CHAPTER 15

APPLICATIONS OF AQUEOUS EQUILIBRIA

Chapter Learning Goals

A. Neutralization Equilibria
1. From the relative strengths of the acid and base in a neutralization reaction, predict whether the pH will be equal to, greater than, or less than 7.00 at the equivalence point.
2. Write balanced net ionic equations for the four types of neutralization reactions.
3. Calculate pH values for a strong acid–strong base titration.
4. Calculate pH values for a weak acid–strong base titration.
5. Given a titration curve, select which indicator(s) could be used to detect the equivalence point.
6. Calculate pH values for a weak base–strong acid titration.
7. Calculate pH values for a diprotic acid–strong base titration.

B. Common–Ion Effect and Buffer Solutions
1. Describe the effect on pH when the conjugate base of a weak acid is added to a solution of the weak acid and the conjugate acid of a weak base is added to a solution of the weak base. Calculate the concentrations of all species present at equilibrium.
2. Given the initial concentration of weak acid (or weak base) and its conjugate base (or weak acid), determine the equilibrium concentrations of all species and the pH of a buffer solution.
3. Calculate the pH of a buffer after the addition of OH^- or H_3O^+.
4. Use the Henderson–Hasselbalch equation to calculate the pH of a buffer.
5. From a table of weak acids and their K_a values, select the weak acid/conjugate base pair that would make the best buffer at a given pH.

C. Solubility Equilibria
1. Write the solubility product expression for a given ionic compound.
2. Given the K_{sp} of an ionic compound, calculate its solubility and vice versa.
3. Given the K_{sp} of an ionic compound, calculate its solubility in the presence of a common ion.
4. Identify ionic compounds that have an enhanced solubility at low pH. Write chemical equations showing why the solubility increases as $[H_3O^+]$ increases.
5. From K_{sp} values, determine whether a precipitate will form on mixing solutions of ionic compounds.
6. Determine which metal sulfides will precipitate from a solution of metal ions on addition of H_2S at a specified pH.

D. Complex Equilibria
1. Given K_f, calculate the concentrations of the species present in a complex–ion equilibrium. Given K_{sp} and K_f, calculate the solubility of a slightly soluble ionic compound in an excess of the complexing agent.
2. Write chemical equations showing how a given oxide or hydroxide exhibits amphoteric behavior.

Chapter in Brief
In the previous chapter, you applied the concept of equilibrium to acid–base chemistry. In this chapter, you will use those concepts to calculate the pH of mixtures of acids and bases and to calculate pH

Chapter 15 – Applications of Aqueous Equilibria

titration curves. You will also apply the concepts of equilibria to the dissolution and precipitation of slightly soluble salts as well as the formation and dissociation of complex ions. This chapter begins with an examination of the chemistry that occurs when solutions of acids and bases of varying strength are mixed together. You will learn about the effect of a common ion in a solution of either an acid or a base and how this effect can be used to prepare buffer solutions. You will also learn how to determine the pH of a buffer solution after a small amount of acid or base has been added. You also will use the concepts of aqueous solution equilibria to determine the K_{sp} of a slightly soluble salt, calculate the solubility of a salt, and determine if a salt will precipitate from solution. Finally, you will learn how the principles of aqueous solution equilibria can be used to separate ions by selective precipitation, which is employed in the qualitative analysis of solutions of unknown ions.

15.1 Neutralization Reactions

A.1, A.2.

A. Strong Acid–Strong Base.
  1. Net ionic equation for the neutralization reaction: H_3O^+ (aq) + OH^- (aq) → 2 H_2O (l).
  2. When equal numbers of moles of acid and base are mixed together, $[H_3O^+]$ and $[OH^-]$ = 1 × 10^{-7} M.
  3. Reaction proceeds far to the right.
     a. equilibrium constant, K_n, for the reaction is the reciprocal of the ion–product constant for water
     b. $K_n = \dfrac{1}{[H_3O^+][OH^-]} = \dfrac{1}{K_w}$
  4. pH = 7.00.
B. Weak Acid–Strong Base.
  1. Net ionic equation for the neutralization reaction involves proton transfer from HA to the strong base, OH^-.
     a. HA (aq) + OH^- (aq) → H_2O + A^- (aq)
  2. $K_n = K_a (1/K_w)$
     a. K_n equals the product of the equilibrium constants for the reactions added.
     b. HA (aq) + H_2O (l) ⇌ H_3O^+ (aq) + A^- (aq) K_a
        H_3O^+ (aq) + OH^- (aq) ⇌ 2 H_2O (l) $1/K_w$
  3. Neutralization of any weak acid by a strong base goes 100% to completion.
     a. OH^- has a great affinity for protons.
  4. pH > 7.00
     a. Anion of a weak acid is a weak base.
C. Strong Acid–Weak Base.
  1. Net ionic equation for the neutralization reaction involves proton transfer from the strong acid, H_3O^+ to the weak base, B.
     a. H_3O^+ (aq) + B (aq) → H_2O (l) + BH^+ (aq)
  2. Equilibrium constant, (K_n), obtained from multiplying the equilibrium constants for the reactions that add to give the net ionic equation.
     a. $K_n = (K_b)(1/K_w)$
     b. B (aq) + H_2O (l) ⇌ BH^+ (aq) + OH^- (aq) K_b
        H_3O^+ (aq) + OH^- (aq) ⇌ 2 H_2O (l) $1/K_w$
  3. Neutralization of any weak base with a strong acid goes 100% to completion.
     a. H_3O^+ is a powerful proton donor.

Chapter 15 – Applications of Aqueous Equilibria

   4. pH < 7.00
      a. Cation of a weak base is a weak acid.
   D. Weak Acid–Weak Base.
      1. Neutralization reaction involves proton transfer from the weak acid to the weak base.
         a. HA (aq) + B (aq) $\rightleftarrows$ BH$^+$ (aq) + A$^-$ (aq)
      2. Obtain K_n from K_a for the weak acid, K_b for the weak base, and $1/K_w$.
         a. HA (aq) + H$_2$O (l) $\rightleftarrows$ H$_3$O$^+$ (aq) + A$^-$ (aq) K_a
            B (aq) + H$_2$O (l) $\rightleftarrows$ BH$^+$ (aq) + OH$^-$ (aq) K_b
            H$_3$O$^+$ (aq) + OH$^-$ (aq) $\rightleftarrows$ 2 H$_2$O (l) $1/K_w$
         b. $K_n = (K_a)(K_b)(1/K_w)$
      3. Less tendency to proceed to completion than neutralizations involving strong acids or strong bases.

EXAMPLE:
Write the net ionic equation, and predict the pH for the following reactions:

HF (aq) + KOH   HClO$_3$ (aq) + CH$_3$NH$_2$ (a weak base)

SOLUTION:
The first equation involves the reaction of a weak acid with a strong base. The net ionic equation is

HF (aq) + OH$^-$ (aq) $\rightleftarrows$ H$_2$O (l) + F$^-$ (aq)

The pH of this solution is expected to be greater than 7 (basic) since the anion of a weak acid is a weak base.

The second equation involves the reaction of a strong acid with a weak base. The net ionic equation is

H$_3$O$^+$ (aq) + CH$_3$NH$_2$ $\rightleftarrows$ H$_2$O (l) + CH$_3$NH$_3^+$ (aq)

The pH of this solution is expected to be less than 7 (acidic) since the cation of a weak base is a weak acid.

Workbook Problem 15.1
Write a balanced net ionic equation for the neutralization of benzoic acid by hydroxylamine. Determine K_n and the position of equilibrium for this neutralization reaction. Predict the pH of the solution.

Strategy: Use Table C.1 and C.3 in the Appendix of your textbook to determine the formulas, K_a and K_b for benzoic acid and hydroxylamine.

Step 1: Write the net neutralization reaction by writing individual reactions for the acid, base, and water.

Chapter 15 – Applications of Aqueous Equilibria

Step 2: Calculate K_n based on the equilibrium constants for the individual reactions in step 2.

Step 3: Determine the position of equilibrium from the value of K_n.

Step 4: Based on the values of K_a and K_b, predict the pH.

What key concept did you use?

15.2 The Common–Ion Effect

A. Common–ion effect — the shift in the position of an equilibrium on addition of a substance that provides an ion in common with one of the ions already involved in the equilibrium.
   1. Example of Le Châtelier's principle.
B. To determine the pH of a solution prepared from a weak acid and a salt of its conjugate base
   1. Identify the acid–base properties of the various species in solution.
   2. Consider the possible proton–transfer reaction these species can undergo.
   3. Principal reaction — dissociation of the weak acid.
   4. Set up a table of concentrations of species involved in the principal reaction.
      a. A^- comes from 2 sources.
         i. salt of the conjugate base, which provides the initial concentration of A^-
         ii. dissociation of weak acid, which determines the change in the concentration of A^-

EXAMPLE:
Calculate the concentration of all species present, the pH, and the percent dissociation of nitrous acid in a solution that is 0.100 M in HNO_2 and 0.050 M in $NaNO_2$.

SOLUTION: Because the salt, $NaNO_2$ is 100% dissociated, the species present initially are HNO_2, NO_2^-, Na^+, and H_2O. Na^+ is inert; HNO_2 is a weak acid ($K_a = 4.5 \times 10^{-4}$); NO_2^- is the conjugate base of a weak acid; and H_2O can be either an acid or a base. Because HNO_2 is the strongest acid present ($K_a > K_w$), the principal reaction is transfer of a proton from HNO_2 to H_2O. We can set up the following table:

263

Chapter 15 – Applications of Aqueous Equilibria

Principal Reaction	$HNO_2\ (aq)\ +\ H_2O\ (l)$	⇌	$H_3O^+\ (aq)$	$NO_2^-\ (aq)$
Initial Concentration (M)	0.100		0	0.050
Change (M)	$-x$		$+x$	$+x$
Eq. Concentration (M)	$0.100 - x$		$+x$	$0.050 + x$

The common ion in this problem is NO_2^-. The equilibrium equation for the principal reaction is

$$K_a = 4.5 \times 10^{-4} = \frac{[H_3O^+][NO_2^-]}{[HNO_2]} = \frac{(x)(0.050+x)}{(0.100-x)} \approx \frac{(x)(0.050)}{(0.100)}$$

x is assumed to be negligible because K_a is so small and the equilibrium is shifted to the left due to the common–ion effect.

$$x = [H_3O^+] = \frac{(4.5 \times 10^{-4})(0.100)}{(0.050)} = 9.0 \times 10^{-4}$$

Note that the assumption concerning the size of x is justified.

pH = $-\log (9.0 \times 10^{-4})$ = 3.05

The percent dissociation of HNO_2 is

$$\text{Percent dissociation} = \frac{[HNO_2]_{dissociated}}{[HNO_2]_{initial}} \times 100 = \frac{9.0 \times 10^{-4}}{0.100} \times 100 = 0.90\%$$

15.3 Buffer Solutions

B.2, B.3

A. Buffer solutions — resist drastic changes in pH.
   1. Contain a weak acid and its conjugate base or a weak base and its conjugate acid.
B. Add a small amount of base to a buffer solution.
   1. Acid component of the solution neutralizes the added base.
C. Add a small amount of acid to a buffer solution.
   1. Base component of the solution neutralizes the added acid.
D. Important in biological systems.
   1. pH of human blood (pH = 7.4) controlled by conjugate acid–base pairs (H_2CO_3/HCO_3^-).
E. For a buffer prepared from a weak acid, HA, and its conjugate base, B.
   1. $[H_3O^+] = K_a \frac{[HA]}{[B]}$.
   2. Can use initial concentrations in the calculations.
      a. K_a is small for commonly used buffer solutions.
      b. Initial concentrations are relatively large.
      c. Change in concentration, x, is generally negligible.
   3. Add acid or base to the buffer solution.
      a. alters the numbers of moles of acid and conjugate base
      b. Concentration ratio, [HA]/[B], changes by only a small amount.
         i. $[H_3O^+]$ changes by only a small amount.
F. Must take neutralization into account before calculating $[H_3O^+]$.
   1. Addition of OH^-.

a. HA (aq) + OH⁻ (aq) ⇌ H₂O (l) + A⁻ (aq)
b. Consume HA and produce A⁻.
   i. [HA] decreases and [A⁻] increases.
3. Addition of H_3O^+.
   a. A⁻ (aq) + H_3O^+ (aq) ⇌ H₂O (l) + HA (aq)
   b. Consume A⁻ and produce HA.
      i. [A⁻] decreases and [HA] increases.
4. Neutralization alters the number of moles of acid and conjugate base.
G. Buffer Capacity — a measure of the amount of acid or base that a buffer solution can absorb without a significant change in pH.
   1. Depends on how much weak acid and conjugate base is present.
   2. For equal volumes of solutions, the more concentrated the solution, the greater the buffer capacity.
   3. For solutions with the same concentration, increasing the volume increases the buffer capacity.

Workbook Problem 15.2

Calculate the pH of a 1.0 L buffer solution containing 0.45 M HCOOH and 0.55 M HCO₂Na. Determine the pH of this solution after the addition of 0.10 mol HCl (assume no volume change).

Strategy: The pH of the initial solution is determined from K_a and the equilibrium expression. The pH of the solution after addition of HCl is determined knowing the changes in concentration that occur after the addition of $[H_3O^+]$.

Step 1: Determine the principal reaction and equilibrium concentrations.

Step 2: Using the equilibrium equation, solve for $[H_3O^+]$.

Step 3: Solve for the pH.

Step 4: Write the neutralization reaction that occurs upon addition of HCl and set up a table showing the number of moles present before and after the addition of HCl.

265

Chapter 15 – Applications of Aqueous Equilibria

Step 5: Determine the concentrations of the buffer components after neutralization occurs.

Step 6: Substitute the concentrations of the buffer components into the equilibrium expression and calculate the pH.

What key concept did you use?

15.4 The Henderson–Hasselbalch Equation
A. $[H_3O^+]$ in a buffer solution depends on:
   1. Dissociation constant of weak acid.
   2. Concentration ratio [HA] : [A⁻]
   3. $[H_3O^+] = K_a \dfrac{[acid]}{[base]}$

B. Henderson-Hasselbalch equation:
   1. $pH = pK_a + \log \dfrac{[base]}{[acid]}$

C. Explains how pH affects the percent dissociation of a weak acid.
   1. $\log \dfrac{[base]}{[acid]} = pH - pK_a$
   2. Ratio of [base] : [acid] tells us how many acid molecules dissociate to produce conjugate base.
   3. If, for example, $pH = pK_a + 2$, then:
      a. $\log \dfrac{[base]}{[acid]} = 2$

Chapter 15 – Applications of Aqueous Equilibria

b. $\dfrac{[\text{base}]}{[\text{acid}]} = 1 \times 10^2 = \dfrac{100}{1}$

   i. indicates you have 100 base molecules and one acid molecule for a total of 101 molecules.

   ii. % dissociation = $\dfrac{100 \text{ base molecules}}{101 \text{ total molecules}} \times 100 = 99\%$

D. Useful for determining how to prepare a buffer solution.   🔑 B.5.
   1. Select a weak acid whose pK_a is close to the desired pH of the buffer solution.
      a. pK_a of the weak acid should be within ±1 pH units of the desired pH.
   2. Adjust the [base]/[acid] ratio.
E. The pH of a buffer solution does not depend on the volume of the solution.
   1. Depends only on pKa and the relative molar amounts of weak acid and conjugate base.

EXAMPLE:
You are performing an experiment that requires your solution be buffered at a pH of 4.32. Suggest an appropriate buffer system based upon the K_a values in Table C.1. in the Appendix of your textbook.

SOLUTION:
Use your knowledge of logarithms to choose your buffer system. Remember

$\log 10^x = x$.

$-\log 10^{-4} = 4$ and the $-\log 10^{-5} = 5$

You want to use an acid whose pK_a is close to 4.32. Therefore, you are looking for an acid whose K_a is close to $10^{-4.32}$ (= 4.79×10^{-5}). The acid with a K_a closest to 4.79×10^{-5} is benzoic acid, with a K_a of 6.5×10^{-5}.

A good buffer system would be benzoic acid and a salt of benzoic acid such as sodium benzoate.

Workbook Problem 15.3
Determine the [sodium benzoate]/[benzoic acid] ratio for a buffer system with a pH = 4.32. How would you prepare this buffer solution?

Strategy: Use the Henderson–Hasselbalch equation to calculate the [base]/[acid] ratio. This information provides the mole ratio of benzoate to benzoic acid.

Step 1: Calculate the mass needed to prepare the buffer.

Chapter 15 – Applications of Aqueous Equilibria

What key concept did you use?

15.5 pH Titration Curves
   A. Acid–base titration — a titrant, a solution containing a known concentration of base (or acid), is slowly added from a buret to a solution containing an unknown concentration of acid (or base).
   B. Equivalence point — the point at which stoichiometrically equivalent quantities of acid and base have been mixed together.
   C. pH titration curve — a plot of the pH of the solution versus the volume of added titrant.
      1. Use to identify the equivalence point in a titration.
      2. Useful in selecting a suitable indicator to signal the equivalence point.
      3. Calculate pH titration curves from the principles of aqueous solution equilibria.

15.6 Strong Acid–Strong Base Titrations   🔑 A.3.
   A. Before addition of any strong base.
      1. Initial $[H_3O^+]$ = concentration of strong acid
   B. Addition of strong base before the equivalence point.
      1. Decrease in $[H_3O^+]$.
         a. Added base will neutralize some of the H_3O^+ present.
      2. mmol H_3O^+ after neutralization = mmol H_3O^+ initial − mmol OH^- added.
         a. $[H_3O^+]$ after neutralization = $\dfrac{\text{mmol } H_3O^+ \text{ after neutralization}}{\text{total volume of acid and base}}$
   C. Addition of strong base at the equivalence point.
      1. Added just enough base to neutralize all the acid initially present.
      2. pH = 7.00
         a. Solution contains water and a salt derived from a strong base and a strong acid.
   D. Addition of strong base after the equivalence point.
      1. Excess of OH^- present.
         a. mmol of excess OH^- = mmol of OH^- added − mmol of acid initially present
      2. $[OH^-]$ after neutralization = $\dfrac{\text{mmol of excess } OH^-}{\text{total volume of acid and base}}$
      3. $[H_3O^+] = \dfrac{K_w}{[OH^-]}$
   E. pH titration curve.
      1. Plot pH versus milliliters of strong base added.
      2. Sharp increase in pH in the region near the equivalence point.
         a. characteristic of the titration curve for any strong acid–strong base titration
         b. can use to identify the equivalence point when the concentration of the acid is unknown

268

Chapter 15 – Applications of Aqueous Equilibria

15.7 Weak Acid–Strong Base Titrations

A. Before addition of strong base.
   1. pH calculated in the same manner as for a solution of a weak acid.
B. Addition of strong base before the equivalence point.
   1. mmol of base, A^-, after neutralization = mL of strong base × [strong base]
   2. $[A^-] = \dfrac{\text{mmol of } A^-}{\text{total mL of acid and base}}$

   3. mmol of acid, HA, after neutralization = mmol of HA initially – mmol of A^-

   4. [HA] after neutralization = $\dfrac{\text{mmol HA after neutralization}}{\text{total volume of acid and base}}$

   5. $\text{pH} = pK_a + \log \dfrac{[A^-]}{[HA]}$

C. Addition of strong base at the equivalence point.
   1. Basic salt solution.
   2. mmol A^- after neutralization = mL of strong base × [strong base]
   3. $[A^-] = \dfrac{\text{mmol of } A^- \text{ after neutralization}}{\text{total mL of acid and base}}$

   4. Calculate pH from the method outlined in section 14.14 in your textbook.
      a. pH > 7.00
      b. Anion of the weak acid is a base.
D. Addition of strong base after the equivalence point.
   1. mmol OH^- added = mL of strong base × [strong base]
   2. mmol A^- present = mmol A^- at the equivalence point
   3. mmol OH^- present = mmol OH^- added – mmol A^- present
   4. $[OH^-]$ present = $\dfrac{\text{mmol } OH^- \text{ present}}{\text{total mL of acid and base}}$

   5. $[H_3O^+] = \dfrac{K_w}{[OH^-]}$

E. Can use pH range at the equivalence point to select a suitable indicator.
F. pH titration curve – see Figure 15.8, pate 612 in your textbook.
   1. Initial rise in pH greater than initial rise in strong acid/strong base titration.
   2. Near equivalence point, pH increase is smaller than strong acid/strong base.
   3. pH at equivalence point > 7 since anion of weak acid is a base.
   4. as value of K_a decreases, the increase in pH at the equivalence point gets smaller.
      a. equivalence point more difficult to detect.

EXAMPLE:
50.0 mL of 0.200 M HF is titrated with 0.100 M NaOH. How many mL of base are required to reach the equivalence point? Calculate the pH at each of the following points: a) after addition of 10.0 mL of base, b) halfway to the equivalence point, c) at the equivalence point, and d) after addition of 130.0 mL of base.

Chapter 15 – Applications of Aqueous Equilibria

SOLUTION: To determine the number of mL of base needed to reach the equivalence point, we need to first calculate the number of mmol of HF present.

$$\text{mmol HF} = 50.0 \text{ mL} \times \frac{0.200 \text{ mmol HF}}{1 \text{ mL}} = 10.0 \text{ mmol}$$

We need 10.0 mmol of NaOH to reach the equivalence point, which means we need 100 mL of 0.100 M NaOH.

To calculate the pH at the different points in the titration, we need to use the steps that are outlined above.

After addition of 10.0 mL of NaOH
mmol F$^-$ = 10.0 mL × 0.100 mmol/mL = 1.00 mmol
[F$^-$] = 1.00 mmol/60.0 mL = 1.67 × 10^{-2} M
mmol HF = 10.0 mmol – 1.00 mmol = 9.0 mmol
[HF] = 9.0 mmol/60.0 mL = 0.15 M

$$\text{pH} = \text{p}K_a + \log\frac{[\text{F}^-]}{[\text{HF}]} \qquad K_a = 3.5 \times 10^{-4} \quad \text{p}K_a = 3.46$$

$$\text{pH} = 3.46 + \log\frac{(1.67 \times 10^{-2})}{(0.15)} = 2.51$$

Halfway to the equivalence point, we have added 50.0 mL of NaOH (since 100 mL is required to reach the equivalence point).

mmol F$^-$ = 50.0 mL × 0.100 mmol/mL = 5.00 mmol
[F$^-$] = 5.00 mmol/100 mL = 0.500 M
mmol HF = 10.0 mmol – 5.00 mmol = 5.00 mmol
[HF] = 5.00 mmol/100 mL = 0.0500 mmol

$$\text{pH} = 3.46 + \log\frac{(0.0500)}{(0.500)} = 3.46$$

At the equivalence point, all the HF has been neutralized, and the pH is determined by the concentration of F$^-$.

mmol F$^-$ = 100 mL × 0.100 mmol/mL = 10.0 mmol
[F$^-$] = 10.0 mmol/150 mL = 6.67 × 10^{-2} M

F$^-$ is the anion of a weak acid and therefore gives a basic solution.

$$K_b = \frac{K_w}{K_a} = \frac{1.0 \times 10^{-14}}{3.5 \times 10^{-4}} = 2.9 \times 10^{-11}$$

Principal reaction: F$^-$ (aq) + H$_2$O (aq) $\rightleftarrows$ HF (aq) + OH$^-$ (aq)

$$K_b = 2.9 \times 10^{-11} = \frac{[\text{HF}][\text{OH}^-]}{[\text{F}^-]} = \frac{(x)(x)}{(6.67 \times 10^{-2})}$$

$x = 1.4 \times 10^{-6} = [OH^-]$

$[H_3O^+] = \dfrac{1.0 \times 10^{-14}}{1.4 \times 10^{-6}} = 7.1 \times 10^{-9}$

$pH = -\log(7.1 \times 10^{-9}) = 8.15$

After addition of 130.0 mL of NaOH
  mmol OH^- added = 130.0 mL × 0.100 mmol/.mL = 13.0 mmol
  mmol F^- = 10.0 mmol
  mmol OH^- present = 13.0 mmol − 10.0 mmol = 3.0 mmol
  $[OH^-]$ = 3.0 mmol/180.0 mL = 1.67×10^{-2} M

$[H_3O^+] = \dfrac{1.0 \times 10^{-14}}{1.67 \times 10^{-2}} = 6.0 \times 10^{-13}$

$pH = -\log(6.0 \times 10^{-13}) = 12.2$

15.8 Weak Base–Strong Acid Titrations  A.6.
  A. Before addition of strong acid.
    1. Calculate pH in the same manner as a weak base.
  B. Addition of strong acid before the equivalence point.
    1. Have a B/BH$^+$ buffer solution.
      a. leveling of the titration curve in the buffer region between the start of the titration and the equivalence point – see Figure 15.10, page 614 in your textbook.
    2. Calculate pH from the Henderson–Hasselbalch equation.
  C. Addition of strong acid at the equivalence point.
    1. Acidic salt solution of BH$^+$.
    2. Calculate pH using the method outlined in section 14.14 of your textbook.
    3. pH < 7.00.
  D. Addition of strong acid after the equivalence point.
    1. Excess H_3O^+ present.
    2. Calculate pH from $[H_3O^+]$ excess.

 Workbook Problem 15.4
50.00 mL of 0.175 M trimethylamine, $(CH_3)_3N$, is titrated with 0.100 M HCl. a) How many mL of acid are required to reach the equivalence point? b) Calculate the initial pH, and the pH after c) the addition of 10.00 mL of acid, d) halfway to the equivalence point, e) at the equivalence point, and f) after addition of 100.00 mL of acid.

Strategy: Use the method outlined above to calculate the pH at various points in the titration.

a) mL of acid required to reach the equivalence point:
Step 1: Write the neutralization reaction that occurs, and determine the number of mmoles of $(CH_3)_3N$ in the initial solution.

Chapter 15 – Applications of Aqueous Equilibria

Step 2: Determine the mmol of HCl needed to react with the $(CH_3)_3N$ and the volume of 0.100 M HCl needed.

b) Initial pH of the $(CH_3)_3N$ solution:

Step 1: Determine the initial pH of the solution, using the method learned in section 15.12 in your textbook.

c) pH of the solution after addition of 10.00 mL of acid:

Step 1: You need to use the Henderson–Hasselbalch equation to calculate the pH at this point. Before proceeding, do a little thinking. The Henderson–Hasselbalch equation has the form

$$pH = pK_a + \log \frac{[\text{base}]}{[\text{acid}]}$$

Keep in mind that the acid in this equation is the conjugate acid of $(CH_3)_3N$ and that the pK_a refers to the K_a of the conjugate acid of $(CH_3)_3N$. (Don't forget that the conjugate acid has one more H^+ than its conjugate base.) You are now armed with the information needed to proceed.

Step 2: Calculate the concentration of $(CH_3)_3N$ and its conjugate acid after the addition of 10.00 mL of HCl.

Step 3: Determine the K_a and pK_a of the conjugate acid of $(CH_3)_3N$.

Step 4: Using the Henderson–Hasselbalch equation, calculate the pH of the solution.

Chapter 15 – Applications of Aqueous Equilibria

d) pH at the equivalence point:
Step 1: Once again, use the Henderson–Hasselbalch equation to calculate the pH. Remember, halfway to the equivalence point. the amount of the $(CH_3)_3N$ that has reacted equals the amount of the conjugate acid that has been produced. That is, [base] = [acid].

e) pH at the equivalence point:
Step 1: You need to do a little more thinking. At the equivalence point, all of the $(CH_3)_3N$ present has now been converted to its conjugate acid. Now consider what is happening to the conjugate acid in an aqueous solution.

Step 2: Write the principal reaction that is occurring at the equivalence point.

Step 3: Calculate the concentration of the conjugate acid.

Step 4: Determine the pH of the solution, using the method learned in section 15.14 in your textbook.

f) pH after the addition of 100.00 mL of 0.100 M HCl:
Step 1: Determine the number of mmoles of excess HCl.

Step 2: Determine the concentration of excess HCl, and calculate the pH.

What key concepts did you use?

15.9 Polyprotic Acid–Strong Base Titrations   A. 7.
A. Diprotic acid — has two dissociable protons and reacts with two molar amounts of OH^-.
B. Amino acids — both acidic and basic and can be protonated by strong acids.

273

Chapter 15 – Applications of Aqueous Equilibria

   1. H_2A^+ — protonated form; neutralized to HA.
      a. $H_2A^+ (aq) + H_2O (l) \rightleftarrows H_3O^+ (aq) + HA (aq)$ K_{a1}
   2. HA — neutralized to A^-.
      a. $HA (aq) + H_2O (l) \rightleftarrows H_3O^+ (aq) + A^- (aq)$ K_{a2}
 C. Before addition of strong base.
   1. Calculate the pH of a diprotic acid.
   2. Principal reaction — dissociation of H_2A^+.
 D. Addition of strong base before the first equivalence point.
   1. H_2A^+ is converted to HA.
   2. H_2A^+/HA buffer solution.
   3. Calculate $[H_2A^+]$ and [HA] of the amino acid in the same way [HA] and $[A^-]$ are calculated in a weak acid–strong base titration.
   4. Use the Henderson–Hasselbalch equation to calculate pH.
 E. Addition of strong base at the first equivalence point.
   1. All of the H_2A^+ is converted to HA.
   2. Principal reaction — proton transfer between HA molecules.
      a. $2 HA (aq) \rightleftarrows H_2A^+ (aq) + A^- (aq)$ $K = K_{a2}/K_{a1}$
   3. pH at the first equivalence point = average of pK_{a1} and pK_{a2}
      a. $pH = \dfrac{pK_{a1} + pK_{a2}}{2}$
      b. isoelectric point — the [HA] is at a maximum and the $[H_2A^+]$ and $[A^-]$ are very small and equal
         i. useful for separating mixtures of amino acids
 F. Addition of strong base between the first and second equivalence points.
   1. HA is converted to A^-.
   2. HA/A^- buffer solution.
   3. Calculate [HA] and $[A^-]$ of the amino acid in the same way [HA] and $[A^-]$ are calculated in a weak acid–strong base titration.
   4. Use the Henderson–Hasselbalch equation to calculate pH.
 G. Addition of strong base at the second equivalence point.
   1. All of the HA is converted to A^-.
   2. $[A^-] = [H_2A^+]$ initially.
   3. Solution of a basic salt.
      a. principal reaction — $A^- (aq) + H_2O (l) \rightleftarrows HA (aq) + OH^- (aq)$ $K_b = K_w/K_{a2}$
      b. calculate $[OH^-]$ from equilibrium equation; calculate $[H_3O^+]$ and pH from $[OH^-]$
 H. Addition of strong base beyond the second equivalence point.
   1. Calculate pH from the excess $[OH^-]$.

15.10 Solubility Equilibria
 A. Dissolution or precipitation of a sparingly soluble ionic compound.
   1. Principles of solubility equilibria — examines quantitative aspects of solubility and precipitation phenomena.

 B. Solubility product constant = K_{sp}
   1. $M_aX_b (s) \rightleftarrows a M^{b+} (aq) + b X^{a-} (aq)$
   2. $K_{sp} = [M^{b+}]^a [X^{a-}]^b$.

15.11 Measuring K_{sp} and Calculating Solubility from K_{sp}
 A. K_{sp} — measured by experiment.
   1. Temperature dependent.
   2. Use to calculate the solubility of a compound.
      a. solubility — amount of compound that dissolves per unit volume of saturated solution

Chapter 15 – Applications of Aqueous Equilibria

3. Can be calculated from the molar solubility of a compound.

EXAMPLE:

The solubility of Ag_2CrO_4 is 6.50×10^{-5} M. Calculate K_{sp} for Ag_2CrO_4.

SOLUTION: The solubility equilibrium for Ag_2CrO_4 is

$$Ag_2CrO_4 \,(s) \rightleftarrows 2\,Ag^+ \,(aq) + CrO_4^{2-} \,(aq)$$

If 6.50×10^{-5} mol is the amount of Ag_2CrO_4 that dissolves in 1.00 L of solution, then the $[Ag^+] = 2(6.50 \times 10^{-5}) = 1.30 \times 10^{-4}$ M and $[CrO_4^{2-}] = 6.50 \times 10^{-5}$ M.

Substituting these values into the equilibrium expression gives

$$K_{sp} = [Ag^+]^2[CrO_4^{2-}] = (1.30 \times 10^{-4})^2 (6.50 \times 10^{-5}) = 1.10 \times 10^{-12}$$

Workbook Problem 15.5

Calculate the K_{sp} for a solution of $Cd(OH)_2$ prepared in pure water, given that the $[Cd^{2+}] = 1.10 \times 10^{-5}$ M.

Strategy: Write the balanced chemical equation for the solubility of $Cd(OH)_2$, and use the stoichiometry and information given to calculate K_{sp}.

Step 1: Write the solubility equilibrium expression for $Cd(OH)_2$.

Step 2: Determine the concentration of Cd^{2+} and OH^- and calculate the K_{sp}.

What key concept did you use?

Chapter 15 – Applications of Aqueous Equilibria

Workbook Problem 15.6

The K_{sp} of $Fe(OH)_3$ is 2.6×10^{-39}. Calculate the molar solubility of $Fe(OH)_3$.

Strategy: Write the solubility equilibrium for $Fe(OH)_3$ and the equilibrium expression. Using the method described in Chapter 13, calculate the molar solubility of $Fe(OH)_3$.

Step 1: Let x be the number of mol/L of $Fe(OH)_3$ that dissolves. The saturated solution then contains x mol/L of Fe^{3+} and $3x$ mol/L of OH^-. Solve for K_{sp}.

What key concept did you use?

15.12 Factors That Affect Solubility

A. Common–ion effect.
   1. Solubility of a slightly soluble ionic compound is decreased by the presence of a common ion in the solution.

B. The pH of the solution.
   1. Solubility of an ionic compound increases with decreasing pH of the solution if the compound contains a basic anion.

C. Formation of complex ions.
   1. Solubility of an ionic compound increases dramatically if the solution contains a Lewis base that can form a coordinate covalent bond to the metal cation.
   2. Complex – ion that contains a metal cation bonded to one or more small molecules or ions.
   3. Formation constant, K_f — equilibrium constant for formation of the complex ion from the hydrated metal cation.
      a. measures the stability of a complex ion

D. Amphoterism.
   1. Amphoteric oxides — soluble both in strongly acidic and in strongly basic solutions.

EXAMPLE:
Determine the molar solubility of $Cu(C_2O_4)$ ($K_{sp} = 2.87 \times 10^{-8}$ at 25°C) in a 0.75 M $CuCl_2$ solution.

SOLUTION: The solubility equilibrium expression is

$$Cu(C_2O_4)\,(s) \rightleftharpoons Cu^{2+}\,(aq) + C_2O_4^{2-}\,(aq)$$

The equilibrium expression for this reaction is

$K_{sp} = [Cu^{2+}][C_2O_4^{2-}]$

Let x be the number of mol/L of $Cu(C_2O_4)$ that dissolves. We can now construct a table showing the equilibrium concentrations:

Solubility Equilibrium	$Cu(C_2O_4)$ (s) ⇌	Cu^{2+} (aq) +	$C_2O_4^{2-}$ (aq)
Initial Concentration (M)		0.75	0
Equilibrium Concentration (M)		$0.75 + x$	$+x$

Given the value of K_{sp}, we can assume that x is negligible. Therefore, the equilibrium concentration of Cu^{2+} is approximately 0.75 M. Substituting these values into the equilibrium expression gives

$2.87 \times 10^{-8} = (0.75)(x)$ $x = 3.8 \times 10^{-8}$

Workbook Problem 15.7

The initial pH of a solution containing $Sn(OH)_2$ was 9.35 after addition of excess NH_3. Determine the molar solubility of $Sn(OH)_2$ in this solution. For $Sn(OH)_2$, $K_{sp} = 5.4 \times 10^{-27}$.

Step 1: Write the solubility equation and solubility expression.

Step 2: Using the pH, determine the [OH⁻] concentration.

Step 3: Let x be the number of mol/L of $Sn(OH)_2$ that dissolves. Construct a table showing the equilibrium concentrations.

Step 4: Calculate the solubility of $Sn(OH)_2$.

What key concept did you use?

EXAMPLE:
Write a balanced net ionic equation for the dissolution reaction between AgBr and $Na_2S_2O_3$, and calculate the equilibrium constant given K_{sp} (AgBr) = 5.4×10^{-13} and K_f {$Ag(S_2O_3)_2^{3-}$} = 4.7×10^{13}.

SOLUTION:
The net ionic equation for the dissolution of AgBr in $Na_2S_2O_3$ is obtained by combining the solubility expression for AgBr and the formation expression for $Ag(S_2O_3)_2^{3-}$.

$AgBr\ (s) \rightleftarrows Ag^+\ (aq) + Br^-\ (aq)$ $K_{sp} = 5.4 \times 10^{-13}$

$Ag^+\ (aq) + 2\ S_2O_3^{2-} \rightleftarrows Ag(S_2O_3)_2^{3-}$ $K_f = 4.7 \times 10^{13}$

Adding the two equations together and canceling out what is common on both the reactant and product sides gives

$AgBr\ (s) \rightleftarrows \cancel{Ag^+}\ (aq) + Br^-\ (aq)$ $K_{sp} = 5.4 \times 10^{-13}$

$\cancel{Ag^+}\ (aq) + 2\ S_2O_3^{2-}\ (aq) \rightleftarrows Ag(S_2O_3)_2^{3-}\ (aq)$ $K_f = 4.7 \times 10^{13}$

$\overline{AgBr\ (s) + 2\ S_2O_3^{2-}\ (aq) \rightleftarrows Ag(S_2O_3)_2^{3-}\ (aq) + Br^-\ (aq) \quad K = K_{sp} \times K_f = 25.3}$

15.13 Precipitation of Ionic Compounds   ☞ C.5.
A. Ion product (IP) — determines if a precipitate will form when solutions containing the constituent ions are mixed.
   1. For the salt, M_aX_b; IP = $[M^{b+}]^a[X^{a-}]^b$.
   2. Defined in the same way as K_{sp}.
   3. Concentrations are initial concentrations, not equilibrium concentrations.
B. IP > K_{sp}
   1. Supersaturated solution.
   2. Precipitation occurs.
C. IP = K_{sp}
   1. Saturated solution.
   2. Equilibrium exists.
D. IP < K_{sp}
   1. Unsaturated solution.
   2. Precipitation will not occur.

Workbook Problem 15.8
Will a precipitate form when 150 mL of 0.75 M $Zn(NO_3)_2$ is mixed with 250 mL of 1.50 M Na_2CO_3?

Chapter 15 – Applications of Aqueous Equilibria

Strategy: Write a metathesis reaction for the reaction between $Zn(NO_3)_2$ and Na_2CO_3, and apply the solubility rules in Chapter 4 to determine if a precipitate will form.

Step 1: Write the solubility equation and IP expression for the precipitate.

Step 2: Calculate the IP, and compare its value to the K_{sp}. Will a precipitate form?

What key concept did you use?

15.15 Separation of Ions by Selective Precipitation
A. To separate a mixture of ions in solution, add a reagent that will precipitate some of the ions but not others.
B. Can separate insoluble metal sulfides from the more soluble metal sulfides. C.6.
  1. Carry out separation in acidic solution.
  2. $MS\,(s) + 2\,H_3O^+\,(aq) \rightleftarrows M^{2+}\,(aq) + H_2S\,(aq) + 2\,H_2O\,(l)$
  3. K_{spa}, solubility product in acid.
     a. $K_{spa} = \dfrac{[M^{2+}][H_2S]}{[H_3O^+]^2}$
  4. Q_c, reaction quotient
     a. $Q_c = \dfrac{[M^{2+}][H_2S]}{[H_3O^+]^2}$ when system is **not** at equilibrium.
  5. $Q_c > K_{spa}$; metal sulfide precipitates out.
  6. Adjust $[H_3O^+]$ so that $Q_c > K_{spa}$ for the more insoluble metal sulfide.
     a. The more insoluble metal sulfide will precipitate out from the more soluble metal sulfide.

15.15 Qualitative Analysis
A. Procedure for identifying the ions present in an unknown solution.
B. Traditional scheme of analysis for metal cations involves the separation of 20 cations into five groups by selective precipitation (see Figure 15.18, page 632 in your textbook).
  1. Determine the presence or absence of the ions in each group with further separations and tests.

Chapter 15 – Applications of Aqueous Equilibria

C. Excellent vehicle for developing laboratory skills and learning about acid–base, solubility, and complex–ion equilibria.

Putting It Together

Aluminum phosphate is formed from the reaction of aluminum chloride and phosphoric acid. a) Write a balanced chemical equation for this reaction. b) If you begin with 125 g of aluminum chloride and 2.50 L of 0.75 M phosphoric acid, how many grams of aluminum phosphate will be produced? c) If you place 25.0 g of aluminum phosphate in water to produce a solution with a volume of 1.00 L, what are the equilibrium concentrations of Al^{3+} and PO_4^{3-}? ($K_{sp} = 1.3 \times 10^{-20}$) d) How does the addition of HCl affect the solubility of aluminum phosphate? e) If you mix 0.75 L of 5.00×10^{-3} M $AlCl_3$ with 1.25 L of 0.750 M Na_3PO_4, will a precipitate of aluminum phosphate form? If so, how many grams of $AlPO_4$ will form?

Self–Test

This section is intended to test your knowledge of the material covered in this chapter. Think through these problems, and make certain you understand what is going on. Ask yourself if your answer makes sense. Many of these questions are linked to the chapter learning goals. Therefore, successful completion of these problems indicates you have mastered the learning goals for this chapter. You will receive the greatest benefit from this section if you use it as a mock exam. You will then discover which topics you have mastered and which topics you need to study in more detail.

True–False

1. When HNO_3 reacts with an equimolar amount of NaOH, the pH of the solution will be equal to 7.00.

2. When HClO reacts with an equimolar amount of NaOH, the pH of the solution will be less than 7.00.

3. When the conjugate base of a weak acid is added to a solution of the weak acid, the equilibrium shifts to the right.

4. The best acid/base conjugate pair to use for preparing a buffer with a pH = 3.35 is HNO_2/NO_2^-.

5. The best indicator to use in the titration of a weak acid with $K_a \approx 1 \times 10^{-6}$ is methyl red.

6. The solubility of FeS increases with increasing pH.

7. The solubility of $PbCl_2$ decreases in a solution that contains both $PbCl_2$ and NaCl.

8. $PbSO_4$ will precipitate from a solution containing 1.5×10^{-3} M $Pb(NO_3)_2$ and 0.05 M Na_2SO_4.

9. The solubility of an ionic compound increases dramatically if the solution contains a substance that can bond to the metal cation.

Fill–in–the–Blank

Chapter 15 – Applications of Aqueous Equilibria

1. In the reaction between a strong acid and weak base, the net ionic equation for the neutralization reaction involves _____ from the _____ to the _____.

2. Neutralization reactions between a _____ and _____ have less tendency to proceed to completion than neutralization reactions involving _____ or _____.

3. The shift in the position of an equilibrium on addition of a substance that provides an ion in common with one of the ions already involved in the equilibrium is referred to as the _____.

4. A buffer solution contains a _____ or a _____ _____ and is able to resist drastic changes in _____.

5. A measure of the amount of acid or base that a buffer solution can absorb without a significant change in pH is referred to as the _____.

6. In preparing a buffer solution with a specific pH, you should select a weak acid whose _____ value is close to the desired pH of the buffer solution.

7. The pH range at the _____ of a titration curve can be used to select a suitable indicator for the titration.

8. The amount of compound that dissolves per unit volume of a saturated solution is the _____.

9. The _____ measures the stability of a complex ion.

10. The _____ determines if a precipitate will form when solutions containing the constituent ions are mixed.

Problems

1. Write balanced net ionic equations, and predict the pH for reactions between
   a. HNO_3 and KOH
   b. HNO_3 and NH_2NH_2
   c. benzoic acid and KOH
   d. benzoic acid and pyridine

2. Calculate the pH of a solution containing 0.15 mol of N_2H_4 and 0.10 mol of N_2H_5Cl in 1.0 L.

3. Determine the pH and concentration of all species present in a buffer solution containing 0.50 mol HClO and 0.35 mol NaClO in 1.0 L of solution. Determine the change in pH upon addition of 0.10 mol NaOH. Determine the change in pH on addition of 0.10 mol HCl.

4. Determine the ratio of lactic acid to lactate ion required for preparing a buffer solution whose pH is 4.25.

5. Determine the change in pH that will occur on addition of 0.150 mol HCl (assuming no volume change) to a 1 L buffer solution prepared from 0.25 M HCOOH and 0.10 M HCO_2Na. How does

Chapter 15 – Applications of Aqueous Equilibria

the buffer capacity of this solution compare to the buffer capacity of the solution in workbook problem 16.2 on page 259?

6. Calculate the pH of 50.0 mL of 0.0500 M HCl after addition of the following volumes of 0.1000M NaOH: a) 0.0 mL; b) 10.0 mL; c) 25.0 mL; d) 30.0 mL.

7. Calculate the pH of 50.0 mL of 0.0500 M NaCN after addition of the following volumes of 0.1000 M HCl: a) 0.0 mL; b) 10.0 mL; c) 25.0 mL; d) 30.0 mL. (**Warning!** This titration should never be attempted in the laboratory!)

8. The molar solubility of $Mg(OH)_2$ is 1.12×10^{-4} M. Calculate the value of K_{sp}.

9. Calculate the molar solubility for $Cu_3(PO_4)_2$, given that the $K_{sp} = 1.4 \times 10^{-37}$.

10. Calculate the solubility of a solution of $BaSO_4$ that contains 0.50 M Na_2SO_4. ($K_{sp} = 1.1 \times 10^{-10}$)

11. Which of the following compounds are more soluble in acidic solution than in pure water?
    a. CuBr   b. Ag_2S c. AgCN   d. MgF_2

12. What are the concentrations of Cu^{2+} and $Cu(NH_3)_4^{2+}$ in a solution prepared by adding 0.25 mol of $Cu(NO_3)_2$ to 1.0 L of 5.0 M NH_3? ($K_f = 5.6 \times 10^{11}$)

13. Will a precipitate form when 250 mL of 1.75 M $CaCl_2$ is mixed with 250 mL of 2.50 M Na_3PO_4?

14. Determine if it is possible to separate Cu^{2+} from Fe^{2+} by bubbling H_2S through a 0.30 M HCl solution that contains 0.005 M Cu^{2+} and 0.005 M Fe^{2+}. (K_{spa} for CuS = 6×10^{-16} and K_{spa} for FeS = 6×10^{2}.)

15. A student is given an unknown solution that may contain metal cations found in the qualitative analysis scheme on page 632 in your textbook. Upon addition of HCl to this solution, a precipitate (ppt A) formed. After separating out the precipitate, the remaining solution (soln a) was treated with H_2S. No visible change occurred. When this same solution (soln a + H_2S) was treated with NH_3, a precipitate (ppt B) formed. After separating out ppt B, $(NH_4)_2CO_3$ was added to the remaining solution (soln b). A white precipitate formed. Flame tests performed on soln b gave a fleeting violet color.

    a. What groups of ions are present? From the information given, what specific ions can be identified?

    b. Ppt A was treated with NH_3. No visible change occurred. The precipitate was separated from the solution (soln c). After treating the precipitate with hot water, the precipitate dissolved. The solution containing the dissolved precipitate was treated with K_2CrO_4, and a yellow precipitate formed. Soln c was treated with excess HNO_3, and a precipitate formed. What ions are present?

Challenge Problem

PbS has $K_{sp} = 3.0 \times 10^{-7}$. Determine the solubility of PbS in a buffer prepared from 0.25 M formic acid and 0.35 M sodium formate.

CHAPTER 16

ENTROPY, FREE ENERGY, AND EQUILIBRIUM

Chapter Learning Goals

A. Molecular Randomness and Chemical and Physical Changes
1. Qualitatively determine whether simple chemical or physical changes are spontaneous.
2. Qualitatively predict whether the sign of ΔS is positive or negative for a chemical or physical change.
3. On the basis of probability, determine which of two states has the higher entropy.
4. Calculate the standard entropy of reaction from the standard molar entropies of products and reactants.
5. Determine whether a reaction is spontaneous by determining the sign of ΔS_{total}.

B. Free Energy Change
1. Use the equation $\Delta G = \Delta H - T\Delta S$ to calculate the free energy of reaction and to determine the temperature at which a nonspontaneous reaction becomes spontaneous.
2. Calculate the standard free energy of reaction from standard free energies of formation.
3. Calculate the free energy of reaction for a system having nonstandard pressures and concentrations.
4. From the standard free energy of reaction, calculate the value of the equilibrium constant.

Chapter in Brief

In Chapter 8, you were introduced to the concepts of entropy and free energy and how the values of these thermodynamic functions can determine the spontaneity of a chemical reaction. In this chapter, you will explore these concepts in greater detail. Your study begins with a look at spontaneous processes as well as a brief review of enthalpy, entropy, and their relationship to spontaneous processes. You will also examine the relationship among entropy, probability, and temperature, and learn how to calculate the standard entropies of reaction. You will examine the second law of thermodynamics and how the thermodynamic properties of enthalpy, entropy, and free energy are interwoven to determine the spontaneity of a reaction. You will learn how to calculate free–energy changes for reactions that take place at both standard and nonstandard–state conditions. Your study of thermodynamics ends by examining the relationship between free energy and chemical equilibrium.

16.1 Spontaneous Processes

A. Spontaneous process — one that proceeds on its own without any external influence.

A.1

B. Spontaneous reaction always moves a system toward equilibrium.
  1. Spontaneity of forward or reverse reaction depends on:
     a. temperature
     b. pressure
     c. composition of the reaction mixture
  2. $Q < K$; reaction proceeds in the forward direction.
  3. $Q > K$; reaction proceeds in the reverse direction.
C. Spontaneity of a reaction is no indication of the speed of the reaction.

EXAMPLE:
Determine which of the following processes are spontaneous and which are nonspontaneous.

Chapter 16 – Entropy, Free Energy, and Equilibrium

   a. The odor of a dead skunk fills your car as you drive by.
   b. The handle of a metal spoon in a pot of hot soup is hot.
   c. The reaction of N_2 and H_2 to produce NH_3 at 300 K and partial pressures of 1.25 atm, 3.00 atm, and 2.50 atm, respectively. $K_p = 4.4 \times 10^5$ at 300 K.

SOLUTION: Both a and b are spontaneous processes. To determine if c is spontaneous or not, we must determine Q for the reaction. If $Q < K$, then the reaction will proceed spontaneously in the forward direction. If $Q < K$, then the reaction will proceed spontaneously in the reverse direction. The equilibrium expression for this reaction is

$$N_2\,(g) + 3\,H_2\,(g) \rightleftarrows 2\,NH_3\,(g)$$

$$Q = \frac{[NH_3]^2}{[N_2][H_2]^3} = \frac{(2.50)^2}{(1.25)(3.00)^3} = 0.185$$

$Q < K$; the reaction is spontaneous in the forward direction.

16.2 Enthalpy, Entropy, and Spontaneous Processes: A Brief Review
   A. Spontaneous reactions.
      1. Enthalpy alone does not account for the direction of spontaneous change.
      2. Some are accompanied by conversion of potential energy to heat.
         a. Reactions are exothermic.
      3. Some spontaneous reactions are endothermic.
         a. System moves spontaneously to a state of higher potential energy by absorbing heat from the surroundings.
   B. Second thermodynamic driving force for spontaneous change.
      1. Molecular systems tend to move spontaneously to a state of maximum randomness or disorder.
   C. Entropy, S — molecular randomness or disorder.
      1. State function.
      2. $\Delta S = S_{final} - S_{initial}$.
      3. ΔS is positive, increase in randomness or disorder of a system.
         a. solid → liquid
         b. liquid → gas
         c. A reaction results in an increase in the number of gaseous molecules.
         d. dissolution of certain solutes
      4. ΔS is negative; decrease in randomness or disorder of a system.

EXAMPLE:
   Predict the sign of ΔS in the system for
   a. $I_2\,(s) \rightarrow I_2\,(g)$
   b. $H_2O\,(g) \rightarrow H_2O\,(l)$
   c. $PbS\,(s) + 2\,HNO_3\,(aq) \rightarrow H_2S\,(g) + Pb(NO_3)_2\,(aq)$

SOLUTION:
   a. ΔS is positive; increase in randomness when the number of moles of gas is increased
   b. ΔS is negative; decrease in randomness when gaseous molecules are condensed into a liquid
   c. ΔS is positive; increase in randomness when the number of moles of gas is increased

Chapter 16 – Entropy, Free Energy, and Equilibrium

16.3 Entropy and Probability
A. Systems tend to a state of maximum randomness because the random state, which is more probable, can be achieved in more ways.
B. Probabilities of ordered and random states are proportional to the number of ways that the state can be achieved.
C. Boltzmann — The entropy of a particular state is related to the number of ways that the state can be achieved.
   1. $S = k \ln W$
      a. $\ln W$ = the number of ways that the state can be achieved
      b. $k = 1.38 \times 10^{-23}$ J/K
D. Gas expands spontaneously because the state of greater volume is more probable.
   1. For an ideal gas.
      a. $\Delta S = R \ln \dfrac{V_{final}}{V_{initial}}$
      b. $P = \dfrac{nRT}{V}$
      c. $\Delta S = R \ln \dfrac{P_{initial}}{P_{final}}$
         i. The entropy of a gas increases when its pressure decreases at constant temperature.
         ii. The entropy of a gas decreases when the pressure increases at constant temperature.

16.4 Entropy and Temperature
A. Entropy — also associated with molecular motion.
B. Random molecular motion increases as the temperature of a substance increases.
   1. Increase in the average kinetic energy of the molecules.
      a. Total energy is distributed among the individual molecules in a number of ways.
      b. Boltzmann — The more ways (W) that the energy can be distributed the greater the randomness of the state and the higher the entropy.
   2. Plot of entropy versus temperature (see Figure 16.8, page 657 in your textbook).
      a. At absolute zero, every substance is a solid where particles are rigidly fixed in a crystalline structure.
C. Third law of thermodynamics — The entropy of a perfectly ordered crystalline substance at 0 K is zero.

16.5 Standard Molar Entropies and Standard Entropies of Reaction
A. Standard molar entropy, $S°$ — the entropy of 1 mole of the pure substance at 1 atm pressure and a specified temperature, usually 25°C.
   1. Absolute entropies — measured with respect to an absolute reference point.
   2. Can compare entropies of different substances under the same conditions.
      a. $S°_{gas} > S°_{liquid} > S°_{solid}$
      b. $S°$ increases with increasing molecular complexity.
B. Standard entropy of reaction, $\Delta S°$.
   1. $\Delta S° = S°(products) - S°(reactants)$.
      a. multiply the $\Delta S°$ value for each substance by the stoichiometric coefficient of that substance in the balanced equation.
   2. Increase in entropy whenever a molecule breaks into two or more pieces.

EXAMPLE:
Calculate the $\Delta S°_{rxn}$ for

$$BaCO_3\ (s) + H_2SO_4\ (l) \rightarrow BaSO_4\ (s) + H_2O\ (l) + CO_2\ (g)$$

Chapter 16 – Entropy, Free Energy, and Equilibrium

SOLUTION: The $\Delta S°_{rxn}$ can be calculated using information found in Appendix B in your textbook.
$\Delta S°$ (BaSO$_4$) = 132 J/mol·K $\Delta S°$ (H$_2$O) = 69.9 J/mol·K
$\Delta S°$ (CO$_2$) = 213.6 J/mol·K $\Delta S°$ (BaCO$_3$) = 112 J/mol·K
$\Delta S°$ (H$_2$SO$_4$) = 156.9 J/mol·K

$\Delta S°_{rxn}$ = [(132 J/mol·K) + (69.9 J/mol·K) + (213.6 J/mol·K)] – [(112 J/mol·K) + (156.9 J/mol·K)]
       = 146.6 J/mol·K

16.6 Entropy and the Second Law of Thermodynamics
A. Value of ΔG — criterion for spontaneity.
  1. $\Delta G = \Delta H - T\Delta S$.
  2. $\Delta G > 0$; nonspontaneous reaction.
  3. $\Delta G < 0$; spontaneous reaction.
  4. $\Delta G = 0$; reaction is at equilibrium.
B. First law of thermodynamics — In any process, spontaneous or nonspontaneous, the total energy of a system and its surroundings is constant.
  1. Helps keep track of energy flow between system and the surroundings.
  2. Does not indicate the spontaneity of the process.
C. Second law of thermodynamics — In any spontaneous process, the total entropy of a system and its surroundings always increases.
  1. Provides a clear cut criterion of spontaneity.
  2. Direction of spontaneous change is always determined by the sign of the total entropy change.
  3. $\Delta S_{total} = \Delta S_{system} + \Delta S_{surroundings}$
     a. $\Delta S_{total} > 0$; spontaneous reaction
     b. $\Delta S_{total} < 0$; nonspontaneous reaction
     c. $\Delta S_{total} = 0$; reaction at equilibrium
D. All reactions proceed spontaneously in the direction that increases the total entropy of the system plus surroundings.
E. To determine ΔS_{total}, need to know ΔS_{system} and $\Delta S_{surroundings}$.
  1. ΔS_{system} — the entropy of reaction.
     a. Calculate from standard molar entropies.
  2. At constant pressure, $\Delta S_{surr} = \dfrac{-\Delta H_{rxn}}{T}$
  3. $\Delta S_{total} = \Delta S°_{rxn} - \dfrac{\Delta H_{rxn}}{T}$

Workbook Problem 16.1
Calculate ΔS_{total} for the following reaction at 298 K, and determine if the reaction is spontaneous under standard state conditions.

AgNO$_3$ (aq) + NaCl (aq) → AgCl (s) + NaNO$_3$ (aq)

Strategy: Calculate the values of ΔS_{system} and ΔS_{surr}.

Step 1: Write the net ionic equation for this reaction.

Step 2: Determine the information you will need to calculate ΔS_{total}

Step 3: Using Appendix B.1 in your textbook, calculate ΔS_{system}.

Step 4: Calculate ΔH_{rxn}.

Step 5: Calculate ΔS_{total}.

What key concept did you use?

16.7 Free Energy
A. Restate the second law in terms of the thermodynamic properties of the system.
   1. Chemists are more interested in the system.
B. Free energy: $G = H - TS$.
   1. TS = part of the system's energy that is already disordered.
   2. $H - TS$ = part of the system's energy that is still ordered and therefore free to cause spontaneous change by becoming disordered.
   3. State function.
      a. $\Delta G = \Delta H - T\Delta S$
C. Relationship between free energy and spontaneity.
   1. $\Delta S_{total} = \Delta S - \dfrac{\Delta H}{T}$ $-T\Delta S_{total} = \Delta H - T\Delta S$.
   2. $-T\Delta S_{total} = \Delta G$
      a. ΔG and ΔS_{total} have opposite signs.
   3. In any spontaneous process at constant temperature and pressure, the free energy of the system decreases.
   4. Temperature acts as a weighting factor that determines the relative importance of the enthalpy and entropy changes (see Table 16.2, page 663 in your textbook).
      a. Can estimate the temperature at which ΔG changes from positive to negative with
         $\Delta G = \Delta H - T\Delta S$

Chapter 16 – Entropy, Free Energy, and Equilibrium

5. To estimate the temperature at which $\Delta G°$ changes from a positive to a negative value, set $\Delta G° = \Delta H° - T\Delta S° = 0$.

   a. $T = \dfrac{\Delta H°}{\Delta S°}$

EXAMPLE:
Pure chromium is obtained by reducing Cr_2O_3 with aluminum.

$Cr_2O_3\ (s)\ +\ 2\ Al\ (s)\ \rightarrow\ 2\ Cr\ (s)\ +\ Al_2O_3\ (s)$

Determine $\Delta H°$ and $\Delta S°$ for the reaction. Is the reaction spontaneous at 25°C?

SOLUTION:
Use Appendix B.1 in your textbook to calculate $\Delta H°_{rxn}$ and $\Delta S°_{rxn}$.

$\Delta H°_{rxn} = \Delta H°_f(Al_2O_3) - \Delta H°_f(Cr_2O_3)$

$\Delta H°_{rxn} = [1\ mol\ Al_2O_3 \times (-1676\ kJ/mol)] - [1\ mol\ Cr_2O_3 \times (-1140\ kJ/mol)]$

$\Delta H°_{rxn} = -536\ kJ$

$\Delta S° = S°_f(Al_2O_3) - S°_f(Cr_2O_3)$

$\Delta S°_{rxn} = [(2\ mol\ Cr \times 23.8\ J/mol·K) + (1\ mol\ Al_2O_3 \times 50.9\ J/mol·K)] - [(2\ mol\ Al \times 28.3\ J/mol·K) + (1\ mol\ Cr_2O_3 \times (-1058\ J/mol·K)]$

$\Delta S°_{rxn} = 1{,}099.9\ J/K$

To determine if the reaction is spontaneous, we need to calculate ΔG for the reaction.

$\Delta G° = \Delta H° - T\Delta S°$

$\Delta G° = -536\ kJ - (298\ K)(1.0999\ kJ/K) = -864\ kJ$

Since the value for $\Delta G°$ is negative, the reaction is spontaneous.

16.8 Standard Free–Energy Changes for Reactions
A. Standard free–energy change, $\Delta G°$ — the change in free energy that occurs when reactants in their standard states are converted to products in their standard states.
  1. $\Delta G°$ — an extensive property; refers to the number of moles indicated in the chemical equation.
  2. Calculate $\Delta G°$ from the standard enthalpy change, $\Delta H°$, and the standard entropy change, $\Delta S°$.
     a. $\Delta G° = \Delta H° - T\Delta S°$
  3. Thermodynamic Standard State
     a. solids, liquids and gases are in the pure form at 1 atm.
     b. solute concentrations = 1 M.
     c. temperature = 25° C.

Chapter 16 – Entropy, Free Energy, and Equilibrium

Workbook Problem 16.2
Determine $\Delta G°$ for the reaction:

$$2\ NaCl\ (s)\ +\ H_2SO_4\ (l)\ \rightarrow\ Na_2SO_4\ (s)\ +\ 2\ HCl\ (g)$$

given the following information:

	$\Delta H°$ (kJ/mol)	$S°$ (J/mol·K)
NaCl (s)	−411.2	72.1
H_2SO_4 (l)	−814.0	156.9
Na_2SO_4 (s)	−1387.1	149.6
HCl (g)	−92.3	186.8

Strategy: From the information given, you can determine $\Delta G°$ from the equation $\Delta G° = \Delta H° - T\Delta S°$.

Step 1: Determine $\Delta H°_{rxn}$

Step 2: Determine $\Delta S°_{rxn}$.

Step 3: Calculate $\Delta G°_{rxn}$.

What key concept did you use?

16.9 Standard Free Energies of Formation
A. Standard free energy of formation, $\Delta G°_f$, the free–energy change for formation of 1 mole of the substance in its standard state from the most stable form of the constituent elements in their standard states.
  1. For an element in its most stable form at 25°C, $\Delta G°_f = 0$.
B. $\Delta G°_f$ measures a substance's thermodynamic stability with respect to its constituent elements.
  1. $\Delta G°_f$ is negative — substance is stable and does not decompose to its constituent elements.
  2. $\Delta G°_f$ is positive — substance is thermodynamically unstable.
    a. may not decompose to its constituent elements if the rate of decomposition is slow
C. Can use $\Delta G°_f$ to calculate standard free–energy changes for reactions.   ⚷ B.2.
  1. $\Delta G° = \Delta G°_f (\text{products}) - \Delta G°_f (\text{reactants})$

Chapter 16 – Entropy, Free Energy, and Equilibrium

EXAMPLE:

Calculate $\Delta G°_{rxn}$ for the following reaction:

$$C_2H_4\,(g) + HBr\,(g) \rightarrow C_2H_5Br\,(g)$$

given the following information:

	$\Delta G°_f$ (kJ/mol)
$C_2H_4\,(g)$	68.4
$HBr\,(g)$	–53.4
C_2H_5Br	–61.9

SOLUTION:

$\Delta G°_{rxn} = [-61.9] - [(68.4) + (-53.4)] = -76.9$ kJ

Workbook Problem 16.3

Determine $\Delta G°_{rxn}$ from $\Delta G°_f$ for the following reaction. Is the reaction spontaneous? If not, determine the temperature at which the reaction becomes spontaneous.

$$NH_4^+\,(aq) + OH^-\,(aq) \rightleftarrows NH_3\,(g) + H_2O\,(l)$$

Strategy: Determine $\Delta G°_{rxn}$, using Appendix B.1 in your textbook.

Step 1: Based on the value of $\Delta G°_{rxn}$, determine if the reaction is spontaneous.

Step 2: Determine the temperature at which the reaction will be spontaneous.

What key concept did you use?

Chapter 16 – Entropy, Free Energy, and Equilibrium

16.10 Free–Energy Changes and Composition of the Reaction Mixture

B.3.

A. To calculate the free–energy change for a reaction when the reactants and products are present at nonstandard–state pressures and concentrations.
   1. $\Delta G = \Delta G° + RT \ln Q$
      a. Q = the reaction quotient — an expression having the same form as the equilibrium constant expression (concentrations are not necessarily at equilibrium)

EXAMPLE:

Calculate ΔG for the reaction in the preceding example if $P(C_2H_4) = 1.57$ atm, $P(HBr) = 2.01$ atm, $P(C_2H_5Br) = 0.83$ atm.

SOLUTION:

Substituting the value for $\Delta G°_{rxn}$ obtained in the preceding example, we have

$$\Delta G = -76{,}900 \frac{J}{mol} + 8.314 \frac{J}{mol \cdot K}(298\ K)\ln\frac{(0.83)}{(1.57)(2.01)} = -80{,}209 \frac{J}{mol} = -80.2 \frac{kJ}{mol}$$

16.11 Free Energy and Chemical Equilibrium

B.4.

A. The total free energy of a reaction mixture changes as the reaction progresses toward equilibrium.
B. $\Delta G = \Delta G° + RT \ln Q$
   1. When the reaction mixture is mostly reactants
      a. $Q \ll 1$
      b. $RT \ln Q \ll 0$
      c. $\Delta G < 0$
      d. total free energy decreases as the reaction proceeds spontaneously in the forward direction
   2. When the reaction mixture is mostly products
      a. $Q \gg 1$
      b. $RT \ln Q > 0$
      c. $\Delta G > 0$
      d. total free energy decreases as the reaction proceeds spontaneously in the reverse direction
C. Free–energy curve — shows how the total free energy of a reaction mixture changes as the reaction progresses. (See Figure 16.10, page 672 in your textbook.)
   1. Goes through a minimum somewhere between pure reactants and pure products.
      a. The system is at equilibrium.
D. Relationship between free energy and the equilibrium constant.
   1. At equilibrium
      a. $\Delta G° = 0$
      b. $Q = K$
   2. $\Delta G° = -RT \ln K$
E. Relationship between $\Delta G°$ and K.
   1. $\Delta G° < 0$, $\ln K > 0$ – equilibrium mixture is mainly products.
   2. $\Delta G° > 0$, $\ln K < 0$ — equilibrium mixture is mainly reactants.
   3. $\Delta G° = 0$, $\ln K = 0$ — equilibrium mixture contains comparable amounts of reactions and products.
F. Properties of nature that determine the direction and extent of a chemical reaction.

Chapter 16 – Entropy, Free Energy, and Equilibrium

1. Value of K.
   a. K determined by $\Delta G°_{rxn}$
   b. $\Delta G°_{rxn}$ determined by $\Delta H°_f$ and $\Delta S°_{rxn}$

EXAMPLE:
Calculate K for the reaction

$$C_2H_4\ (g) + HBr\ (g) \rightarrow C_2H_5Br\ (g)$$

SOLUTION:

Using the $\Delta G°_{rxn}$ calculated earlier and rearranging the above equation, we have

$$\ln K = \frac{-76{,}900\ \frac{J}{mol}}{-\left[8.314\ \frac{J}{mol \cdot K}(298\ K)\right]} = 31.0$$

Taking the antilog of both sides gives $K = 3 \times 10^{13}$

 **Putting It Together**

At 37°C, $\Delta G° = -14.0$ kJ for the reaction

$$Hb-O_2\ (aq) + CO\ (g) \rightleftarrows Hb-CO\ (aq) + O_2\ (g)$$

This reaction represents the reaction that occurs when a person is exposed to carbon monoxide gas. Hb represents the hemoglobin molecule in blood, which transports O_2. Hb–O_2 is the oxygenated form of this molecule. When exposed to carbon monoxide, the CO and O_2 exchange places. If $[O_2] = [CO]$, what is the ratio of $[Hb-CO]/[Hb-O_2]$?

 **Self–Test**

This section is intended to test your knowledge of the material covered in this chapter. Think through these problems, and make certain you understand what is going on. Ask yourself if your answer makes sense. Many of these questions are linked to the chapter learning goals. Therefore, successful completion of these problems indicates you have mastered the learning goals for this chapter. You will receive the greatest benefit from this section if you use it as a mock exam. You will then discover which topics you have mastered and which topics you need to study in more detail.

True–False

1. The spontaneity of a reaction also gives an indication of the speed of a reaction.

2. All spontaneous reactions are exothermic reactions.

3. The spontaneity of a reaction is determined by the enthalpy of the reaction.

Chapter 16 – Entropy, Free Energy, and Equilibrium

4. Molecular systems tend to move spontaneously to a state of maximum randomness or disorder.

5. The entropy of a particular state is related to the number of ways that the state can be achieved.

6. Whenever a molecule breaks into two or more pieces, there is a decrease in entropy.

7. The first law of thermodynamics helps keep track of the energy flow between the system and surroundings and also indicates the spontaneity of the process.

8. The direction of spontaneous change is always determined by the sign of the total entropy change.

9. In determining the spontaneity of a reaction, the temperature is the weighting factor that determines the relative importance of the enthalpy and entropy changes.

10. When a reaction mixture is mostly reactants, the total free energy increases as the reaction proceeds spontaneously in the forward direction.

Matching

Spontaneous process	a. In any process, spontaneous or nonspontaneous, the total energy of a system and its surroundings is constant.
Entropy	b. the free–energy change for formation of 1 mole of the substance in its standard state from the most stable form of the constituent elements in their standard states
Third law of thermodynamics	c. the entropy of 1 mol of the pure substance at 1 atm pressure and a specified temperature, usually 25°C
Standard molar entropy	d. a process that proceeds on its own without any external influence
First law of thermodynamics	e. In any spontaneous process, the total entropy of a system and its surroundings always increases.
Second law of thermodynamics	f. molecular randomness or disorder
Standard free energy of formation	g. The entropy of a perfectly ordered crystalline substance at 0 K is zero.

Fill–in–the–Blank

1. A spontaneous reaction always moves a system toward _____.
2. Molecular systems tend to move spontaneously to a state of maximum _____.
3. The probabilities of ordered and random states are proportional to _____.
4. Random molecular motion increases as the _____ of a substance increases.

Chapter 16 – Entropy, Free Energy, and Equilibrium

5. The entropy of a gas _____ when its pressure decreases at constant temperature.
6. The direction of spontaneous change is always determined by the _____ of the total entropy change.
7. All reactions proceed spontaneously in the direction that increases the _____.
8. In any spontaneous process at constant temperature and pressure, the free energy of the system _____.
9. _____ measures a substance's thermodynamic stability with respect to its constituent elements.
10. An equilibrium mixture is mainly products when $\Delta G°$ is _____ and $\ln K$ _____.

Problems

1. Explain, in terms of probability, which state has the higher entropy.
   a. a file cabinet neatly organized in alphabetical order or a three–year–old's toy box
   b. a perfectly ordered crystal of salt or frozen slush (after the road was salted)
   c. 1 mol of CO_2 gas at STP or 1 mol of CO_2 gas at 273 K in a volume of 15.5 L

2. Determine the sign of ΔS for the following processes or reactions:
   a. an increase in the volume of a gas at constant temperature
   b. formation of gaseous products from solid reactants
   c. $CaO\ (s)\ +\ 2\ NH_4Cl\ (s) \rightarrow 2\ NH_3\ (g)\ +\ CaCl_2\ (s)$
   d. $4\ NH_3\ (g)\ +\ 3\ O_2\ (g) \rightarrow 2\ N_2\ (g)\ +\ 6\ H_2O\ (g)$
   e. $^{235}_{92}U$ is separated from a mixture of $^{235}_{92}U$ and $^{238}_{92}U$

3. Calculate the standard entropy of reaction for

   a. $Na_2CO_3\ (s)\ +\ 2\ HCl\ (aq) \rightarrow 2\ NaCl\ (aq)\ +\ CO_2\ (g)\ +\ H_2O\ (g)$

   b. $4\ NH_3\ (g)\ +\ 3\ O_2(g) \rightarrow 2\ N_2\ (g)\ +\ 6\ H_2O\ (g)$

4. Given $\Delta S°_{total} = 1.814 \times 10^4$ J/K and $\Delta S°_{rxn} = 310.8$ J/K, calculate $\Delta H°_{rxn}$ for the following reaction

   $2\ C_4H_{10}\ (g)\ +\ 13\ O_2\ (g) \rightarrow 8\ CO_2\ (g)\ +\ 10\ H_2O\ (g)$

5. Calculate ΔS if 1 mol of methane gas at STP expands to a volume of 30.5 L at 273 K and 1 atm of pressure.

6. Determine $\Delta G°$ for the following reaction, using the equation $\Delta G° = \Delta H° - T\Delta S°$:

   $Pb\ (s)\ +\ PbO_2\ (s)\ +\ 2\ H_2SO_4\ (l) \rightarrow 2\ PbSO_4\ (s)\ +\ 2\ H_2O\ (l)$

   Determine the temperature at which the reverse reaction becomes spontaneous.

7. Calculate the standard free energy of reaction from the standard free energy of formation for the following reaction:

$$2\,C_2H_2\,(g) + 5\,O_2\,(g) \rightarrow 4\,CO_2\,(g) + 2\,H_2O\,(l)$$

8. Calculate the free-energy change for the following reaction if the partial pressures are 3.0 atm for SO_2 and SO_3 and 1.5 atm for O_2.

$$2\,SO_2\,(g) + O_2\,(g) \rightleftarrows 2\,SO_3\,(g)$$

9. Determine the equilibrium constant for the following reaction, using the information found in Appendix B.

$$Na_2CO_3\,(s) + 2\,HCl\,(aq) \rightleftarrows 2\,NaCl\,(aq) + CO_2\,(g) + H_2O\,(l)$$

10. For the reaction

$$2\,KCl\,(s) \rightleftarrows Cl_2\,(g) + 2\,K\,(s)$$

   a. calculate the ΔG°_{rxn} from ΔG°_f
   b. calculate the temperature at which the reaction becomes spontaneous
   c. determine K for the reaction.

Challenge Problem

Determine the value of K at 25°C for the reaction

$$N_2\,(g) + O_2\,(g) \rightleftarrows 2\,NO\,(g)$$

Assume that ΔH and ΔS remain constant with changing temperature. Calculate K for this reaction at 700° C. Determine the partial pressures at equilibrium if you mix 5.00 mol of N_2 and 5.00 mol of O_2 in a 10.0 L vessel.

CHAPTER 17

ELECTROCHEMISTRY

Chapter Learning Goals

A. Galvanic Cells: Spontaneous Oxidation–Reduction Reactions
1. Sketch a galvanic cell, identifying the anode and cathode half–reactions, the sign of each electrode, and the direction of electron and ion flow.
2. Write balanced chemical equations for reactions occurring in a galvanic cell.
3. Write and interpret shorthand notations for galvanic cells.
4. Write balanced chemical equations for reactions occurring in common batteries.
5. Describe the reactions that occur when iron rusts.

B. Galvanic Cells: Cell Potentials
1. Use a table of standard reduction potentials to calculate standard cell potentials.
2. Use a table of standard reduction potentials to rank substances in order of increasing oxidizing strength or reducing strength and to determine whether a reaction is spontaneous.
3. Use the Nernst equation to calculate cell potentials for reactions occurring under nonstandard–state conditions.
4. From a measured cell potential for a reaction involving hydrogen ion and a reference cell potential, calculate the pH of the solution.

C. Galvanic Cells: Free–Energy Changes
1. Interconvert cell potential and free–energy change for a reaction.
2. Calculate equilibrium constants from standard cell potentials and vice versa.

D. Electrolytic Cells
1. Describe half–cell and overall reactions occurring in electrolytic processes.
2. Perform electrolytic cell calculations interconverting current and time, charge, moles of electrons, and moles (or grams) of product.

Chapter in Brief

Electrochemistry is the area of chemistry concerned with the interconversion of chemical and electrical energy. An electrochemical cell is the device used for this interconversion. In this chapter, you will examine the principles involved in the design and operation of electrochemical cells, as well as some of the important connections between electrochemistry and thermodynamics. You will learn how the table of standard reduction potentials was derived and how this table can be used to obtain an enormous amount of chemical information. You will also learn how to use the Nernst equation to calculate cell potentials under nonstandard–state conditions and how this equation is used to determine the pH of a solution. You will then apply your knowledge of galvanic cells to the study of batteries and the process of corrosion. Finally, you will examine the commercial applications and the quantitative aspects of electrolysis and electrolytic cells.

Chapter 17 – Electrochemistry

17.1 Galvanic Cells   🔑 A.1.

A. Galvanic cell — a spontaneous chemical reaction generates an electric current.
B. Electrolytic cell — an electric current drives a nonspontaneous reaction.
C. Review of redox reaction.
   1. Oxidation — a loss of electrons (an increase in oxidation number).
   2. Reduction — a gain of electrons (a decrease in oxidation number).
   3. Represent oxidation and reduction aspects of the reaction with half–reactions.
   4. Oxidizing agent — species that causes oxidation to occur and is itself reduced.
   5. Reducing agent — species that causes reduction to occur and is itself oxidized.
   6. If a spontaneous reaction is carried out in a beaker:
      a. Oxidizing agent and reducing agent are in direct contact
      b. Electrons are directly transferred.
      c. Enthalpy of reaction is lost to the surroundings for an exothermic reaction.
   7. If a spontaneous reaction is carried out in a galvanic cell
      a. Chemical energy released by the reaction is converted to electrical energy.
D. For the reaction: $Zn\ (s) + Cu^{2+}\ (aq) \rightarrow Zn^{2+}\ (aq) + Cu\ (s)$.
   1. Use a Daniell cell, a type of galvanic cell, to carry out the reaction. (See Figure 17.2, page 689 in your textbook.)
      a. consists of two half–cells
         i.. a beaker with a strip of Zn in a solution of $ZnSO_4$
         ii. a beaker with a strip of Cu in a solution of $CuSO_4$
      b. electrodes — strips of zinc and copper
      c. salt bridge — a U–shaped tube that contains a gel permeated with a solution of an inert electrolyte
   2. The electrons can be transferred only through the wire.
      a. oxidation and reduction half–reactions occur at separate electrodes
      b. electric current flows through the wire
   3. Anode — the electrode at which oxidation takes place.
      a. the negative (–) electrode
      b. produces electrons
   4. Cathode — the electrode at which reduction takes place.
      a. the positive (+) electrode
      b. consumes electrons

   5. Anode and cathode half–reaction must add to give the overall cell reaction.   🔑 A.2.
   6. Salt bridge maintains electrical neutrality by a flow of ions.
      a. Anions flow through the salt bridge from the cathode to the anode compartment.
      b. Cations migrate through the salt bridge from the anode to the cathode compartment.
   7. Electrons move through the external circuit from the anode to the cathode.

EXAMPLE:
Describe how you would construct a galvanic cell based on the following reaction:

$$Pb^{2+}\ (aq) + Zn\ (s) \rightarrow Pb\ (s) + Zn^{2+}\ (aq)$$

SOLUTION: Let's start by taking the overall cell reaction and breaking it into two half–reactions.

$$Pb^{2+}\ (aq) + 2\ e^- \rightarrow Pb\ (s)$$

$$Zn\ (s) \rightarrow Zn^{2+}\ (aq) + 2\ e^-$$

Looking at the two half reactions, we find that the Pb^{2+} is being reduced, and the Zn is being

Chapter 17 – Electrochemistry

oxidized. Therefore, the anode compartment of our cell would consist of a strip of zinc metal immersed in a solution containing Zn^{2+} ions (such as zinc nitrate). The cathode compartment would consist of a strip of lead immersed in a solution containing Pb^{2+} ions (such as lead(II) nitrate). The two half–cells would be connected to each other with a salt bridge and an external wire. Electrons flow through the wire from the zinc anode to the lead cathode. Anions move from the cathode compartment toward anode the, while cations migrate from the anode compartment toward the cathode.

17.2 Shorthand Notation for Galvanic Cells
A. Single vertical line, |, represents a phase boundary.
B. Double vertical line, ||, represents a salt bridge.
C. Shorthand for the anode half–cell is always written on the left of the salt–bridge symbol, followed on the right of the salt-bridge symbol by the shorthand for the cathode half–cell.
   1. Reactants in each half–cell are written first, followed by products.
   2. Electrons move through the external circuit from left to right.
   3. For $Zn\ (s) + Cu^{2+}\ (aq) \rightarrow Zn^{2+}\ (aq) + Cu\ (s)$:
      $Zn\ (s)\,|\,Zn^{2+}\ (aq)\,||\,Cu^{2+}\ (aq)\,|\,Cu\ (s)$.
D. Cell involving a gas.
   1. Additional vertical line due to presence of additional phase.
   2. List the gas immediately adjacent to the appropriate electrode.
E. Detailed notation includes ion concentrations and gas pressures.

EXAMPLE:
Give the shorthand notation for a galvanic cell that employs the overall reaction

$Pb(NO_3)_2\ (aq) + Ni\ (s) \rightarrow Pb\ (s) + Ni(NO_3)_2\ (aq)$

Give a brief description of the cell.

SOLUTION: The two half–reactions for this overall reaction are

$Pb^{2+}\ (aq) + 2\ e^- \rightarrow Pb\ (s)$

$Ni\ (s) \rightarrow Ni^{2+}\ (aq) + 2\ e^-$

From these half–reactions, we know that lead is being reduced and nickel is being oxidized. Therefore, Ni is the anode and Pb is the cathode. The cell notation is

$Ni\ (s)\,|\,Ni^{2+}\ (aq)\,||\,Pb^{2+}\ (aq)\,|\,Pb\ (s)$

This cell would consist of a strip of nickel as the anode dipping into an aqueous solution of $Ni(NO_3)_2$ and a strip of Pb as the cathode dipping into an aqueous solution of $Pb(NO_3)_2$. The two half–cells would be connected by a salt bridge and a wire.

Workbook Problem 17.1
Given the shorthand notation below, determine the half–reactions occurring at the anode and cathode. Write the overall reaction for the cell and give a brief description of the cell's construction.

$Al\ (s)\ |\ Al^{3+}\ (aq)\ |\ |\ Co^{2+}\ (aq)\ |\ Co\ (s)$

Strategy: Based on the shorthand notation above, determine which species is the cathode and which is the anode, and write the half–reaction that occurs at each electrode.

Step 1: Write the half–reaction occurring at the anode. Remember, the reactant in each half–cell is written first, followed by the product.

Step 2: Write the half–reaction occurring at the cathode.

Step 3: Combine the two half–reactions to obtain the overall cell reaction. Don't forget that the number of electrons lost in an oxidation half–reaction must equal the number of electrons gained in the reduction half–reaction (Chapter 4).

Step 4: Describe the cell indicated in the shorthand notation.

What key concept did you use?

17.3 Cell Potentials and Free–Energy Changes for Cell Reactions
A. Electromotive force (emf) — the driving force (electrical potential) that pushes the negatively charged electrons away from the anode and pulls them toward the cathode.
   1. Also called the cell potential (E) or the cell voltage.
   2. Potential of a galvanic cell is a positive quantity.
B. Coulomb (C) — the amount of charge transferred when a current of 1 ampere (A) flows for 1 s.
   1. 1 C = 1 A × 1 s
   1. 1 J = 1 C × 1 V
C. Cell potential — measured with a voltmeter.
   1. Gives a positive reading when the + and − terminals of the voltmeter are connected to cathode (+) and anode (−), respectively.
      a. can use voltmeter–cell connections to determine which electrode is the anode and which is the cathode
D. Two driving forces of a chemical reaction: cell potential, E; and free–energy change, ΔG.
   1. Related by $\Delta G = -nFE$.
      a. n = number of moles of electrons transferred in the reaction

C.1.

Chapter 17 – Electrochemistry

        b. F (faraday) — the electrical charge on 1 mol of electrons
           i. 1 F = 96,500 C/mol e^-
        c. ΔG and E have opposite signs.
           i. Spontaneous reaction has a positive cell potential but negative ΔG.
   E. Standard cell potential, $E°$ – the cell potential when both reactants and products are in their standard states.
      1. Solutes at 1 M concentration.
      2. Gases at a partial pressure of 1 atm.
      3. Solids and liquids in pure form.
      4. $T = 25°C$
   F. $\Delta G° = -nFE°$

17.4 Standard Reduction Potentials

   A. $E°_{cell} = E°_{ox} + E°_{red}$
      1. Can't measure potential of a single electrode.
      2. Measure a potential difference by placing a voltmeter between two electrodes.
      3. Develop a set of standard half–cell potentials.
         a. Choose an arbitrary standard half–cell as a reference point and assign an arbitrary potential.
         b. Express the potential of all other half–cells relative to the reference half–cell.
   B. Standard hydrogen electrode (S.H.E.) — reference half–cell (see Figure 17.4, page 696 in your textbook).
      1. Corresponding half–reaction — assigned an arbitrary potential of exactly 0 V.
         a. $2 H^+ (aq, 1 M) + 2 e^- \rightarrow H_2 (g, 1 atm)$ $E° = 0.00$ V
      2. Shorthand notation for S. H. E.
         a. H^+ (1 M) | H_2 (1 atm) | Pt (s)
           i. Pt electrode in contact with H_2 gas and H^+ (aq) at standard conditions.
   C. Determine standard potentials for half–cells by constructing a galvanic cell in which the half–cell of interest is paired up with the standard hydrogen electrode.
      1. Standard oxidation potential — the corresponding half–cell potential for an oxidation half–reaction.
      2. Standard reduction potential — the corresponding half–cell potential for a reduction half–reaction.
      3. Whenever the direction of a half–reaction is reversed, the sign of $E°$ must be reversed.
         a. The standard oxidation potential and the standard reduction potential always have the same magnitude, but they have opposite signs
      4. Construct a table of standard reduction potentials (Appendix D in your textbook).
   D. Conventions used in constructing a table of half–cell potentials.
      1. The half–reactions are written as reductions.
         a. Oxidizing agents and electrons are on the reactant side.
         b. Reducing agents are on the product side.
      2. The half–cell potentials are standard reduction potentials.
         a. also known as standard electrode potentials

      3. The half–reactions are listed in order of decreasing standard reduction potential.
         a. Strongest oxidizing agents are located in the upper left of the table.
         b. Strongest reducing agents are in the lower right of the table.
      4. Ordering of half–reactions correspond to ordering of the oxidation reactions in the activity series.
         a. The more active metals at the top of the activity series have the more positive oxidation potential (more negative reduction potential).

17.5 Using Standard Reduction Potentials

A. Table of standard reduction potentials — summarizes an enormous amount of chemical information.
   1. Can arrange any two or more oxidizing or reducing agents in order of increasing strength.
   2. Predict the spontaneity or nonspontaneity of thousands of redox reactions.   *B.2.*
      a. combine half-reactions of interest and use $E°_{cell} = E°_{anode} + E°_{cathode}$
      b. may need to multiply half-reactions by some factor to ensure that electrons cancel
         i. **Do not** multiply values of $E°$ for the half-reactions by that factor.
B. $E°$ values are independent of the amount of reaction.
   1. $\Delta G° = -nFE°$
      a. $\Delta G°$ is an extensive property because it depends on the amount of substance.
      b. change the amount of substance that reacts, $\Delta G°$ changes by the same amount as does n, the number of electrons transferred
      c. $E° = -\Delta G°/nF$ remains constant
C. Can predict the spontaneity of a reaction by knowing the location of the oxidizing and reducing agents in the table.
   1. An oxidizing agent can oxidize any reducing agent that lies below it in the table.   *B.2.*
      a. $E°$ for overall reaction must be positive.

EXAMPLE:
Write the balanced net ionic equation, and calculate $E°$ for the following galvanic cell:

$$Al\,(s)\,|\,Al^{3+}\,(aq)\,||\,Cu^{2+}\,(aq)\,|\,Cu\,(s)$$

SOLUTION: Al (s) is the anode and therefore undergoes oxidation, while Cu (s) is the cathode and Cu^{2+} therefore undergoes reduction. The half-reactions and their cell potentials are

$$Al\,(s) \rightarrow Al^{3+}\,(aq) + 3\,e^{-} \quad E° = +1.66\text{ V}$$

$$Cu^{2+}\,(aq) + 2\,e^{-} \rightarrow Cu\,(s) \quad E° = +0.34\text{ V}$$

(Notice that the sign for $E°$ for the Al/Al^{3+} half-reactions has been reversed.) To write the balanced net ionic equation, we need to make sure that the electrons cancel out on both sides. Therefore, we need to multiply the top reaction by 2 and the bottom reaction by 3.

$$2\,Al\,(s) \rightarrow 2\,Al^{3+}\,(aq) + 6\,e^{-} \qquad E° = +1.66\text{ V}$$

$$3\,Cu^{2+}\,(aq) + 6\,e^{-} \rightarrow 3\,Cu\,(s) \qquad E° = +0.34\text{ V}$$

Notice that although we multiplied the coefficients in both half-reactions by the factor of 2 and 3 respectively, we did not multiply the values of $E°$ by these factors. This is because $E°$ values are independent of the amount of reaction.

$$E°_{cell} = E°_{Al \rightarrow Al^{3+}} + E°_{Cu^{2+} \rightarrow Cu} = +1.66\text{ V} + 0.34\text{ V} = 2.00\text{ V}$$

Chapter 17 – Electrochemistry

Workbook Problem 17.2
Using the table of standard reduction potentials and without calculating the cell potential, determine if the following reactions are spontaneous.

Ag (s) + Cu^{2+} (aq) → 2 Ag^{+} (aq) + Cu (s)

MnO$_4^-$ (aq) + 8H$^+$ (aq) + 10 Br$^-$ (aq) → Mn^{2+} (aq) + 4 H$_2$O (l) + 5 Br$_2$ (l)

Strategy: Determine the position of the reducing agent relative to the position of the oxidizing agent, and based on those results, determine the spontaneity of the reactions.

What key concept did you use?

17.6 Cell Potentials and Composition of the Reaction Mixture: The Nernst Equation   B.3.
A. Cell potentials depend on temperature and on the composition of the reaction mixture.
  1. $\Delta G = \Delta G° + RT \ln Q$.
     a. $\Delta G = -nFE$; $\Delta G° = -nFE°$
  2. $-nFE = -nFE° + RT \ln Q$.

B. Nernst equation: $E = E° - \dfrac{0.0592}{n} \log Q$ (in volts at 25°C).
  1. Enables us to calculate cell potentials under nonstandard-state conditions.

EXAMPLE:
Calculate E_{cell} for the following cell reaction:

2 Cr (s) + 3 Pb^{2+} (aq) → 2 Cr^{3+} (aq) + 3 Pb (s)

[Pb^{2+}] = 0.15 M   [Cr^{3+}] = 0.50 M

SOLUTION: The half-reactions for this equation are

2 Cr (s) → 2 Cr^{3+} + 6 e$^-$ $E° = +0.74$ V

3 Pb^{2+} (aq) + 6 e$^-$ → 3 Pb (s) $E° = -0.13$ V

$E°_{cell} = +0.74 + (-0.13) = +0.61$ V

The Nernst equation for this reaction is:

$$E_{cell} = E°_{cell} - \frac{0.0592}{6}\log\frac{[Cr^{3+}]^2}{[Pb^{2+}]^3}$$

Substituting the information given and solving for E_{cell} gives:

$$E_{cell} = +0.61 - \frac{0.0592}{6}\log\frac{(0.5)^2}{(0.15)^3} = 0.59$$

Workbook Problem 17.3

Determine the pH for the following cell reaction:

$$Ca\,(s) + ClO^-\,(aq) + H_2O\,(l) \rightleftarrows Cl^-\,(aq) + 2\,OH^-\,(aq) + Ca^{2+}\,(aq)$$

if the concentrations of ClO^-, Cl^-, and Ca^{2+} are 1.00 M and $E_{cell} = 4.00$ V.

Strategy: Calculate the $E°_{cell}$ for the cell reaction, substitute the information given into the Nernst equation, and solve for $[OH^-]$.

Step 1: Calculate the pH.

What key concept did you use?

17.7 Electrochemical Determination of pH

B.4.

A. Important application of Nernst equation — electrochemical determination of pH, using a pH meter.

B. Consider a cell with a hydrogen electrode as the anode and a second reference electrode as the cathode.

1. $Pt\,(s)\,|\,H_2\,(1\text{ atm})\,|\,H^+\,(?\text{ M})\,||\,$reference cathode

Chapter 17 – Electrochemistry

2. $E_{cell} = 0.0592\, pH + E_{ref}$
3. pH is a linear function of the cell potential.
   a. $pH = \dfrac{E_{cell} - E_{ref}}{0.0592}$
   b. can measure the pH of a solution by measuring E_{cell}
C. Actual pH measurements use a glass electrode with a calomel electrode as the reference.

EXAMPLE:
The following cell has a potential of 0.49 V. Calculate the pH of the solution in the anode compartment.

Pt (s) | H$_2$ (g) (1 atm) | H$^+$ (pH = ?) || Cl$^-$ (aq) (1M) | Hg$_2$Cl$_2$ (s) | Hg (l)

SOLUTION: The cell reaction is

Hg$_2$Cl$_2$ (s) + H$_2$ (g) → 2 Hg (l) + 2 Cl$^-$ (aq) + 2 H$^+$ (aq)

and the standard cell potential can be calculated from the data in Appendix D.

$$E° = E°_{H_2 \to H^+} + E°_{Hg_2Cl_2 \to Hg, Cl^-} = 0.00\, V + 0.28\, V = 0.28\, V$$

The pH can be calculated using the equation

$$pH = \dfrac{E_{cell} - E_{ref}}{0.0592}$$

$$pH = \dfrac{0.49\, V - 0.28\, V}{0.0592} = 3.55$$

17.8 Standard Cell Potentials and Equilibrium Constants
A. Standard free–energy change for a reaction is related to both the standard cell potential and the equilibrium constant.
   1. $\Delta G° = -nFE°$
   2. $\Delta G° = -RT \ln K$
B. Can combine the two equations.
   1. $E° = \dfrac{RT}{nF} \ln K = \dfrac{2.303 RT}{nF} \log K$
   2. Simplified form

   $$E° = \dfrac{0.0592}{n} \log K$$

   2. Most common use — calculating equilibrium constants from standard cell potentials.
C. Equilibrium constants for redox reactions tend to be either very large or very small in comparison with equilibrium constants for acid–base reactions.
   1. Positive value of $E°$ corresponds to $K > 1$.
   2. Negative value of $E°$ corresponds to $K < 1$.
D. Three different ways to determine the value of an equilibrium constant K.
   1. K from concentration data: $K = \dfrac{[C]^c [D]^d}{[A]^a [B]^b}$.

2. K from thermochemical data: $\ln K = \dfrac{-\Delta G°}{RT}$.

3. K from electrochemical data: $\ln K = \dfrac{nFE°}{RT}$.

EXAMPLE:
Calculate the equilibrium constant for the following reaction at 25°C.

$$5\ S_2O_8^{2-}\ (aq) + I_2\ (s) + 6\ H_2O\ (l) \rightarrow 10\ SO_4^{2-}\ (aq) + 2\ IO_3^-\ (aq) + 12\ H^+\ (aq)$$

SOLUTION: The half–reactions for this reaction are:

$S_2O_8^{2-}\ (aq) + 2\ e^- \rightarrow 2\ SO_4^{2-}\ (aq)$ $\qquad\qquad\qquad\qquad E° = +2.01\ V$

$I_2\ (s) + 6\ H_2O\ (l) \rightarrow 2\ IO_3^-\ (aq) + 12\ H^+\ (aq) + 10\ e^-$ $\qquad E° = -1.20\ V$

$E°_{cell} = 2.01 + (-1.20) = +0.81\ V$

The value of n for this reaction is 10. We can now solve for K.

$$\log K = \dfrac{(10)(0.81)}{(0.0592)} = 137 \qquad K = 6.67 \times 10^{136}$$

17.9 Batteries

A. Most important practical application of galvanic cells is their use as batteries.
B. Features required in a battery depend on the application.
C. General features.
  1. Compact and lightweight.
  2. Physically rugged and inexpensive.
  3. Provide a stable source of power for relatively long periods of time.
D. Lead storage battery.
  1. Used as a reliable source of power for starting automobiles for more than three–quarters of a century.
  2. 12 V battery — six 2 V cells connected in series.
  3. Anode — a series of grids packed with spongy lead.
  4. Cathode — a series of grids packed with lead dioxide, dipped into an aqueous solution of H_2SO_4 (38% w/w).
  5. Electrode half–reactions and the overall cell reaction.

Anode: $Pb\ (s) + HSO_4^-\ (aq) \rightarrow PbSO_4\ (s) + H^+\ (aq) + 2\ e^-$ $\qquad\qquad E° = 0.296\ V$

Cathode: $PbO_2\ (s) + 3\ H^+\ (aq) + HSO_4^-\ (aq) + 2\ e^- \rightarrow PbSO_4\ (s) + 2\ H_2O\ (l)$ $\qquad E° = 1.628\ V$

Overall: $Pb\ (s) + PbO_2\ (s) + 2\ H^+\ (aq) + 2\ HSO_4^-\ (aq) \rightarrow 2\ PbSO_4\ (s) + 2\ H_2O\ (l)$ $\qquad E° = 1.924\ V$

  6. $PbSO_4$ adheres to the surface of the electrodes.
    a. recharge by using an external source of direct current to drive the cell reaction in the reverse, nonspontaneous direction
E. Dry–Cell Batteries (Leclanché cells) — common household batteries.
  1. Anode — Zn metal can.

Chapter 17 – Electrochemistry

      2. Cathode — inert graphite rod surrounded by a paste of solid MnO_2 and carbon black.
      3. Electrolyte — a moist paste of NH_4Cl and $ZnCl_2$ in starch.
         a. surrounds the MnO_2 containing paste
         b. acidic — causes corrosion of the Zn anode ($Zn \rightarrow Zn^{2+}$)
      4. Alkaline dry cell — modified version of Leclanché.
         a. replace NH_4Cl (acidic) with NaOH or KOH
         b. electrode reactions — oxidation of zinc and reduction of manganese dioxide
            i. produces ZnO due to basic conditions
            ii. zinc corrodes more slowly
            iii. battery has a longer life
         c. produces higher power and more stable current and voltage
            i. more efficient ion transport in the alkaline electrolyte
      5. Mercury battery — used in watches, heart pacemakers, and other devices.
         a. small size
         b. anode — Zn (same as dry cell)
         c. cathode — steel in contact with HgO in an alkaline medium of KOH and $Zn(OH)_2$
   F. Nickel–Cadmium batteries — used in calculators and portable power tools.
      1. Rechargeable.
      2. Anode — cadmium metal.
      3. Cathode — NiO(OH) supported on nickel metal
      4. Solid products of electrode reaction adhere to the surface of the electrodes.
         a. allows battery to be recharged
   G. Nickel-Metal Hydride Batteries – NiMH
      1. Replacement for Ni-Cad batteries
         a. Cd – expensive, toxic, heavy metal.
      2. More environmentally friendly than Ni-Cad.
      3. Save voltage as NiCad battery, but twice the energy density.
         a. energy density – the amount of energy stored per unit mass.
      4. Anode – special metal alloy that can absorb and release large amounts of H_2 at ordinary temperature.
      5. Cell reaction:

Anode: $MH_{ab}(s) + OH^-(aq) \rightarrow M(s) + H_2O(l) + e^-$

Cathode: $NiO(OH)(s) + H_2O(l) + e^- \rightarrow Ni(OH)_2(s) + OH^-(aq)$

Overall: $MH_{ab}(s) + NiO(OH)(s) \rightarrow M(s) + Ni(OH)_2(s)$

         a. Metal hydride – oxidized at anode.
         b. NiO(OH) (s) – reduced at cathode
         c. Overall reaction transfers hydrogen from anode to the cathode.
      6. Used in hybrid gas-electric automobiles powered by a gasoline engine and electric motor.
         a. battery packs:
            i. supply energy to electric motor.
            ii. capture and store energy from braking.
         b. hit the brakes:
            i. electric motor recharges the batteries.
            ii. some kinetic energy converted to electrical energy.
   G. Lithium batteries – light weight, high–voltage, rechargeable.
      1. Anode — lithium metal.
         a. highest standard oxidizing potential
      2. Cathode — metal oxide or sulfide that can incorporate Li^+.
      3. Electrolyte — lithium salt in an organic solvent.
      4. Cell reaction:

Anode: $x\,Li\,(s) \rightarrow x\,Li^+\,(soln) + x\,e^-$

Cathode: $MnO_2\,(s) + Li^+\,(soln) + x\,e^- \rightarrow Li_xMnO_2\,(s)$

Overall: $x\,Li\,(s) + MnO_2\,(s) \rightarrow Li_xMnO_2\,(s)$

    5. Used in cell phones, laptop computers, and cameras.
I. Lithium Ion Battery
    1. Anode – graphite anode with lithium atoms inserted between layers of carbon atoms.
        a. lithiated graphite, Li_xC_6.
    2. Cathode – CoO_2
        a. also can incorporate Li^+ into its structure.
    3. Electrolyte –
        a. lithium salt in an organic solvent
        b. solid-state polymer that can transport Li^+.
    4. Cell Reaction:

Anode: $Li_xC_6\,(s) \rightarrow x\,Li^+\,(soln) + 6\,C\,(s) + x\,e^-$

Cathode: $Li_{1-x}CoO_2\,(s) + xLi^+\,(soln) + x\,e^- \rightarrow LiCoO_2\,(s)$

    5. Used in cell phones, laptops, digital cameras, power tools and Tesla Motors electric cars.

17.10 Fuel cells
    A. A galvanic cell in which one of the reactants is a fuel such as hydrogen or methanol.
    B. Reactants are not self-contained within the cell.
        1. supplied from an external reservoir
    C. Hydrogen–oxygen fuel cell – best known fuel cell.
        1. Used in space vehicles as a source of electric power.
        2. Porous carbon electrodes impregnated with metallic catalyst.
        3. Electrolyte – hot, aqueous KOH.
        4. H_2, the fuel, and O_2, the oxidizing agent, flow into separate cell compartments.
            a. H_2 oxidized at anode.
            b. O_2 reduced at cathode.
        5. Overall cell reaction – conversion of hydrogen and oxygen to water.
    D. Proton-exchange membrane (PEM Fuel Cells).
        1. Electrolyte – special plastic membrane that conducts protons but not electrons.
        2. Electric motor driven by:
            a. Protons that pass through the membrane from the anode to the cathode.
            b. Electrons move through the external circuit from the anode to the cathode.
        3. Doesn't produce environmental pollutants or CO_2.
            a. only reaction product is water.
        4. Commercialization of fuel-cell vehicles requires
            a. reduced cost and improved performance PEM's.
            b. newer and safer methods of storing hydrogen
            c. development of hydrogen-fuel infrastructure.
    E. Direct Methanol Fuel Cell (DMFC)
        1. Similar to PEM fuel cell.
        2. Uses aqueous methanol, CH_3OH, as the fuel
        3. Cell reaction:

Anode: $2 CH_3OH (aq) + 2 H_2O (l) \rightarrow 2 CO_2 (g) + 12 H^+ (aq) + 12 e^-$

Cathode: $3 O_2 (g) + 12 H^+ (aq) + 12 e^- \rightarrow 6 H_2O (l)$

Overall: $2 CH_3OH (aq) + 3 O_3 (g) \rightarrow 2 CO_2 (g) + 4 H_2O (l)$

    4. Advantages:
       a. lighter than conventional batteries
       b. high energy density
    5. Produces CO_2 (g) – not as environmentally friendly as PEM.
    6. Uses – small electronics.

17.11 Corrosion
  A. Corrosion — the oxidative deterioration of a metal.
  B. Well–known example of corrosion — conversion of iron to rust.
    1. Requires both oxygen and water.
    2. Involves pitting of the metal surface.
       a. rust is deposited at a location physically separated from the pits

  C. Proposed mechanism for formation of rust — an electrochemical process in which iron is oxidized in one region of the surface and oxygen is reduced in another region (see Figure 17.12, page 714 in your textbook.)
    1. Anode region: $Fe (s) \rightarrow Fe^{2+} (aq) + 2 e^-$ $E° = 0.45$ V
    2. Cathode region: $O_2 (g) + 4 H^+ (aq) + 4 e^- \rightarrow 2 H_2O (l)$ $E° = 1.23$ V
    3. Electrons flow from the anode to the cathode through the metal.
    4. Ions migrate through the water droplets.
       a. Fe^{2+} reacts with O_2 and is oxidized to Fe^{3+}
       b. Fe^{3+} reacts with H_2O to form $Fe_2O_3 \cdot xH_2O$ (s) (rust)
    5. Explains why cars rust more rapidly when road salt is used to melt snow and ice.
       a. Dissolved salt in water greatly increases the conductivity of the electrolyte.
  D. O_2 able to oxidize all metals except a few.
    1. O_2/H_2O half./reaction lies above the M^{n+}/M half–reaction
    2. Oxidation of Al:
       a. creates a very hard, protective Al_2O_3 surface.
          i. prevents further oxidation.
  E. Prevention of corrosion — shield the metal surface from oxygen and moisture.
    1. Durable surface coating — metals such as chromium, tin, or zinc.
    2. Galvanizing — coating by dipping into a bath of molten zinc.
    3. Cathodic protection — protecting a metal from corrosion by connecting it to a second metal that is more easily oxidized.

17.12 Electrolysis and Electrolytic Cells
  A. Electrolytic cell — an electric current is used to drive a nonspontaneous reaction.
    1. Processes occurring in galvanic and electrolytic cells are the reverse of each other.
  B. Electrolysis — the process of using an electric current to bring about chemical change.
  C. Electrolytic cell.
    1. Two electrodes that dip into an electrolyte and are connected to a battery or some other source of direct electric current.
       a. battery — an electron pump, pushing electrons into one electrode and pulling them out of the other electrode
    2. Anode — electrode where oxidation takes place.

Chapter 17 – Electrochemistry

    a. positive sign
    b. the battery pulls electrons out of it.
  3. Cathode — electrode where reduction takes place.
    a. negative sign
    b. the battery pushes electrons into it.
D. Electrolysis of molten NaCl.
  1. Cathode — attracts Na^+.
    a. $Na^+ + e^- \rightarrow Na\ (l)$
  2. Anode — attracts Cl^-.
    a. $2\ Cl^- \rightarrow Cl_2\ (g) + e^-$
E. Electrolysis of aqueous NaCl.
  1. Electrode reactions in an aqueous solution can differ from those for a molten salt.
  2. Cathode reaction can involve the reduction of Na^+ or the reduction of water.
    a. reduction of water preferred.
      i. $E°$ less negative for H_2O than Na^+.
    b. $2\ H_2O\ (l) + 2\ e^- \rightarrow H_2\ (g) + 2\ OH^-\ (aq)$
  3. Anode reaction can involve the oxidation of Cl^- or the oxidation of water.
    a. actual reaction — the oxidation of Cl^- due to overvoltage
    b. $2\ Cl^-\ (aq) \rightarrow Cl_2\ (g) + 2\ e^-$
  4. Overvoltage — amount of voltage needed above the calculated standard reduction (or oxidation) potential for electrolysis to occur.
    a. needed when the half–reaction has a substantial barrier for electron transfer (slow rate).
      i. surmounts barrier
      ii. Reaction proceeds at satisfactory rate.
    b. small overvoltage needed for solution or deposition of metals
    c. large overvoltage needed for formation of O_2 or H_2
    d. can't predict; need experimental evidence if cell potentials are similar
  5. Overall cell reaction.
    $2\ Cl^-\ (aq) + 2\ H_2O\ (l) \rightarrow Cl_2\ (g) + 2\ H_2\ (g) + 2\ OH^-\ (aq)$
    a. Na^+ is a spectator ion and reacts with the OH^- to form NaOH
F. Electrolysis of water.
  1. Electrolysis of any aqueous solution requires the presence of an electrolyte to carry the current in solution.
  2. If the ions of the electrolyte are less easily oxidized and reduced than water is, then water will react at both electrodes.
  3. Anode: $2\ H_2O\ (l) \rightarrow O_2\ (g) + 4\ H^+\ (aq) + 4\ e^-$.
  4. Cathode: $4\ H_2O\ (l) + 4\ e^- \rightarrow 2\ H_2\ (g) + 4\ OH^-\ (aq)$.
  5. Overall cell reaction:
    $2\ H_2O\ (l) \rightarrow O_2\ (g) + 2\ H_2\ (g)$

17.13 Commercial Applications of Electrolysis
A. Manufacture of sodium — produced commercially in a Downs cell by electrolysis of a molten mixture of NaCl and $CaCl_2$.
  1. Liquid Na produced at the cylindrical steel cathode is less dense than the molten salt and thus floats to the top part of the cell, where it is drawn off into a suitable container.
B. Manufacture of chlorine and sodium hydroxide — electrolysis of aqueous NaCl.
  1. Basis of chlor–alkali industry.
  2. Anode and cathode reactions for electrolysis of aqueous NaCl carried out in membrane cell.
    a. Membrane keeps Cl_2 and OH^- apart but allows a current of Na^+ to flow.
C. Manufacture of aluminum — Hall–Heroult process.
  1. Electrolysis of a molten mixture of Al_2O_3 and cryolite (Na_3AlF_6) at $1000°$ C in a cell with graphite electrodes.
    a. success of process — the use of cryolite as a solvent

Chapter 17 – Electrochemistry

   2. Electrode reactions involve the formation of complex ions.
      a. Ions are reduced at the cathode to produce Al (l).
      b. Ions are oxidized at the anode to produce O_2 (g).
         i. O_2 (g) reacts with the graphite electrode to produce CO_2 (g)
         ii. requires frequent replacement of the anodes
   3. Largest single consumer of electricity in the U.S.
      a.
         One mole of electrons produces only 9 g of Al.
D. Electrorefining — the purification of a metal by means of electrolysis.
E. Electroplating — the coating of one metal on the surface of another, using electrolysis.
   1. Cathode — object to be plated (carefully cleaned).
   2. Electrolytic cell contains a solution of ions of the metal to be deposited.

17.14 Quantitative Aspects of Electrolysis

A. The amount of substance produced at an electrode by electrolysis depends on the quantity of charge passed through the cell.
   1. Follows directly from the stoichiometry of the reaction and the atomic mass of the product.
B. Moles of electrons passed through a cell are determined from the electric current and the time that the current flows.

$$\text{Moles of } e^- = \text{charge(C)} \times \frac{1 \text{ mol } e^-}{96,500 \text{ C}}$$

C. Sequence of conversion used to calculate the mass or volume of product produced by passing a known current for a fixed period of time (see Figure 17.19, page 723).

Current and time → Charge → Moles of e^- → Moles of product → Grams or liters of product

D. Think of electrons as reactants in a balanced equation, and proceed as with any other stoichiometry problem.

EXAMPLE:
How many grams of Cl_2 would be produced in the electrolysis of molten NaCl by a current of 4.25 A for 35.0 min?

SOLUTION: (Remember that a coulomb is an A·s or that an ampere is C/s.)

$2 Cl^- \rightarrow Cl_2 + 2 e^-$

Moles of electrons = 2

$$4.25 \frac{C}{s} \times 35.0 \text{ min} \times \frac{60 \text{ s}}{1 \text{ min}} \times \frac{1 \text{ mol } e^-}{96,500 \text{ C}} \times \frac{1 \text{ mol } Cl_2}{2 \text{ mol } e^-} \times \frac{70.9 \text{ g } Cl_2}{1 \text{ mol } Cl^-} = 3.28 \text{ g } Cl_2$$

 **Workbook Problem 17.4**

The Dow process isolates Mg (s) from seawater. The final step in this process involves the electrolysis of molten $MgCl_2$ to the metal. How long would it take to produce 25 lb of Mg (s) at a current of 20 A?

Chapter 17 – Electrochemistry

Strategy: First, write the electrolysis reaction and consider the conversion process to calculate time.

Step 1: Treat the electrons as reactants in the chemical equation, and solve for the time required to produce 25 lbs. of magnesium. Remember, the conversion factor of 96,500 C (or A·s) per mole of electrons.

What key concept did you use?

Putting It Together

Determine $E°$ for the following reaction:

$$O_2\,(g) + 4\,H^+\,(aq) + 4\,Br^-\,(aq) \rightarrow 2\,H_2O\,(l) + 2\,Br_2\,(l)$$

Is this reaction spontaneous? A buffer containing 0.25 M sodium formate ($NaCHO_2$) and 0.35 M formic acid is added to adjust the pH of the reaction. Assume $[Br^-] = 1.0$ M and P_{O_2}. Determine E after the addition of the buffer. Now is the reaction spontaneous?

Self–Test

This section is intended to test your knowledge of the material covered in this chapter. Think through these problems, and make certain you understand what is going on. Ask yourself if your answer makes sense. Many of these questions are linked to the chapter learning goals. Therefore, successful completion of these problems indicates you have mastered the learning goals for this chapter. You will receive the greatest benefit from this section if you use it as a mock exam. You will then discover which topics you have mastered and which topics you need to study in more detail.

True–False

1. An electrolytic cell is one in which a spontaneous chemical reaction generates an electric current.

2. The reducing agent in a redox reaction is the species that causes reduction to occur and it itself is oxidized.

3. The cathode is the electrode where oxidation takes place.

4. The cell potential of a galvanic cell is positive.

5. Whenever the direction of a half–reaction is reversed, the sign of $E°$ must be reversed.

6. In constructing a table of half–cell potentials, the half–reactions are written as reduction reactions.

Chapter 17 – Electrochemistry

7. An oxidizing agent can oxidize any reducing agent that lies below it in the table of standard reduction potentials.

8. Rust is deposited in the same location where pitting of the metal surface has occurred.

9. The amount of substance produced at an electrode by electrolysis depends on the quantity of charge passed through the cell.

10. In an electrolytic cell, the anode is the electrode where reduction takes place.

Multiple Choice

1. For the following galvanic cell Ni (s) | Ni^{2+} (aq) || Br^- (aq) | Br_2 (l) | Pt (s)
   a. the cathode is Ni (s)
   b. electrons flow from the Pt (s) electrode to the Ni (s) electrode
   c. the Ni^{2+} ions flow to the anode
   d. the electrons flow from the Ni (s) electrode to the Pt (s) electrode

2. The reaction carried out in a galvanic cell
   a. is spontaneous, and therefore has a negative $E°$ value
   b. is nonspontaneous, and therefore has a negative $E°$ value.
   c. is spontaneous, and therefore has a positive $E°$ value
   d. is nonspontaneous, and therefore has a positive $E°$ value.

3. For the net reaction Cr (s) + Fe^{3+} (aq) → Cr^{3+} (aq) + Fe (s)
   a. Cr (s) is the oxidizing agent
   b. Fe (s) is the cathode
   c. Fe^{3+} (aq) is the reducing agent
   d. Cr (s) undergoes reduction

4. Given the following reduction half–cells:
   PbO_2 (s) + 3 H^+ (aq) + HSO_4^- (aq) + 2 e^- → $PbSO_4$ (s) + 2 H_2O (l) $E° = 1.628$ V
   $Cr_2O_7^{2-}$ (aq) + 14 H^+ (aq) + 6 e^- → 2 Cr^{3+} (aq) + 7 H_2O (l) $E° = 1.33$ V
   SO_4^{2-} (aq) + 4 H^+ (aq) + 4 H^+ (aq) + 2 e^- → H_2SO_3 (aq) + H_2O (l) $E° = 0.17$ V
   2 CO_2 (g) + 2 H^+ (aq) + 2 e^- → $H_2C_2O_4$ (aq) $E° = -0.45$ V

   a. the strongest oxidizing agent is $PbSO_4$ (s)
   b. $PbSO_4$ (s) will spontaneously react with CO_2 (g)
   c. the weakest oxidizing agent is PbO_2 (s)
   d. $H_2C_2O_4$ (aq) will spontaneously react with PbO_2 (s)

5. In the electrolysis of molten BaI_2
   a. the Ba^{2+} ions migrate toward the cathode
   b. the I^- ions migrate toward the cathode
   c. water undergoes oxidation at the anode
   d. the Ba^{2+} ions migrate toward the anode

6. In the table of standard reduction potentials
   a. the strongest reducing agents are located in the bottom left of the table
   b. the strongest oxidizing agents are located in the top left of the table
   c. the strongest reducing agents are located in the top left of the table

d. the strongest oxidizing agents are located in the bottom left of the table

7. In an electrolytic cell
   a. the cathode has a positive sign
   b. reduction occurs at the anode
   c. the anode has a negative sign
   d. reduction occurs at the cathode

8. In an electrolytic cell
   a. anions migrate towards the cathode
   b. cations migrate towards the anode
   c. ions migrate through a salt bridge
   d. anions migrate towards the anode

9. Alkaline dry cells
   a. are rechargeable nickel–cadmium batteries
   b. have a cathode in which steel is in contact with HgO in an alkaline medium
   c. contain an electrolyte of a moist paste of NH_4Cl and $ZnCl_2$
   d. contain an electrolyte of a moist paste of NaOH and $ZnCl_2$

10. The amount of substance produced at an electrode by electrolysis depends
    a. on the quantity of reactant present
    b. on the quantity of charge passed through the cell
    c. on the spontaneity of the reaction
    d. on the size of the electrolytic cell

Fill–in–the–Blank

1. _____ is the area of chemistry concerned with the interconversion of chemical and electrical energy.

2. It's convenient to separate overall cell reactions into _____ because oxidation and reduction occur at separate _____.

3. The standard cell potential is the sum of the _____ for the anode half–reaction and the _____ for the cathode half–reaction.

4. Standard half–cell potentials are defined relative to an arbitrary value of 0 V for the _____.

5. _____ are used to arrange oxidizing and reducing agents in order of increasing strength.

6. A _____ is a convenient portable source of electrical energy consisting of one or more galvanic cells.

7. A _____ differs from an ordinary battery in that the reactants are continuously supplied to the cell.

8. The process of _____ involves covering iron with another metal, such as zinc in order to prevent corrosion.

Chapter 17 – Electrochemistry

9. The _____ is used to produce aluminum metal from a mixture of Al$_2$O$_3$ and cryolite.

10. The _____ is the amount of voltage needed above the calculated standard reduction (or oxidation) potential for electrolysis to occur.

Matching

Galvanic cell	a. the oxidative deterioration of a metal
Electrolytic cell	b. a U–shaped tube that contains a gel permeated with a solution of an inert electrolyte
Oxidizing agent	c. an equation used to calculate cell potentials under nonstandard–state conditions
Reducing agent	d. an electrochemical cell in which a spontaneous chemical reaction generates an electric current
Cathode	e. the coating of one metal on the surface of another, using electrolysis
Anode	f. the species that causes oxidation to occur and is itself reduced
Salt bridge	g. the cell potential when both reactants and products are in their standard states
Electromotive force	h. the amount of charge transferred when a current of 1 ampere flows for 1 second
Coulomb	i. the purification of a metal by means of electrolysis
Standard cell potential	j. the process of using an electric current to bring about chemical change
Nernst equation	k. the electrode at which oxidation takes place
Corrosion	l. an electrochemical cell in which an electric current drives a nonspontaneous reaction
Electrolysis	m. the species that causes reduction to occur and is itself oxidized
Electrorefining	n. the electrode at which reduction takes place
Electroplating	o. the driving force that pushes the negatively charged electrons away from the anode and pulls them toward the cathode

Problems

1. Write the cell notation for the galvanic cells formed by using the following pairs of half–reactions.

   a. $NO_3^- (aq) + 4 H^+ (aq) + 3 e^- \rightarrow NO (g) + 2 H_2O (l)$
   $MnO_4^- (aq) + 8 H^+ (aq) + 5 e^- \rightarrow Mn^{2+} (aq) + 4 H_2O (l)$

   b. $Fe^{3+} (aq) + 3 e^- \rightarrow Fe (s)$
   $Ag^+ (aq) + e^- \rightarrow Ag (s)$

2. Write the overall balanced reaction and the half–reactions for MnO_4^- (in acid) reacting with $FeCl_2$ to produce Mn^{2+} and Fe^{3+}. Describe the galvanic cell you would use to carry out this reaction, and give the shorthand notation for that cell.

3. Calculate $E°$ for the following cell:

   $Hg (l) | Hg_2Cl_2 (s) | Cl^- (aq) || Hg_2^{2+} | Hg(l)$

4. Predict if the following reactions occur spontaneously in aqueous solution.

   a. $Ca (s) + Cd^{2+} (aq) \rightarrow Ca^{2+} (aq) + Cd (s)$

   b. $2 Ag (s) + Ni^{2+} (aq) \rightarrow 2 Ag^+ (aq) + Ni (s)$

   c. $SO_4^{2-} (aq) + 4 H^+ (aq) + 2 I^- (aq) \rightarrow H_2SO_3 (aq) + I_2 (s) + H_2O (l)$

5. Calculate $\Delta G°$ and K for the following reaction:

   $O_2 (g) + 4 H^+ (aq) + 4 Fe^{2+} (aq) \rightarrow 2 H_2O (l) + 4 Fe^{3+}$

6. $E°$ for a galvanic cell in which Sn^{2+} is reduced to $Sn (s)$ is +1.04 V. What is the potential at the anode? What metal is oxidized at the anode? Write the half–reaction that occurs at the anode.

7. Determine the Cl^- concentration if the following cell has $E = -0.30$ V.

   $C (s) | Cl_2 (g, 1 atm) | Cl^- (aq) || MnO_4^- (aq, 0.010 M), H^+ (pH = 3.87), Mn^{2+} (aq, 0.10 M) | Pt (s)$

8. Calculate E and ΔG for the following reaction:

   $3 Zn (s) + 2 Cr^{3+} (aq) \rightarrow 3 Zn^{2+} (aq) + 2 Cr (s)$
   $[Cr^{3+}] = 0.050$ M; $[Zn^{2+}] = 0.035$ M

9. $E°_{cell} = 2.48$ V for the following reaction:

   $NiO_2 (s) + 4 H^+ (aq) + 2 Ag (s) \rightarrow Ni^{2+} (aq) + 2 H_2O (l) + 2 Ag^+ (aq)$

   Calculate the pH of the solution if $E_{cell} = 2.06$ V and $[Ag^+]$ and $[Ni^{2+}] = 0.010$ M.

10. Calculate the K_{sp} for AgBr (s) using the table of standard reduction potentials.

11. Determine $\Delta G°$ and $E°$ given that $K = 2.35 \times 10^{-8}$ and $n = 2$.

12. Calculate the amount of product produced at the cathode in the electrolysis of molten NaBr if a current of 35 A is applied for 6.0 hours.

13. How many liters of O_2 are produced when 1.19×10^3 C are passed through water at a pressure of 755 mm Hg and 25°C?

14. How many grams of $Fe(OH)_2$ are produced at an iron anode when a basic solution undergoes electrolysis at a current of 5.00 A for 3 hours?

15. Silver tarnish is due to the build–up of Ag_2S. It can be removed by heating a piece of silver in Na_2CO_3 in an aluminum pan. The reaction that occurs is:

$$3\ Ag_2S\ (s) + 2\ Al\ (s) \rightleftarrows 6\ Ag^+\ (aq) + 3\ S^{2-}\ (aq) + 2\ Al^{3+}\ (aq)$$

Calculate $E°$ and K for this reaction given that $\Delta G_f° = -480.6$ kJ/mol for Al^{3+} (aq) and $\Delta G_f° = 86.0$ kJ/mol for S^{2-} (aq).

Challenge Problem

The K_{sp} for lead(II) chromate is 2.8×10^{-13}. The standard reduction potential for Pb^{2+}/Pb is $E° = -0.126$ V. Use this information to determine $E°$ for the reaction:

$$PbC_2O_4 + 2\ e^- \rightarrow Pb\ (s) + C_2O_4^{2-}\ (aq)$$

CHAPTER 18

HYDROGEN, OXYGEN, AND WATER

Chapter Learning Goals

🔑 A. *Properties of Hydrogen*
1. Describe the properties of elemental hydrogen, including its appearance, molecular structure, occurrence in nature, laboratory synthesis, industrial synthesis, and industrial use.
2. Describe the isotopes of hydrogen, and compare and contrast the properties of the isotopes and compounds containing the isotopes.

🔑 B. *Reactions of Hydrogen*
1. Apply the gas laws to problems involving hydrogen, including calculation of its density and calculation of the number of grams of solid reactants needed to produce a certain volume of hydrogen.
2. Assign oxidation numbers and identify the oxidizing agent and reducing agent for redox reactions of hydrogen. Balance equations for these reactions, using either the oxidation–number method or the half–reaction method.
3. Classify binary hydrides as ionic, covalent, or metallic.

🔑 C. *Properties of Oxygen*
1. Describe the properties of elemental oxygen, including its appearance, molecular structure, occurrence in nature, laboratory synthesis, industrial synthesis, and industrial use.
2. Classify oxides as basic, acidic, or amphoteric.
3. Identify compounds as oxides, peroxides, and superoxides.
4. Use structure and bonding concepts to explain the physical and chemical properties of elemental oxygen and ozone.

🔑 D. *Reactions of Oxygen*
1. Apply the gas laws to problems involving oxygen, including the calculation of its density and the calculation of the number of grams of solid reactants needed to produce a certain volume of oxygen.
2. Assign oxidation numbers and identify the oxidizing agent and reducing agent for redox reactions of oxygen. Balance equations for these reactions, using either the oxidation–number method or the half–reaction method.

🔑 E. *Properties and Reactions of Water*
1. Write balanced net ionic equations for the reaction of water with alkali metals, alkaline earth metals, and halogens.
2. Identify and describe the general properties of a hydrate.
3. Determine the empirical formula of a hydrate.

Chapter in Brief
In this chapter, you will continue your study of descriptive chemistry by examining hydrogen, oxygen, and water. This study will include the properties, synthesis, and use of the elements as well as their redox chemistry. You will investigate the properties of ozone, an allotrope of oxygen, the isotopes of

Chapter 18 – Hydrogen, Oxygen, and Water

hydrogen, and the hydrides and oxides. Finally, you will examine the reactivity and the properties of water, along with the properties of hydrates.

18.1 Hydrogen
A. Henry Cavendish — English chemist who isolated hydrogen in pure form.
   1. Acid + Metal → H_2 (g) + Metal ion
B. Lavoisier — French chemist who named hydrogen.
   1. Hydrogen — water former.
      a. combines with oxygen to form water
C. Properties of hydrogen.
   1. Colorless, odorless, and tasteless gas.
   2. Nonpolar, diatomic molecule.
   3. Weak intermolecular forces.
      a. low boiling and melting points
   4. Bond dissociation energy = 436 kJ/mol
D. Natural occurrence.
   1. 75% of the mass of the universe.
   2. Atmospheric abundance — 0.53 ppm by volume.
   3. Ninth most abundant element in terms of mass in the earth's crust and oceans.

18.2 Isotopes of Hydrogen
A. Three isotopes of hydrogen.
   1. Protium — ordinary hydrogen ($_1^1H$).
      a. 99.985% of atoms in naturally occurring hydrogen
   2. Deuterium — heavy hydrogen ($_1^2H$ or D).
   3. Tritium — ($_1^3H$ or T).
B. Similar properties (see Table 18.1, page 739 in your textbook).
   1. All three isotopes have the same electron configuration.
      a. determines chemical behavior
   2. Isotope effects — arise from the differences in the mass of the isotopes.
      a. greater for hydrogen than for any other element
         i. Percentage differences between the masses of the isotopes are larger for hydrogen.
      b. D_2 — higher boiling and melting point and greater heat of dissociation than H_2
      c. D_2O — higher boiling and melting point than H_2O
         i. smaller equilibrium constant for dissociation
         ii. used as a coolant and a moderator in nuclear reactors
C. D_2 and D_2O manufactured by electrolysis of an inert electrolyte in ordinary water.
   1. Electrolysis — the process of using an electric current to bring about chemical change.
   2. Kinetic isotope effect — effect of isotopic mass on the rate of a chemical reaction.
      a. heavier isotopes — slower reaction
      b. 2 H_2O (l) → 2 H_2 (g) + O_2 (g)
         2 D_2O (l) → 2 D_2 (g) + O_2 (g)   slower

18.3 Preparation and Uses of Hydrogen
A. Purest hydrogen prepared by electrolysis of water.
   1. Requires a large amount of energy.
   2. Not economical for large–scale production.
   3. 2 H_2O (l) → 2 H_2 (g) + O_2 (g) $\Delta H° = +572$ kJ
B. Laboratory synthesis — reaction of dilute acid with an electropositive metal.

1. M (s) + $2n$ H$^+$ → n H$_2$ (g) + M^{n+}
C. Large–scale industrial preparation methods — reducing agent extracts the oxygen from steam.
   1. Steam–hydrocarbon re–forming process.
      a. first step – produce synthesis gas
         i. H$_2$O (g) + CH$_4$ (g) → CO (g) + 3 H$_2$ (g)
         ii. high temperature, moderately high pressure, and a nickel catalyst
      b. second step — water–gas shift reaction
         i. shifts the composition of synthesis gas
         ii. removes CO and produces more H$_2$
         iii. CO (g) + H$_2$O (g) → CO$_2$ (g) + H$_2$ (g)
         iv. metal oxide catalyst, T = 400°C
      c. third step – removal of CO$_2$ (g)
         i. CO$_2$ (g) + 2 OH$^-$ (aq) → CO$_3^{2-}$ (aq) + H$_2$O (l)
         ii. CO$_3^{2-}$ remains in the aqueous phase
D. Uses of hydrogen
   1. 95% of H$_2$ produced — synthesized and consumed in industrial plants.
   2. Largest single consumer – Haber process.
      a. N$_2$ (g) + 3 H$_2$ (g) ⇌ 2 NH$_3$ (g)
   3. Synthesis of methanol.
      a. CO (g) + H$_2$ (g) → CH$_3$OH (l)

18.4 Reactivity of Hydrogen

A.1.

A. Hydrogen has properties similar to both the alkali metals and the halogens.
   1. Lose electron — H$^+$.
      a. E_i = 1312 kJ/mol
      b. doesn't completely transfer its valence electron in ordinary chemical reactions
      c. shares electron with a nonmetallic element to produce a covalent compound
      d. gas phase — complete ionization of a hydrogen atom to produce a bare proton
      e. liquids and solids — bare proton is too reactive to exist by itself
   2. Gain electron — H$^-$.
      a. E_{ea} = –73 kJ/mol
      b. will accept an electron from an active metal to give an ionic hydride
B. H$_2$ — relatively unreactive.
   1. Due to strong H–H bond.
   2. Reaction of H$_2$ and O$_2$ – highly exothermic
      a. ΔH = –572 kJ
      b. mixtures of H$_2$ in O$_2$ – highly flammable and can explode

18.5 Binary Hydrides

B.3.

A. Binary hydride — a compound that contains hydrogen and just one other element.
B. Ionic hydride — formed by the alkali metals and the heavier alkaline–earth metals.
   1. Saltlike, high–melting, white crystalline compounds.
   2. Prepared by direct reaction of the elements.
   3. Alkali–metal hydrides — face–centered cubic crystal structure.
      a. ionic in the liquid state
         i. Molten compounds conduct electricity.
   4. H$^-$ — good proton acceptor.
      a. ionic hydrides react with water to give H$_2$ gas and OH$^-$ ions
         i. redox reaction
   5. Good reducing agents

Chapter 18 – Hydrogen, Oxygen, and Water

   C. Covalent hydrides — hydrogen is attached to another element by a covalent bond.
      1. Most common — hydrides of nonmetallic elements.
      2. Discrete, small molecules with relatively weak intermolecular forces.
         a. gases or volatile liquids at ordinary temperatures
   D. Metallic hydrides — lanthanide, actinide and certain d–block transition metals.
      1. Interstitial hydrides — crystal lattice of metal atoms with the smaller hydrogen atoms occupying holes (interstices) between the larger metal atoms.
      2. Nonstoichiometric compounds — atomic composition can't be expressed as a ratio of small whole numbers.
      3. Properties depend on composition.
         a. function of the partial pressure of H_2 gas in the surroundings
      4. Potential hydrogen–storage devices.
         a. can contain large amounts of hydrogen

18.6 Oxygen
   A. Joseph Priestley (English) and Karl Wilhelm Scheele (Swedish) — first isolated and characterized oxygen..
      1. Odorless, tasteless gas.
      2. Supports combustion better than air.
      3. Priestley — called gas "dephlogisticated air"
      4. Lavoisier — named element oxygen.
   B. Properties of oxygen.
      1. Pale blue liquid and solid.
      2. Paramagnetic in all three phases.
      3. O_2 — double bonded.
         a. bond length — 121 pm
         b. Bond dissociation energy = 498 kJ
   C. Additional information.
      1. Most abundant element on the planet's surface.
      2. Crucial to human life.
      3. Oxidizing agent in metabolic "burning" of food
      4. Occurrence in nature.
         a. 23% of atmosphere — primarily as O_2
         b. 46% of lithosphere
         c. 85% of hydrosphere — in the form of H_2O
   D. Photosynthesis process — green plants use solar energy to produce O_2 and glucose from carbon dioxide and water.
      1. Replaces O_2 removed by combustion and respiration processes.
      2. $6\,CO_2 + 6\,H_2O \rightarrow 6\,O_2 + C_6H_{12}O_6$
   E. Metabolism of carbohydrates.
      1. Reverse of photosynthesis.

18.7 Preparation and Uses of Oxygen
   A. Laboratory preparation.
      1. Electrolysis of water.
      2. Decomposition of aqueous hydrogen peroxide in the presence of a catalyst.
      3. Thermal decomposition of an oxoacid salt.
   B. Industrial synthesis — fractional distillation of liquefied air.
      1. Third ranking industrial chemical produced in the U.S.
   C. Industrial uses.
      1. Steelmaking — removes impurities from iron by forming oxides.

Chapter 18 – Hydrogen, Oxygen, and Water

2. Sewage treatment — destroys malodorous compounds by oxidation.
3. Paper bleaching — oxidizes compounds that cause unwanted colors.
4. Oxyacetylene torches — provide the high temperatures needed for cutting and welding metals.
5. Inexpensive and readily available oxidizing agent.

18.8 Reactivity of Oxygen

C.1.

A. High electronegativity.
   1. Needs only two electrons to achieve an octet.
B. Reacts with active metals to produce an ionic oxide.
   1. Gains two electrons and achieves a noble gas configuration.
C. Forms covalent oxides with nonmetals.
   1. Sharing of two additional electrons achieves a noble gas configuration.
   2. Forms two single bonds or one double bond.
      a. Double bonds are formed between oxygen and other small atoms.
         i. good overlap between oxygen $p\pi$ orbital and the $p\pi$ orbitals of small atoms
D. Reacts directly with nearly all the elements in the periodic table.
   1. Doesn't react with noble gases, platinum, or gold.
   2. Reactions are slow at room temperature.
   3. Extremely reactive at higher temperatures.

18.9 Oxides

C.2.

A. Classified on the basis of the oxygen's oxidation state.
B. Oxidation number = –2, oxides.
   1. Categorize in terms of their acid–base character.
      a. amphoteric – both basic and acidic
   2. Basic oxides.
      a. ionic
      b. formed by the metals on the left side of the periodic table
   3. Acidic oxides (acid anhydrides).
      a. covalent
      b. formed by the nonmetals on the right side of the periodic table
   4. Amphoteric oxides.
      a. exhibit both acidic and basic properties
      b. formed by elements with intermediate electronegativity
      c. strongly polar covalent character
      d. bonds have intermediate ionic–covalent character
   5. Acid–base properties and ionic–covalent character depend on
      a. position of element in the periodic table
         i. Acidic and covalent character increases left to right across the periodic table
         ii. Basic and ionic character increases top to bottom down the periodic table.
      b. oxidation state of element
         i. Acidic and covalent character increases with increasing oxidation state.
   6. As the bonding in oxides changes from ionic to covalent, a change in structure and physical properties occurs.
C. Uses determined by properties
   1. High thermal stability, mechanical strength, and electrical resistance:
      a. high–temperature electrical insulators
   2. Amphoteric oxides — SiO_2.
      a. optical fibers
   3. Acidic oxides.

Chapter 18 – Hydrogen, Oxygen, and Water

   a. precursors to industrial acids

18.10 Peroxides and Superoxides
  A. Oxidation number = –1, peroxides (O_2^{2-}).
  B. Oxidation number = $-\frac{1}{2}$, superoxides (O_2^-).
  C. Formed by heating the heavier group 1A and 2A metals in an excess of air.
  D. Ionic solids.
  E. MO Theory – explains trends in bond lengths and magnetic properties of O_2, O_2^{2-}, and O_2^-.
    1. due to number of electrons in π_{2p}^* orbital (see Table 18.2, page 754 in your textbook)
  F. Peroxide ion.
    1. Diamagnetic.
    2. O–O single bond.
    3. Has basic properties.
      a. reacts with strong acids to form hydrogen peroxide, H_2O_2
  G. Superoxide ion.
    1. One unpaired electron.
    2. Paramagnetic.
    3. Bond order = 1.5.
    4. Evolves oxygen when dissolved in water.
      a. disproportionation reaction — a substance is both oxidized and reduced in a reaction
      b. $2\ KO_2\ (s) + H_2O\ (l) \rightarrow O_2\ (g) + 2\ K^+\ (aq) + HO_2^-\ (aq) + OH^-\ (aq)$
         ↑   ↑   ↑
         –1/2   0   –1

18.11 Hydrogen Peroxide
  A. Uses (due to oxidizing properties).
    1. Antiseptic, bleach, starting material for synthesis of other peroxide containing compounds.
  B. Properties.
    1. Colorless, syrupy liquid.
    2. mp = –0.4°C; bp = 150°C.
    3. Explodes when heated.
    4. Strong hydrogen bonding present in liquid H_2O_2 (indicated by high bp).
    5. Weak acid in aqueous solutions.
  C. Both an oxidizing and a reducing agent.
    1. Oxidizing agent: $O_2^{2-} + 2\ e^- \rightarrow 2\ O^{2-}$.
    2. Reducing agent: $O_2^{2-} \rightarrow O_2 + 2\ e^-$.
    3. Can oxidize and reduce itself (unstable with respect to disproportionation to water and oxygen).
      a. $2\ H_2O_2\ (l) \rightarrow 2\ H_2O\ (l) + O_2\ (g) \quad \Delta H° = -196\ kJ$
      b. slow at room temperature
      c. heat, light, and many catalysts can cause rapid, exothermic, and potentially explosive decomposition

18.12 Ozone
  A. Two allotropes of oxygen.
    1. Ordinary dioxygen, O_2.
    2. Ozone, O_3.
  B. Ozone.
    1. Toxic, pale blue gas.
    2. Sharp, penetrating odor.

3. Produced by passing an electric discharge through O_2.
  C. Bonding in ozone.
    1. Two resonance structures that are bent.
    2. Central atom surrounded by two σ bonds and one lone pair of electrons.
    3. π bond is delocalized over all three oxygen atoms.
    4. Net bond order = 1.5.
  D. Powerful oxidizing agent.
    1. Used to kill bacteria in drinking water.

18.13 Water
  A. Most familiar and abundant compound on earth.
  B. Natural occurrence
    1. 97.3% — oceans
       a. volume of oceans = 1.35×10^{24} mL (twice Avogadro's number)
    2. 2.0% — polar ice caps and glaciers
    3. 0.6% — underground fresh water
    4. 0.01% — freshwater lakes and rivers
  C Seawater.
    1. Major constituent — sodium chloride.
    2. Unlimited source of chemicals.
    3. Three substances obtained from seawater commercially: NaCl, Mg, and Br_2
  D. Freshwater lakes.
    1. Purification process.
       a. preliminary filtration
       b. sedimentation
       c. sand filtration
       d. aeration
       e. sterilization
  E. Hard water – contains appreciable concentrations of doubly charged cations.
    1. Cations react to form soap scum and boiler scale.
    2. Softened by ion exchange.
       a. process by which Na^+ replaces doubly charged ions.

18.14 Reactivity of Water   ☞ E.1
  A. Reacts with alkali metals, heavier alkaline earth metals, and the halogens.
    1. Alkali metals and heavier alkaline earth metals + H_2O → H_2 + metal hydroxide.
    2. Fluorine + H_2O → O_2 + HF.
    3. Chlorine, bromine, and iodine disproportionate.
       a. extent of disproportionation decreases down the table
       b. $X_2 + H_2O$ → HOX + X^- (X = Cl, Br, I)

18.15 Hydrates
  A. Solid compounds that contain water molecules.
  B. Complex or unknown structures.
    1. Use a dot in the formula to specify the composition.
  C. Formation is common with salts that contain +2 and +3 cations.
    1. Bonding interactions between water and a metal cation increase with increasing charge on the cation.
  D. Hygroscopic compounds — absorb water from the air.
    1. Anhydrous compounds that have a tendency to form hydrates.
    2. Useful as drying agents.

Putting It Together

When sodium hydride reacts with sulfur dioxide liquid, sodium dithionate ($Na_2S_2O_4$) solid and hydrogen gas are produced. (Sodium dithionate is used to bleach paper pulp.) Write a balanced equation for this reaction. What volume of hydrogen gas is produced when 32.83 g of sodium hydride reacts with 125.25 mL of sulfur dioxide at a temperature of 25° C and pressure of 1 atm?

Self-Test

This section is intended to test your knowledge of the material covered in this chapter. Think through these problems, and make certain you understand what is going on. Ask yourself if your answer makes sense. Many of these questions are linked to the chapter learning goals. Therefore, successful completion of these problems indicates you have mastered the learning goals for this chapter. You will receive the greatest benefit from this section if you use it as a mock exam. You will then discover which topics you have mastered and which topics you need to study in more detail.

True–False

1. Hydrogen transfers a valence electron to a nonmetallic element to produce an ionic compound.

2. Hydrogen is the most abundant element on the planet's surface.

3. Oxygen is an inexpensive and readily available reducing agent.

4. Oxygen is paramagnetic in all three phases.

5. The largest single consumer of hydrogen is the Haber process.

6. The three isotopes of hydrogen have different chemical behavior.

7. Interstitial hydride compounds are considered to be ionic hydrides.

8. The acidic and covalent character of oxides increases from left to right across the periodic table.

9. The peroxide ion has basic properties

10. Hydrogen peroxide is stable with respect to disproportionation.

11. Ozone is a powerful oxidizing agent.

12. The formation of hydrates is common with salts that contain +2 and +3 cations.

Chapter 18 – Hydrogen, Oxygen, and Water

Matching

Isotope effect	a. a compound in which the structure consists of a crystal lattice of metal atoms with the smaller hydrogen atoms occupying holes between the larger metal atoms
Electrolysis	b. a compound that contains hydrogen and just one other element
Binary hydride	c. a solid compound that contains water molecules
Interstitial hydride	d. a reaction in which a substance is both oxidized and reduced
Nonstoichiometric compound	e. differences in properties due to the differences in the mass of isotopes
Disproportionation reaction	f. the process of using an electric current to bring about chemical change
Hydrate	g. compound in which the atomic composition can't be expressed as a ratio of small, whole numbers

Fill–in–the–Blank

1. The low boiling and melting points of hydrogen are due to _____.
2. Hydrogen is synthesized in the laboratory by _____
   _____.
3. The water–gas shift reaction shifts the composition of _____ to remove _____ and produce more _____.
4. Ionic hydrides are formed between hydrogen and _____ and are prepared by _____.
5. The properties of interstitial hydrides depend on _____ and are a function of _____.
6. Oxygen does not react with _____.
7. Amphoteric oxides are formed between oxygen and _____ .
   _____.
8. Peroxides and superoxides can be formed between oxygen and the _____ _____ and are prepared by _____.
9. Ozone is produced by _____.

325

Chapter 18 – Hydrogen, Oxygen, and Water

10. Hygroscopic compounds are _____ compounds that have a tendency to form _____ and are useful as _____ since they absorb _____.

Problems

1. How many grams of zinc are required to generate 750 mL of hydrogen at 22°C and 740 mm Hg by the reaction between zinc and hydrochloric acid?

2. Give a chemical equation for the largest industrial use of hydrogen.

3. Write balanced equations for the following reactions, using the half-reaction method. Assign oxidation numbers, and identify the oxidizing agent and the reducing agent.

   a. $Ag^+ (aq) + AsH_3 \rightarrow H_3AsO_4 (aq) + Ag (s)$ (Acidic solution)

   b. $Al (s) + H_2SO_4 (aq) \rightarrow Al_2(SO_4)_3 (aq) + H_2 (g)$

   c. $SnO_2 (s) + H_2 (g) \rightarrow Sn (s) + H_2O (l)$

   d. $WO_3 (s) + H_2 (g) \rightarrow W (s) + H_2O (g)$

4. Write the chemical equation for the reaction between LiH and H_2O. Identify the reducing and oxidizing agents in this reaction. This reaction can also be classified as an acid–base reaction. Identify the acid and the base.

5. Using chemical equations, explain why the price of ammonia depends on the price of natural gas.

6. Identify the following hydrides as ionic, covalent, or metallic:
   a. MgH_2 b. AsH_3 c. B_2H_6 d. $ZrH_{1.9}$ e. UH_3

7. Oxygen can be prepared in the laboratory by the thermal decomposition of $KClO_3$ in the presence of MnO_2. How many grams of $KClO_3$ are needed to produce 500 mL of O_2 at 22°C and 758 mm Hg?

8. Write balanced equations for the following reactions, using the half-reaction method. Assign oxidation numbers, and identify the oxidizing agent and the reducing agent.
   a. $VO^{2+} (aq) + Cr_2O_7^{2-} (aq) \rightarrow Cr^{3+} (aq) + VO_2^+ (aq)$ (Acidic solution)

   b. $O_2 (g) + H_2O (l) + Mg (s) \rightarrow Mg(OH)_2 (s)$

   c. $O_2 (g) + 4 H^+ (aq) + Cu (s) \rightarrow Cu^{2+} (aq) + H_2O (l)$

9. Classify the following oxides as basic, acidic, or amphoteric.
   a. CaO   b. P_4O_{10} c. Al_2O_3 d. SO_3

Chapter 18 – Hydrogen, Oxygen, and Water

10. Complete and balance the following equations:
    a. CaH_2 (s) + H_2O (l) →

    b. KH (s) + O_2 (g) →

    c. Li (s) + O_2 (g) →

    d. P_4 (s) + O_2 (g) →

    e. Na_2O (s) + H_2O (l) →

    f. BaO_2 (s) + H_2SO_4 (aq) →

    g. N_2O_5 (s) + H_2O (l) →

    h. CsO_2 (s) + H_2O (l) →

    i. H_2O_2 (aq) + MnO_4^- (aq) → (Acidic solution)

    j. O_3 (g) + I^- (aq) → (Basic solution)

11. Calculate the density of H_2 if 0.2348 g of CaH_2 reacts with an excess of H_2O at 0°C 1 atm of pressure.   *B.1.*

12. Calculate the volume of 0.7500 M $KMnO_4$ needed to react with H_2O_2 to produce 3.000 g of O_2 at 0°C and 1 atm pressure. What volume of O_2 is produced? From this information, calculate the density of O_2.   *D.1.*

13. Identify the following as oxides, peroxides, or superoxides:
    a. K_2O_2 b. RbO_2 c. CaO   d. SrO_2 e. Al_2O_3

14. When 2.879 g of the hydrate $MgSO_4 \cdot XH_2O$ was heated, 1.406 g of $MgSO_4$ was obtained. What was the formula of the hydrate?   *E.3.*

15. A 2.75 g sample of Na_2CO_3 was exposed to moist air. If 7.42 g of a hydrate of Na_2CO_3 was obtained, what was the formula of the hydrate?   *E.3.*

Challenge Problem

Hydrogen peroxide can act as either an oxidizing agent or a reducing agent.

a. When H_2O_2 reacts with KI (aq), solid I_2 is formed. Write a balanced equation for this reaction. If necessary, use either the oxidation–number method or half–reaction method to balance the equation. Identify the oxidizing and reducing agents.

Chapter 18 – Hydrogen, Oxygen, and Water

b. When H_2O_2 reacts with potassium permanganate, the purple color of the solution disappears and Mn^{2+} is formed. Write a balanced equation for this reaction. If necessary, use either oxidation–number method or half–reaction method to balance the equation. Identify the oxidizing and reducing agents.

CHAPTER 19

THE MAIN–GROUP ELEMENTS

Chapter Learning Goals

A. Periodic Properties
1. Determine which of two main–group elements has: a) the more metallic character, b) the higher ionization energy, c) the larger atomic radius, d) the higher electronegativity e) the more acidic oxide, f) the more ionic hydride, and g) the more ionic oxide.
2. Contrast the chemical and physical properties of the second–row main–group elements with the properties of the heavier members in the same groups.

B. Boranes
1. Compare the properties of the group 3A elements. Include valence electron configurations, common oxidation states, and trends in atomic radii, ionic radii, first ionization energies, and electronegativities.
2. Draw the structure of diborane and describe its bonding.

C. Group 4A
1. Compare the properties of the group 4A elements. Include valence electron configurations, common oxidation states, and trends in atomic radii, first ionization energies, and electronegativities. Describe how each is found in nature, and give one method of preparation and one commercial use.
2. Describe the carbon allotropes.
3. Briefly describe the chemistry of carbon oxides, carbonates, cyanides, and carbides, including commercial uses of these compounds.
4. Given the formula of a silicate–containing mineral, determine the charge, the number of shared oxygens, and the structure of the silicate.

D. Group 5A
1. Compare the properties of the group 5A elements. Include valence electron configurations, common oxidation states, and trends in atomic radii, first ionization energies, and electronegativities. Describe how each is found in nature, and give one commercial use.
2. Give an example of a nitrogen–containing compound for each common oxidation state exhibited by nitrogen. For each compound, sketch its electron–dot structure and describe its geometry.
3. Briefly describe the chemistry of ammonia, hydrazine, and the nitrogen oxides.
4. Give an example of a phosphorus–containing compound for each common oxidation state exhibited by phosphorus. For each compound, sketch its electron–dot structure and describe its geometry.
5. Show how phosphoric acids can be interconverted by removing or adding water molecules.

E. Group 6A
1. Compare the properties of the group 6A elements. Include valence electron configurations, common oxidation states, and trends in atomic radii, ionic radii, first ionization energies, electron affinities, electronegativities, and redox potentials for $X + 2 H^+ + 2 e^- \rightarrow H_2X$.

Chapter 19 – The Main-Group Elements

2. Give an example of a sulfur–containing compound for each common oxidation state exhibited by sulfur. For each compound, write a balanced chemical equation for its preparation, sketch its Lewis electron–dot structure, and describe its geometry.

F. Halogen Oxoacids
1. Give an example of a halogen oxoacid, HXO_n, for $n = 1, 2, 3,$ and 4. Name each acid, sketch its Lewis electron–dot structure, and describe its geometry.

Chapter in Brief
In this chapter, you will explore the chemistry of the groups 3A through 7A elements, paying particular attention to boron, carbon, silicon, nitrogen, phosphorus, and sulfur. You will examine how the chemistry of the elements in each group is determined by the valence electron configurations and how this configuration affects the common oxidation states, atomic radii, ionic radii, first ionization energies, and electronegativities of the elements. You will also examine how the size and electronegativities of the second–row elements changes their chemistry relative to the other elements in the group. You will then see how the chemical concepts you have been studying throughout this study guide can be applied in understanding the chemistry of specific compounds of the elements.

19.1 A Review of General Properties and Periodic Trends
A. Z_{eff} increases across the periodic table.
  1. Each additional valence electron does not completely shield the additional nuclear charge.
     a. Atom's electrons are more strongly attracted to the nucleus.
     b. Ionization energy increases, atomic radius decreases.
     c. Eectronegativity increases.
     d. Metallic character decreases, and nonmetallic character increases across the table.
B. Atomic radius increases from the top to the bottom of a group in the periodic table.
  1. Additional shells of electrons are occupied.
  2. Valence electrons are farther from the nucleus.
     a. Ionization energy and electronegativity generally decrease.
     b. Metallic character increases.
     c. Nonmetallic character decreases.
C. Metals form ionic compounds with nonmetals.
  1. Metallic hydrides — ionic solids with high melting points.
D. Nonmetals tend to form covalent molecular compounds with each other.
  1. Nonmetallic hydrides — covalent molecules that or either gases or volatile liquids at room temperature.
E. Comparison of properties of metallic and nonmetallic elements – see Table 19.2, page 771 in your textbook.

19.2 Distinctive Properties of the Second–Row Elements
A. Properties of second–row elements differ markedly from those of heavier elements in the same periodic group.
B. Second–row atoms have especially small sizes and especially high electronegativities.
  1. Accentuate nonmetallic behavior.
     a. BeO — amphoteric (other group 2A element oxides are basic)
     b. Boron forms mainly covalent molecular compounds.
     c. hydrogen bonding interactions — restricted to compounds of the highly electronegative second–row elements N, O, and F
C. Second–row elements lack valence d orbitals.
  1. Have only four valence orbitals.
     a. form a maximum of four covalent bonds

D. Small size of second–row atoms — allows formation of multiple bonds involving π overlap of the 2p orbitals.
  1. 3p orbitals are more diffuse.
     a. longer bond distances
     b. poor π overlap
  2. π bonds involving p orbitals are rare for elements of the third and higher rows.

19.3 The Group 3A Elements

🔑 B.1.

A. Gallium — the largest liquid range of any metal.
  1. Used in making GaAs, a semiconductor.
B. Indium — also used in making semiconductor devices.
C. Thallium — extremely toxic and has no commercial uses.
D. Valence electron configuration: ns^2np^1.
  1. Most stable oxidation state for Ga and In: +3.
  2. Most stable oxidation state for Tl: +1.
E. Properties — consistent with increasing metallic character down the group.
  1. Metal (except B).
  2. Boron — much higher electronegativity and much smaller atomic radius.
     a. shares valence electrons
     b. has nonmetallic character

19.4 Boron

A. Occurs in concentrated deposits of borate minerals.
B. Obtained from BBr_3.
  1. $2 BBr_3 (g) + 3 H_2 (g) \rightarrow 2 B (s) + 6 HBr (g)$.
C. Crystalline boron — strong, hard, high–melting substance.
  1. Chemically inert at room temperature.
  2. Desirable component in high–strength composite materials.
D. Boron halide — highly reactive, volatile covalent compounds.
  1. BX_3 molecules.
  2. Behave as Lewis acids.
     a. uses its vacant 2p orbital in accepting a share in a pair of electrons from a Lewis base
E. Boron hydrides (boranes) — volatile, molecular compounds with formulas B_nH_m.
  1. Simplest — diborane (B_2H_6).
  2. Two BH_2 groups are connected by two bridging H atoms.
     a. geometry around B atoms — tetrahedral
     b. Bridging B–H bonds are significantly longer than the terminal B–H bonds.

  3. Diborane — 12 valence electrons; electron deficient.

🔑 B.2.

     a. B atoms use sp^3 hybrid orbitals to bond to four neighboring H atoms.
        i. formed from overlap of a boron sp^3 hybrid orbital and hydrogen 1s orbital
     b. three–center, two–electron bond — joins each bridging H atom to both B atoms
     c. Electron density between adjacent atoms is less than in an ordinary 2c–2e bond.
     d. Two electrons in B–H–B bridge are spread out over three atoms.

Workbook Problem 19.1

A solid compound containing boron and hydrogen has the empirical formula B_5H_7. When 11.4 g of this compound was dissolved in 50.0 g of benzene, $T_f = -0.87°$ C. Determine the molecular formula of the compound. T_f (benzene = 5.5° C.)

Strategy: Determine the molar mass of the compound, using the freezing–point depression information.

Step 1: Determine the number of moles of the compound that are presen,t using the freezing–point depression.

Step 2: Calculate the molar mass from the number of moles and the mass of the compound.

Step 3: Determine the ratio of the molecular molar mass to the empirical molar mass. Multiply the empirical formula by the result.

What key concept did you use?

19.5 The Group 4A Elements
A. Especially important, both in industry and in living organisms.
B. Increase in metallic character down a group in the periodic table.
C. Valence electron configuration: ns^2np^2.
   1. Most common oxidation state: +4.
      a. covalent
   2. +2 oxidation state occurs for Sn and Pb.
      a. most stable for Pb
      b. ionic compounds
   3. No simple M^{4+} (aq) ions for any of the group 4A elements.

19.6 Carbon
A Uncombined form, carbon is found as diamond and graphite.
B. Diamond.
   1. Has a covalent network structure in which each C atom uses sp^3 hybrid orbitals to form a tetrahedral array of σ bonds.
      a. interlocking three–dimensional network of strong bonds
      b. hardest known substance
      c. highest known melting point
   2. Electrical insulator.

        a. Valence electrons are localized in the σ bonds.
   C. Graphite.
      1. Two-dimensional sheetlike structure.
      2. Uses sp^2 hybrid orbitals.
         a. forms trigonal planar σ bonds to three neighboring C atoms
         b. remaining p orbital (⊥ plane of sheet) to form a π bond
         c. π bonds delocalized and free to move in plane of sheet
         d. electrical conductivity parallel to the sheets 10^{20} greater than diamond
      3. Useful as an electrode material.
      4. Carbon sheets held together by London dispersion forces.
         a. can easily slide over each other
         b. slippery feel
         c. used as a lubricant
         d. difficult for electrons to hop from one sheet to another
   D. Fullerene — third crystalline allotrope of carbon; found in the soot formed by vaporizing graphite.
      1. Spherical C_{60} molecules with the shape of a soccer ball.
      2. Prepared by electrically heating a graphite rod in a helium atmosphere.
      3. Molecular substance.
         a. soluble in nonpolar, organic solvents
      4. Derivatives of fullerene.
         a. compounds in which other atoms are attached to C_{60} cage
            i. example — $C_{60}F_{36}$
         b. compounds in which metal atoms are trapped in the C_{60} cage.
            i. example — $La@C_{60}$
      5. Other carbon clusters.
         a. egg shaped — C_{70}
         b. nanotubes — tube-shaped molecules.
   E. Forty amorphous forms that resemble graphite.
      1. Coke — heat coal in the absence of air
         a. used as reducing agent in the manufacture of steel
      2. Charcoal — wood heated in absence of air.
         a. porous spongelike structure
         b. strong adsorbent properties
         c. used in filters for removing foul-smelling molecules from water
      3. Carbon black — heat hydrocarbons in a limited supply of oxygen.
         a. used in manufacturing of tires and printing ink
   F. Oxides of carbon.
      1. CO: colorless, odorless, toxic gas.
         a. forms when C or hydrocarbon fuels are burned in a limited supply of oxygen
         b. used for the industrial synthesis of methanol
         c. toxicity — due to its ability to bond strongly to the Fe^{2+} atom of hemoglobin
            i. impairs the ability of hemoglobin to carry O_2 to the tissues
      2. CO_2: colorless, odorless, nonpoisonous gas.
         a. formed when fuels burn in an excess of O_2
         b. by-product of yeast-catalyzed fermentation of sugar in the manufacture of alcoholic beverages
         c. produced when metal carbonates are reacted with acids
         d. used in beverages
            i. CO_2 solutions — mildly acidic
            ii. $CO_2\ (g) + H_2O\ (l) \rightleftarrows H^+\ (aq) + HCO_3^-\ (aq)$
            iii. gives bite to carbonated beverages
         e. used in fire extinguishers

Chapter 19 – The Main-Group Elements

        i. nonflammable and 1.5 times more dense than air
        ii. settles over fire and cuts off source of O_2
  G. Carbonates.
    1. Carbonic acid — forms two series of salts: carbonates and hydrogen carbonates.
    2. Na_2CO_3 — soda ash.
      a. used in making glass
      b. $Na_2CO_3 \cdot 10\ H_2O$ — washing soda
        i. used in laundering textiles
        ii. CO_3^{2-} removes cations from hard water
        iii. produces OH^-, which removes grease from fabrics
    3. $NaHCO_3$ — baking soda.
      a. reacts with acidic substances in food
      b. produces CO_2
        i. causes dough to rise
  H. Hydrogen cyanide and cyanides.
    1. HCN — highly toxic, volatile substance.
      a. $CN^- (aq) + H^+ (aq) \rightarrow HCN (aq)$
    2. CN^- (pseudohalide ion) — behaves like halides.
      a. forms insoluble AgCN
    3. CN^- acts as a Lewis base.
      a. bonds through the lone pair of electrons on carbon
      b. toxicity of HCN and cyanides — due to the strong bonding of CN^- to Fe^{3+} in cytochrome oxidase (important enzyme in food metabolism)
        i. enzyme is unable to function
        ii. halts cellular energy production
        iii. rapid death
    4. CN^- used to extract Au and Ag from their ores.
      a. produces $M(CN)_2^-$ ion
      b. $M(CN)_2^-$ reduced by Zn to produce M
  I. Carbides — binary compounds of carbon.
    1. Carbon atom has a negative oxidation state.
      a. carbides of active metals
      b. interstitial carbides of transition metals
      c. covalent network carbides

19.7 Silicon

  A. Hard, gray, semiconducting solid that melts at 1410°C.
    1. Diamondlike structure.
    2. Relatively poor overlap of π orbitals.
      a. no graphitelike allotrope
    3. Naturally found combined with oxygen (SiO_2 and various silicate minerals).
  B. Ultrapure silicon needed for making solid–state semiconductor devices. Purification methods involve
    1. Convert to $SiCl_4$: $Si (s) + 2\ Cl_2 (g) \rightarrow SiCl_4 (l)$.
      a. Separate by fractional distillation.
    2. Convert $SiCl_4$ back to elemental silicon by reduction with hydrogen.
      a. $SiCl_4 (g) + 2\ H_2 (g) \rightarrow Si (s) + 4\ HCl (g)$
    3. Further purification — zone refining. (See figure 19.4, page 782 in your textbook.)
  C. Silicates — 90% of Earth's crust.
    1. Silicon oxoanions.
    2. Basic structural building block — SiO_4 tetrahedron.
      a. occurs as simple orthosilicate ion, SiO_4^{4-}.

Chapter 19 – The Main-Group Elements

3. Common minerals — anions in which two or more O atoms bridge between Si atoms to give rings, chains, layers, and extended three–dimensional structures.

C.4.
   D. Sharing of two oxygen atoms per SiO_4 tetrahedron → cyclic anions or infinitely extended chain anions (repeating unit $Si_2O_6^{4-}$). See Figure 19.5, page 783 in your textbook.
      1. $Si_6O_{18}^{12-}$ — cyclic anion.
         a. present in beryl, $Be_3Al_2Si_6O_{18}$
         b. emerald gemstone — 2% of Al^{3+} in beryl is replaced with Cr
   E. Additional sharing of O atoms gives
      1. The $(Si_4O_{11}^{6-})_n$ — double–stranded chain anions.
         a. found in asbestos minerals
            i. tremolite — $Ca_2Mg_5(Si_4O_{11})_2(OH)_2$
         b. Asbestos is fibrous material.
            i. Bonds between chains are relatively weak and easily broken.
      2. The $(Si_4O_{10}^{4-})_n$ — infinitely extended two–dimensional layer anions.
         a. found in clay minerals, micas, and talc
         b. talc — $Mg_3(OH)_2(Si_4O_{10})$
         c. Mica is a sheetlike material.
            i. Bonds between two–dimensional layers are relatively weak and easily broken.
   F. Silica — an infinitely extended three–dimensional structure in which the layer anions $(Si_4O_{10}^{4-})_n$ are stacked on top of one another, and SiO_4 and AlO_4 tetrahedra share all four of their corners with neighboring tetrahedra.
      1. Quartz — crystalline form of SiO_2.
      2. Aluminosilicates (feldspars) — most abundant of all minerals.
         a. partially substitute the Si^{4+} in SiO_2 with Al^{3+}
         b. orthoclase — $KAlSi_3O_8$
         c. zeolites – SiO_4 and AlO_4 tetrahedra are joined together in an open structure that has a three–dimensional network of cavities linked by channels
            i. act as molecular sieves for separating small molecules from larger ones
            ii. used as catalyst in the manufacture of gasoline

Workbook Problem 19.2

Cl_2 will react with Mg_2Si to produce a silicon chloride compound with the empirical formula $SiCl_3$ when the correct conditions are met. The molecular mass of this compound is 269 g/mol. What is the molecular formula of this compound? Propose a structure for this compound.

Strategy: From the information provided, determine the molecular formula. Propose a structure based upon the descriptive chemistry of silicon.

Step 1: Determine the molecular formula.

Step 2: Propose a structure.

What key concept did you use?

19.8 Germanium, Tin, and Lead
A. Relatively low abundances in the Earth's crust.
   1. Sn and Pb.
      a. concentrated in workable deposits
      b. soft, malleable, low–melting metals
B. Sn — obtained from purified cassiterite (SnO_2) by reduction with carbon.
   1. Protective coating over steel in making tin cans.
   2. Important component of alloys.
   3. Two allotropic forms.
      a. silver–white metallic form (white tin)
      b. brittle, semiconducting form with diamond structure (gray tin)
   4. Tin disease — the conversion of white tin to gray tin below 13°C.
C. Pb — obtained from galena (PbS) by roasting in air ($\rightarrow$ PbO) and reducing PbO with CO.
   1. Used in making pipes, cables, pigments, and electrodes for storage batteries.
D. Germanium — one of the holes in Mendeleev's periodic table.
   1. Used in making transistors and special glasses for infrared devices.
   2. High–melting semiconductor.
   3. Same crystal structure as diamond and silicon.

19.9 The Group 5A Elements
A. Down the periodic group — increasing atomic size, decreasing ionization energy, and decreasing electronegativity.
B. Increasing metallic character of heavier elements — evident in the acid–base properties of their oxides.
   1. N and P oxides — acidic.
   2. As and Sb oxides — amphoteric.
   3. Bi_2O_3 oxides — basic.
C. Electron configuration: ns^2np^3.
   1. Maximum oxidation state: +5.
   2. Minimum oxidation state: –3.
   3. N and P exhibit all oxidation states between –3 and +5.
   4. As and Sb — most important oxidation states: +3 and +5.
D. Metallic character increases down the group.
   1. Sb^{3+} and Bi^{3+} — found in salts.
   2. No simple cations in compounds of N or P.
E. As, Sb, and Bi are found in sulfide ores and are used in making various metal alloys.

19.10 Nitrogen
A. Colorless, odorless, tasteless gas.
   1. Makes up 78% of Earth's atmosphere.
B. Separated out of liquid air by fractional distillation.
C. Uses of N_2.
   1. Gas — protective inert atmosphere in manufacturing processes.
   2. Liquid — refrigerant.
   3. Most important use — Haber process (manufacture of NH_3).
D. N_2 unreactive due to N–N triple bond.
   1. Reactions involving N_2 have a high activation energy and/or an unfavorable equilibrium constant.

2. Haber process requires moderately high temperatures, high pressures, and a catalyst.
   a. $N_2\,(g)\ +\ 3\,H_2\,(g)\ \rightarrow\ 2\,NH_3\,(g)$

E. Classify nitrogen compounds by oxidation state (see Table 19.6, page 788 in your textbook).　　*D.2.*

F. Ammonia — gateway to nitrogen chemistry.　　*D.3.*
   1. Starting material for industrial synthesis of other important nitrogen compounds.
   2. Colorless, pungent–smelling gas.
   3. Polar, trigonal pyramidal NH_3 molecules.
   4. Very soluble in water (due to hydrogen bonding).
   5. Easily condensed.
   6. Excellent solvent for ionic compounds.
   7. Reacts with acids to yield ammonium salts.
      a. resemble alkali metal salts in their solubility
   8. Brønsted–Lowry base
      a. produces weakly alkaline aqueous solutions.

G. Hydrazine (H_2NNH_2) — derivative of NH_3.　　*D.3.*
   1. Replace one H atom with an NH_2 group.
   2. Prepared by reaction of ammonia with OCl^-.
   3. Poisonous, colorless liquid.
   4. Explosive in the presence of air or other oxidizing agents.
   5. Used as a rocket fuel.
   6. Weak base and versatile reducing agent in aqueous solutions.

H. Oxides of nitrogen.　　*D.3.*
   1. N_2O — colorless, sweet–smelling gas.
      a. laughing gas — small doses are mildly intoxicating
      b. used as a dental anesthetic
      c. propellant for dispensing whipped cream
   2. NO — colorless gas.
      a. produced in laboratory by reacting Cu with dilute HNO_3
      b. large quantities prepared by catalytic oxidation of NH_3
      c. important in biological processes
   3. NO_2 — highly toxic, reddish brown gas that forms rapidly when NO is exposed to air
      a. paramagnetic
      b. dimerizes to form N_2O_4
         i. N_2O_4 predominates at lower temperatures and NO_2 predominates at higher temperatures
   4. HNO_2 — produced when NO_2 reacts with water.
      a. disproportionation reaction — NO_2 undergoes both oxidation and reduction
      b. $2\,NO_2\,(g)\ +\ H_2O\,(l)\ \rightarrow\ HNO_2\,(aq)\ +\ H^+\,(aq)\ +\ NO_3^-\,(aq)$

I. Nitric acid — one of the most important inorganic acids.
   1. Used in making ammonium nitrate for fertilizers.
   2. Used in manufacturing explosives, plastics, and dyes.
   3. Produced by the Ostwald process.
      a. air oxidation of NH_3 to nitric oxide
      b. oxidation of nitric oxide to nitrogen dioxide
      c. disproportionation of NO_2 in water
   4. Yellow color of concentrated HNO_3.
      a. due to presence of NO_2 produced by a slight amount of decomposition
   5. Strong acid — 100% dissociated in water.

6. Strong oxidizing agent.
   a. stronger oxidizing agent than H^+ (aq)
   b. can oxidize inactive metals
   c. Product depends upon the nature of the reducing agent and the reaction conditions.
7. Aqua regia — mixture of concentrated HCl and concentrated HNO_3 (3:1).
   a. more potent oxidizing agent than HNO_3
   b. NO_3^- — serves as the oxidizing agent
   c. Cl^- — converts Au^{3+} to $AuCl_4^-$.

Workbook Problem 19.3
Upon heating, ammonium nitrate decomposes to produce nitrous oxide (N_2O) and water. What volume of nitrous oxide, collected over water, at a total pressure of 705 mm Hg and 22°C, can be produced from 3.5 g ammonium nitrate? (The vapor pressure of water at 22°C is 24 mm Hg.)

Strategy: After writing the balanced chemical reaction, determine the volume of N_2O from the pressure of N_2O using the ideal gas law. You must determine the pressure of N_2O using Dalton's law.

Step 1: Write the balanced chemical equation.

Step 2: Calculate the pressure of N_2O, using the total pressure and the vapor pressure of water.

Step 3: Calculate the number of moles of N_2O produced from 3.5 g NH_4NO_3.

Step 4: Calculate the volume of N_2O using the ideal gas law.

What key concept did you use?

19.11 Phosphorus
A. Found in phosphate rock, $Ca_3(PO_4)_2$, and in fluorapatite, $Ca_5(PO_4)_3F$.
   1. Apatites — phosphate minerals with the formula 3 $Ca_3(PO_4)_2 \cdot CaX_2$ (X^- = F^- or OH^-).
   2. Sixth most abundant element in human body.

a. bones — $Ca_3(PO_4)_2$
b. tooth enamel — $Ca_5(PO_4)_3OH$
c. found in DNA and RNA
B. Exists in two common allotropic forms.
1. White phosphorus — toxic, waxy, white solid.
a. discrete tetrahedral P_4 molecules
b. low melting point
c. soluble in nonpolar solvents
d. highly reactive — due to unusual bonding
i. P–P bonds are "bent," relatively weak, and highly reactive
2. Red phosphorus — nontoxic; has a polymeric structure.

C. Phosphorus compounds — most important oxidation states: +3 and +5.   ☞ *D.4.*
1. More likely to be found in a positive oxidation state because of its electronegativity.
D. Phosphine (PH_3) — colorless, extremely poisonous gas.
1. Most important hydride of phosphorus.
2. Neutral aqueous solutions.
a. poor proton acceptor
3. Easily oxidized: burns in air to form H_3PO_4.
E. Phosphorous halides (PX_3 or PX_5).
1. Product depends on the relative amounts of the reactants.
2. Gases, volatile liquids, or low–melting solids.

F. Oxides and oxoacids of phosphorus.   ☞ *D.5.*
1. Burn P in presence of oxygen → P_4O_6 or P_4O_{10}.
a. molecular compounds with a tetrahedral array of P atoms
b. acidic oxides — react with water → H_3PO_3 or H_3PO_4
c. P_4O_{10} — used as a drying agent
2. H_3PO_3 — weak diprotic acid.
a. only two of the three H atoms are bonded to oxygen.
3. H_3PO_4 — low–melting, colorless, crystalline solid.
a. phosphoric acid used in the laboratory — syrupy aqueous solution (82% H_3PO_4 by mass)
b. Manufacturing method depends on use.
i. food additive — begin with molten phosphorus
ii. fertilizers — begin with $Ca_3(PO_4)_2$
c. sometimes called orthophosphoric acid
4. Other phosphoric acids.
a. diphosphoric acid (pyrophosphoric acid) $H_4P_2O_7$ – combine 2 molecules of H_3PO_4 and eliminate H_2O.
b. triphosphoric acid, $H_5P_3O_{10}$
c. polymetaphosphoric acid $(HPO_3)_n$ — infinitely long chain of phosphate groups
5. $Na_5P_3O_{10}$ — component of some synthetic detergents.
a. $P_3O_{10}^{3-}$ acts as a water softener

19.12 The Group 6A Elements   ☞ *E.1.*
A. Periodic trends.
1. Oxygen and sulfur — typical nonmetals.
2. Gray selenium — most stable allotrope of selenium.
a. lustrous semiconducting solid
3. Tellurium — semiconductor; classified as a semimetal.

Chapter 19 – The Main-Group Elements

    4. Polonium — radioactive element that occurs in trace amounts in uranium ores, is a silvery white metal.
- B. Electron configuration: ns^2np^4.
   1. Common oxidation state: –2.
      a. stability of state decreases with increasing metallic character
         i. oxygen — powerful oxidizing agent
         ii. H_2Se and H_2Te — reducing agents
   2. S, Se and Te — less electronegative than oxygen.
      a. found in positive oxidation states
      b. common oxidation states: +4 and +6
- C. Commercial uses of Se, Te, and Po — limited.
   1. Se — making red-colored glass and in photocopiers.
   2. Te — used in alloys.
   3. Po — heat source in space equipment and a source of alpha particles in research.

19.13 Sulfur
- A. Many allotropic forms.
   1. Rhombic sulfur — yellow crystalline solid.
      a. most stable form of sulfur
      b. contains crown-shaped S_8 rings
      c. above 95°C convert to monoclinic sulfur (cyclic S_8 molecules pack differently in the crystal)
      d. melts at 113°C when heated at an ordinary rate
   2. Above melting point, sulfur is a fluid, straw-colored liquid.
   3. Between 160°C and 195°C — dark, reddish brown, viscous liquid.
      a. S_8 ring opens and forms S_8 chains that form long polymers with more than 200,000 S atoms in the chain.
   4. Above 195°C — more fluid liquid.
      a. Chains fragment.
   5. Boils at 445°C.
      a. pour boiling liquid into water — forms plastic sulfur
         i. Chains freeze in disordered array.
- B. Hydrogen sulfide — colorless gas with the strong, foul odor of rotten eggs.
   1. Extremely toxic.
   2. Prepared by treating FeS with dilute H_2SO_4.
   3. Generated in solution by hydrolysis of thioacetamide for use in qualitative analysis.
   4. Weak diprotic acid and a mild reducing agent.
- C. Oxides and oxoacids of sulfur.
   1. SO_2 — colorless, toxic gas formed when sulfur burns in air.
      a. slowly oxidized in the atmosphere to SO_3
      b. used for sterilizing wine and dried fruit
      c. SO_3 dissolves in rainwater to give sulfuric acid
      d. Aqueous solutions contain dissolved SO_2 and little H_2SO_3.
- D. Sulfuric acid — world's most important industrial chemical.
   1. Manufactured by contact process.
      a. S burns in air to give SO_2.
      b. SO_2 is oxidized to SO_3.
      c. SO_3 reacts with water to give H_2SO_4.
   2. Strong acid in dissociation of first proton.
   3. Forms two series of salts: sulfates and hydrogen sulfates.
   4. Oxidizing properties depend on its concentration and the temperature.

a. dilute solutions at room temperatures — behaves like HCl
b. Hot, concentrated H_2SO_4 can oxidize metals not oxidized by H^+.
5. Used in manufacturing soluble phosphate and ammonium sulfate fertilizers.

Workbook Problem 19.4
When lead(II) sulfide is treated with hydrogen peroxide, the possible products are either a) lead(IV) oxide and sulfur dioxide or b) lead(II) sulfate. Which product(s) are favored?

Strategy: To determine the more favorable products we need to calculate the free energy of the proposed reactions. The reaction that has the more negative free energy will be the favored one.

Step 1: Write a balanced chemical equation for the two proposed reactions.

Step 2: Determine the $\Delta G°$ for each reaction.

Step 3: From your results, determine which reaction is more favorable.

What key concept did you use?

19.14 The Halogens: Oxoacids and Oxoacid Salts
A. Electron configuration: ns^2np^5.
 1. Most electronegative group of elements in the periodic table.
 2. Gain an electron when forming ionic compounds.
 3. Share an electron with nonmetals to form molecular compounds.

B. Oxoacids and oxoacid salts — most important compounds of halogens in positive oxidation states.
 1. General formula — HXO_n.
 2. Oxidation state = +1, +3, +5, or +7 (depends on value of n).

C. Acid strength increases with increasing oxidation state of the halogen.
   1. Acidic hydrogen bonded to oxygen.
   2. Strong oxidizing agents.
D. Hypohalous acids formed from halogen disproportionation reactions.
   1. X_2 (g, l, or s) + H_2O (l) $\rightleftarrows$ HOX (aq) + H^+ (aq) + X^- (aq)
      a. Equilibrium lies to the left.
      b. shift right in basic solution
   2. NaOCl — strong oxidizing agent.
      a. 5% solution — chlorine bleach
E. React Cl_2 with hot NaOH → $NaClO_3$.
   1. Chlorate slats
      a. used as weed–killers and strong oxidizing agent
F. Anhydrous perchloric acid — colorless, shock–sensitive liquid that decomposes explosively on heating.
   1. Powerful and dangerous oxidizing agent.
   2. Perchlorate salts — strong oxidants.
G. Iodine forms more than one perhalic acid.
   1. Paraperiodic acid, H_5IO_6 — white crystals obtained from evaporating HIO_4 solutions.
      a. weak, polyprotic acid
   2. Metaperiodic acid, HIO_4 (s) — produced when H_5IO_6 loses water.
      a. strong monoprotic acid

Workbook Problem 19.5
The oxidation of iodine by nitric acid produces iodic acid and NO_2. What volume of 16 M nitric acid is needed to produce 15.0 g of iodic acid?

Strategy: Use solution stoichiometry to calculate the volume of nitric acid from the balanced chemical equation.

Step 1: Write the balanced chemical equation recognizing that this is a redox reaction.

Step 2: Determine the number of moles of nitric acid needed from the mass of iodic acid and the balanced chemical equation.

Step 3: Determine the volume of nitric acid from the moles of nitric acid and the concentration of nitric acid.

What key concept did you use?

 **Putting It Together**

When sulfur dioxide reacts with aqueous hydrogen sulfide, elemental sulfur is produced. a) Write a balanced equation for the reaction. b) If one ton of coal, containing 3.5% by mass sulfur, is burned, what volume of hydrogen sulfide is needed to remove the sulfur dioxide at 27°C and 740 mm Hg? c) How much elemental sulfur is produced?

 **Self–Test**

This section is intended to test your knowledge of the material covered in this chapter. Think through these problems, and make certain you understand what is going on. Ask yourself if your answer makes sense. Many of these questions are linked to the chapter learning goals. Therefore, successful completion of these problems indicates you have mastered the learning goals for this chapter. You will receive the greatest benefit from this section if you use it as a mock exam. You will then discover which topics you have mastered and which topics you need to study in more detail.

Fill–in–the–Blank

1. The second–row elements are unable to form more than four bonds because they lack _____.

2. Boron forms _____ compounds.

3. Boranes are electron–deficient molecules that contain _____ bonds.

4. The Ostwald process is used to produce _____.

5. Sulfuric acid is manufactured by the _____.

6. The three primary allotropes of carbon are _____.

7. The basic structural building block for silicates is _____.

8. Binary compounds of carbon with a negative oxidation state are referred to as _____.

9. Ammonia is produced by the _____.

10. The two common allotropic forms of phosphorus are _____.

Chapter 19 – The Main-Group Elements

General Questions

1. For the following pairs of elements, determine which has the more metallic character.
   a. Ca or Ba
   b. Al or Ga
   c. Rb or Sr

2. For the following pairs of elements, determine which has the more ionic hydride.
   a. Li or Na
   b. Sn or Sb
   c. Sb or Bi

3. What are the general trends down a group in the periodic table for ionization energy, atomic radius, electronegativity, and basicity of oxides?

4. Carbon has two common allotropes: diamond and graphite. Silicon is found only in the diamond structure. Explain this difference.

5. With the exception of boron, the group 3A elements are metals. Why does boron exhibit nonmetallic character?

6. Why are boron halides able to behave as Lewis acids?

7. Give the valence electron configuration for the group 3A elements. Give the stable oxidation states for these elements.

8. Describe the bonding in diborane.

9. Give the valence electron configuration and oxidation states for the group 4A elements.

10. Explain why graphite has a slippery feel and can be used as a lubricant.

11. Give some of the industrial uses of CO_2.

12. Describe the steps used for the purification of silicon. Include chemical equations in your description.

13. Determine the charge, number of shared oxygens per silicon and the structure of the silicates in the following minerals:
    a. beryl
    b. asbestos
    c. talc
    d. quartz
    e. orthoclase

14. Describe the acid–base properties of the oxides of group 5A.

15. What are the possible oxidation states for the group 5A elements?

16. Give an example of a nitrogen–containing compound for each common oxidation state exhibited by nitrogen.

17. Using chemical equations, show how the phosphoric acids can be interconverted by removing or adding water molecules.

18. What are the common oxidation states for the group 6A elements?

19. Give the name of the halogen oxoacid, HXO_n (X = Cl, Br, or I and n = 1, 2, 3, and 4).

20. Write chemical equations describing the actions of washing soda and baking soda.

21. Write a balanced equation for the reaction of Ag(CN)$_2^-$ by zinc.

22. Give the chemical equation for the production of NO$_2$ from nitric acid and copper.

23. Write the chemical reaction for each step in the Ostwald process.

24. Write chemical equations for each step in the contact process.

25. Write the chemical equation for the formation of hypoiodous acid from I$_2$ (s).

Challenge Problem

For the reaction

XeF$_2$ (aq) + 2 H$^+$ (aq) + 2 e$^-$ → Xe (g) + 2 HF (aq)

$E°$ = +2.32 V. When XeF$_2$ decomposes in aqueous solution, O$_2$ is produced. Is this reaction spontaneous?

CHAPTER 20

TRANSITION ELEMENTS AND COORDINATION CHEMISTRY

Chapter Learning Goals

A. d–Block Elements: Properties
1. Write valence electron configurations for transition metal atoms and ions.
2. Compare the properties of the first–transition–series elements. Include appearance, valence electron configurations for atoms and ions, common oxidation states, and trends in melting points, atomic radii, densities, ionization energies, and standard oxidation potentials.
3. Balance equations for redox reactions of chromium, iron, and copper.

B. Coordination Compounds: Ligands
1. Write formulas of coordination complexes. Identify the ligands and their donor atoms. Determine the coordination number and the oxidation state of the metal and the charge on any complex ion.
2. Given the electron dot structure of a molecule or ion, determine whether it can serve as a chelate ligand.
3. Name coordination compounds.

C. Coordination Compounds: Constitutional Isomers and Stereoisomers
1. Identify linkage isomers and ionization isomers.
2. Determine the number and structures of diastereoisomers possible for a given coordination complex.
3. Determine which coordination complexes are chiral, and draw the structure of the enantiomers.

D. Coordination Compounds: Valence Bond Theory
1. Give a valence bond theory description of a coordination complex, and show the number of unpaired electrons and the hybrid orbitals used by the metal ion.

E. Coordination Compounds: Crystal Field Theory
1. Give a crystal field theory description of a coordination complex, and show the number of unpaired electrons.

Chapter in Brief
This chapter examines the properties and chemical behavior of transition-metal compounds, with particular emphasis on coordination compounds. You begin with a review of the electronic configurations of the transition elements and how these configurations affect the properties and oxidation states. These concepts are then illustrated with the chemistry of chromium, iron, and copper. You will then explore the area of coordination chemistry, including the structural properties, color, and magnetism of these compounds. Finally, you will examine the different bonding theories that are used to explain the color and magnetism of these complexes.

Chapter 20 – Transition Elements and Coordination Chemistry

20.1 Electron Configurations
A. 4s subshell filled first; 3d subshell filled according to Hund's rule.
   1. Add one electron to each of the five 3d orbitals before adding a second electron.
   2. Two exceptions: Cr and Cu.
      a. Cr: $4s^1 3d^5$
      b. Cu: $4s^1 3d^{10}$
B. Electron configurations depend on both orbital energies and electron–electron repulsions.
   1. If two valence subshells have similar energies, can't always predict the configurations.
   2. Exceptions from the expected orbital–filling pattern result in either half–filled or completely filled subshells.
   3. For Cr and Cu: 3d and 4s have similar energies and as a consequence, one electron shifts from 4s to 3d.
      a. Shift decreases the electron–electron repulsions.
      b. gives either a filled or half–filled subshell
C. Easy to predict electron configurations of transition metal cations.
   1. All the valence electrons occupy the d orbitals.
   2. Neutral atom loses one or more electrons; Z_{eff} increases.
      a. remaining electrons more strongly attracted to the nucleus
      b. orbital energies decrease
      c. 3d orbitals experience a steeper drop in energy with increasing Z_{eff} than does the 4s
         i. 3d orbitals in cations lower in energy than the 4s orbitals
D. Groups 1B – 8B
   1. Valence electron configuration similar to analogous elements in groups 1A – 8A.
      a. groups 1B, 2B — electron configuration similar to groups 1A, 2A
      b. groups 3B – 8B — electron configuration similar to groups 3A – 8A

20.2 Properties of Transition Elements
A. Properties can be understood in terms of electron configurations.
B. Metallic properties.
   1. Malleable, ductile, lustrous, and good conductors of heat and electricity.
   2. Sharing of d, as well as s, electrons gives rise to stronger metallic bonding.
      a. Transition metals are harder, have higher melting and boiling points, and are more dense than the group 1A and 2A metals
   3. Melting points increase as the number of unpaired d electrons available for metallic bonding increases and then decrease as the d electrons pair up and become less available for bonding.
C. Atomic radii and densities.
   1. Decrease in radii with increasing atomic number.
      a. added d electrons only partially shield the added nuclear charge
      b. increase in radii toward the end of each series
         i. due to more effective shielding and increasing electron–electron repulsion as d orbitals become doubly occupied
   2. Similar radii — accounts for ability to blend together in forming alloys.
   3. Second and third series — group 4B on have nearly identical radii.
      a. smaller than expected size of the third–series atoms due to lanthanide contraction
   4. Lanthanide contraction — the general decrease in atomic radii of the f–block lanthanide elements between the second and third transition series.
      a. increase in effective nuclear charge as 4f subshell is filled
      b. Size decrease due to a larger Z_{eff} almost exactly compensates for the expected size increase due to an added quantum shell of electrons.
      c. Atoms of third series have radii very similar to those of the second series.

Chapter 20 – Transition Elements and Coordination Chemistry

   5. Densities — inversely related to atomic radii.
      a. Second and third series have nearly the same atomic volume.
         i. Third series has unusually high densities.
   D. Ionization energies and standard oxidation potentials.
      1. Ionization energies — increase left to right across a transition series.
         a. due to increase in Z_{eff} and a decrease in atomic radius
      2. $E°$ for **oxidation** potential of first series are positive (except Cu).
         a. The solid metal is oxidized to the aqueous cation more readily than H_2 gas is oxidized to H^+
         b. $M(s) + 2 H^+(aq) \rightarrow M^{2+}(aq) + H_2(g)$
         c. better reducing agents than H_2
         d. can be oxidized by "nonoxidizing" acids (lack an oxidizing anion)
         e. Cu (s) requires a stronger oxidizing agent.
      3. General trend for $E°$ correlates with the general trend in ionization energies.
         a. Ease of oxidation of the metal decreases as the ionization energies increase across the first series.

20.3 Oxidation States of Transition Elements

   A. Exhibit a variety of oxidation states.
      1. Have oxidation states less than their group number.
   B. First series form +2 cation (except Sc).
      1. Due to loss of the two 4s electrons.
   C. First series — can also lose 3d electrons.
      1. 3d and 4s energies are similar.
      2. Energy required to remove the third electron is provided by the larger $\Delta G°$ of hydration of the more highly charged +3 cation.
   D. Highest oxidation state for the groups 3B–7B metals is the group number.
      1. Corresponds to loss of all the valence s and d electrons.
      2. Later transition metals — loss of all valence electrons is prohibited by the increasing value of Z_{eff}.
   E. Transition metal ions in a high oxidation state — good oxidizing agents.
   F. Early transition metal ions in low oxidation state — good reducing agents.
   G. Divalent ions of the later metals — poor reducing agents because of the larger value of Z_{eff}.
   H. Stability of the higher oxidation states increases down a periodic group.

20.4 Chemistry of Selected Transition Metals

   A. Chromium.
      1. Obtained from chromite ($FeO \cdot Cr_2O_3$).
         a. reduce chromite with carbon $\rightarrow$ ferrochrome (Fe (s) + Cr (s))
            i. used in making stainless steel
      2. Pure form obtained from reduction of Cr_2O_3 with Al.
      3. Used to electroplate metallic objects with an attractive, protective coating.
      4. Hard, lustrous, takes a high polish.
      5. Resistant to corrosion.
         a. microscopic film of Cr_2O_3 protects surface
      6. Systemize aqueous chemistry by oxidation states and the species that exist under acidic and basic conditions. (See Table 20.3, page 822 in your textbook.)
      7. Cr^{2+}.
         a. in aqueous solution — $Cr(H_2O)_6^{2+}$ is beautiful blue solution
         b. rapidly oxidized by O_2 to Cr^{3+}

348

8. Cr^{3+}.
    a. $Cr(H_2O)_6^{3+}$ — violet
    b. Cr^{3+} solutions are green
        i. anions replace some of the bound water molecules to give green complex ions
9. $Cr(OH)_n$ — acid strength increases with increasing polarity of O–H bond.
    a. polarity increases with increasing oxidation state of Cr atom
        i. $Cr(OH)_2$ — basic
        ii. $Cr(OH)_3$ — amphoteric
        iii. $CrO_2(OH)_2$ — chromic acid (H_2CrO_4)
10. +6 oxidation state.
    a. acidic solution — $Cr_2O_7^{2-}$
        i. powerful oxidizing agent
        ii. used as an oxidant in analytical chemistry
    b. basic solution — CrO_4^{2-}
        i. weaker oxidizing agent
B. Iron — immensely important in human civilization and in living systems.
    1. Relatively soft and easily corroded.
        a. combine with carbon and other metals to make alloys
    2. Most important ores — hematite (Fe_2O_3) and magnetite (Fe_3O_4).
        a. reduce with coke in a blast furnace → Fe (s)
    3. Most important oxidation states: +2 and +3.
    4. Oxidize with an acid that lacks an oxidizing anion in absence of air:
       Fe (s) → Fe^{2+} (aq).
        a. Oxidation of Fe^{2+} has negative standard potential.
        b. in presence of air: Fe^{2+} oxidized to Fe^{3+}
    5. Oxidize with an acid that has an oxidizing anion: Fe (s) → Fe^{3+} (aq).
    6. Fe^{3+} (aq) + base → $Fe(OH)_3$ (s)
        a. very insoluble
        b. forms if pH > 2
        c. reaction accounts for red–brown rust stains in sinks
C. Copper — found in the elemental state.
    1. Most important ores — sulfides.
        a. chalcopyrite, $CuFeS_2$
    2. High electrical conductivity and negative oxidation potential.
        a. used to make electrical wiring and corrosion–resistant water pipes
    3. Less reactive than other first–series transition metals.
    4. Prolonged exposure to moist air and CO_2: Cu (s) → $Cu_2(OH)_2CO_3$.
        a. continues to react with acid rain forming $Cu_2(OH)_2SO_4$, the green patina seen on bronze monuments
    5. Two common oxidation states: +1 and +2.
    6. $E°_{Cu^+/Cu^{2+}}$ less negative than $E°_{Cu/Cu^+}$.
        a. Any oxidizing agent strong enough to oxidize copper to the Cu(I) ion is also able to oxidize the Cu(I) ion to the Cu(II) ion
        b. Cu^+ (aq) can undergo a disproportionation reaction
            i. 2 Cu^+ (aq) → Cu (s) + Cu^{2+} (aq) $E° = +0.37$ V
            ii. large equilibrium constant, $K = 1.8 \times 10^6$
        c. Cu^+ — not an important species in aqueous solution
    7. Cu^+ exists in solid compounds.
        a. disproportionation equilibrium is reversed in presence of Cl^-
            i. CuCl precipitation shifts the preceding reaction left
    8. Cu^{2+} — more common.
        a. most common compound — blue–colored $CuSO_4 \cdot 5\ H_2O$

Chapter 20 – Transition Elements and Coordination Chemistry

i. Four H$_2$O's bound to the Cu^{2+}; fifth H$_2$O hydrogen bonded to SO$_4^{2-}$
ii. loses color on heating — suggests that blue color is due to the bonding of Cu^{2+} to the water molecules

20.5 Coordination Compounds
A. Coordination compound — a compound in which a central metal ion is attached to a group of surrounding molecules or ions by coordinate covalent bonds.
B. Ligands — the molecule or ions that surround the central metal ion in a complex.
C. Donor atoms — the atoms that are attached directly to the metal ion.
D. Complex formation is a Lewis acid–base interaction in which the ligands act as Lewis bases and the central metal ion behaves as a Lewis acid.
E. Some coordination compounds are salts which contain a complex cation or anion along with enough ions of opposite charge to give a compound that is electrically neutral overall.
   1. Enclose the complex ion in brackets in the formula.
      a. indicates that the complex ion is a discrete structural unit
F. Metal complex — refers both to neutral molecules and to complex ions.
G. Coordination number — the number of ligand donor atoms that surround a central metal ion in a complex.
   1. Most common coordination numbers: 4 and 6.
   2. Depends on the metal ion's size, charge, and electron configuration and the size and shape of the ligands.
H. Characteristic shape of metal complex is determined by the metal ion's coordination number.
   1. Two–coordinate complexes — linear.
   2. Four–coordinate complexes — tetrahedral and square planar.
   3. Six–coordinate complexes — octahedral.
I. Charge on a metal complex is equal to the charge on the metal ion plus the sum of the charges on the ligands.

20.6 Ligands
A. All ligands are Lewis bases.
   1. Have at least one unshared pair of electrons.
      a. forms a coordinate covalent bond to a metal ion

   2. Classified as monodentate or polydentate.
      a. depends on number of ligand donor atoms that bond to the metal
      b. use the electron pair of a single donor atom — monodentate
      c. use electron pairs on more than one donor atom — polydentate

B. Chelating agents — polydentate ligands.
   1. Multipoint attachment to a metal ion resembles the grasping of an object by the claws of a crab.
      a. forms a chelate ring
   2. Metal chelate — a complex that contains one or more chelate rings.

20.7 Naming Coordination Compounds
A. If the compound is a salt, name the cation first and then the anion.
B. In naming a complex ion or a neutral complex, name the ligands first, in alphabetical order, and then the metal.
   1. Anionic ligands end in –o. (See Table 20.5, page 832 in your textbook.)

Chapter 20 – Transition Elements and Coordination Chemistry

    2. Complex name is one word.
  C. More than one ligand of a particular type, use Greek prefixes to indicate number.
  D. If the name of a ligand itself contains a Greek prefix, put the ligand name in parenthesis and use an alternate prefix.
  E. Use a Roman numeral in parenthesis immediately following the name of the metal, to indicate the metal's oxidation state.
  F. In naming the metal, use the ending –*ate* if the metal is in an anionic complex.

EXAMPLE:
Name the following compounds:

$[Co(NH_3)_4Cl_2]NO_3$ $K_2[OsCl_5N]$ $[Co(en)_2CO_3]Br$

SOLUTION: Using the rules for naming coordination complexes, we find

$[Co(NH_3)_4Cl_2]NO_3$ — The ligands are ammine and chloro. Since we have 2 Cl^- and 1 NO_3^-, we know that the charge on cobalt is +3. The name is tetraamminedichlorocobalt(III) nitrate.

$K_2[OsCl_5N]$ — The ligands in this compound are chloro and nitrido. Since we have 2 K^+, 5 Cl^-, and 1 N^{3-}, we know that the charge on osmium must be +6. Since the complex is an anion, the ending on osmium is changed to *ate*. The compound's name is potassium pentachloronitridoosmate(VI).

$[Co(en)_2CO_3]Br$ — The ligands in this compound are ethylenediamine and carbonato. Since ethylenediamine contains a Greek prefix, we will use the prefix *bis* to indicate the number of en ligands present. Since we have CO_3^{2-} and Br^-, we know the charge on cobalt is +3. The compound's name is carbonatobis(ethylenediamine)cobalt(III) bromide.

EXAMPLE:
Write formulas for the following compounds: triammineaquadichlorocobalt(III) chloride, potassium trioxalatochromate(III), sodium hexacyanoferrate(III).

SOLUTION: Using Table 20.5 on page 882 of your textbook, we can determine the formula for the ligands.

The ammine ligand is NH_3 and the aqua ligand is H_2O. The formula for this compound is $[CoCl_2(NH_3)_3(H_2O)]Cl$.

The oxalato ligand has the formula $C_2O_4^{2-}$. Since there are three oxalato ligands with a total charge of –6 and the chromium has a total charge of +3, we know we need 3 K^+ to balance the charge. The formula for this compound is $K_3[Cr(C_2O_4)_3]$.

The cyano ligand has the formula CN^-. Since there are six cyano ligands with a total charge of –6 and the iron has a total charge of +3, we know we need 3 Na^+ to balance the charge. The formula for this compound is $Na_3[Fe(CN)_6]$.

20.8 Isomers
  A. Isomers — compounds that have the same formula but a different arrangement of their constituent atoms.
    1. Different compounds with different physical and chemical properties.

Chapter 20 – Transition Elements and Coordination Chemistry

B. Constitutional isomers — isomers that have different connections among their constituent atoms.   C.1.
   1. Linkage isomers — arise when a ligand can bond to a metal through either of two different donor atoms.
   2. Ionization isomers — isomers that differ in the anion that is bonded to the metal ion.
C. Stereoisomers — isomers that have the same connections among atoms but have a different arrangement of the atoms in space.

   1. Diastereoisomers (geometric isomers) — have different relative orientations of their metal–ligand bonds.   C.2.
      a. *cis* isomer — identical ligands occupy adjacent corners of the square in a square planar complex
      b. *trans* isomer — identical ligands are across from one another in a square planar complex
      c. have different properties
         i. *cis*–$Pt(NH_3)_2Cl_2$ — polar molecule; more soluble in water
         ii. *trans*–$Pt(NH_3)_2Cl_2$ — nonpolar; two Pt–Cl and Pt–NH_3 dipoles point in opposite directions and cancel out
      d. Square planar complexes of the type MA_2B_2 and MA_2BC (M= metal ion; A, B, C = ligands) can exist as *cis–trans* isomers.
      e. no *cis–trans* isomers for four coordinate tetrahedral complexes
         i. All four corners of a tetrahedron are adjacent to one another.
      f. Octahedral complexes of the type MA_4B_2 can also exist as diastereoisomers.
         i. Two B ligands can be either on adjacent or on opposite corners of the octahedron.
      g. easy to distinguish
         i. Various bonds in the *cis* and *trans* isomers point in different directions.

20.9 Enantiomers and Molecular Handedness   C.3

A. Enantiomers — molecules or ions that are nonidentical mirror images of one another.
   1. Differ from one another because of their handedness.
      a. chiral — objects that have a handedness to them
      b. achiral — objects that lack handedness
B. An object is not chiral if it has a symmetry plane cutting through its middle so that one half of the object is a mirror image of the other half.
C. Certain molecules and ions are chiral.
   1. Tris(ethylenediamine)cobalt(III) ion, $[Co(en)_3]^{3+}$ (see Figure 20.21, page 841 in your textbook) — two nonidentical mirror image forms.
      a. "right–handed" enantiomer — the three ethylenediamine ligands spiral to the right (clockwise)
      b. "left–handed" enantiomer — the ethylenediamine ligands spiral to the left (counterclockwise)
   2. $[Co(NH_3)_6]^{3+}$ — achiral; has several symmetry planes.
D. Enantiomers have identical properties except for their reactions with other chiral substances and their effect on plane–polarized light.
   1. Plane–polarized light — light in which the electric vibrations of the light wave are restricted to a single plane.
   2. Pass ordinary light through a polarizing filter then through a solution of one of the enantiomers.
      a. Plane of polarization is rotated either to the right or to the left
      b. pass through the solution of the other enantiomer — plane of polarization is rotated through an equal angle, but in the opposite direction

Chapter 20 – Transition Elements and Coordination Chemistry

3. Optical isomers — another term used for enantiomers because of their effect on plane–polarized light.
4. Label enantiomers (+) or (–) depending on the direction of rotation of the plane of polarization.
5. Racemic mixture — 50:50 mixture of the (+) and (–) isomers that produces no net optical rotation.
   a. Rotations produced by the individual enantiomers exactly cancel.

20.10 Color of Transition Metal Complexes
A. Color of transition metal complexes depend on the identity of the metal and the ligands.
B. Metal complex can absorb light by undergoing an electronic transition from its lowest energy state (E_1) to a higher energy state (E_2).
   1. Wavelength of absorbed light depends on the energy separation $\Delta E = E_2 - E_1$ between the two states.
   2. Absorbance — the measure of the amount of light absorbed by a substance.
      a. absorption spectrum — plot of absorbance versus wavelength
      b. Color that we see is complementary to the color absorbed (see figure 20.26, page 845 in your textbook).

20.11 Bonding in Complexes: Valence Bond Theory
A. Valence bond theory — bonding results when a filled ligand orbital containing a pair of electrons overlaps a vacant hybrid orbital on the metal ion.
   1. Produces a coordinate covalent bond.
B. If you know the geometry of a complex, you know which hybrid orbitals the metal ion uses. (See Table 20.7, page 845 in your textbook.)
C. Magnetic behavior of transition metal complexes.
   1. Paramagnetic — substance that contains unpaired electrons and is attracted by magnetic fields.
   2. Diamagnetic substance — contains only paired electrons and is weakly repelled by magnetic fields.
   3. Number of unpaired electrons in a transition metal complex can be determined by quantitative measurement of the force exerted on the complex by a magnetic field.
D. High–spin complex — one in which the d electrons are arranged according to Hund's rule to give the maximum number of unpaired electrons.
E. Low–spin complex — one in which the d electrons are paired up to give a maximum number of doubly occupied d orbitals and a minimum number of unpaired electrons.

20.12 Crystal Field Theory
A. Crystal field theory — a model that views the bonding in complexes as arising from electrostatic interactions and considers the effect of the ligand charges on the energies of the metal ion d orbitals.
   1. Accounts for color and magnetic properties of transition metal complexes.
   2. No covalent bonds, no shared electrons, and no hybrid orbitals.
   3. Electrostatic interactions within an array of ions.
   4. Neutral, dipolar ligand — the electrostatic interactions are of the ion–dipole type.
B. Octahedral complexes.
   1. Metal is positively charged, and the ligands are negatively charged.
      a. Ligands repel each other.
      b. minimize repulsion by getting as far apart from one another as possible
      c. metal–ligand attractions > ligand–ligand repulsions
         i. Complex is more stable than the separated ions.

Chapter 20 – Transition Elements and Coordination Chemistry

C. Effect of ligand charges on the energies of d orbitals.
  1. Negatively charged d electrons are repelled by the negatively charged ligands.
     a. Orbital energies are higher in the complex.
     b. not all raised in energy by the same amount
  2. d_{z^2}, $d_{x^2-y^2}$ are raised higher in energy than the d_{xy}, d_{xz}, d_{yz} (point between ligands).
  3. Crystal field splitting (Δ) — the energy splitting between the two sets of d orbitals. (See Figure 20.31, page 851 in your textbook.)
D. Crystal field splitting energy, Δ — corresponds to λ's in the visible region of the spectrum.
  1. Color of complexes due to electronic transitions between the lower and higher energy sets of d orbitals.
  2. Size depends on the nature of the ligands.
     a. spectrochemical series — changes in the crystal field splitting as the ligand varies
     b. weak–field ligands — produce a relatively small value of Δ
        i. high–spin complexes
     c. strong–field ligands — produce a relatively large value of Δ
        i. low–spin complexes
E. Spin–state depends on the relative values of Δ and P, the spin–pairing energy.
  1. Spin–pairing energy — energy needed to overcome repulsion that exists when an electron is placed into an orbital that already contains another electron.
  2. $\Delta > P$; low–spin arrangement has lower energy.
  3. $\Delta < P$; high–spin arrangement has lower energy.
F. Choice of spin–states arises only for complexes d^4–d^7 complexes.
G. Tetrahedral and square planar complexes — have different energy splitting for the d orbitals. (See Figure 20.32, page 853 in your textbook.)
H. Energy splitting in tetrahedral complexes is just the opposite of that in octahedral complexes.
  1. Δ(tetrahedral complexes) $\approx \frac{1}{2}\Delta$(octahedral complexes).
     a. none of the orbitals points directly at the ligands
     b. four ligands instead of six
     c. $\Delta < P$
     d. all complexes are high–spin
I. Square planar complexes — two trans ligands along the z–axis are missing.
  1. Large energy gap between the x^2–y^2 orbital and the four lower–energy orbitals.
  2. Most common for metal ions with d^8 electronic configurations.
     a. favors low–spin complexes in which all four lower–energy orbitals are filled and the higher energy x^2–y^2 orbital is vacant

 **Putting It Together**

The amount of iron in ore can be determined by dissolving a sample in a nonoxidizing acid, reducing all of the Fe^{3+} to Fe^{2+} and titrating with potassium dichromate. The reaction is the oxidation of the Fe^{2+} to Fe^{3+} and the reduction of the dichromate ion to Cr^{2+} in an acidic environment. Determine the mass percent of iron in a 1.213 g sample of ore if 22.05 mL of 0.105 M potassium dichromate is needed to reach the end point in a titration.

 **Self–Test**

This section is intended to test your knowledge of the material covered in this chapter. Think through these problems, and make certain you understand what is going on. Ask yourself if your answer makes sense. Many of these questions are linked to the chapter learning goals. Therefore, successful completion of these problems indicates you have mastered the learning goals for this chapter. You will receive the greatest benefit from this section if you use it as a mock exam. You will then discover which topics you have mastered and which topics you need to study in more detail.

Fill–in–the–Blank

1. Objects that have a handedness are called _____.
2. The _____ isomer has identical ligands that occupy adjacent corners of the square in a square planar complex.
3. For the first–series transition metals, the solid metal is oxidized to the aqueous cation _____ than H_2 gas is oxidized to H^+.
4. Melting points of transition elements _____ as the number of _____ _____ available for metallic bonding increases.
5. Melting points of the transition elements _____ as the d electrons _____ and become less available for bonding.
6. Early transition metal ions in low oxidation states are good _____ agents.
7. For $Cr(OH)_n$ compounds, acid strength increases with _____ _____.
8. $Cr_2O_7^{2-}$ is a _____ agent in _____ solution.
9. The most important oxidation states of iron are _____.
10. Cu^+ undergoes _____ reaction to produce Cu^{2+} and Cu.
11. Complex formation is a _____ interaction in which the ligands act as a _____ and the metal ion behaves as a _____.
12. The most common coordination numbers are _____.
13. _____ ligands use electron pairs on more than one donor atom.
14. A _____ is formed by multipoint attachment of a ligand to a metal ion.
15. Ligands form a _____ bond to a metal ion.
16. _____ are compounds that have the same formula but different arrangement of constituent atoms.
17. Square planar complexes are most common for metal ions with _____ electron configurations.
18. Objects that lack handedness are _____.
19. Light in which the electric vibrations of the light wave are restricted to a single plane is called _____.
20. When naming a coordination complex, the number of ligands of a particular type is indicated with a _____.
21. Another term used for enantiomers is _____ due to their effect on plane–polarized light.

Chapter 20 – Transition Elements and Coordination Chemistry

22. A plot of absorbance versus wavelength is called an _____.
23. _____ substances contain unpaired electrons and are attracted by magnetic fields.
24. According to _____, bonding results when a filled ligand orbital containing a pair of electrons overlaps a vacant hybrid orbital on the metal ion.
25. A _____ complex is one in which the d electrons are arranged according to Hund's rule to give the maximum number of unpaired electrons.
26. _____ is a model that views the bonding in complexes as arising from electrostatic interactions and considers the effect of the ligand charges on the energies of the metal ion d orbitals.
27. Low–spin complexes result from a _____ value of Δ.
28. Tetrahedral complexes are _____ spin complexes due to _____.
29. Square planar complexes are _____ spin complexes.
30. The color of the transition metal complexes is due to the _____ between the lower and higher energy sets of d orbitals.

General Questions

1. Write the valence electron configurations for the following atoms and ions: Mn, Pt, Co^{3+}, V^{2+}.
2. How do the d electrons affect the metallic properties of the transition metals?
3. Why do the third series transition metals have unusually high densities?
4. Using only the periodic table, predict which is the better reducing agent, Mn or Fe.
5. Is FeO_4^{2-} a good oxidizing or reducing agent? Why?
6. Which is the more stable oxidation state, Fe^{6+} or Os^{6+}?
7. Write a balanced net ionic equation in which $Cr_2O_7^{2-}$ in acidic solution reacts with Pb (s) and HSO_4^- (aq) to produce $PbSO_4$ and Cr^{3+}.
8. Write a balanced net ionic equation showing what happens when Fe(II) is oxidized in the presence of air and a base.
9. Define
    a. coordination compound
    b. ligand
    c. donor atom
    d. coordination number
    e. chelating agent
    f. linkage isomer
    g. ionization isomers
    h. diastereoisomers

i. enantiomers
j. racemic mixture

10. For the following compounds identify the ligands and their donor atoms. Determine the coordination number and the oxidation state of the metal and the charge on any complex ion. Determine if the compound can display geometric, linkage, or optical isomerism.
   a. $[Co(NH_3)_4(H_2O)_2]Br_3$
   b. $[Co(NH_3)_5(H_2O)]Cl_2$
   c. $[Co(en)_2(Cl)(SCN)]NO_2$
   d. $Na_4[Fe(CN)_6]$
   e. $[Co(en)_3]Cl_3$

11. Name the compounds in question 10.

12. Write formulas for the following compounds:
   a. tetraquadichlorochromium(III) chloride
   b. pentaamminedinitrogenruthenium(II) chloride
   c. sodium hexachloropalladate(IV)
   d. sodium pentacyanoiodoferrate(II) dihydrate
   e. tetraamminecopper(II) sulfate

13. Use valence bond theory to explain the bonding in $[FeCl_4]^-$ (tetrahedral complex) and $[Ni(CN)_4]^{2-}$ (square planar). What hybrid orbitals are used by the metal? Are the complexes low–spin or high–spin complexes?

14. Using the spectrochemical series in your textbook (page 899), determine the number of unpaired electrons in $[CoF_6]^{3-}$ and $[Fe(CN)_6]^{3-}$.

15. Why is the crystal field splitting in a tetrahedral complex approximately half of the crystal field splitting in an octahedral complex?

Challenge Problem

Three different isomers with the formula $CrCl_3 \cdot 6\,H_2O$ exist. Two are green in color; one is violet. These isomers are labeled *A*, *B* and *C*. When a 0.248 g sample of isomer *A* reacts with a dehydrating agent, the resulting mass is 0.212 g. When 100 mL of a 0.10 M solution of isomer *A* is titrated with excess $AgNO_3$, 1.430 g of AgCl are produced. When a 0.248 g sample of isomer *B* reacts with a dehydrating agent, the resulting mass is 0.230 g. Reaction of a 100 mL solution of 0.100 M isomer *B* with excess $AgNO_3$ produced 2.860 g of AgCl. When a 0.248 g sample of isomer *C* reacts with a dehydrating agent, the resulting mass is 0.248 g. When 100 mL of a 0.10 M solution of isomer *C* is titrated with $AgNO_3$, 4.290 g of AgCl are produced. Determine the structural formula of each isomer.

CHAPTER 21

METALS AND SOLID–STATE MATERIALS

Chapter Learning Goals

A. Minerals and Free Metals
1. Predict whether a given metal is likely to be found in nature as an oxide, a sulfide, a carbonate, a silicate, a chloride, or a free metal.
2. Describe the three steps in the metallurgical processes for isolating and purifying a metal from its ore.
3. Describe the Mond process.
4. Describe the processes by which iron ore is converted to steel.

B. Bonding Descriptions
1. Explain the properties of a metal according to the electron–sea and the band theory models.
2. Use band theory to explain transition metal melting–point trends.

C. Semiconductors
1. Classify doped semiconductors as n–type or p–type.

D. Ceramics and Composites
1. Define superconducting transition temperature (T_c) and the Meissner effect.
2. Define the terms ceramics, sintering, sol, and gel.
3. Describe the sol–gel method for the production of ceramic powders.
4. Classify composite materials as ceramic–ceramic, ceramic–metal, or ceramic–polymer.

Chapter in Brief
This chapter explores the chemistry behind metals and solid–state materials. You will examine the natural sources of the metallic elements, the methods used to obtain metals from their ores, and the models used to describe the bonding in metals. You will also study the structure, bonding, properties, and applications of semiconductors, superconductors, ceramics, and composites.

21.1 Sources of the Metallic Elements
A. Minerals — the crystalline, inorganic constituents of the rocks that make up the Earth's crust.
   1. Source of most metals.
B. Most abundant minerals — silicates and aluminosilicates.
   1. Difficult to concentrate and reduce.
   2. Not important sources of metals.
C. More important minerals — sulfides and oxides.
   1. Hematite (Fe_2O_3), rutile (TiO_2), and cinnabar (HgS).
   2. Ores — mineral deposits from which metals can be produced economically.

D. Chemical composition of the most common ores correlates with the location of the metal in the periodic table.
   1. Early transition metals — occur as oxides.
      a. Less electronegative metals form compounds by losing electrons to highly electronegative nonmetals.
   2. Late transition metals — occur as sulfides.

Chapter 21 – Metals and Solid–State Materials

   a. The more electronegative metals tend to form compounds with more covalent character by bonding to the less electronegative nonmetals.
  3. Electronegative *p*–block metals — sulfide ores.
  4. *s*–Block metals — occur as carbonates, silicates, and chlorides (for Na and K).
   a. *s*–Block oxides are strongly basic and too reactive to exist with acidic oxides.

21.2 Metallurgy

A. Ore — a complex mixture of a metal–containing mineral and economically worthless material called gangue.
  1. Gangue — consists of sand, clay, and other impurities.
B. Metallurgy — the science and technology of extracting metals from their ores.
  1. Three–step process.   *A.2.*
   a. concentration of the ore; chemical treatment prior to reduction
   b. reduction of the mineral to the free metal
   c. refining or purification of the metal
  2. Used for making alloys — metallic materials composed of two or more elements.
C. Concentration and chemical treatment of ores.
  1. Concentrate by separating the mineral from the gangue.
   a. have different properties that are exploited in various separation methods
  2. Metal sulfides ores — concentrated by flotation.
   a. Process exploits the differences in the ability of water and oil to wet the surfaces of the mineral and the gangue.
  3. Bayer process — uses chemical treatment to concentrate the mineral.
   a. Separate Al_2O_3 in bauxite from Fe_2O_3 impurities by treatment with hot aqueous NaOH.
  4. Roasting — a process that involves heating the mineral in air.
   a. used to convert minerals to compounds that are more easily reduced
D. Reduction — the concentrated ore is reduced to the free metal, either by chemical reduction or by electrolysis.
  1. Method used depends on the activity of the metal as measured by $E°$ for reduction (see Table 21.2, page 869 in your textbook).
   a. most active metals — most negative standard reduction potentials
    i. the most difficult to reduce
   b. the least active metals have the most positive standard reduction potentials
    i. easiest to reduce
  2. Au and Pt — inactive; found in nature in uncombined form.
  3. Cu, Ag, and Hg — occur as sulfide ores.
   a. easily reduced by roasting
  4. Cr, Zn, and W — more active metals.
   a. reduce metal oxides with a chemical reducing agent (C, H_2, or a more electropositive metal)
  5. Most active metals — produced by electrolytic reduction.
   a. no chemical reducing agent strong enough to reduce compounds
E. Refining — purification methods used include distillation, chemical purification, and electrorefining.
  1. Zn — volatile enough to be refined by distillation.
  2. Ni — purified by the Mond process.   *A.3.*
   a. a chemical method involving formation and subsequent decomposition of $Ni(CO)_4$
  3. Cu — purified by electrorefining.
   a. an electrolytic process in which Cu is oxidized to Cu^{2+} at an impure Cu anode, and Cu^{2+} from aqueous $CuSO_4$ is reduced to Cu at a pure Cu cathode

Chapter 21 – Metals and Solid–State Materials

21.3 Iron and Steel
A. Metallurgy of iron.
  1. Special technological importance.
    a. major constituent of steel
B. Production of iron — carbon monoxide reduction of Fe ore in a blast furnace (see Figure 21.5, page 871 in your textbook).
  1. Overall reaction: $Fe_2O_3\ (s)\ +\ 3\ CO\ (g)\ \rightarrow\ 2\ Fe(l)\ +\ 3\ CO_2\ (g)$.
C. Cast iron (pig iron) obtained from a blast furnace; brittle material containing 4% elemental carbon and smaller amounts of other impurities.

D. Basic oxygen process — used for the purification and conversion of iron to steel.
  1. Expose molten iron from the blast furnace to a jet of pure oxygen gas in a furnace lined with basic oxides.
    a. oxidizes impurities
    b. Acidic oxides produced react with CaO and form a slag that is poured off.
  2. Produces steels with about 1% carbon and very small amounts of P and S.
E. Composition of liquid steel — monitored by chemical analysis.

21.4 Bonding in Metals
A. Metals — malleable, ductile, lustrous, and good conductors of heat and electricity.
  1. Malleable — can be hammered into sheets.
  2. Ductile — can be drawn into wires.
  3. Lustrous — shiny appearance.
B. Can understand properties by examining the bonding in metals.
C. Valence electrons can't be localized in a bond between any particular pair of atoms.
  1. Delocalized and belong to the crystal as a whole.

D. Electron–sea model — the crystal is a three–dimensional array of metal cations immersed in a sea of delocalized electrons that are free to move throughout the crystal.
  1. Delocalized, mobile valence electrons act as electrostatic glue that holds the metal cations together.
  2. Simple qualitative explanation for the electrical and thermal conductivity of metals.
    a. Mobile electrons are free to move from a negative electrode to a positive electrode when subjected to an electrical potential
    b. Mobile electrons conduct heat by carrying kinetic energy from one part of the crystal to another.
    c. Electrons extend in all directions, therefore, when a metallic crystal is deformed, no localized bonds are broken.
      i. explains malleability and ductility

E. Molecular orbital theory for metals.
  1. The number of molecular orbitals formed is the same as the number of atomic orbitals combined.
  2. The difference in energy between successive MOs decreases as the number of atoms increases.
    a. MOs merge into an almost continuous band of energy levels.
    b. MO theory for metals is called band theory.
    c. Bottom half of band consists of bonding MOs.
    d. Top half of band consists of antibonding MOs.
  3. Each electron in a metal has a discrete kinetic energy and a discrete velocity.
    a. depends on the particular MO
    b. increases from the bottom to the top of a band
  4. One–dimensional metal wire.

a. Energy levels within a band occur in degenerate pairs (see figure 21.8, page 875 in your textbook).
      i. one set — electrons moving to the right
      ii. other set — electrons moving to the left
   b. no electrical potential — no net electric current in either direction
      i. Two sets are equally populated.
   c. Electrical potential — electrons moving right (toward the + battery terminal) are accelerated while electrons moving left (toward the − battery terminal) are decelerated.
      i. Electrons moving very slowly left, change direction.
      ii. number of electrons moving right > number of electrons moving left
      iii. net electric current
5. Electrical potential can shift electrons from one set of energy levels to the other only if the band is partially filled.
   a. electrical insulators — materials that have only completely filled bands
   b. metals — materials that have partially filled bands
   c. can have s and p subshells sufficiently close in energy so that the s and p bands overlap, resulting in a partially filled composite band
6. Transition metals — d band overlaps the s band.
   a. Composite band has six MOs per metal atom.
   b. maximum bonding for metals with six valence electrons per metal atom
      i. near group 6B
   c. causes melting points of transition metals to be at a maximum at group 6B ☞ *B.2.*

21.5 Semiconductors
A. Semiconductor — a material that has an electrical conductivity intermediate between that of a metal and that of an insulator.
B. Valence band — the bonding MOs.
C. Conduction band — the higher–energy, antibonding MOs.
D. Band gap — the separation of the valence band and the conduction band in terms of energy.
E. Insulator — no vacant MOs in the valence band; large band gap.
   1. Can't excite electrons within the valence band.
   2. Can't excite electrons to vacant MOs in the conduction band due to the large band gap.
F. Metal conductors — no energy gap between the highest occupied and lowest unoccupied MOs.
G. Semiconductor — smaller band gap than insulators.
   1. A few electrons can jump the gap and occupy the higher–energy conduction band.
   2. Partially filled conduction band and valence band .
      a. Unoccupied MOs in valence band result from electrons that jumped the gap.
   3. Electrical potential can accelerate the electrons in the partially filled bands.
      a. Semiconductor conducts a small current.
   4. Conductivity increases with increasing temperature.
      a. electrons with enough energy to jump the band gap increase

H. Doping — a process in which the conductivity of a semiconductor is increased by adding certain impurities in small amounts. ☞ *C.1.*
   1. Add an impurity with more electrons than the semiconductor — n–type semiconductor.
      a. Charge carriers are negative electrons.
   2. Example of n–type semiconductor: Si doped with P.
      a. Each P atom occupies a Si position and introduces an extra electron.
      b. Extra electron occupies the conduction band.
      c. More electrons in conduction band cause the conductivity to be higher.
   3. Add an impurity with fewer electrons than the semiconductor — p–type semiconductor.

Chapter 21 – Metals and Solid–State Materials

   a. Creates positive holes in the valence band.
4. Essential components in modern solid–state electronic devices.

21.6 Semiconductor Applications
A. Doped semiconductors – essential ocmponents in modern solid-state electronic devices.
B. Diode – permits flow of electrons in one directions but is highy resistant to current flow in the opposite direction.
   1. Consists of a *p*-type semiconductor in contact with a *n*-type semiconductor.
      a. creates a *p-n* junction.
      b. see figure 21.12, page 881 in your textbook.
   2. Forward bias – negative battery terminal on the *n*-type side.
      a. current flows through *p-n* junction.
         i. The charge carriers combine in the region of the junction.
   3. Reverse bias – negative battery terminal on the *p*-type side.
      a. No current flows.
         i. Electrons and holes move away from each other.
C. Light Emitting Diodes (LEDs)
   1. Electronic transition from conduction band to valence band under a forward bias.
      a. energy released as light.
   2. 3-5 Semiconductors – 1:1 compounds of group 3A and 5A elements.
      a. mixtures form a continuous series of solid solutions.
      b. vary the composition of mixtures
         i. tune the band gap and the color of emitted light.
   3. Advantages:
      a. smaller and brighter
      b. longer lived.
      c. more energy efficient.
      d. faster switching times.
D. Diode Lasers
   1. Laser – Light Amplification by Stimulated Emission of Radiation.
   2. Produces light due to combination of holes and electrons in *p-n* junction.
   3. Laser light
      a. more intense and highly directional.
      b. all of the same frequency and phase.
   4. Essential features:
      a. high forward bias
      b. laser cavity that allows emitted light to bounce back and forth
         i. stimulates cascade of electrons and holes
         ii. amplifies amount of light used
E. Photovoltaic Cells (Solar Cells)
   1. Converts light to electricity.
   2. E(light) $> E$ (band gap)
   3. Light shines on *p-n* junctions.
      a. electrons excited from valence band of *p*-type semiconductor into conduction band of *n*-type conductor.
   4. *p-n* junction part of an electrical circuit.
      a. electrons flow through junction from *p*-side to *n*-side.
F. Transistors – consists of *n-p-n* or *p-n-p* junctions that control or amplify electrical signals.
   1. large number can be packed into a very small space.

Chapter 21 – Metals and Solid–State Materials

21.7 Superconductors

A. Superconductor — a material that loses all electrical resistance below a characteristic temperature, the superconducting transition temperature (T_c).
   1. Below T_c, once an electric current is started, it will flow indefinitely without loss of energy.

B. 1986 — $Ba_xLa_{2-x}CuO_4$ was discovered to have a $T_c = 35$ K.
   1. Soon after, other copper–containing oxides were discovered with even higher T_c.
   2. Ceramics — nonmetallic inorganic solids.
      a. Most are electrical insulators.

C. No generally accepted theory of superconductivity in ceramic superconductors.
   1. Infinitely extended layers of Cu and O and the fractional oxidation number of Cu appear to play a role in the current flow.

D. Superconductors can levitate a magnet.
   1. Cool a superconductor below T_c and lower a magnet toward it, the magnet and superconductor repel each other.

E. Present applications of conventional superconductors require the use of liquid helium (expensive and requires cryogenic equipment).
   1. Can use liquid N_2 with new higher–temperature superconductors.
      a. abundant refrigerant that is cheaper than milk

F. Problems with high–temperature superconductors.
   1. Brittle powders with high melting points.
   2. Not easily made into wires and coils needed for electrical equipment.
   3. Currents carried at 77 K (boiling point of liquid N_2) are too low for practical applications.

G. High–temperature superconductors based on fullerene — the allotrope of carbon that contains C_{60} molecules.
   1. K_3C_{60} — metallic conductor at room temperature; superconductor at 18 K.
   2. Fullerides may prove to be better materials than the Cu–O ceramics for making superconducting wires.
      a. Fullerides are three–dimensional superconductors.

D.1.

21.8 Ceramics

A. Ceramics — inorganic, nonmetallic, nonmolecular solids, including both crystalline and amorphous materials, such as glasses.
   1. Traditional silicate ceramics — made by heating aluminosilicate clays to high temperatures.
   2. Advanced ceramics — materials that have high–tech engineering, electronic, and biomedical applications.
      a. oxide ceramics — alumina (Al_2O_3)
      b. nonoxide ceramics — silicon carbide and silicon nitride

B. Properties of ceramics — superior to those of metals.
   1. Higher melting points.
   2. Stiffer, harder, and more resistant to wear and corrosion.
   3. Maintain their strength at high temperatures.
   4. Less dense than steel.
      a. attractive lightweight, high–temperature materials for replacing metal components in aircraft, space vehicles, and automobiles

C. Strong chemical bonding.
   1. Covalent network solid.
      a. Highly directional covalent bonds prevent planes of atoms from sliding over one another when the solid is subjected to the stress of a load or an impact.
      b. Solid can't deform to relieve the stress.
      c. maintains its shape up to a point, but then the bonds give way suddenly
   2. Oxide ceramics — bonding is largely ionic.

D.2.

Chapter 21 – Metals and Solid–State Materials

- a. behave similarly to covalent network solids
- 3. Metals — deform under stress because their planes of metal cations can slide easily in the electron sea.
- 4. Metals dent; ceramics shatter.
- D. Ceramic processing — the series of steps that leads from raw material to the finished ceramic object.
    1. Determines the strength and the resistance to fracture of the product.
    2. Sintering — a process in which the particles of the powder are welded together without completely melting.
        - a. occurs below the melting point
        - b. Crystal grains grow larger and the density of the material increases as the void spaces between particles disappear.
        - c. need high–purity, fine powders that are tightly compacted prior to sintering
            - i. Impurities and remaining voids can lead to microscopic cracks
- E. Sol–gel method — method used for preparing high–purity, fine powders that can be tightly compacted.
    1. Synthesis of a metal oxide powder from a metal alkoxide (a compound derived from a metal and an alcohol).
    2. Sol — a colloidal dispersion consisting of extremely fine particles having a diameter of only 0.001 to 0.1 µm.
    3. Gel — a rigid, gelatin–like material.

21.9 Composites

- A. Ceramic composite — a hybrid material in which a ceramic powder is mixed, prior to sintering, with fibers of a second ceramic material.
    1. Combines the advantageous properties of both components.
    2. Whiskers — tiny fiber–shaped particles that are very strong because they are single crystals.
- B. Fibers and whiskers increase the strength and fracture toughness of composite materials.
    1. Most of the chemical bonds are aligned along the fiber axis, giving the fibers great strength.
    2. Can deflect cracks.
        - a. prevents cracks from moving cleanly in one direction
    3. Can bridge cracks.
        - a. holds two sides of a crack together
- C. Composites in which the two phases are different types of materials.
    1. Ceramic–metal composites (cermets).
    2. Ceramic–polymer composites.
    3. High strength–to–weight ratios.
- D. Ceramic fibers used in composites are usually made by high–temperature methods.

Putting It Together

The presence of manganese in steel can be determined by the following titration procedure:
1. A steel sample is dissolved in an acidic solution that oxidizes the manganese to the permanganate ion.
2. The permanganate ion is reduced to Mn^{2+} by reaction with an excess of iron(II) sulfate.
3. The unreacted Fe^{2+} is oxidized to Fe^{3+} by reaction with potassium dichromate.

Determine the percent by mass of manganese in a 0.450 g sample of steel if 22.4 mL of 0.0100 M potassium dichromate is needed to react with the Fe^{2+} remaining after 50.0 mL of 0.0800 M iron (II) sulfate reacts with the steel sample.

 Self–Test

This section is intended to test your knowledge of the material covered in this chapter. Think through these problems, and make certain you understand what is going on. Ask yourself if your answer makes sense. Many of these questions are linked to the chapter learning goals. Therefore, successful completion of these problems indicates you have mastered the learning goals for this chapter. You will receive the greatest benefit from this section if you use it as a mock exam. You will then discover which topics you have mastered and which topics you need to study in more detail.

True–False

1. The early transition metals occur as sulfides in nature.

2. The *s*–block elements occur as oxides in nature.

3. The most active metals have the most negative standard reduction potentials and are the most difficult to reduce.

4. Insulators have small band gaps.

5. The electrons in a metal extend in all directions, and therefore the metal breaks when it is hammered.

6. The difference in energy between successive MOs in a metal decreases as the number of atoms increases.

7. The valence band in a semiconductor contains the higher–energy, antibonding MOs.

8. When an impurity with more electrons than the semiconductor is added to the semiconductor, an *n*–type semiconductor is created.

9. Most ceramics are electrical conductors.

10. One of the advantages of the new superconductors that have been synthesized is the ability of these materials to conduct at liquid nitrogen temperatures.

Matching

Minerals	a. the science and technology of extracting metals from their ores
Ore	b. a material that loses all electrical resistance below a characteristic temperature
Metallurgy	c. inorganic, nonmetallic, nonmolecular solids, including both crystalline and amorphous materials

Chapter 21 – Metals and Solid–State Materials

        Bayer process   d. tiny fiber–shaped particles that are very strong because they are single crystals

        Semiconductor   e. a hybrid material in which a ceramic powder is mixed, prior to sintering

        Doping   f. a process in which the particles of the powder are welded together without completely melting

        Superconductor   g. method used for preparing high–purity, fine powders that can be tightly compacted

        Ceramics   h. a material that has an electrical conductivity intermediate between that of a metal and that of an insulator

        Sintering   i. the crystalline, inorganic constituents of the rocks that make up the earth's crust

        Sol–gel method   j. a process in which the conductivity of a semiconductor is increased by adding certain impurities in small amounts

        Composite   k. a complex mixture of a metal–containing mineral and economically worthless material called gangue

        Whisker   l. a chemical treatment used to separate Al_2O_3 in bauxite from impurities by treatment with hot aqueous NaOH

Fill–in–the–Blank

1. The chemical composition of the most common ores correlates with the _____ _____.

2. Roasting is a process that involves _____ and is used to convert minerals to compounds that are _____.

3. The three steps of the metallurgical process for producing metals are _____ _____.

4. The most active metals are produced by electrolytic reduction because _____ _____.

5. The overall reaction for the production of iron using a blast furnace is _____ _____.

6. The electron–sea model for bonding in metals assumes that the crystal is a 3–dimensional array of _____ immersed in a sea of _____ that are free to _____.

7. Based on the electron–sea model, thermal conductivity of metals results from the mobile electrons _____.

8. According to the band theory for bonding in metals, an electrical potential can shift electrons from one set of energy levels to the other only if the _____.

9. Electrical insulators are materials that have only _____.

10. Metals are materials that have _____.

11. The separation of the valence band and the conduction band in terms of energy is called the _____.

12. An insulator is unable to conduct electricity because it can't excite electrons within the filled _____ and the band gap is too _____ to excite electrons to the vacant MOs in the _____.

13. The conductivity of a semiconductor increases with increasing temperature because the number of electrons with enough energy to jump the _____ increases.

14. *p*–Type semiconductors are _____ with impurities that create _____ in the valence band.

15. The fullerides may prove to be better materials than the Cu–O ceramics for making superconducting wires because _____.

16. A ceramic breaks when subjected to a stress because _____ _____.

17. _____ increase the strength and fracture toughness of composite materials.

Challenge Problem

A concentration technique used to extract gold from–low grade ores is cyanidation. This technique involves reaction of the cyanide ion with the low–grade ore to produce the Au(CN)$_2^-$ ion. This ion is then reduced by zinc to produce gold in its pure form. Is it possible to use cyanidation to remove silver, which exists as Ag$_2$S, from the ore argentite? K_f [(Ag(CN)$_2^-$] = 1 × 10^{21}. K_{sp} (Ag$_2$S) = 6 × 10^{-51} Could cyanidation be used to remove silver, which exists as AgCl, from the ore horn silver?

CHAPTER 22

NUCLEAR CHEMISTRY

Chapter Learning Goals

A. Atomic Nuclei
1. Summarize the differences between nuclear reactions and chemical reactions.
2. Use the neutron/proton plot for stable isotopes to determine whether a given nuclide is expected to be stable or unstable.

B. Nuclear Reactions
1. Write balanced equations for nuclear reactions, identifying the types of radiation and nuclides involved.
2. Use the integrated first–order rate law, solving for half–life, decay constant, or ratio of nuclei initially present to nuclei present at time t.
3. Classify nuclear reactions as fission or fusion. Calculate the energy released by a nuclear fission or fusion reaction.
4. Write balanced equations for nuclear transmutations.

C. Energy Changes
1. Calculate mass defects and binding energies for nuclides. Use values of binding energy per nucleon to compare the relative stabilities of two nuclides.

D. Radionuclides
1. Show how radiocarbon dating is used to determine the age of an object.

Chapter in Brief
Nuclear chemistry is the study of the properties and reactions of atomic nuclei. In this chapter, you will learn about the characteristics of nuclear reactions, the different types of radioactive particles involved in these reactions, the first–order decay rates for these reactions, as well as the energy changes that occur during reaction. You will examine the factors that tend to induce nuclear stability, the phenomena of nuclear fission and fusion, and nuclear transmutation. You will explore how radioactivity is detected and measured, along with the biological effects of radiation. Finally, you will study some of the applications of nuclear chemistry.

22.1 Nuclear Reactions and Their Characteristics
A. An atom is characterized by its atomic number, Z and its mass number, A.
  1. Z — written as a subscript to the left of the element symbol; gives the number of protons in the nucleus.
  2. A — written as a superscript to the left of the element symbol, gives the total number of nucleons.
     a. nucleons — a general term for both protons (p) and neutrons (n)
B. Isotopes — atoms with identical atomic numbers, but different mass numbers.
  1. Nuclide — nucleus of a specific isotope.
C. Nuclear reactions — reactions that change the nucleus.

Chapter 22 – Nuclear Chemistry

1. Indicate the electrons as $_{-1}^{0}e$.
   a. superscript 0 — the mass of an electron is essentially zero when compared with that of a proton or neutron
   b. subscript (–1) — the charge of an electron is –1

D. Differences between nuclear reactions and chemical reactions.
   1. Nuclear reactions — change in an atom's nucleus; chemical reaction — a change in distribution of the outer shell electrons around an atom.
   2. Different nuclides of an element have essentially the same behavior in chemical reactions, but have different behavior in nuclear reactions.
   3. Changes in temperature or pressures or the addition of a catalyst, do not affect the rate of a nuclear reaction.
   4. The nuclear reaction of an atom is essentially the same regardless of whether the atom is in an element or a compound.
   5. The energy changes accompanying nuclear reactions are far greater than those accompanying chemical reactions.

22.2 Nuclear Reactions and Radioactivity

A. Radionuclides — radioactive nuclei
   1. Radioactive — the spontaneous emission of radiation.
B. Three common types of radiation with different properties. (See Table 22.1, page 906 in your textbook.)
   1. Alpha (α).
   2. Beta (β).
   3. Gamma (γ).
C. Alpha (α) radiation — a stream of particles that consists of two protons and two neutrons.
   1. Repelled by a positively charged electrode.
   2. Attracted by a negatively charged electrode.
   3. Mass–to–charge ratio: $_{2}^{4}He^{2+}$.
   4. Emission of an α particle reduces the mass number of the nucleus by four and reduces the atomic number by two.
      a. $_{92}^{238}U \rightarrow {_{2}^{4}He} + {_{90}^{234}Th}$
   5. Common for heavy radioactive isotopes (radionuclides).
D. Balanced nuclear reaction.
   1. Not balanced in usual chemical sense.
      a. Nuclei are not the same on both sides of the reaction.
   2. Sums of the nucleons on both sides are equal.
   3. Sums of the charges on the nuclei and any elementary particles on both sides are equal.
   4. Not concerned with ionic charges on atoms.
      a. irrelevant to nuclear disintegration
      b. disappear
E. Beta (β) radiation — occurs when a neutron in the nucleus spontaneously decays into a proton plus an electron ($_{-1}^{0}e$, or β^-), which is then ejected.
   1. Product nucleus — same mass number but a higher atomic number (creates a proton).
F. Gamma (γ) radiation — electromagnetic radiation of very high energy and short wavelength.
   1. Stream of high–energy photons.
   2. Almost always accompanies α and β emission.
      a. mechanism for release of energy
   3. γ emission is often not shown in nuclear equations.
      a. doesn't change either the mass number or the atomic number of the product nucleus

Chapter 22 – Nuclear Chemistry

G. Positron emission and electron capture.
   1. Positron emission — the conversion of a proton in the nucleus into a neutron plus an ejected positron.
      a. positron ($^{0}_{1}e$, or β^{+}) — a particle with the same mass as an electron but opposite charge
      b. decrease in atomic number of the product nucleus
      c. no change in the mass number
   2. Electron capture — a proton in the nucleus captures an inner–shell electron, which is converted into a neutron.
      a. no change in the mass number of the product nucleus
      b. atomic number decreases by one

22.3 Radioactive Decay Rates

B.2

A. Radioactive decay is kinetically a first–order process whose rate is proportional to the number of radioactive nuclei N in a sample:
   1. Decay rate $= k \times N$.
   2. k = first–order rate constant called the decay constant.
   3. Integrated rate law

   $$\ln\left(\frac{N}{N_o}\right) = -kt.$$

   a. N_o = number of radioactive nuclei originally present; N = number remaining at time t
B. Radioactive decay is characterized by a half–life, $t_{1/2}$.
   1. Time required for the number of radioactive nuclei in a sample to drop to one–half of the initial value.
   2. Each passage of a half–life causes the decay of one–half of whatever sample remains.
   3. The half–life is the same no matter what the size of the sample, the temperature, or any other external condition.
   4. $t_{1/2} = \dfrac{0.693}{k}$ and $k = \dfrac{0.693}{t_{1/2}}$.
   5. To calculate the ratio of remaining and initial amounts of radioactive sample N/N_o at any time t.

   a. $\ln\left(\dfrac{N}{N_o}\right) = -0.693\left(\dfrac{t}{t_{1/2}}\right)$

EXAMPLE:
The radioactive decay of Tl–206 to Pb–206 has a half–life of 4.20 min. Starting with 0.250 g of Tl–206, calculate the grams of Tl–206 left after 60.00 min.

SOLUTION:

We can use the equation $\ln\left(\dfrac{N}{N_o}\right) = -0.693\left(\dfrac{t}{t_{1/2}}\right)$, and solve for the mass of Tl–206 by realizing that the mass of a sample is proportional to the number of nuclei.

$$\frac{m}{m_o} = \frac{N}{N_o}$$

$$\ln\left(\frac{m}{0.250\ \text{g}}\right) = -0.693\left(\frac{60.00}{4.20}\right) = -9.900 \qquad \frac{m}{0.250\ \text{g}} = 5.017 \times 10^{-5} \qquad N = 1.25 \times 10^{-5}\ \text{g}$$

Workbook Problem 22.1
Technecium–99m is utilized in bone scans, and is an important tool in the diagnosis of cancer and other pathological conditions.

$^{99m}_{43}$Tc has a $t_{1/2}$=6.01 hours, thereby minimizing a patient's exposure to harmful effects. What percent of $^{99m}_{43}$Tc remains 2.50 hours after treatment? How long will a patient have to wait for the $^{99m}_{43}$Tc to be reduced to 15%?

Strategy: Use the first–order integrated rate law to solve (assume that N_o = 100).

Step 1: Determine the amount of $^{99m}_{43}$Tc (N) remaining after 2.50 hours.

Step 2: Solve for the percentage of $^{99m}_{43}$Tc remaining.

Step 3: Determine the time needed for $^{99m}_{43}$Tc to be reduced to 15%.

What key concept did you use?

22.4 Nuclear Stability
A. Radioactive isotopes
 1. Stable — one that can be prepared and whose half–life can be measured.
 2. Unstable — those isotopes that can't be prepared or that decay too rapidly for their half–lives to be measured.
 3. Nonradioactive (stable indefinitely) — isotopes that do not undergo radioactive decay.
B. Neutron/proton ratio in the nucleus determines if an isotope is radioactive.
C. Plot of number of neutrons (y–axis) versus number of protons (x–axis) (see Figure 22.3, page 911 in your textbook).
 1. Stable nuclides fall in a curved band called the band of nuclear stability.
 2. Sea of instability — area on either side of the band of nuclear stability.
   a. represents the large number of unstable neutron/proton combinations
 3. Island of stability — area predicted to exist for a few superheavy nuclides near 114 protons and 184 neutrons.

Chapter 22 – Nuclear Chemistry

D. Generalizations from plot.
  1. Every element has at least one radioactive isotope.
  2. Hydrogen is the only element whose most abundant stable isotope contains more protons than neutrons.
  3. The ratio of neutrons to protons gradually increases for elements heavier than calcium.
  4. All isotopes beyond bismuth–209 are radioactive.
  5. Nonradioactive isotopes generally have an even number of neutrons.
E. Neutrons function as a kind of nuclear glue that holds nuclei together by overcoming proton–proton repulsions.
F. Magic numbers of protons or neutrons — 2, 8, 20, 28, 50, 82, 126.
  1. Give rise to particularly stable nuclei.
  2. Nucleus with a magic number of either protons or neutrons is unusually stable.
  3. Analogous to chemical stability brought about by an octet of electrons.
G. Trends that are apparent from the band of nuclear stability.
  1. Elements with an even atomic number have a larger number of nonradioactive nuclides than do elements with an odd atomic number.
  2. Radioactive nuclei on the right side of the band undergo nuclear disintegration by positron emission, electron capture, or alpha emission.
     a. lower neutron/proton ratio
     b. Processes increase the neutron/proton ratio.
  3. Nuclei on the left side of the band emit beta particles.
     a. higher neutron/proton ratios
     b. Process decreases the neutron/proton ratio.
H. Some nuclides can't reach a nonradioactive nucleus in a single emission.
  1. Undergo a decay series of disintegrations.

22.5 Energy Changes During Nuclear Reactions

A. Can calculate the energy change on formation of the nucleus from isolated protons and neutrons.
  1. $\Delta E = \Delta mc^2$.
     a. relates the energy change of a nuclear process to a corresponding mass change
B. Mass defect — the loss in mass that occurs when protons and neutrons combine to form a nucleus.
  1. Lost mass is converted into energy known as the binding energy.
     a. binding energy — energy that holds the nucleons together
     b. calculate from the Einstein equation
     c. usually expressed on a per–nucleon basis using the electron volts as the energy unit
        i. $1 \text{ eV} = 1.60 \times 10^{-19}$ J
C. Plot binding energy per nucleon.
  1. Higher binding energy corresponds to higher stability.
D. Mass and energy are interconvertible.
  1. Laws of conservation of mass and conservation of energy must be combined.
     a. Neither mass nor energy is conserved separately.

EXAMPLE:

Calculate the nuclear binding energy in MeV/nucleon for $^{174}_{77}\text{Ir}$, atomic mass = 173.966 66 amu. (Mass of $e^- = 5.486 \times 10^{-4}$.)

SOLUTION: First, calculate the total mass of the nucleons (97 n + 77 p)

Chapter 22 – Nuclear Chemistry

Mass of 97 neutrons = (97) (1.008 66 amu) = 97.840 02 amu
Mass of 77 protons = (77)(1.007 28 amu) = 77.560 56 amu
_____
Mass of 97 n + 77 p = 175.400 58 amu

Determine the mass of the $^{174}_{77}$Ir nucleus by subtracting the mass of 77 e⁻ from the atomic mass.

Mass of nucleus = 173.966 66 – 0.042 24 = 173.924 42

Next, solve for the mass defect by subtracting the mass of the ^{174}Ir nucleus from the ^{174}Ir atom.

Mass defect = Mass of nucleons – Mass of nucleus
= (175.40058 amu) – (173.924 42 amu)
= 1.476 16 amu

Now convert the mass defect from amu to gram to get the mass defect in grams per mole.

$$(1.476\,16\,\frac{\text{amu}}{\text{atom}})\left(1.660\,54\times10^{-24}\,\frac{\text{g}}{\text{amu}}\right)\left(6.022\times10^{23}\,\frac{\text{atom}}{\text{mol}}\right) = 1.476\,\frac{\text{g}}{\text{mol}}$$

Now, use the Einstein equation to convert the mass defect into the binding energy.

$$\Delta E = \Delta mc^2 = \left(1.476\,\frac{\text{g}}{\text{mol}}\right)\left(10^{-3}\,\frac{\text{kg}}{\text{g}}\right)\left(3.00\times10^8\,\frac{\text{m}}{\text{s}}\right)^2 = 1.33\times10^{14}\,\frac{\text{kg}\cdot\text{m}^2}{\text{s}^2\cdot\text{mol}} = 1.33\times10^{14}\,\text{J/mol}$$

We now convert to MeV/nucleon

$$\frac{1.33\times10^{14}\,\text{J}}{1\,\text{mol}} \times \frac{1\,\text{mol}}{6.022\times10^{23}\,\text{nuclei}} \times \frac{1\,\text{MeV}}{1.60\times10^{-13}\,\text{J}} \times \frac{1\,\text{nucleus}}{174\,\text{nucleons}} = 7.93\,\text{MeV/nucleon}$$

Workbook Problem 22.2

Calculate the nuclear binding energy for $^{241}_{95}$Am in MeV/nucleon. The mass of $^{241}_{95}$Am is 241.056 823 amu.

Strategy: First, calculate the mass defect, Δm. Then calculate the binding energy from $\Delta E = \Delta mc^2$.

Step 1: First, calculate the total mass of the nucleons (146 n + 95 p)

Step 2: Determine the mass of the $^{241}_{95}$Am nucleus by subtracting the mass of 95 e⁻ from the atomic mass.

Chapter 22 – Nuclear Chemistry

Step 3: Determine the mass defect.

Step 4: Convert the mass defect to grams per mole.

Step 5: Convert the mass defect into the binding energy.

Step 6: Convert to MeV/nucleon.

What key concept did you use?

22.6 Nuclear Fission and Fusion
   A. Lighter and heavier elements are less stable than mid–mass elements near iron–56.
      1. Heavy nuclei — gain stability and release energy if they fragment to yield mid–mass elements.
      2. Light nuclei can gain stability and release energy if they fuse together.
   B. Fission — the fragmentation of heavy nuclei.
      1. Nuclei break into fragments when struck by neutrons.
      2. Doesn't occur in exactly the same way each time.
      3. Uranium–235 — more than 100 different fission pathways.
         a. more frequently occurring pathways
            $${}^{1}_{0}n + {}^{235}_{92}U \rightarrow {}^{142}_{56}Ba + {}^{91}_{36}Kr + 3\, {}^{1}_{0}n$$
         b. three neutrons released induce three more fissions → 9 neutrons
         c. chain reaction — a reaction that continues to occur even if the supply of neutrons from outside is cut off
         d. small sample size — many of the neutrons escape before initiating additional fission events
         e. critical mass — a sufficient amount of radioactive nuclide that allows the chain reaction to become self–sustaining
      4. Can calculate the amount of energy released during nuclear fission, using the mass defect and Einstein equation.

a. Use the masses of the atoms corresponding to the relevant nuclei.
C. Nuclear reactors.
1. Principle of a nuclear reactor — a subcritical amount of uranium fuel is placed in a containment vessel surrounded by circulating coolant, and control rods are added.
   a. Control rods are made of boron and cadmium.
      i. absorb and regulate the flow of neutrons
      ii. raised and lowered to maintain fission at a controlled rate
   b. Energy from the controlled fission heats the circulating coolant.
   c. Heated circulating coolant produces steam to drive a turbine and produce electricity.
D. Nuclear fusion — the joining together of light nuclei.
1. Release enormous amounts of energy.
2. Fusion of hydrogen nuclei — a potential power source.
   a. Hydrogen isotopes are cheap and plentiful.
   b. Fusion products are nonradioactive and nonpolluting.
3. Technical problems to achieving a practical and controllable fusion method.
   a. to initiate the process, $T = 40 \times 10^6$ K

Workbook Problem 22.3

Determine how much energy is released by the fission reaction

$$^{241}_{95}\text{Am} \rightarrow {}^{237}_{93}\text{Np} + {}^{4}_{2}\text{He}$$

The masses are

$^{241}_{95}\text{Am} = 241.056\,823$ g/mol

$^{237}_{93}\text{Np} = 237.048\,167$ g/mol

$^{4}_{2}\text{He} = 4.002\,60$ g/mol

Strategy: Use the mass defect and the Einstein equation to calculate ΔE.

Step 1: Calculate the mass change in the reaction.

Step 2: Substitute the mass change into the Einstein equation and solve for the energy.

Chapter 22 – Nuclear Chemistry

What key concept did you use?

22.7 Nuclear Transmutation
A. Nuclear transmutation — the change of one element into another, brought about by bombardment of an atom with a high–energy particle.
  1. Creates an unstable nucleus, which causes a nuclear change to occur.
B. Can lead to the synthesis of entirely new elements.
  1. All transuranium elements have been produced by bombardment reactions.

22.8 Detecting and Measuring Radioactivity
A. Detect radiation by measuring its ionizing properties.
B. Ionizing radiation — high–energy radiation of all kinds.
  1. Interaction of the radiation with a molecule knocks an electron from the molecule.
    a. Molecule $\xrightarrow{radiation}$ ion + e$^-$
  2. Includes α particles, β particles, γ particles, as well as X rays and cosmic rays.
  3. X rays — high–energy photons.
  4. Cosmic rays — energetic particles from interstellar space.
    a. primarily protons
C. Devices used for measuring radiation.
  1. Photographic film badge.
    a. badge fogs when struck by radiation.
  2. Geiger counter.
    a. Radiation ionizes Argon in the tube.
      i. Ar produces electrons that conduct a current.
      ii. produces a clicking sound.
  3. Scintillation counter.
    a. phosphor emits a flash of light when hit by radiation.
D. Expression of radiation intensity depends on what is being measured.
  1. Becquerel (Bq) — SI unit for measuring the number of radioactive disintegrations occurring each second in a sample.
    a. 1 Bq = 1 disintegration/s
  2. Curie (Ci): 1 Ci = 3.7×10^{10} disintegrations/s.
    a. larger units than Bq.
    b. decay rate of 1 g of radium.
  3. Gray (Gy) — SI unit for measuring the amount of energy absorbed per kilogram of tissue exposed to a radiation source.
    a. 1 Gy = 1 J/kg
    b. 1 rad = 0.01 Gy
  4. Sievert (Sv) — SI unit that measures the amount of tissue damage caused by radiation.
    a. 1 rem (roentgen equivalent for man) = 0.01 Sv

22.9 Biological Effects of Radiation
A. The effects of ionizing radiation on the human body depend on:
  1. The kind of radiation and its energy.
  2. The length of exposure.
  3. Whether the source is outside or inside the body.

Chapter 22 – Nuclear Chemistry

22.10 Applications of Nuclear Chemistry

A. Radiocarbon dating — depends on the slow and constant production of radioactive carbon–14 in the upper atmosphere by neutron bombardment of nitrogen atoms.
   1. $^{14}_{7}N + ^{1}_{0}n \rightarrow ^{14}_{6}C + ^{1}_{1}H$
   2. $^{14}C \rightarrow ^{14}CO_2$ which mixes with $^{12}CO_2$ and is taken up by plants during photosynthesis.
   3. Ratio of ^{14}C to ^{12}C in a living organism is the same as that in the atmosphere.
   4. $^{14}C/^{12}C$ decreases when the organism dies.
      a. ^{14}C undergoes radioactive decay.
         $^{14}_{6}C \rightarrow ^{14}_{7}N + ^{0}_{-1}e$
   5. For ^{14}C, $t_{1/2} = 5715$ y.
   6. Can determine how long ago the organism died by measuring the amount of ^{14}C remaining.
B. Use other radionuclides for dating rocks.
   1. ^{238}U
      a. $t_{1/2} = 4.46 \times 10^9$ y for ^{238}U
      b. ^{238}U decays through a series to give ^{206}Pb
      c. Measurement of $^{238}U/^{206}Pb$ ratio can determine the age of uranium–containing rock.
   2. ^{40}K.
      a. $t_{1/2} = 1.28 \times 10^9$ y
      b. $^{40}_{19}K + ^{0}_{-1}e^- \rightarrow ^{40}_{18}Ar$
      c. $^{40}_{19}K \rightarrow ^{18}_{40}Ar + ^{0}_{1}e^-$
      d. compare amount of ^{40}Ar to ^{40}K remaining in sample
C. Medical uses of radioactivity — grouped into four classes.
   1. In vivo procedures.
   2. In vitro procedures.
   3. Radiation therapy.
   4. Imaging procedures.
D. In vivo procedures — studies that take place inside the body.
   1. Assess the functioning of a particular organ or body system.
   2. A radiopharmaceutical agent is administered, and its path in the body is determined by analysis of blood or urine samples.
E. In vitro procedures — studies that take place outside the body.
   1. Radioimmunoassay — techniques which measure small concentrations of substances in body fluids.
   2. Takes advantage of antibodies to bind to antigens.
F. Therapeutic procedures — radiation is used as a weapon to kill diseased tissue.
   1. Involve either external or internal sources of radiation.
   2. Internal radiation therapy is a much more selective technique than external therapy.
G. Imaging procedures — give diagnostic information about the health of body organs by analyzing the distribution pattern of radionuclides introduced into the body.
   1. A radiopharmaceutical agent known to concentrate in a specific tissue or organ is injected into the body and its distribution pattern is monitored by external radiation detectors.
H. Magnetic resonance imaging (MRI) uses radio waves to stimulate certain nuclei in the presence of an extremely powerful magnetic field.
   1. Stimulated nuclei give off a signal that can be measured, interpreted, and correlated with their environment in the body.

Chapter 22 – Nuclear Chemistry

 **Putting It Together**

31.0 % of the chlorine atoms in a 37.8 mg sample of sodium perchlorate are ^{36}Cl, which is a radioactive isotope of chlorine. The half–life of ^{36}Cl is 3.0×10^5 yr. How many disintegrations per second are produced by this sample?

 **Self–Test**

This section is intended to test your knowledge of the material covered in this chapter. Think through these problems, and make certain you understand what is going on. Ask yourself if your answer makes sense. Many of these questions are linked to the chapter learning goals. Therefore, successful completion of these problems indicates you have mastered the learning goals for this chapter. You will receive the greatest benefit from this section if you use it as a mock exam. You will then discover which topics you have mastered and which topics you need to study in more detail.

General Questions

1. Summarize the differences between nuclear reactions and chemical reactions.

2. Balance the following nuclear reactions:

    a. $^{81}_{36}\text{Kr} + ^{0}_{-1}\text{e} \rightarrow$

    b. $^{104}_{47}\text{Ag} \rightarrow ^{0}_{1}\text{e} +$

    c. $^{73}_{31}\text{Ga} \rightarrow ^{0}_{-1}\text{e} +$

    d. $^{104}_{48}\text{Cd} \rightarrow ^{104}_{47}\text{Ag} +$

3. Write nuclear equations for:

    a. alpha emission by $^{11}_{5}\text{B}$

    b. beta emission by $^{121}_{51}\text{Sb}$

    c. neutron emission by $^{70}_{35}\text{Br}$

    d. proton emission by $^{41}_{19}\text{K}$

4. What is the age of a bone fragment that shows an average of 3.5 disintegrations per minute per gram of carbon? The carbon in living organisms undergoes an average of 15.3 disintegrations per minute per gram, and the half–life of ^{14}C is 5715 y.

5. The half–life for selenium–75 is 120.0 days. What is the decay constant for this radioactive element?

6. The half–life of strontium–90 is 28.1 years. How much of a 0.500 mg sample of strontium–90 remains after 5.00 years?

Chapter 22 – Nuclear Chemistry

7. From the neutron–proton plot for stable isotopes, determine if an element with 48 protons and 42 neutrons will be stable. If this element is unstable, what type of nuclear processes can increase the stability?

8. Calculate the mass defect and binding energy for ^{92}Mo (atomic mass = 91.906 91 amu).

9. How much energy in kJ/mol is released in the following reaction:
$$^1_0n + ^{235}_{92}U \rightarrow ^{137}_{52}Te + ^{97}_{40}Zr + 2\,^1_0n$$

The atomic masses are ^{235}U (235.0439 amu); ^{137}Te (136.9254 amu); ^{97}Zr (96.991 10 amu); and n (1.008 66 amu); 1 amu = 1.6605×10^{-27} kg

10. Write reactions for the following nuclear transmutations

    a. Curium–242 reacts with an alpha particle to produce californium–245.

    b. ^{14}N reacts with an alpha particle to produce ^{17}O.

11. What is the age of a piece of wood that shows an average of 2.2 disintegrations per minute per gram of carbon? The carbon in living organisms undergoes an average of 15.3 disintegrations per minute per gram, and the half–life of ^{14}C is 5715 y.

Challenge Problem

^{51}Cr decays by electron capture, emitting γ radiation. ^{14}C emits β particles during the decay process. A sample of $K_2^{51}Cr_2O_7$ with an activity of 843 cpm/g is mixed with a sample of $H_2^{14}C_2O_4$ with an activity of 345 cpm/g (cpm = counts per minute). The resulting chromium oxalate complex gives a γ count of 165 cpm and a β count of 83 cpm. How many oxalate ions are bound to the chromium ion?

CHAPTER 23

ORGANIC CHEMISTRY

Chapter Learning Goals

A. *Saturated Hydrocarbons*
1. Draw electron dot structures for isomers of simple alkanes.
2. Write condensed structures of organic molecules.
3. Determine which structures represent different molecules and which are merely different conformations of the same molecule.
4. Given the structure of an alkane, determine its IUPAC name and vice versa.
5. Given the structure of a cycloalkane, determine its IUPAC name and vice versa.
6. Predict the products of and write balanced chemical equations for reactions of alkanes with chlorine.
7. Identify functional groups in molecules. Draw structures of molecules containing functional groups listed in Table 23.1 on page 951 of your textbook.

B. *Unsaturated Hydrocarbons*
1. Given the structure of an alkene or alkyne, determine its IUPAC name and vice versa.
2. Predict the products and write balanced chemical equations for alkene addition reactions.

C. *Aromatic Compounds*
1. Given the structure of an aromatic compound, determine its IUPAC name and vice versa.
2. Predict the products and write balanced chemical equations for aromatic substitution reactions.

D. *Alcohol, Ethers, and Amines*
1. Classify molecules as alcohols, ethers, or amines. State the general properties of these classes of molecules.

E. *Carbonyl Containing Compounds*
1. Classify molecules as ketones or aldehydes.
2. Classify molecules as carboxylic acids, esters, or amides. State the general properties of these classes of molecules.
3. Give systematic names of carboxylic acids, esters, and amides.
4. Predict the products and write balanced chemical equations for carbonyl substitution reaction and reactions of carboxylic acids, esters, and amides with water.

F. *Polymers*
1. Give the formula for a segment of the polymer formed as a result of an alkene polymerization and as a result of a polymerization involving molecules with two different functional groups.

Chapter in Brief
Organic chemistry, once defined as the study of compounds from living organisms, is now defined as the study of carbon compounds. In this chapter, you will begin to explore the ability of carbon atoms to bond together, forming a vast array of compounds such as long chains and rings. You will learn to draw the structures of and apply the IUPAC rules of nomenclature for the simple hydrocarbons. You will classify molecules based on the functional groups present and learn the general properties of the

Chapter 23 – Organic Chemistry

different classes of molecules. Finally, you will learn how to predict the products and write balanced chemical equations for reactions involving the different classes of molecules.

23.1 The Nature of Organic Molecules
A. Carbon is tetravalent.
B. Organic molecules have covalent bonds.
C. Organic molecules have polar covalent bonds when carbon bonds to an element on the right or left side of the periodic table.
D. Carbon can form multiple covalent bonds by sharing more than two electrons with a neighboring atom.
E. Organic molecules have specific three-dimensional shapes, as predicted by the VSEPR model.   🔑 *A.1.*
   1. C bonded to four atoms — bonds point towards corner of a tetrahedron.
   2. C bonded to three atoms — bond angle = 120°.
   3. C bonded to two atoms — bond angle = 180°.
F. Carbon uses hybrid atomic orbitals for bonding to other atoms.
   1. Single bonded carbon (C bonded to four atoms) — uses sp^3 hybrids.
   2. Double-bonded carbons (C bonded to three atoms) — uses sp^2 hybrids that point toward the corners of an equilateral triangle.
     a. Unhybridized p orbital is perpendicular to the plane of sp^2 hybrid orbitals.
     b. Overlap of unhybridized p orbitals on two C's forms π bond.
   3. Triply bonded carbons (C bonded to 2 atoms) – uses sp hybrid orbitals which are 180° to each other.
     a. Two unhybridized p orbitals
       i. oriented 90° from each other.
       ii. overlap other unhybridized orbitals on two C's to form two π bonds.
G. Covalent bonds give organic compounds properties that are different from ionic salts.
   1. Weak intermolecular forces in organic compounds lead to
     a. lower melting and boiling points
     b. insolubility in water and lack of electrical conductivity

23.2 Alkanes and Their Isomers
A. Hydrocarbons — molecules that contain only carbon and hydrogen.
B. Alkanes — hydrocarbons that have only single bonds.
   1. Saturated hydrocarbons.
C. Straight-chain alkanes — compounds with all their carbons connected in a row.
D. Branched-chain alkanes — those with a branching connection of carbons.

E. Isomers — molecules that have the same molecular formula but different structures.   🔑 *A.3.*
   1. Different isomers are different chemical compounds.
     a. have different structures and different physical properties

23.3 Drawing Organic Structures   🔑 *A.2.*
A. Condensed structures — carbon–hydrogen and carbon–carbon single bonds aren't shown.
   1. Bonds are understood.
     a. carbon with 3 H's bonded to it — CH_3
     b. carbon with 2 H's bonded to it — CH_2
   2. The horizontal bonds between carbons aren't shown.
   3. Vertical bonds shown for clarity.

Chapter 23 – Organic Chemistry

A.2, A.3.   **EXAMPLE:**

Draw all possible isomers for C_6H_{14}.

SOLUTION:

We begin by drawing the simplest structure of this compound, one in which there are six carbons bonded to each other (a six–carbon chain) with only hydrogen atoms bonded to each carbon.

$CH_3CH_2CH_2CH_2CH_2CH_3$

Next, we draw as many different structures as possible in which there are five carbons bonded to each other (a five–carbon chain) with a CH_3 bonded to a carbon in the chain.

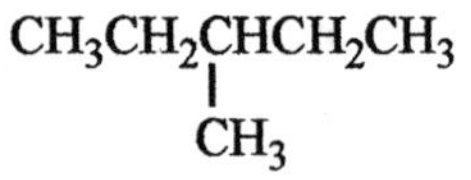

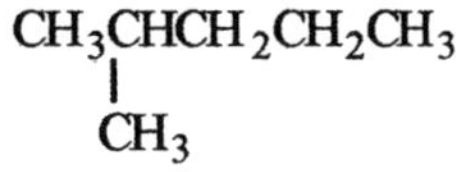

$CH_3CH_2\underset{\underset{CH_3}{|}}{C}HCH_2CH_3$

$CH_3\underset{\underset{CH_3}{|}}{C}HCH_2CH_2CH_3$

These two structures are the only different five–carbon chain structures that can be drawn with the formula C_6H_{14}. If we were to draw another five–carbon chain with the CH_3 bonded to the second carbon from the right, it would be the same structure with the CH_3 bonded to the second carbon from the left, only flipped over.

We now draw as many different structures as possible in which there are four carbons bonded to each other (a four–carbon chain) with two CH_3's hanging off the chain.

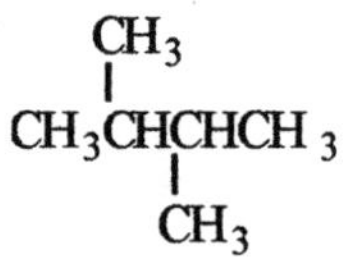

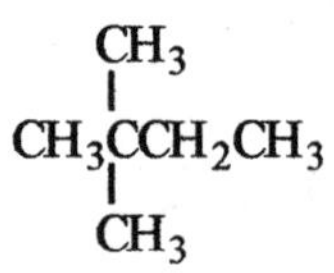

$CH_3\underset{\underset{CH_3}{|}}{\overset{\overset{CH_3}{|}}{C}}HCHCH_3$

$CH_3\underset{\underset{CH_3}{|}}{\overset{\overset{CH_3}{|}}{C}}CH_2CH_3$

Again, if we were to draw the structure with both of the CH_3's bonded to the second carbon on the right, we would have the same molecule as the one with the CH_3's bonded to the second carbon from the left.

23.4 The Shapes of Organic Molecules

A.

A. A molecule can be arbitrarily shown in a great many ways.
B. Rotation around carbon–carbon single bonds is possible.
  1. Conformations — a number of possible three–dimensional structures resulting from the ability of parts of a molecule to spin around a carbon–carbon single bond.

EXAMPLE:
The following molecules have the same formula, C_7H_{16}. Which are isomers, and which are the same molecule?

a.
$$CH_3CH_2CH_2CH_2CH_2CH_2CH_3$$

b.
$$CH_3CHCH_2CH_2CH_3$$
$$|$$
$$CH_2CH_3$$

c.
$$CH_3CH_2CH_2CHCH_3$$
$$|$$
$$CH_2CH_3$$

d.
$$CH_3CH_2CH_2CH_2$$
$$|$$
$$CH_2$$
$$|$$
$$CH_2$$
$$|$$
$$CH_3$$

e.
$$CH_3$$
$$|$$
$$CH_3CH_2CHCHCH_3$$
$$|$$
$$CH_3$$

SOLUTION:
Because rotation around carbon–carbon single bonds is possible, we need to determine if any of the molecules have the same number of carbon atoms bonded to each other in a chain. Structures *a* and *d* have seven carbon atoms bonded to each other in a chain, and structures *b* and *c* have six carbon atoms bonded to each other in a chain. Also notice that the CH_2CH_3 is bonded to the second carbon from the end in both structures *b* and *c*.

Structures *a* and *d* are the same molecule, and structures *b* and *c* are the same molecule. The isomers are *a*, *b*, and *e*.

23.5 Naming Alkanes
    A. Three parts to a chemical name.
        1. Parent name — specifies the overall size of the molecule by telling how many carbon atoms are present in the longest continuous chain.
        2. Suffix — identifies what family the molecule belongs to.
        3. The prefix — specifies the location of various substituent groups attached to the parent chain.
    B. Straight–chain alkanes — named by counting the number of carbon atoms in the chain and adding the family suffix *–ane*.

Chapter 23 – Organic Chemistry

1. First four compounds.
   a. methane — CH_4
   b. ethane — CH_3CH_3
   c. propane — $CH_3CH_2CH_3$
   d. butane — $CH_3CH_2CH_2CH_3$
2. All other alkanes are named from Greek numbers according to the number of carbons present.
C. Four steps to naming branched–chain alkanes.
   1. Name the main chain.
      a. the longest continuous chain of carbons present
      b. use the name of that chain as the parent name
   2. Number the carbon atoms in the main chain.
      a. begin at the end near the first branch point
         i. branches should have the lowest number combination
   3. Identify and number the branching substituent.
      a. alkyl group — the part of an alkane that remains when a hydrogen is removed
         i. name by replacing the *–ane* ending of the parent alkane with an *–yl* ending
   4. Write the name as a single word.
      a. Separate different prefixes with hyphens.
      b. Separate numbers with commas.
      c. Cite substituent names in alphabetical order.
         i. Don't use the prefixes for alphabetizing.
D. Alkyl groups.
   1. Hydrogens in methane and ethane are equivalent.
      a. can form only one alkyl group
   2. Hydrogens in propane are not equivalent.
      a. six H's on each end and two H's in the middle
      b. forms two different propyl groups
      c. remove an end H — *n*–propyl
      d. remove a middle H – isopropyl
   3. Four different kinds of butyl groups.
      a. butyl and *sec*–butyl
         i. are derived from straight–chain alkanes
         ii. *n*–butyl: remove H from end C
         iii. *sec*–butyl: remove H from a C atom bonded to two other C atoms
      b. isobutyl and *tert*–butyl
         i. derived from branched–chain isobutane
         ii. isobutyl: remove H from end C on the branched–chain isobutane
         iii. *tert*–butyl: remove H from a C bonded to three other C atoms
   4. Alkyl groups are not compounds.

EXAMPLE:
Name the compounds in the previous example.

SOLUTION:
a and *d*: no branching, seven carbons in a chain — *n*–heptane.

b and *c*: six carbons in a chain with a methyl branch — 3–methylhexane. (Remember, look for the largest number of carbon atoms bonded to each other.)

e: five carbons in a chain with two methyl groups — 2,3–dimethylpentane.

Chapter 23 – Organic Chemistry

Workbook Problem 23.1
Draw structures for the following compounds:

a. 2–methyloctane b. 3–ethylnonane c. 5–isobutyldecane

d. 5–ethyl–2,3–dimethylheptane e. 2,2,4–trimethylhexane

Step 1: For each compound, determine the number of carbons in the chain based on the parent name.

Step 2: Draw the parent chain, and place the branches on the chain starting from either end. (Don't forget that carbon only can have **four** bonds. When adding branches, be sure to subtract the correct number of hydrogens.)

What key concept did you use?

23.6 Cycloalkanes
A. Acylic alkanes — open–chain alkanes.
B. Cycloalkanes — alkanes that contain rings of carbon atoms.
C. Structures are represented by polygons.
   1. Carbon atom is understood to be at every junction of lines.
   2. Mentally supply the proper number of hydrogen atoms needed to fill out carbon's valency.
D. Name substituted cycloalkanes, using the cycloalkane as the parent name and identifying the positions on the ring where substituents are attached.   *A. 5.*
   1. Start numbering at the group that has alphabetical priority.
   2. Proceed in the direction that gives the second substituent the lowest possible number.

EXAMPLE:
Name the following compounds:

a. 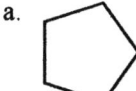   b.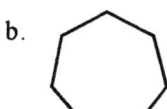

Chapter 23 – Organic Chemistry

c.

SOLUTION:
a. Five junctions representing five carbons bonded to each other, no substituents — cyclopentane.

b. Seven junctions representing seven carbons bonded to each other, no substituents — cycloheptane.

c. Six junctions representing six carbons bonded to each other, with three substituents. Cl — has first position (due to alphabetical priority); lower possible number of next substituent (CH_3) is two; continues in a counter clockwise manner — 1–chloro–5–isopropyl–2–methylcyclohexane.

23.7 Reactions of Alkanes
A. Alkanes have relatively low chemical reactivity.
B. React with oxygen and with halogens.
 1. Alkanes react with oxygen during combustion.
  a. CO_2 and H_2O are products

 2. Alkanes react with Cl_2 or Br_2 when irradiated with UV light.
  a. sequential substitution of alkane hydrogen by halogen

23.8 Families of Organic Molecules: Functional Groups
A. Can classify organic compounds into families according to their structural features.
 1. Chemical behavior of the members of a family is predictable.
B. Functional group — a part of a larger molecule that is composed of an atom or group of atoms that has characteristic chemical behavior.
 1. A given functional group undergoes the same kinds of reactions in every molecule it's a part of.
C. The chemistry of an organic molecule, regardless of its size and complexity, is determined by the functional groups it contains.
D. Most common functional groups — Table 23.1 (page 981 in your textbook.)

EXAMPLE:
Give the functional group for alcohols, ketones, and amines.

SOLUTION:
See Table 23.1, page 981 in your textbook.

Chapter 23 – Organic Chemistry

—C—O—H	—C—C(=O)—C—	—C—N—
Alcohols	Ketones	Amines

Workbook Problem 23.2
Determine which functional groups are present in the following molecules:

a. $CH_3CH_2CH_2\overset{\overset{O}{\|}}{C}H$

b. $CH_3CH_2-O-CH_2CH_3$

c. $CH_3\overset{\overset{O}{\|}}{C}-O-CH_2CH_2CH_3$

d. $CH_3CH_2CH_2\overset{\overset{O}{\|}}{C}-NH_2$

Step 1: Use Table 23.1 on page 981 in your textbook to identify the functional groups.

What key concept did you use?

23.9 Alkenes and Alkynes
A. Alkenes — hydrocarbons that contain a carbon–carbon double bond.
B. Alkynes — hydrocarbons that contain a carbon–carbon triple bond.
C. Unsaturated hydrocarbons — hydrocarbons that have fewer hydrogens per carbon than the related alkanes.
   1. Alkenes and alkynes are unsaturated.
D. Name alkenes by counting the longest chain of carbons that contains the double bond, and add the family suffix *–ene*.

Chapter 23 – Organic Chemistry

E. Alkene isomers that exist because of double–bond position.
   1. For butene and higher — specify the position of the double bond with a numerical prefix.
      a. Start numbering from the chain end nearer the double bond.
      b. Cite only the first of the double–bonded carbons.
      c. Identify and number the position of any substituent present.
      d. If the double bond is equidistant from both ends of the chain, numbering starts at the end nearer the substituent
F. Alkene isomers that exist because of double–bond geometry.
   1. *cis* isomer — functional groups are on the same side of the double bond.
   2. *trans* isomer — function groups are on different sides of the double bond.
G. Alkynes are similar to alkenes.
   1. Names using the family suffix *–yne*.
   2. Simplest alkyne — acetylene.
   3. Isomers are possible depending on the position of the triple bond in the chain.

EXAMPLE:
Name the following compounds:

a. $CH_2=CHCH_2CH_3$ b. $CH_3-C\equiv CCH_3$

c. $CH_3-C(=CH_2)-CH_3$ d. $Br-CH_2-C\equiv C-CH(CH_3)CH_3$ (with CH_3 on the fourth carbon)

e. $CH_3C(Cl)=C(Cl)CH_3$

SOLUTION:
a. Four–carbon chain with a double bond between the first and second carbon: 1–butene.
b. Four–carbon chain with a triple bond between the second and third carbon: 2–butyne.
c. Three–carbon chain with a methyl group on the second carbon: 2–methyl–1–propene
d. Five–carbon chain with the triple bond between the second and third carbon, a Br group on the first carbon and a methyl group on the fourth: 1–bromo–4–methyl–2–pentyne.
e. Four carbon chain with a double bond between the second and third carbon, 2 Cl groups on the second and third carbon which are on different sides of the double bond: *trans*–2,3–dichloro–2–butene

Workbook Problem 23.3
Draw structures for:
a. 2,4–dibromo–6–methyl–3–octene
b. 1–bromo–3,4–dimethyl–1–pentyne

c. 2,4–dimethyl–1–pentene

Step 1: Identify the parent chain in each molecule.

Step 2: Draw the structure, placing the double or triple bond in the proper position. The positions of the branches are determined relative to the double or triple bond. Take into consideration any geometrical isomers.

What key concept did you use?

23.10 Reactions of Alkenes and Alkynes

🔑 B.2.

A. Addition reactions — X–Y adds to the multiple bond of the unsaturated reactant to yield a saturated product.
B. Hydrogenation — addition of hydrogen.
   1. Alkene + H_2 → Alkane
C. Halogenation — addition of Cl_2 and Br_2.
   1. Alkene + X_2 → 1,2–Dihaloalkane
D. Hydrolysis — alkenes react with water in the presence of H_2SO_4 to produce an alcohol.
   1. Adding an H^+ and OH^-.
   2. H^+ goes to the carbon containing the most H's.

EXAMPLE:
Show the products of the reaction of propene with a) H_2, b) Cl_2, and c) H_2O (H_2SO_4 catalyst).

SOLUTION:

a. $CH_3CH_2CH_3$

Chapter 23 – Organic Chemistry

b. ClCH$_2$CHCH$_3$
        |
        Cl

c. CH$_3$CHCH$_3$
        |
        OH

23.11 Aromatic Compounds and Their Reactions
A. Aromatic compounds — a class of compounds containing a six–membered ring with three double bonds.
  1. Simplest aromatic — benzene.
B. Benzene and other aromatic compounds are much less reactive than alkenes and don't normally undergo addition reaction.
C. Benzene's stability is a consequence of its electronic structure.
  1. six π electrons are spread around the entire ring.
  2. 2 resonance structures

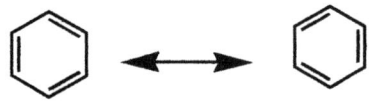

   a. often depicted as:

D. Name substituted aromatic compounds by using the suffix –*benzene*.
E. Disubstituted aromatic compounds — use one of the prefixes *ortho–*, *meta–*, or *para–*.
  1. *ortho* (*o*) — two substituents in a 1,2 relationship on the ring.
  2. *meta* (*m*) — two substituents in a 1,3 relationship on the ring.
  3. *para* (*p*) — two substituents in a 1,4 relationship on the ring.
F. Phenyl — the benzene ring itself is a substituent.
G. Substitution reactions — a group substitutes for one of the hydrogen atoms on the aromatic ring without changing the ring itself.
  1. All six H's are equivalent.
  2. Nitration — substitution of a nitro group (–NO$_2$) in the presence of H$_2$SO$_4$.
    a. key step in synthesis of explosives
    b. important pharmaceutical agents
    c. nitrobenzene used as starting material for dyes
  3. Halogenation — substitution of either Br or Cl in the presence of FeBr$_3$ or FeCl$_3$.
    a. step used in the synthesis of numerous pharmaceutical agents
  4. Sulfonation — substitution of a sulfonic acid group (–SO$_3$H).
    a. result of benzene reacting with H$_2$SO$_4$ and SO$_3$
    b. key step in synthesis of aspirin and the sulfa–drug family of antibiotics

EXAMPLE:
Name the following compounds:

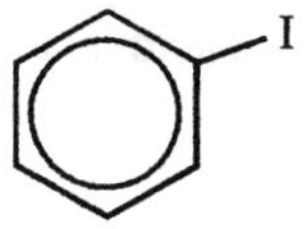

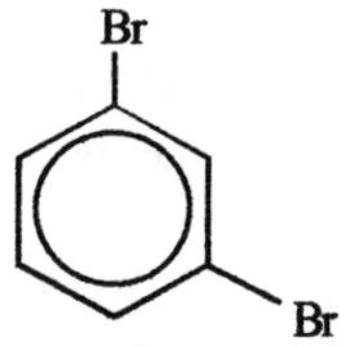

SOLUTION:
Determine the substituents on the benzene ring. If the ring is disubstituted, indicate the position of the substituents relative to each other.

iodobenzene, *meta*–dibromobenzene

Workbook Problem 23.4
Draw structures for the following molecules:

a. 4–ethyl–1,2–dimethylbenzene
b. *para*–bromochlorobenzene
c. The product form the reaction of benzene with Cl_2 in the presence of $FeCl_3$.
d. 2–phenyl–pentane

Step 1: Draw either the ring or the parent chain placing the substituents in the correct position. For ring structures, the actual placement of the substituents is not important, only that they are in the correct position relative to each other.

What key concept did you use?

23.12 Alcohols, Ethers, and Amines

A. Alcohols can be considered in two ways.
1. Derivatives of water in which one of the H's is replaced by an organic substituent.
2. Derivatives of alkanes in which one of the H's is replaced by a –OH group.
3. Structural resemblance to water.
   a. Simple alcohols are water soluble.
4. Name by
   a. specifying the point of attachment of the –OH group to the hydrocarbon chain

Chapter 23 – Organic Chemistry

  b. Use the suffix –*ol* to replace the terminal –*e* in the alkane name.
  c. Number chain at the end nearer to the –OH group.
 5. Simple alcohols — most important and commonly encountered organic chemicals.
 6. Methanol — wood alcohol.
  a. produced by catalytic reduction of CO with hydrogen gas
  b. important industrial starting material for preparing formaldehyde, acetic acid, and other chemicals
 7. Ethanol — one of the oldest known pure organic chemicals.
  a. alcohol present in wine, beer, and distilled liquors
 8. 2–Propanol (rubbing alcohol) — used primarily as a solvent.
 9. Other important alcohols.
  a. ethylene glycol — principal constituent of automobile antifreeze
  b. glycerol — moisturizing agent in many foods and cosmetics
  c. phenol — used in preparing nylon, epoxy adhesives, and heat–setting resins
B. Ethers — compounds that have two organic groups bonded to the same oxygen atom.
 1. Inert chemically — used as reaction solvents.
 2. Diethyl ether — used for many years as a surgical anesthetic agent.
C. Amines — organic derivatives of NH_3.
 1. One or more of the NH_3 hydrogens is replaced by an organic substituent.
 2. Use suffix –*amine* in naming these compounds.
 3. Amines are bases.
  a. can use the lone pair of electrons on nitrogen to accept H^+ from an acid
   i. produce ammonium salts
 4. Ammonium salts are more soluble in water than are neutral amines.
  a. practical consequences in drug delivery

EXAMPLE:
Name the following compounds:

a. $CH_3CHCH_2CH_2\text{-}OH$ b. $CH_3CH_2NH_2$
      |
      CH_3

SOLUTION:
 a. Functional group is OH; therefore, name as alcohol; four carbons in the chain, OH is on the first carbon, methyl group is on the third carbon: 3–methyl–1–butanol.
 b. Functional group is NH_2; therefore, name as an amine; alkyl group is ethyl: ethylamine.

23.13 Aldehydes and Ketones
A. Carbonyl group — group with a carbon–oxygen double bond.
 1. C=O — polar bond.
 2. Classify into two categories.
  a. Based on their chemical properties.
B. Aldehydes and ketones — CO group is bonded to atoms (H and C) that are not strongly electronegative.
 1. Aldehyde — hydrogen atom bonded to CO group.
 2. Ketone — two carbon atoms bonded to CO group.
 3. Two groups of compounds have similar properties.
 4. Industrial preparation of simple aldehydes and ketones involves oxidation of the related alcohol.

5. Aldehyde and ketone functional groups are present in many biologically important compounds.
C. Naming aldehydes and ketones.
   1. Aldehydes – use suffix *al*.
   2. Ketones – use suffix *one*.

23.14 Carboxylic Acids, Esters, and Amides

A. CO group is bonded to an atom (O or N) that is strongly electronegative.

B. Carboxylic acids contain

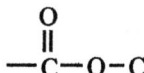

   1. Occur widely throughout the plant and animal kingdoms.
   2. Long-chain carboxylic acids – constituents of all animal fats and vegetable oils.
   3. Name by replacing the *e* ending of the corresponding alkane with *–oic acid*.
   4. Dissociate slightly in aqueous solution to give H_3O^+ and a carboxylate anion.
   5. Undergo acid–catalyzed reaction with an alcohol to yield an ester.

C. Esters contain

   $$-\overset{\overset{O}{\|}}{C}-O-C$$

   1. Many uses in medicine, in industry, and in living systems.
      a. aspirin and benzocaine
      b. Dacron and Mylar
      c. fragrant odors of fruits and flowers
   2. Most important reaction – carbonyl-group substitution reaction.
      a. hydrolysis — splits the ester molecule into a carboxylic acid and an alcohol.
         i. can be either acid or base catalyzed
   3 Saponification — base–catalyzed ester hydrolysis.
      a. soap — mixture of sodium salts of long–chain carboxylic acids
      b. produced by hydrolysis with aqueous NaOH of the naturally occurring esters in animal fat
   4. Named by identifying the alcohol–related part and the acid–related part using the *–ate* ending.

D. Amides contain

   $$-\overset{\overset{O}{\|}}{C}-N$$

   1. Amide bond between nitrogen and a carbonyl–group carbon
      a. fundamental link used by organisms for forming proteins.
      b. also found in synthetic fibers and pharmaceutical agents.
   2. Neutral — do not act as proton acceptors and do not form ammonium salts.
   3. Can be prepared by the reaction of a carboxylic acid with ammonia or an amine.
   4. Name by first citing the *N*–alkyl group on the amine part and then identifying the carboxylic acid part using the *–amide* ending.

Chapter 23 – Organic Chemistry

   a. *N*-group is attached to nitrogen.
   b. use *amide* as the ending.
5. Acid or base hydrolysis reaction with water.
   a. produces a carboxylic acid and amine.
   b. key process that occurs in the stomach during digestion of proteins.

EXAMPLE:
Name the following compounds:

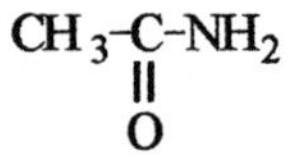

a.
$$CH_3CH_2-O-\underset{\underset{O}{\|}}{C}-CH_3$$

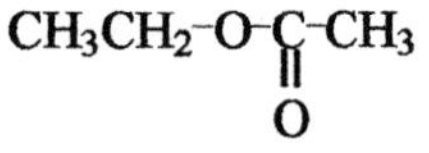

b. $CH_3-\underset{\underset{O}{\|}}{C}-NH_2$

c. $CH_3\underset{\underset{}{|}}{\overset{CH_3}{C}}HC\overset{=O}{\underset{OH}{\diagdown}}$

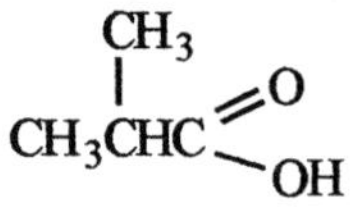

SOLUTION:
a. alcohol–related part – ethanol; carboxylic acid part acetic acid: ethyl acetate
b. carboxylic acid part – acetic acid: acetamide
c. carboxylic acid – isopropanoic acid

EXAMPLE:
Give the products for the following reaction:

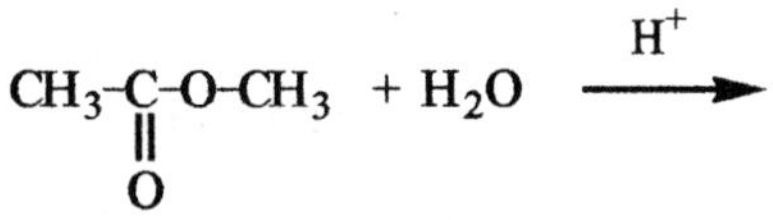

$$CH_3-\underset{\underset{O}{\|}}{C}-O-CH_3 + H_2O \xrightarrow{H^+}$$

SOLUTION:

$$CH_3-\underset{\underset{O}{\|}}{C}-OH + CH_3OH$$

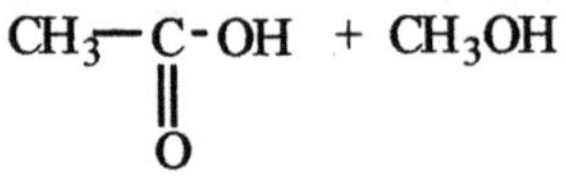

23.15 Synthetic Polymers

A. Polymers — large molecules formed by the repetitive bonding together of many smaller molecules, called monomers.
   1. Simple alkenes — vinyl monomers undergo polymerization reactions.
      a. important polymers — Table 23.3, page 967 in your textbook.
B. Fundamental process — addition reaction to the double bond
   1. Initiator – adds to double bond of alkene.
      a. produces reactive intermediate.
   2. Reactive intermediate adds to another alkene.

Chapter 23 – Organic Chemistry

  a. creates another reactive intermediate.
 C. Second kind of polymerization process — molecules with two functional groups react.
  1. Diacids and diamines react to give polyamides.
  2. Diacids and dialcohols react to give polyesters.

 **Putting It Together**

Ethyl caprate is often referred to as "cognac essence" since it is used in the manufacture of wine bouquets. The percent composition of this ester is 71.89% carbon, 12.13% hydrogen, and 15.98% oxygen. Reaction with water yields ethanol and an acid. The molar mass of the acid is 172 g/mol. What is the molecular formula of ethyl caprate?

 **Self–Test**

This section is intended to test your knowledge of the material covered in this chapter. Think through these problems, and make certain you understand what is going on. Ask yourself if your answer makes sense. Many of these questions are linked to the chapter learning goals. Therefore, successful completion of these problems indicates you have mastered the learning goals for this chapter. You will receive the greatest benefit from this section if you use it as a mock exam. You will then discover which topics you have mastered and which topics you need to study in more detail.

True–False

1. Organic molecules have polar covalent bonds when carbon bonds to hydrogen.
2. Doubly bonded carbons use sp^2 hybrid orbitals to form σ bonds.
3. Hydrocarbons are molecules that contain only C and H.
4. Different isomers have different structures but the same physical properties.
5. Alkyl groups are not compounds.
6. Alkanes are very reactive.
7. Benzene's stability is a consequence of the six π electrons that are spread around the ring.
8. Amines are organic derivatives of H_2O.
9. Aldehyde and ketone functional groups are present in many biologically important compounds.
10. Amides are neutral.

Fill–in–the–Blank

1. Carbon is _____ in terms of its bonding.
2. Molecules that have the same molecular formula but different structures are called _____.
3. A number of possible three–dimensional structures resulting from the ability of parts of a molecule to spin around a carbon–carbon single bond are referred to as _____.
4. Alkanes that contain rings of carbon atoms are known as _____.

Chapter 23 – Organic Chemistry

5. The chemistry of an organic molecule, regardless of its size or complexity, is determined by the _____ it contains.
6. The simplest aromatic compound is _____.
7. _____ are compounds that have two organic groups bonded to the same oxygen atom and whose reactivity is _____.
8. Organic derivatives of NH_3 are _____.
9. Upon hydrolysis, an ester molecule splits into an _____ and a _____.
10. Polyesters are polymers formed by the reaction of _____ and _____.

General Questions

1. Name the following compounds:

    a.
    $$\begin{array}{c} CH_3 \\ | \\ CH_2 \\ | \\ CH_3-CH-CH-CH_3 \\ | \\ CH_2 \\ | \\ CH_3 \end{array}$$

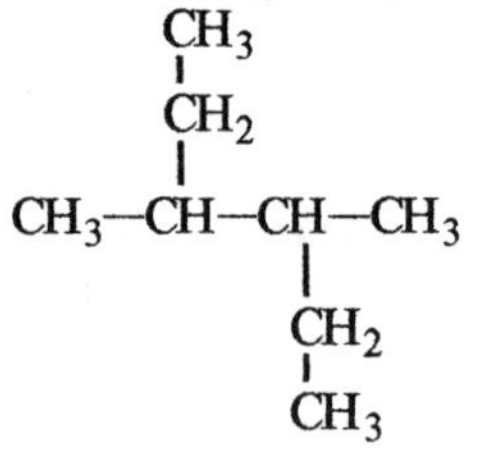

    b. 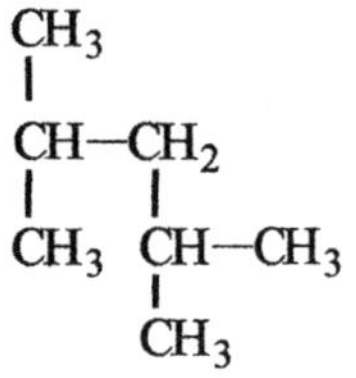
    $$\begin{array}{c} CH_3 \\ | \\ CH-CH_2 \\ | \quad\; | \\ CH_3 \;\; CH-CH_3 \\ | \\ CH_3 \end{array}$$

    c. $CH_2\!=\!C-CH_2CH-CH_3$ with CH_3 on the CH and CH_2-CH_3 on the C.

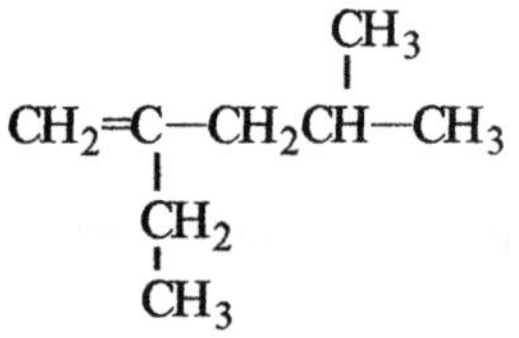

    d. cyclopentene with a CH_3 substituent

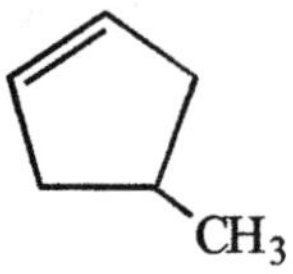

    e. $Br-CH_2-C\equiv C-CH_2-CH_3$

    f. bromobenzene

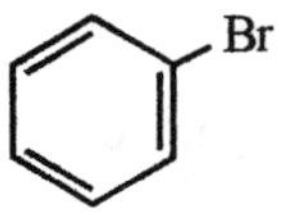

g. 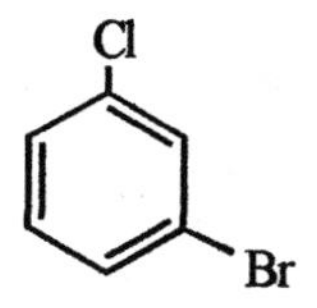 [structure: benzene ring with Cl and Br in meta positions]

h. 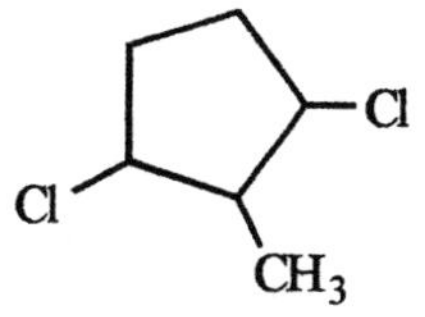 [structure: cyclopentane with Cl, Cl, and CH₃ substituents]

i. 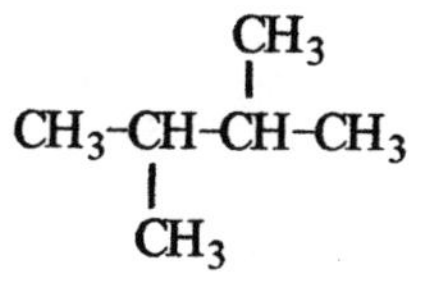 $CH_3-CH-CH-CH_3$ with CH_3 above middle right C and CH_3 below middle left C

j. 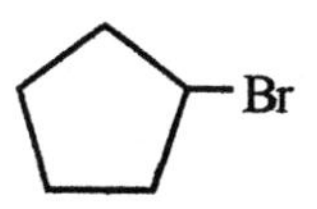 [structure: cyclopentane with Br substituent]

k. 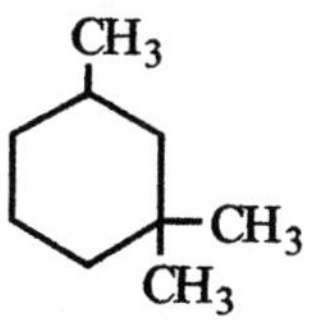 [structure: cyclohexane with CH₃ at one position and two CH₃ groups (gem-dimethyl) at another position]

l. 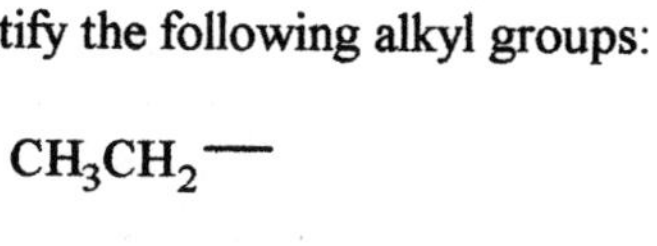 $CH_3-\underset{Br}{C}=\underset{Br}{C}-CH_3$

2. Identify the following alkyl groups:

   a. CH_3CH_2-

   b. CH_3CH-
       $|$
       CH_3

c. 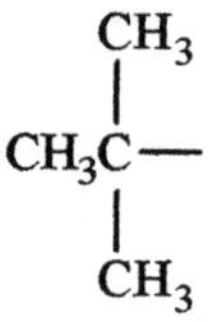
$$CH_3C\begin{array}{c}CH_3\\|\\-\\|\\CH_3\end{array}$$

3. Write structures for the following compounds:
   a. 4–methylcyclohexene
   b. 2–pentyne
   c. *para*–chloroiodobenzene
   d. 1–chloro–3–ethyl–3–hexene
   e. 2,3–dimethylhexane
   f. 1,3–dichloro–1–butyne
   g. chlorobenzene

4. Give the first chemical equation for the reaction between ethane and Br_2.

5. Predict the outcome of the following reactions. Name the organic product.
   a. propene and bromine
   b. propene and hydrogen
   c. cyclohexene and water

6. Determine if the following isomers are *cis* or *trans*.

a. 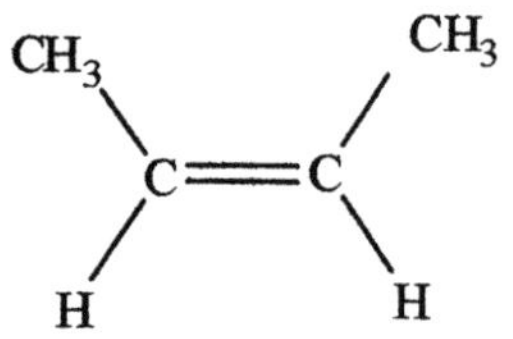
$$\begin{array}{c}CH_3\\ \diagdown\\ C=C\\ \diagup\quad\diagdown\\ H\qquad H\end{array}\begin{array}{c}CH_3\\ \diagup\\ \\ \end{array}$$

b. 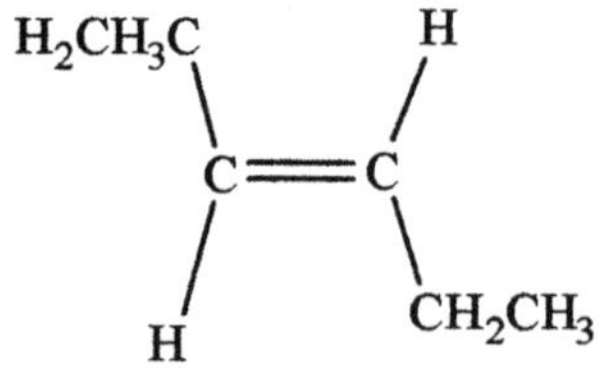
$$\begin{array}{c}H_2CH_3C\\ \diagdown\\ C=C\\ \diagup\quad\diagdown\\ H\qquad CH_2CH_3\end{array}\begin{array}{c}H\\ \diagup\\ \\ \end{array}$$

7. How would you prepare the following compounds, beginning with benzene?

a. 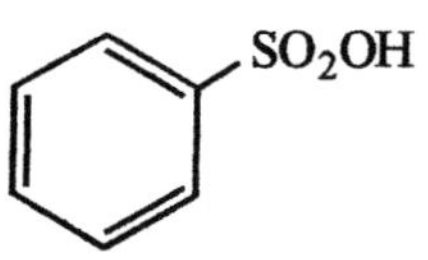 [benzene ring]—SO₂OH

b. 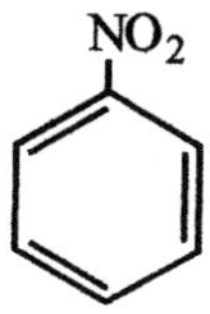 [benzene ring]—NO₂

8. Classify the following compounds based on the functional group present. Name these compounds.

a. $CH_3CH_2CH_2CH_2COOH$

b. $CH_3CH_2\underset{\underset{O}{\|}}{C}CH_2CH_3$

c. $CH_3CH_2CH_2CH_2CH_2\underset{\underset{O}{\|}}{C}H$

d. $CH_3CH_2OCH_2CH_2CH_3$

e. $CH_3CH_2CH_2\underset{\underset{O}{\|}}{C}OCH_2CH_2CH_3$

f. $CH_3CH_2\underset{\underset{OH}{|}}{C}HCH_2CH_3$

g. $CH_3\underset{\underset{CH_3}{|}}{C}HNH_2$

h. $CH_3\underset{\underset{O}{\|}}{C}-\underset{\underset{CH_3}{|}}{N}CH_3$

9. Predict the outcome of the following reactions. Name the organic product.

a. $CH_3CH_2CH_2-O-\underset{\underset{O}{\|}}{C}H-CH_2CH_2CH_3 + H_2O \longrightarrow$

b. $CH_3\underset{\underset{CH_3}{|}}{C}HCH_2COOH + CH_3CH_2\underset{\underset{CH_3}{|}}{C}H\underset{\underset{}{|}}{C}H_2\overset{OH}{|} \longrightarrow$

c. $CH_3CH_2-\underset{\underset{O}{\|}}{C}-OH + H-\underset{\underset{H}{|}}{N}-CH_2CH_3 \longrightarrow$

10. Draw structural formulas for the polymers made from:

   a. $CH_2=CH-Br$

   b. $HO-\underset{O}{\underset{\|}{C}}-C_6H_4-\underset{O}{\underset{\|}{C}}-OH \; + \; HO-CH_2CH_2-OH$

Challenge Problem

Maleic acid is an unsaturated, dicarboxylic acid prepared by the catalytic oxidation of benzene. When 0.175 g of the acid undergoes combustion, 0.266 g of CO_2 and 0.0544 g of H_2O are produced. When a 0.242 g sample of maleic acid is titrated with sodium hydroxide, 27.80 mL of 0.150 M NaOH is needed for complete titration. Determine the molecular formula of the acid.

CHAPTER 24

BIOCHEMISTRY

Chapter Learning Goals

A. Biochemical Energy Changes
1. Explain the metabolic reactions between a molecule containing the –OH functional group and ATP to form a phosphate.

B. Proteins
1. Identify functional groups in α–amino acids. Draw structures that show the geometry about each atom.
2. Classify amino acid side chains as acidic, basic, or neutral, hydrophilic or hydrophobic.
3. Determine whether a molecule is chiral or achiral.
4. Draw the structure and give the three–letter shorthand notation for simple proteins.

C. Carbohydrates
1. Classify a carbohydrate as simple or complex. Classify a monosaccharide as a ketose or aldose.
2. Draw tetrahedral representations of simple carbohydrates, and determine the number of chiral carbon atoms and the maximum number of isomers of the molecule.

D. Fats and Oils
1. Draw the structures of simple fats and oils.

E. Nucleic Acids
1. Draw the structures of simple nucleic acids.
2. Show the complementary base pairs in strands of DNA.
3. Show the RNA base sequence complementary to a given DNA base sequence.

Chapter in Brief
This chapter begins by looking at biochemical energetics. You then begin an examination of the main classes of biomolecules: proteins, carbohydrates, lipids, and nucleic acids. Your detailed study of these molecules includes the molecular handedness of amino acids and carbohydrates, the levels of protein structure, and the cyclic structures of monosaccharides. You will also examine the lock–and–key model of enzyme action and the Watson–Crick model for base pairing in DNA.

24.1 Biochemical Energy

A.1.

A. Metabolism — the sum of the many organic reactions that go on in cells.
  1. Extract and release energy from food.
  2. End products — carbon dioxide, water and energy.
  3. Occur in either linear or cyclic long sequences.
     a. linear sequence — product of one reaction serves as the starting material for the next
     b. cyclic sequence — a series of reactions regenerates the first reactant
B. Catabolism — reaction sequences that break molecules apart.
  1. Generally releases energy that is used to power living organisms.
C. Four stages of catabolism – Figure 24.1, page 980 in your textbook.
  1. First stage — digestion.

Chapter 24 – Biochemistry

   a. bulk food is broken down into small molecules
 2. Second stage — small molecules are broken down into two–carbon acetyl groups attached to coenzyme A.
   a. produces acetyl coenzyme A — intermediate in the breakdown of all main classes of food molecules
 3. Third stage — citric acid cycle.
   a. acetyl groups are oxidized; produces carbon dioxide and water
   b. releases a great deal of energy, which is used in the fourth stage
 4. Fourth stage — the respiratory chain.
   a. produces adenosine triphosphate (ATP)
   b. plays a pivotal role in the production of biological energy
D. Anabolism — reaction sequences that put building blocks back together to assemble larger molecules.
 1. Absorb energy.
E. The entire process of energy production revolves around the ATP ⇌ ADP interconversion.
 1. Catabolic reactions "pay off" in ATP by synthesizing it from adenosine diphosphate (ADP) plus hydrogen phosphate ion.
 2. Anabolic reactions "spend" ATP by transferring a phosphate group to other molecules, thereby regenerating ADP.
F. ATP releases a large amount of energy when its P–O–P bonds are broken and a phosphate group is transferred.
 1. Usefulness is due to ATP's ability to drive otherwise unfavorable reactions.
 2. Couple with energetically unfavorable reaction.
   a. Overall free–energy change for the two reactions together is favorable.
 3. Transfer of phosphate group from ATP.
   a. single most important reaction in life.

24.2 Amino Acids and Peptides
A. Protein — biological polymers made up of many amino acids linked together to form long chains.
 1. A group of biological molecules that are of primary importance to all living organisms.
 2. Different biological functions.
   a. Some serve a structural purpose.
   b. Some act as hormones to regulate specific body processes.
   c. Some are enzymes.
     i. biological catalysts that carry out body chemistry

B. Amino acids — molecules that contain two functional groups.
 1. A basic amino group (–NH_2).
 2. An acidic group (–COOH).
C. Peptide bond — an amide bond that forms when two amino acids link together.
D. Dipeptide — the molecule that results when two amino acids are linked together by formation of a peptide bond between the –NH_2 group of one and the –COOH group of the second.
 1. Tripeptide — linkage of three amino acids by two peptide bonds.
 2. Polypeptides — chains of up to 100 amino acids.
 3. Protein — chains with more than 100 amino acids.
E. 20 different amino acids commonly found in proteins.
 1. Refer to each by a three–letter shorthand code (Figure 24.2 page 984 in your textbook).
 2. α–amino acids — the amino group is connected to the carbon atom alpha to (next to) the carboxylic acid group.
 3. Differ in the nature of the group (side chain, R) attached to the α–carbon.
 4. Classified as neutral, basic, or acidic.

Chapter 24 – Biochemistry

   a. depends on side chain
   b. neutral — divide into hydrophobic (nonpolar side chains) and hydrophilic (polar) side chains
   c. acidic — an additional carboxylic acid on side chain
   d. basic — additional amine function on side chain

24.3 Amino Acids and Molecular Handedness
   A. Chiral molecule — a molecule that is not identical to its reflected mirror image.
      1. Right-handed molecule (D).
      2. Left-handed molecule (L).
   B. A carbon atom bonded to four different atoms or groups of atoms is chiral. 🔑 B.3.
      1. Achiral — carbon atom is bonded to two or more of the same groups.
   C. Enantiomers — two mirror-image forms of a chiral molecule.
      1. Nineteen of common amino acids are chiral because they have four different groups bonded to their α-carbons.
      2. Only L-amino acids are found in proteins.

24.4 Proteins
   A. Residues — individual amino acids that are linked together by peptide bonds to form proteins.
   B. Backbone — the repeating chain of amide linkages to which the side chains are attached.
   C. The number of possible isomeric peptides increases rapidly as the number of amino acid residues increases.
   D. All noncyclic proteins have an N-terminal amino acid and a C-terminal amino acid. 🔑 B.4.
      1. N-terminal amino acid — has a free –NH$_2$ group on one end.
      2. C-terminal amino acid — has a free –COOH group on the other end.
      3. Protein is written with the N-terminal residue on the left and the C-terminal residue on the right.
      4. Use three-letter abbreviations to indicate the name. (See Figure 24.2, page 984 in your textbook.)
   E. Fibrous proteins — consist of polypeptide chains arranged side by side in long filaments.
   F. Globular proteins — polypeptide chains coiled into compact, nearly spherical shapes.
   G. Proteins are also classified by their biological functions. (Table 24.1, page 989 in textbook.)

24.5 Levels of Protein Structure
   A. Four levels of structure used to describe proteins.
      1. Primary structure — specifies the sequence in which the various amino acids are linked together.
      2. Secondary structure — specifies how segments of the protein chain are oriented into a regular pattern.
      3. Tertiary structure — specifies how the entire protein chain is coiled and folded into a specific three-dimensional shape.
      4. Quaternary structure — how several protein chains aggregate to form a larger unit.
   B. Primary protein structure — most important of the four structural levels.
      1. Amino acid sequence determines the overall shape and functions.
         a. change one amino acid — alter biological properties
   C. Secondary protein structure — the regular patterns that result from the orientation of segments of folded proteins.

Chapter 24 – Biochemistry

   1. α–helix — the protein wraps into a helical coil, much like the cord on a telephone.
      a. stabilized by formation of hydrogen bonds between the N–H group of one amino acid and the C=O group of another amino acid four residues away
   2. β–pleated sheet — polypeptide chains line up in a parallel arrangement held together by hydrogen bonds.
      a. not as common as the α–helix
      b. found in proteins where sections of peptide chains double back on themselves
  D. Tertiary protein structure — structures which result primarily from interactions of side–chain R groups in the protein.
   1. Stabilization results from the hydrophobic interactions of the hydrocarbon side chains on amino acids.
   2. Neutral, nonpolar side chains congregate on the hydrocarbon–like interior of a protein molecule.
   3. Polar side chains are found on the exterior of the protein.
   4. Other stabilizing factors.
      a. disulfide bridges — covalent S–S bonds formed between nearby cysteine residues
      b. salt bridges — ionic attractions between positively and negatively charged sites on the protein
      c. hydrogen bonds between nearby amino acids

24.6 Carbohydrates

  A. Carbohydrates — a large class of polyhydroxylated aldehydes and ketones.
   1. Monosaccharides — carbohydrates that can't be broken down into smaller molecules by hydrolysis with aqueous acid.
   2. Polysaccharides — compounds that are made of many simple sugars linked together and that cleave into many molecules of simple sugars upon hydrolysis.
  B. Monosaccharides.
   1. Aldose — contains an aldehyde carbonyl group.
   2. Ketose — contains a ketone carbonyl group.
   3. The "–ose" suffix — used to indicate a sugar.
   4. Use one of the Greek prefixes to indicate the number of carbon atoms in the sugar.

24.7 Handedness of Carbohydrates

  A. Compounds are chiral if they have a carbon atom bonded to four different atoms or groups of atoms.
   1. Lack a plane of symmetry.
   2. Can exist as a pair of enantiomers.
      a. right–handed — D form
      b. left–handed — L form
  B. A compound with n chiral carbon atoms has a maximum of 2^n possible forms.

24.8 Cyclic Structures of Monosaccharides

  A. Monosaccharides are shown as having open–chain structures.
  B. Exist as cyclic molecules.
   1. –OH group near the bottom of the chain adds to the –C=O group near the top of the chain.
   2. Cyclic α form — the C1 –OH group is on the bottom side of the ring.

24.9 Some Common Disaccharides and Polysaccharides

  A. Lactose — major carbohydrate present in mammalian milk.

1. Disaccharide whose hydrolysis with aqueous acids yields one molecule of glucose and one molecule of galactose.
2. Two sugars bonded together by a 1,4 link.
   a. bridging oxygen atom between C1 of β–galactose and C4 of β–glucose
B. Sucrose — most common pure organic chemical in the world.
   1. Hydrolysis yields one molecule of glucose and one molecule of fructose.
   2. Invert sugar — 50:50 mixture of sugar produced from the hydrolysis of sucrose.
      a. commonly used as a food additive
C. Cellulose — consists of several thousand β–glucose molecules joined together by 1,4 links to form an immense polysaccharide.
   1. Used as a structural material in plants.
D. Starch — consists of several thousand α–glucose molecules joined together by 1,4 links.
   1. More structurally complex than cellulose.
   2. Two types.
      a. amylose — several hundred to 1000 α–glucose units joined together in a long chain by 1,4 links
      b. amylopectin — larger than amylose and has branches every 25 units along its chain
E. Glycogen — a long polymer of α–glucose units with branch points in its chain.
   1. Breaks down into simple glucose units.
      a. some used as fuel
      b. some is stored for later use.

24.10 Lipids

🔑 D.1.

A. Lipids — naturally occurring organic molecules that dissolve in nonpolar organic solvents.
   1. Defined by solubility rather than by chemical structure.
   2. Contain large hydrocarbon portions.
      a. accounts for their solubility behavior
B. Fats and oils — most plentiful lipids in nature.
   1. Triacylglycerols (triglycerides) — triesters of glycerol with three long–chain carboxylic acids (fatty acids).
      a. fatty acids — unbranched and have an even number of carbon atoms
      b. Three fatty acids can be different.
   2. Monounsaturated — fatty acid with only one double bond.
   3. Polyunsaturated fatty acids — have more than one carbon–carbon double bond.
C. Steroids — a lipid whose structure is based on the tetracyclic system.
   1. Three rings are six–membered, while the fourth is five–membered.
D. Cholesterol — most abundant animal steroid; serves two important functions.
   1. Minor component of cell membranes.
   2. Serves as the body's starting material for the synthesis of all other steroids.

24.11 Nucleic Acids

A. Deoxyribonucleic acid (DNA) and ribonucleic acid (RNA) are the chemical carriers of an organism's genetic information.

B. Nucleic acids — polymers made up of nucleotide units linked together to form a long chain.

🔑 E.1.

   1. Nucleotide — composed of nucleoside plus phosphoric acid.
      a. nucleoside — composed of an aldopentose sugar plus an amine base.
         i. RNA – sugar = ribose
         ii. DNA – sugar = 2–deoxyribose
C. Four different cyclic amine bases in DNA: adenine, guanine, cytosine, and thymine.
D. Four different cyclic amine bases in RNA: adenine, guanine, cytosine and uracil.

Chapter 24 – Biochemistry

E. Cyclic amine base is bonded to C1' of the sugar, and the phosphoric acid is bonded to the C5' sugar position.
  1. Numbers with a prime superscript refer to positions on the sugar component of a nucleotide.
  2. Numbers without a prime superscript refer to positions on the cyclic amine base.
F. Nucleotides link through a phosphate ester bond between the phosphate group at the 5' end of one nucleotide and the hydroxyl group on the sugar component at the 3' end of another nucleotide.
G. Structure of a nucleic acid depends on its sequence of individual nucleotides.
  1. An alternating sugar–phosphate backbone with different amine base side chains attached.
H. Describe the sequence of nucleotides by starting at the 5' phosphate and identifying the bases in order.
  1. Use first letter of the base name as an abbreviation for each nucleotide.

24.12 Base Pairing in DNA: The Watson–Crick Model

A. DNA consists of two polynucleotide strands coiled around each other in a double helix.
  1. Sugar–phosphate backbone is on the outside of the helix.
  2. Heterocyclic bases are on the inside of the helix.
  3. Base on one strand points directly at a base on the other strand.
  4. Two strands run in opposite directions.
     a. held together by hydrogen bonds between pairs of bases
        i. A and T
        ii. G and C
B. Two strands are complementary.
  1. Whenever a G base occurs in one strand, a C base occurs opposite it in the other strand.

24.13 Nucleic Acids and Heredity

A. Chromosomes — threadlike strands that are coated with proteins and wound into complex assemblies.
B. Gene — segment of a DNA chain that contains the instructions necessary to make a specific protein.
C. DNA functions as a storage medium for an organism's genetic information.
D. RNA reads, decodes, and uses the information received from DNA to make proteins.
E. Three processes in the transfer and use of genetic information.
  1. Replication — the means by which identical copies of DNA are made.
     a. forms additional molecules
     b. preserves genetic information for passing on to offspring
  2. Transcription — means by which information in the DNA is transferred to and decoded by RNA.
  3. Translation — means by which RNA uses the information to build proteins.
F. Replication — an enzyme-catalyzed process that begins with a partial unwinding of the double helix.
  1. New nucleotides line up on each strand in a complementary manner.
  2. Two new strands begin to grow.
G. Transcription — similar to replication except that new ribonucleotides line up.
  1. Uracil lines up opposite adenine.
H. Translation — protein biosynthesis; directed by messenger RNA (mRNA).
  1. Occurs on ribosomes — knobby protuberances within a cell.
  2. Ribonucleotide sequence in mRNA specifies the order of different amino acid residues.
     a. Three ribonucleotides that are specific for a given amino acid

Chapter 24 – Biochemistry

3. Read by transfer RNA (tRNA) — contains a complementary base sequence that allows it to recognize a three–letter word on mRNA.
   a. acts as a carrier to bring a specific amino acid into place for transfer to the peptide chain

 Putting It Together

The monoanion of adenosine monophosphate (AMP) is an intermediate in phosphate metabolism

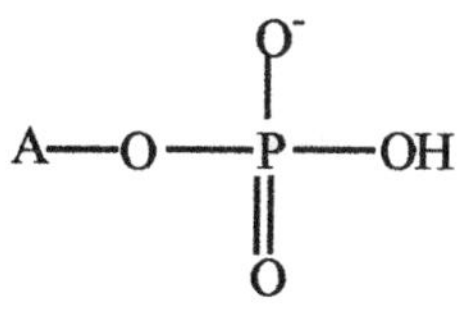

$$A-O-\overset{\overset{O^-}{|}}{\underset{\underset{O}{\|}}{P}}-OH$$

where A = adenosine. If the pK_a for this anion is 7.21, what is the ratio of [AMP—OH$^-$] to [AMP—O^{2-}] in blood at pH 7.40?

 **Self–Test**

This section is intended to test your knowledge of the material covered in this chapter. Think through these problems, and make certain you understand what is going on. Ask yourself if your answer makes sense. Many of these questions are linked to the chapter learning goals. Therefore, successful completion of these problems indicates you have mastered the learning goals for this chapter. You will receive the greatest benefit from this section if you use it as a mock exam. You will then discover which topics you have mastered and which topics you need to study in more detail.

Matching

Metabolism	a. a large class of polyhydroxylated aldehydes and ketones
Protein	b. large proteins that act as catalysts for biological reactions.
Dipeptide	c. proteins that consist of polypeptide chains arranged side by side in long filaments
Residue	d. the means by which RNA uses the information to build proteins.
Fibrous proteins	e. molecules composed of an aldopentose sugar plus an amine base.

Chapter 24 – Biochemistry

    Globular proteins   f. threadlike DNA strands that are coated with proteins and wound into complex assemblies.

    Enzyme   g. an enzyme–catalyzed process that begins with a partial unwinding of the double helix.

    Cofactor   h. proteins that consist of polypeptide chains coiled into compact, nearly spherical shapes

    Carbohydrates   i. the sum of the many organic reactions that go on in cells

    Lipids   j. small nonprotein portions of an enzyme, either a metal ion or a small organic molecule

    Nucleoside   k. polypeptide chain with more than one hundred amino acids

    Chromosomes   l. the molecule that results when two amino acids are linked together by formation of a peptide bond between the –NH$_2$ group of one and the –COOH group of the second

    Replication   m. naturally occurring organic molecules that dissolve in nonpolar organic solvents

    Translation   n. individual amino acids that are linked together by peptide bonds to form proteins

Fill–in–the–Blank

1. In the second stage of catabolism, small molecules are broken down into _____ _____ and attached to _____.
2. The entire process of energy production revolves around the _____ interconversion.
3. The two functional groups in amino acids are _____ _____.
4. α–Amino acids differ in the nature of the _____.
5. An acidic α–amino acid has an additional _____ on its side chain.
6. A protein is written with the _____ residue on the left and the _____ on the right.
7. The α–helix structure of a protein is stabilized by the formation of _____ between the _____ group of one amino acid and the _____ group of another amino acid _____.
8. The number of substrate molecules acted on by one molecule of enzyme per unit time is the _____.

9. The difference between an aldose and a ketose is that an aldose contains _____ _____ and a ketose contains a _____.

10. _____ consists of several thousand β–glucose molecules joined together by 1,4 links to form an immense polysaccharide.

11. Lipids are defined by _____ rather than _____.

12. _____ and _____ are the chemical carriers of an organism's genetic information.

13. The four different cyclic amine bases in DNA are _____.

14. The base _____ replaces the base _____ in RNA.

15. Transcription is similar to replication except that _____ line up and _____ lines up opposite adenine.

Problems

1. State the amino acids represented by the following abbreviations.

    a. Trp   b. Glu   c. His   d. Arg

2. Draw the structure of the amino acid that contains

    a. an isobutyl group   b. a *sec*–butyl group

3. Identify the amino acids present in the following tetrapeptide:

4. Which of the following amino acids are likely to be found on the outside and which are likely to be found on the inside of a globular protein?

    a. Tyr   b. Met   c. Lys   d. Glu

Chapter 24 – Biochemistry

5. If the sequence A–G–T–A–A–T appeared on one strand of DNA, what sequence would appear opposite it on the other strand?

6. What RNA sequence would be complementary to the DNA sequence in #5?

7. Draw the structure of the dinucleotide A–G.

Challenge Problem

Explain why a graph of rate versus temperature of most enzyme–controlled reactions is bell–shaped.

WORKBOOK PROBLEMS SOLUTIONS

Chapter 1

WP 1.1

Strategy: Consider the differences in the degree size and zero point adjustments for the Fahrenheit and Celsius scales.

Step 1: A Fahrenheit degree is smaller than a Celsius degree. Just think about the melting (freezing) point of water. Water melts at 0° on the Celsius scale and at 32° on the Fahrenheit scale. Therefore, when converting from Fahrenheit to Celsius, the temperature should be lower.

Step 2: Apply the formula for the conversion of °F to °C. (Remember, for this conversion, you do a zero–point correction followed by a size correction.)

$$°C = \frac{5}{9} \times (102.7 - 32) = 39.3°C$$

Step 3: Does your answer in step 2 agree with your answer in step 1?
We predicted that the Celsius temperature should be lower, and it is.

Step 4: Apply the formula for the conversion of °C to K.

$$K = 39.3 + 273.15 = 312.5 \text{ K}$$

Key Concept: Experimentation involving measurement

WP 1.2

Strategy: Think about the information you have been given. Set up a mathematical equation that allows you to solve for the mass of iron. Remember that the definition of density is mass per unit volume (or mass divided by volume).

$$\text{Density}\left(\frac{g}{mL}\right) = \frac{\text{Mass of sample (g)}}{\text{Volume of sample (mL)}}$$

Step 1: Substitute the information given in the problem into the mathematical equation you set up.

$$7.87 \frac{g}{mL} = \frac{\text{mass of iron (g)}}{28.3 \text{ mL}}$$

Step 2: Solve for the mass of iron.

$$7.87 \frac{g}{mL} \times 28.3 \text{ mL} = 223 \text{ g}$$

Key Concept: Matter: understanding properties

Appendix A – Workbook Problems Solutions

WP 1.3

Strategy: Following the example and workbook problem 1.2, set up an equation to determine the volume of the cylinder.

$$\text{Density}\left(\frac{g}{mL}\right) = \frac{\text{Mass of sample (g)}}{\text{Volume of sample (mL)}}$$

Step 1: Solve for the volume of the cylinder.

$$3.678\,\frac{g}{mL} = \frac{98.075\,g}{\text{volume of sample (mL)}} \qquad \text{volume of sample (mL)} = \frac{98.075\,g}{3.678\,\frac{g}{mL}} = 26.665\,307\,23\ mL$$

Step 2: Apply the rule for multiplication and division to determine the number of significant numbers in your answer.

When carrying out either multiplication or division, your answer cannot have more significant figures than either of the original numbers. 3.678 has four significant figures; 98.075 has five significant figures. Your answer can't have more than four significant figures.

Step 3: If necessary, use the rules for rounding off numbers.

We must round off 26.665 307 23 to four significant figures. The first digit we must remove is a 5 with more non–zero digits following, therefore, we must round up. As a consequence, the final answer is 26.67 mL.

Key Concept: Experimentation involving measurement.

WP 1.4

Strategy: Using the conversion table on the back cover of your book, determine the conversion factor for meters to miles and convert time to one unit.

1 mi = 1.6093 km

$$4\ min \times \frac{60\ s}{1\ min} = 240\ s; \quad 240\ s + 38\ s = 278\ s$$

Step 1: Calculate the amount of time it will take for the athlete to run that distance.

$$3200\ m \times \frac{1\ km}{1000\ m} \times \frac{1\ mi}{1.6093\ km} \times \frac{278\ s}{1\ mi} = 553\ s$$

Key Concept: Experimentation involving measurement

Chapter 2

WP 2.1

Step 1: Determine the chlorine–to–phosphorus ratio for the first compound.

$$\text{First compound: Cl:P mass ratio} = \frac{2.58\ g\ Cl}{0.75\ g\ P} = 3.44$$

Step 2: Determine the chlorine-to-phosphorus ratio for the second compound.

$$\text{Second compound: Cl:P mass ratio} = \frac{7.73 \text{ g Cl}}{1.35 \text{ g P}} = 5.73$$

Step 3: Divide the ratio found for the first compound by the ratio for the second compound.

$$\frac{\text{Cl : P mass ratio in 1st compound}}{\text{Cl : P mass ratio in 2nd compound}} = \frac{3.44}{5.73} = 0.6$$

Step 4: Can your answer be converted to a small whole-number ratio?

Yes. $0.6 = \frac{6}{10} = \frac{3}{5}$

Key Concept: Law of Multiple Proportions

WP 2.2

Step 1: First, it's necessary to know the chemical symbol for iron.

The chemical symbol for iron is Fe.

Step 2: Use the periodic table to determine the atomic number for iron.

From the periodic table, we find that the atomic number for Fe is 26.

Step 3: The number of neutrons can be determined from the definition for mass number.

A (mass number) = Z (atomic number) + number of neutrons.
number of neutrons = $A - Z$
For iron – 56: number of neutrons = 56 – 26 = 30
For iron – 58: number of neutrons = 58 – 26 = 32

Step 4: The standard symbol is written with the mass number as a superscript and the atomic number as a subscript, both to the left of the symbol.

$^{56}_{26}\text{Fe}$ $^{58}_{26}\text{Fe}$

Key Concepts: Atoms and isotopes

WP 2.3

Strategy: Determine the atomic number from the information given.

We are told that element X has 29 protons; therefore, the atomic number of the element is 29.

Step 1: Knowing the atomic number, identify the element by using the periodic table.

From the periodic table, we find that the element with $Z = 29$ is Cu (copper).

Appendix A – Workbook Problems Solutions

Step 2: The standard symbol is written with the mass number as a superscript and the atomic number as a subscript, both to the left of the symbol.

The mass number is determined by adding the number of protons and the number of neutrons.

$A = 29 + 34 = 63$.

The standard symbol is $^{63}_{29}\text{Cu}$.

Key Concepts: Atoms and isotopes

WP 2.4

Strategy: Determine the conversion factors needed to convert from grams of sample to number of atoms.

We are beginning with grams and want to end up with atoms. The first conversion factor we will need is one that will allow us to cancel out the unit grams.

$$\frac{1\,\text{amu}}{1.660\,54 \times 10^{-24}\,\text{g}}$$

Using this conversion factor leaves us with the unit, amu. We now need a conversion factor that will cancel out the unit, amu.

$$\frac{1\,\text{atom Cd}}{112.4\,\text{amu}}$$

(112.4 amu is the atomic mass for cadmium.) This conversion factor leaves us with the desired unit.

Step 1: Set up a mathematical equation such that all of the units except number of atoms cancel.

$$25.2\,\text{g} \times \frac{1\,\text{amu}}{1.660\,54 \times 10^{-24}\,\text{g}} \times \frac{1\,\text{atom Cd}}{112.4\,\text{amu}} = 1.35 \times 10^{23} \text{ atoms of Cd}$$

Key Concept: Atomic mass units

WP 2.5

Strategy: Remember, the atomic mass of an element = Σ(mass of each isotope × the abundance of the isotope).

Step 1: Use the information given to determine the average atomic mass.

$(0.7899 \times 23.985\,\text{amu}) + (0.1000 \times 24.986\,\text{amu}) + (0.1101 \times 25.983\,\text{amu}) = 24.31\,\text{amu}$

Key Concepts: Atomic mass units and isotopes

WP 2.6

Step 1: Determine whether the formula contains both metals and nonmetals or only nonmetals.

a) HI – 2 nonmetals;   b) CO_2 – 2 nonmetals
c) $Mg(OH)_2$ – metal and nonmetals   d) $FeCl_3$ – metal and nonmetal

Step 2: Identify the compounds as either molecular or ionic, based on your conclusions in step 1.

a) and b) – molecular compounds   c) and d) – ionic compounds

Appendix A – Workbook Problems Solutions

Step 3: Determine if the substance is capable of providing either an H^+ or OH^- ion in water.

a) HI – produces H^+ when dissolved in water.
b) CO_2 – cannot produce either H^+ or OH^- when dissolved in water.
c) $Mg(OH)_2$ – produces OH^- when dissolved in water.
d) $FeCl_3$ – cannot produce either H^+ or OH^- when dissolved in water.

Step 4: Identify the compounds as either acids or bases (if applicable) based on your conclusions in step 3.

a) HI – acid   c) $Mg(OH)_2$ – base

Key Concepts: Molecules, ions and chemical bonds

WP 2.7

Step 1: When naming compounds, determine if the metal is a representative metal or a transition metal. If the metal is a transition metal, you must indicate the charge on the metal when naming the compound. The charge on the metal is determined from the number and charge on the anion. (Remember, you must maintain electrical neutrality.)

$CaCl_2$ — Calcium chloride. (Ca^{2+} is a group 2A metal and has only one charge.)

VO_2 — Vanadium is a transition metal; therefore, we need to include the charge on vanadium when naming the compound. The oxide anion has a –2 charge, and there are two oxide anions. The total negative charge is –4. To maintain electrical neutrality, the charge on vanadium must be a +4. The name of the compound is vanadium(IV) oxide.

$TiCl_4$ — Titanium also is a transition metal; therefore, we need to follow the same procedure for VO_2. The name of the compound is titanium(IV) chloride.

Co_2S_3 — Cobalt is a transition metal. We use the same procedure for the previous two compounds. However, this compound requires a bit more math. The charge on the sulfide anion is –2. We have three sulfides for a total charge of –6. To maintain electrical neutrality, the total charge on the two cobalts must be +6. The charge on each cobalt cation is $\frac{+6}{2} = +3$. The name of the compound is cobalt(III) sulfide.

MgSe — Magnesium selenide. (Mg^{2+} is a group 2A metal and has only one charge.)

Step 2: When writing formulas, use the periodic table or the name of the compound to determine the charge on the metal. Use the periodic table to determine the charge on the anion. (Remember, you must maintain electrical neutrality.)

strontium bromide — Strontium is a group 2A metal and forms only Sr^{2+}. Bromide has a –1 charge. To maintain electrical neutrality, we need two bromide ions. The formula is $SrBr_2$.

chromium (II) oxide — The name tells us that the charge on the chromium cation is +2. Using the periodic table, we find that the charge on the oxide anion is –2. The charges on each ion are balanced; therefore, the formula is CrO.

Appendix A – Workbook Problems Solutions

magnesium nitride — Magnesium is a group 2A metal and forms only Mg^{2+}. Using the periodic table, we find that the charge on the nitride anion is –3. Using the hint in the *Example*, let's use the number in the charge on magnesium (2) as the subscript for the nitride anion and the number in the charge on the nitride anion (3) as the subscript for magnesium. This would give us the formula Mg_3N_2. A check for electrical neutrality gives $[3\times(+2)] + [2\times(-3)] = 0$.

aluminum oxide — Aluminum is a group 3A metal and forms the Al^{3+} cation. The oxide anion has –2 charge. Using the same procedure for magnesium nitride gives Al_2O_3. A check for electrical neutrality gives $[2\times(+3)] + [3\times(-2)] = 0$.

Step 3: Make sure the formula contains the smallest whole–number ratio of cation to anion.

The formulas above all have the smallest whole–number ratios of cation to anion.

Key Concept: Naming binary ionic compounds

WP 2.8

Step 1: When naming the molecules, remember that the element that is more anionlike uses the *–ide* suffix. Also, remember to use the numerical prefixes found in Table 2.2 (p. 61) in your textbook.

SF_6 — Sulfur hexafluoride.

ICl_3 — Iodine trichloride.

HBr (*g*) — The formula was written to indicate that this binary hydrogen compound is a gas. Therefore, it is named as hydrogen bromide gas.

Step 2: When writing formulas, refer to Table 2.2 in your textbook for the meaning of the numerical prefixes. (Eventually, you'll need to know these numerical prefixes by heart.) Remember to indicate if a binary hydrogen compound is a gas or is in aqueous solution.

dihyrogen selenide gas — H_2Se (*g*)

iodine pentafluoride — IF_5

Key Concept: Naming binary molecular compounds

WP 2.9

Step 1: When naming the compounds, refer to Table 2.3 on page 62 in your book. (Eventually, you will need to know the names of these polyatomic ions by heart. Flashcards will certainly come in handy when learning these names.) Also, don't forget that you must indicate the charge on certain metal cations.

$Co(NO_3)_3$ — NO_3 is the nitrate ion and has a charge of –1. Since we have three nitrate ions, we have an overall charge of –3. Therefore, the charge on the cobalt must be +3. Cobalt is a transition metal, and we must include the charge on the metal when naming the compound. The name of the compound is cobalt(III) nitrate.

Appendix A – Workbook Problems Solutions

Fe$_2$(SO$_4$)$_3$ — SO$_4$ is the sulfate ion and has a charge of –2. Since we have three nitrate ions, we have an overall charge of –6. Therefore, the overall charge on the iron must be +6. The charge on each iron is determined from $\frac{+6}{2} = +3$. Iron is a transition metal, and we must include the charge on the metal when naming the compound. The name of the compound is iron(III) sulfate.

Step 2: When writing formulas, use the periodic table or the name of the compound to determine the charge on the metal. To determine the formula for a polyatomic ion, refer to Table 2.3 on page 62 in your book. Remember, you must maintain electrical neutrality.

sodium sulfite — The formula for sulfite is SO$_3^{2-}$. Sodium is a group 1A metal and has a charge of +1. To maintain electrical neutrality, we need to have two sodium cations for every one SO$_3^{2-}$ anion. The formula is Na$_2$SO$_3$.

barium hydroxide — The formula for hydroxide is OH$^-$. Barium is a group 2A metal and has a charge of +2. To maintain electrical neutrality, we need to have to hydroxide ions for every one barium ion. The formula is Ba(OH)$_2$.

ammonium hypochlorite — The formula for ammonium is NH$_4^+$. The formula for hypochlorite is ClO$^-$. The formula for the compound is NH$_4$ClO.

Key Concept: Naming compounds with polyatomic ions

WP 2.10

Strategy: First, determine the type of compound. Is it a simple binary ionic compound, a molecule, an ionic compound containing a polyatomic ion, or an acid (either binary acid or oxoacid)? Once you have determined the type of compound, apply the appropriate nomenclature rules.

MnS — This compound contains a metal and a nonmetal. Therefore, it is a simple binary ionic compound. Manganese is a transition metal, and therefore, we must indicate the charge on the cation. From the periodic table, we know that the sulfide ion has a –2 charge. There is only one sulfide ion. To maintain electrical neutrality, the manganese ion must have a +2 charge. The name of the compound is manganese(II) sulfide.

N$_2$O$_4$ — This compound contains two nonmetals and therefore, is a molecule. We use the numerical prefixes in Table 2.2 on page 61 in your textbook to indicate the number of atoms of each element present. The name of this compound is dinitrogen tetroxide.

GaCl$_3$ — This compound contains a metal and a nonmetal, making it a simple binary ionic compound. Gallium is a group 3A metal. Therefore, we do not need to indicate the charge on the metal cation. The name of the compound is gallium chloride.

hydrobromic acid — From the name of the compound, we know that it is a binary acid containing hydrogen and bromide. The bromide ion has a –2 charge. The formula for this compound is HBr (*aq*). (Remember, when writing formulas for binary hydrogen compounds, we must indicate whether the compound is a gas or in aqueous solution.)

sulfurous acid — This compound is an oxoacid containing the sulfite ion (SO$_3^{2-}$) and hydrogen. The charge on the sulfite ion is –2. To maintain electrical neutrality, we must have two hydrogen ions. The formula is H$_2$SO$_3$.

Appendix A – Workbook Problems Solutions

calcium carbonate	This compound contains a metal and a polyatomic anion. It is an ionic compound. Calcium is a group 2A metal and has a +2 charge. The carbonate anion, CO_3^{2-}, has a –2 charge. The formula for this compound is $CaCO_3$.
strontium nitrate	This compound contains a metal and a polyatomic anion. It is an ionic compound. Strontium is a group 2A metal and has a +2 charge. The nitrate anion, NO_3^-, has a –1 charge. To maintain electrical neutrality, we need two nitrates. The formula for this compound is $Sr(NO_3)_2$.
aluminum sulfide	This compound contains a metal and a nonmetal. It is a simple binary ionic compound. Aluminum is a group 3A metal with a +3 charge. The sulfide anion has a –2 charge. Let's use the number in the charge on aluminum (3) as the subscript for the sulfide anion and the number in the charge on the sulfide anion (2) as the subscript for aluminum. The formula will then be Al_2S_3. A check for electrical neutrality gives $[2\times(+3)]+[3\times(-2)]=0$.

All formulas for the ionic compounds have the smallest– whole number ratio of cation to anion.

Key Concept: Naming inorganic compounds

Chapter 3

WP 3.1:

Step 1: Write the unbalanced chemical equation. (Remember, the term *combustion reaction* is used to indicate reaction with oxygen. When hydrocarbons (compounds containing primarily C and H) undergo combustion reaction, carbon dioxide and water are produced.

$$C_8H_{18} + O_2 \rightarrow CO_2 + H_2O$$

Step 2: Use coefficients to balance the equation. (Remember, it helps to save oxygen for last.)

Begin with carbon. There are eight C's on the reactant side, but only one C on the product side. Place an 8 in front of CO_2.

$$C_8H_{18} + O_2 \rightarrow 8\,CO_2 + H_2O$$

There are 18 H's on the reactant side, but only two H's on the product side. Place a 9 in front of H_2O.

$$C_8H_{18} + O_2 \rightarrow 8\,CO_2 + 9\,H_2O$$

Now balance the oxygens. There are two oxygens on the reactant side and a total of 25 on the product side. For the time being, we will use a fraction (25/2) to balance the oxygens. (Remember, to multiply the coefficient by the subscript on oxygen. This will then give a total of 17 oxygens.)

$$C_8H_{18} + 25/2\,O_2 \rightarrow 8\,CO_2 + 9\,H_2O$$

Step 3: Reduce the coefficients to their smallest whole–number ratio.

The ratio is 1:25/2: 8: 9. This is **not** the smallest whole–number ratio. To achieve the smallest **whole**–number ratio, we need to multiply each coefficient by 2. This gives us the ratio of 2:17:8:9. The balanced equation then becomes

$2\ C_8H_{18}\ +\ 25\ O_2\ \rightarrow\ 16\ CO_2\ +\ 18\ H_2O$

Step 4: Check your answer.

Reactant side	Product Side
16 C	16 C
36 H	36 H
50 O	50 O

Key Concept: Balanced chemical equations (Law of Conservation of Mass)

WP 3.2
Part a
Step 1: Write an unbalanced chemical equation based on the information given in the problem.

$N_2H_4\ +\ N_2O_4\ \rightarrow\ N_2\ +\ H_2O$

Step 2: Use coefficients to balance the equation. (Remember to save oxygen for last.)

There are four N's on the reactant side and only two N's on the product side. Place a 2 in front of N_2.

$N_2H_4\ +\ N_2O_4\ \rightarrow\ 2\ N_2\ +\ H_2O$

There are four H's on the reactant side and only two H's on the product side. Place a 2 in front of the water.

$N_2H_4\ +\ N_2O_4\ \rightarrow\ 2\ N_2\ +\ 2\ H_2O$

There are four O's on the reactant side and only two O's on the product side. Change the 2 in front of the water to a 4.

$N_2H_4\ +\ N_2O_4\ \rightarrow\ 2\ N_2\ +\ 4\ H_2O$

The H's are once again unbalanced. We now need to place a 2 in front of the hydrazine.

$2\ N_2H_4\ +\ N_2O_4\ \rightarrow\ 2\ N_2\ +\ 4\ H_2O$

The O's and H's are now balanced. However, the N's are not. We now have six N's on the reactant side and only four N's on the product side. However, if we change the 2 in front of N_2 to a 3, the N's will be balanced.

$2\ N_2H_4\ +\ N_2O_4\ \rightarrow\ 3\ N_2\ +\ 4\ H_2O$

Step 3: Reduce the coefficients to their smallest whole number ratio if necessary.

The ratio, 2:1:3:4 already is the smallest whole number ratio.

Step 4: Check your answer.

Reactant side	Product side
6 N	6 N
8 H	8 H
4 O	4 O

Part b

Appendix A – Workbook Problems Solutions

Step 1: Determine the mole ratio for hydrazine and dinitrogen tetroxide from your balanced chemical equation.

There are 2 mol N_2H_4 for every 1 mol of N_2O_4.

Step 2: Use the mole ratio from above as a conversion factor, and calculate the number of moles of hydrazine needed to react with 0.25 mol of dinitrogen tetroxide.

We are starting with 0.25 mol N_2O_4, and we want to determine the number of moles of hydrazine. Therefore, the number of moles of N_2O_4 need to cancel out. To accomplish this, we need to put the number of moles of N_2O_4 on the bottom of our conversion factor. Remember, when using dimensional analysis, start with the information given and write your conversion factor(s) so that everything cancels out except the desired unit.

$$0.25 \text{ mol } N_2O_4 \times \frac{2 \text{ mol } N_2H_4}{1 \text{ mol } N_2O_4} = 0.50 \text{ mol } N_2H_4$$

Ballpark check: Your answer should be two times the amount you start with since you have a 2:1 mole ratio.

Part c

Step 1: Determine the mole ratio for dinitrogen tetroxide and nitrogen from your balanced chemical equation.

There are 3 mol of N_2 for every 1 mol of N_2O_4.

Step 2: Use the mole ratio from above as a conversion factor, and calculate the number of moles of nitrogen produced from 0.25 mol of dinitrogen tetroxide.

$$0.25 \text{ mol } N_2O_4 \times \frac{3 \text{ mol } N_2}{1 \text{ mol } N_2O_4} = 0.75 \text{ mol } N_2$$

Ballpark Check: Your answer should be three times the amount you start with since you have a 3:1 mole ratio.

Step 3: Determine the mole ratio for dinitrogen tetroxide and water from your balanced chemical equation.

There are 4 mol of water for every 1 mol of N_2O_4.

Step 4: Use the mole ratio from above as a conversion factor, and calculate the number of moles of water produced form 0.125 mol of dinitrogen tetroxide.

$$0.25 \text{ mol } N_2O_4 \times \frac{4 \text{ mol } H_2O}{1 \text{ mol } N_2O_4} = 1.0 \text{ mol } H_2O$$

Ballpark Check: Your answer should be four times the amount you start with since you have a 4:1 mole ratio.

Key concept: Formula units and moles (Law of Conservation of Mass)

WP 3.3

Appendix A – Workbook Problems Solutions

Step 1: Write an unbalanced chemical equation from the information given.

$$Ca_3(PO_4)_2 + H_2SO_4 \rightarrow H_3PO_4 + CaSO_4$$

Step 2: Balance the chemical equation, using the procedure you have learned.

$$Ca_3(PO_4)_2 + 3\,H_2SO_4 \rightarrow 2\,H_3PO_4 + 3\,CaSO_4$$

Step 3: From the balanced chemical equation, determine the mole ratio of phosphoric acid to calcium phosphate.

There are 3 mol of phosphoric acid for every 1 mol of calcium phosphate.

Step 4: Determine the formula mass for calcium phosphate.

$Ca_3(PO_4)_2$ formula mass = (3 × 40.1 g/mol) + (2 × 31.0 g/mol) + (8 × 16.0 g/mol) = 310.3 g/mol

Step 5: Determine the molecular mass for phosphoric acid.

H_3PO_4 molecular mass = (3 × 1.00 g/mol) + 31.0 g/mol = (4 × 16.0 g/mol) = 98.0 g/mol

Step 6: From the information in steps 3, 4, and 5, create conversion factors. Use these conversion factors and follow the flow diagram in Figure 3.2 on page 84 of your textbook so that you proceed from grams of calcium phosphate to moles of calcium phosphate to moles of phosphoric acid to grams of phosphoric acid.

$$6.75\ \text{g}\ Ca_3(PO_4)_2 \times \frac{1\ \text{mol}\ Ca_3(PO_4)_2}{310.3\ \text{g}\ Ca_3(PO_4)_2} \times \frac{2\ \text{mol}\ H_3PO_4}{1\ \text{mol}\ Ca_3(PO_4)_2} \times \frac{98.0\ \text{g}\ H_3PO_4}{1\ \text{mol}\ H_3PO_4} = 4.26\ \text{g}\ H_3PO_4$$

Ballpark Check: The molar mass of $Ca_3(PO_4)_2$ is approximately three times the molar mass of H_3PO_4. However, you have a mole ratio of 1 mol $Ca_3(PO_4)_2$:3 mol H_3PO_4. Therefore, you would expect the answer you get in grams of H_3PO_4 to be similar to the grams of $Ca_3(PO_4)_2$.

Key Concepts: Law of Conservation of Mass, stoichiometry.

WP 3.4

Step 1: Determine the theoretical amount of H_3PO_4 produced when starting with 13.5 g of $Ca_3(PO)_2$. (Follow the same procedure as in WP 3.3.)

$$Ca_3(PO_4)_2 + 3\,H_2SO_4 \rightarrow 2\,H_3PO_4 + 3\,CaSO_4$$

There are 3 mol of phosphoric acid for every 1 mol of calcium phosphate.

$Ca_3(PO_4)_2$ formula mass = (3 × 40.1 g/mol) + (2 × 31.0 g/mol) + (8 × 16.0 g/mol) = 310.3 g/mol

H_3PO_4 molecular mass = (3 × 1.00 g/mol) + 31.0 g/mol = (4 × 16.0 g/mol) = 98.0 g/mol

From the information in steps 3, 4, and 5, create conversion factors. Use these conversion factors and follow the flow diagram in Figure 3.2 on page 84 of your textbook so that you proceed from grams of calcium phosphate to moles of calcium phosphate to moles of phosphoric acid to grams of phosphoric acid.

Appendix A – Workbook Problems Solutions

$$13.5 \text{ g Ca}_3(\text{PO}_4)_2 \times \frac{1 \text{ mol Ca}_3(\text{PO}_4)_2}{310.3 \text{ g Ca}_3(\text{PO}_4)_2} \times \frac{2 \text{ mol H}_3\text{PO}_4}{1 \text{ mol Ca}_3(\text{PO}_4)_2} \times \frac{98.0 \text{ g H}_3\text{PO}_4}{1 \text{ mol H}_3\text{PO}_4} = 8.53 \text{ g H}_3\text{PO}_4$$

Step 2: Calculate the percent yield using the actual yield stated in the problem and the theoretical yield just calculated.

$$\text{Percent yield} = \frac{6.82 \text{ g H}_3\text{PO}_4}{8.53 \text{ g H}_3\text{PO}_4} \times 100\% = 80.0\%$$

Key concepts: Stoichiometry, percent yield

WP 3.5

Step 1: Using the information given, write an unbalanced chemical equation.

$$\text{Cr}_2\text{O}_3 + \text{Al} \rightarrow \text{Al}_2\text{O}_3 + \text{Cr}$$

Step 2: Balance the chemical equation.

$$\text{Cr}_2\text{O}_3 + 2 \text{ Al} \rightarrow \text{Al}_2\text{O}_3 + 2 \text{ Cr}$$

Step 3: Determine the mole ratio for chromium(III) oxide and aluminum.

There are 2 mol of Al for every 1 mol of Cr_2O_3.

Step 4: Determine the number of moles of each reactant present.

Formula mass of Cr_2O_3 = (2 × 52.0 g/mol) + (3 × 16.0 g/mol) = 152 g/mol Cr_2O_3

Atomic mass of Al = 27.0 g/mol

$$4.27 \text{ g Cr}_2\text{O}_3 \times \frac{1 \text{ mol Cr}_2\text{O}_3}{152 \text{ g Cr}_2\text{O}_3} = 0.0281 \text{ mol Cr}_2\text{O}_3$$

$$7.48 \text{ g Al} \times \frac{1 \text{ mol Al}}{27.0 \text{ g Al}} = 0.277 \text{ mol Al}$$

Step 5: Compare the mole ratio in step 3 to the mole ratio in step 4.

From step 3, we know that we need 2 mol of Al for every 1 mol of Cr_2O_3. If we start with 0.0218 mol Cr_2O_3, then we need 0.0562 mol Al.

Step 6: Compare the number of theoretical moles of aluminum needed to the actual number of moles of aluminum present. Are there enough moles of aluminum present to react? Is aluminum the limiting reactant or the excess reactant?

0.0562 mol Al are needed to react with 0.0281 mol Cr_2O_3. There are 0.277 mol Al present. There is more than enough Al present. Therefore, Al is the excess reactant, and Cr_2O_3 is the limiting reactant.

Step 7: Based upon the number of moles of excess reactant consumed and the number of moles of excess reactant present, calculate the number of moles and the number of grams of excess reactant left over.

mol Al left over = 0.277 mol Al present − 0.0562 mol Al needed for reaction = 0.221 mol Al left over

Appendix A – Workbook Problems Solutions

$$0.221 \text{ mol Al} \times \frac{27.0 \text{ g Al}}{1 \text{ mol Al}} = 5.97 \text{ g Al}$$

Step 8: Calculate the number of moles of chromium produced based on the number of moles of limiting reactant and the balanced chemical equation.

$$0.0281 \text{ mol Cr}_2\text{O}_3 \times \frac{2 \text{ mol Cr}}{1 \text{ mol Cr}_2\text{O}_3} \times \frac{52.0 \text{ g Cr}}{1 \text{ mol Cr}} = 2.92 \text{ g Cr}$$

Step 9: Determine the percent yield of the reaction.

$$\text{Percent yield} = \frac{1.89 \text{ g Cr}}{2.92 \text{ g Cr}} \times 100\% = 64.7\%$$

Key Concepts: Law of Conservation of Mass, stoichiometry, and limiting reactants.

WP 3.6

Step 1: First, determine the number of moles of Pb(NO$_3$)$_2$ found in 250 mL of a 0.10 M solution.

$$0.250 \text{ L} \times \frac{0.10 \text{ mol Pb(NO}_3)_2}{1 \text{ L}} = 0.025 \text{ mol Pb(NO}_3)_2$$

Step 2: Determine the number of grams of Pb(NO$_3$)$_2$, knowing the moles of Pb(NO$_3$)$_2$.

$$0.025 \text{ mol Pb(NO}_3)_2 \times \frac{331.2 \text{ g Pb(NO}_3)_2}{1 \text{ mol Pb(NO}_3)_2} = 8.3 \text{ g Pb(NO}_3)_2$$

Key Concept: Concentration (Molarity)

WP 3.7

Step 1: Rearrange the equation $M_i \times V_i = M_f \times V_f$ to solve for the initial volume.

$$V_i = \frac{M_f \times V_f}{M_i} = \frac{0.75 \frac{\text{mol HCl}}{\text{L HCl}} \times 0.500 \text{ L HCl}}{6.0 \frac{\text{mol HCl}}{\text{L HCl}}} = 0.0625 \text{ L HCl} = 62.5 \text{ L HCl}$$

Ballpark Check: You want to decrease the concentration by 1/8, $\left(\frac{0.75 \text{ M}}{6 \text{ M}}\right)$. Therefore, the volume should also decrease by 1/8.

Key Concepts: Solution concentration and dilution

WP 3.8

Step 1: Once again, we begin by writing a balanced chemical equation.

$$\text{Zn (s)} + 2 \text{ HCl (aq)} \rightarrow \text{ZnCl}_2 \text{ (aq)} + \text{H}_2 \text{ (g)}$$

Appendix A – Workbook Problems Solutions

Step 2: Determine the mole ratio of zinc to hydrochloric acid.

1 mol Zn:2 mol HCl

Step 3: Determine the number of moles of zinc present.

$$\text{mol Zn} = 8.75 \text{ g Zn} \times \frac{1 \text{ mol Zn}}{65.4 \text{ g Zn}} = 0.134 \text{ mol Zn}$$

Step 4: Determine the number of moles of HCl.

From the balanced equation, we know that there are 2 mol of HCl for every 1 mol Zn. Therefore, the number of moles of HCl is 0.268.

Step 5: Calculate the volume of 12 M HCl needed for this reaction.

$$\text{Volume HCl} = 0.268 \text{ mol HCl} \times \frac{1 \text{ L soln}}{12 \text{ mol HCl}} = 0.0223 \text{ L} = 22.3 \text{ mL soln}$$

Step 6: To calculate the volume of H_2 gas produced, we first need to calculate the grams of H_2 gas produced. This is easily accomplished by beginning with the moles of zinc present.

$$0.134 \text{ mol Zn} \times \frac{1 \text{ mol } H_2}{1 \text{ mol Zn}} \times \frac{2.00 \text{ g } H_2}{1 \text{ mol } H_2} = 0.268 \text{ g } H_2$$

Step 7: Determine the volume of H_2 gas produced, using the grams of H_2 produced and the density of H_2 gas.

$$\text{Volume of } H_2 = 0.268 \text{ g } H_2 \times \frac{1 \text{ L } H_2}{0.0899 \text{ g } H_2} = 2.98 \text{ L } H_2$$

Key Concepts: Law of Conservation of Mass, solution stoichiometry, understanding properties of matter (Chapter 1)

WP 3.9
Step 1: Write a balanced chemical equation for the reaction.

2 HNO_3 + $Mg(OH)_2$ → $Mg(NO_3)_2$ + 2 H_2O

Step 2: Determine the mole ratio of nitric acid to $Mg(OH)_2$.

There are 2 mol of HNO_3 for every 1 mole of $Mg(OH)_2$

Step 3: Calculate the number of moles of $Mg(OH)_2$ present.

$$0.0316 \text{ g Mg(OH)}_2 \times \frac{1 \text{ mol Mg(OH)}_2}{58.3 \text{ g Mg(OH)}_2} = 5.42 \times 10^{-4} \text{ mol Mg(OH)}_2$$

Step 4: Determine the number of moles of nitric acid that will react with the Mg(OH)$_2$.

$$5.42 \times 10^{-4} \text{ mol Mg(OH)}_2 \times \frac{2 \text{ mol HNO}_3}{1 \text{ mol Mg(OH)}_2} = 1.08 \times 10^{-3} \text{ mol HNO}_3$$

Step 5: Calculate the molarity of the nitric acid solution.

$$\frac{1.08 \times 10^{-3} \text{ mol HNO}_3}{0.04335 \text{ L HNO}_3} = 0.0250 \frac{\text{mol HNO}_3}{\text{L}}$$

Key Concepts: Law of Conservation of Mass, solution stoichiometry

WP 3.10

Step 1: Convert the grams of each element into moles of each element.

32.4 g Na, 0.70 g H, 21.8 g P, and 45.1 g O.

$$32.4 \text{ g Na} \times \frac{1 \text{ mol Na}}{23.0 \text{ g Na}} = 1.41 \text{ mol Na}$$

$$0.70 \text{ g H} \times \frac{1 \text{ mol H}}{1.00 \text{ g H}} = 0.70 \text{ mol H}$$

$$21.8 \text{ g P} \times \frac{1 \text{ mol P}}{31.0 \text{ g P}} = 0.703 \text{ mol P}$$

$$45.1 \text{ g O} \times \frac{1 \text{ mol O}}{16.0 \text{ g O}} = 2.82 \text{ mol O}$$

Step 2: Knowing the relative number of moles, find the ratio of moles by dividing the three larger numbers by the smallest number.

The smallest number is 0.70.

$$\frac{1.41}{0.70} = 2.01 \qquad \frac{0.703}{0.70} = 1.00 \qquad \frac{2.82}{0.70} = 4.02$$

Step 3: If necessary, multiply the subscript by a small integer to find whole numbers for the formula and write the formula of the compound.

The empirical formula is Na$_2$HPO$_4$.

Key Concepts: Law of Definite Proportions, empirical formula

WP 3.11

Step 1: Write the formula for aluminum bromate.

Appendix A – Workbook Problems Solutions

$Al(BrO_3)_3$

Step 2: Determine the mole ratio of the elements in the compound.

1 mol Al:3 mol Br:9 mol O

Step 3: Assume a 1 mole sample is present, and convert the mole ratio into a mass ratio.

$$1 \text{ mol Al} \times \frac{27.0 \text{ g Al}}{1 \text{ mol Al}} = 27.0 \text{ g Al}$$

$$3 \text{ mol Br} \times \frac{79.9 \text{ g Br}}{1 \text{ mol Br}} = 240 \text{ g Br}$$

$$9 \text{ mol O} \times \frac{16.0 \text{ g O}}{1 \text{ mol O}} = 144 \text{ g O}$$

Step 4: Determine the percent composition by dividing the mass of each element present by the total mass of the compound and multiplying by 100.

Formula mass of $Al(BrO_3)_3$ = 27.0 g/mol + (3 × 79.9 g/mol) + (9 × 16.0 g/mol) = 410.7 g/mol

$$\%Al = \frac{27.0 \text{ g Al}}{410.7 \text{ g}} \times 100\% = 6.57\%$$

$$\%Br = \frac{240 \text{ g Br}}{410.7 \text{ g}} \times 100\% = 58.4\%$$

$$\%O = \frac{144 \text{ g O}}{410.7 \text{ g}} \times 100\% = 35.1\%$$

(A good check is to add up all of the percentages. The sum should equal 100%.)

Key Concepts: Law of Definite Proportions, percent composition.

WP 3.12

Step 1: Find the molar amounts of C and H in CO_2 and H_2O.

$$7.29 \text{ g CO}_2 \times \frac{1 \text{ mol CO}_2}{44.0 \text{ g CO}_2} \times \frac{1 \text{ mol C}}{1 \text{ mol CO}_2} = 0.166 \text{ mol C}$$

$$2.98 \text{ g H}_2\text{O} \times \frac{1 \text{ mol H}_2\text{O}}{18.0 \text{ g H}_2\text{O}} \times \frac{2 \text{ mol H}}{1 \text{ mol H}_2\text{O}} = 0.331 \text{ mol H}$$

Step 2: Carry out mole–to–gram conversions to find the number of grams of C and H in the original sample.

$$0.166 \text{ mol C} \times \frac{12.0 \text{ g C}}{1 \text{ mol C}} = 1.99 \text{ g C}$$

$$0.331 \text{ mol H} \times \frac{1.00 \text{ g H}}{1 \text{ mol H}} = 0.331 \text{ g H}$$

Step 3: Subtract the masses of C and H form the mass of the starting sample to determine the mass of O.

2.85 g sample − 1.99 g C − 0.331 g H = 0.529 g O

Step 4: Convert the mass of O to moles of O.

$$0.529 \text{ g O} \times \frac{1 \text{ mol O}}{16.0 \text{ g O}} = 0.00.0331 \text{ mol O}$$

Step 5: Find the ratio of the number of moles by dividing the larger numbers of moles by the smallest number of moles.

$$\frac{0.166}{0.0331} = 5.01 \qquad \frac{0.331}{0.0331} = 10.0$$

Step 6: Use the ratio above to write the empirical formula of the compound.

$C_5H_{10}O$

Step 7: Determine the multiple for the empirical formula.

Empirical mass = (5× 12.0 g/mol) + (10 ×1.0 g/mol) + 16.0 g/mol = 86.0 g/mol

$$\text{Multiplier} = \frac{172.0 \text{ g/mol}}{86.0 \text{ g/mol}} = 2$$

Step 8: Write the molecular formula.

$C_{10}H_{20}O_2$

Key Concepts: Law of Definite Proportions and *Law of Conservation of Mass*

Chapter 4

WP 4.1
Strategy: $Ca_3(PO_3)_2$ is a strong electrolyte and, therefore, completely dissociates.

Step 1: Determine the total number of moles of ions formed when Ca_3P_2 completely dissociates in water.

Ca_3P_2 forms 3 moles of Ca^{2+} and 2 moles of P^{3-} when it is dissolved in water. Therefore, the total number of ions formed is 5.

Step 2: Create a conversion factor comparing the total number of moles of ions in solution to 1 mol of Ca_3P_2.

Appendix A – Workbook Problems Solutions

The conversion factor is $\dfrac{5 \text{ mol ions}}{1 \text{ mol Ca}_3\text{P}_2}$.

Step 3: Use the conversion factor to calculate the molar concentration of ions in solution.

$$\dfrac{0.275 \text{ mol Ca}_3\text{P}_2}{\text{L soln}} \times \dfrac{5 \text{ mol ions}}{1 \text{ mol Ca}_3\text{P}_2} = \dfrac{1.38 \text{ mol ions}}{\text{L soln}}$$

Key Concepts: Law of Conservation of Mass, strong electrolytes

WP 4.2

Strategy: From the information given, write a molecular equation and determine if any of the reactants or products are strong electrolytes.

Step 1: Write an unbalanced chemical equation.

$Na_2S\ (aq)\ +\ HCl\ (aq)\ \rightarrow\ NaCl\ (aq)\ +\ H_2S\ (g)$

Step 2: Balance the molecular equation.

$Na_2S\ (aq)\ +\ 2\ HCl\ (s)\ \rightarrow\ 2\ NaCl\ (aq)\ +\ H_2S\ (g)$

Step 3: Determine if any of the reactants or products are strong electrolytes. If so, write the strong electrolytes in terms of their free ions to obtain the ionic equation.

Na_2S, HCl, and NaCl are strong electrolytes.

$2\ Na^+\ (aq)\ +\ S^{2-}\ (aq)\ +\ 2\ H^+\ (aq)\ +\ 2\ Cl^-\ (aq)\ \rightarrow\ 2\ Na^+\ (aq)\ +\ 2\ Cl^-\ (aq)\ +\ H_2S\ (g)$

Step 4: Determine if any spectator ions are present, and write the net ionic equation.

$S^{2-}\ (aq)\ +\ 2\ H^+\ (aq)\ \rightarrow\ H_2S\ (g)$

Key Concept: Net ionic equations

WP 4.3

Strategy: Using the solubility guidelines and solution stoichiometry, determine the reactants to use in the preparation of $Ba_3(PO_4)_3$ and the amount of reactants needed.

Step 1: Determine the reactants that are soluble and will produce the insoluble $Ba_3(PO_4)_2$ and another soluble product.

From the solubility guidelines given in your textbook, we know that the group 1A cations and NH_4^+ are soluble. We also know that the halide anions, NO_3^-, ClO_4^-, $CH_3CO_2^-$, and SO_4^{2-}, with the exception of $BaSO_4$, Hg_2SO_4, and $PbSO_4$, are soluble. We need to use a soluble salt that contains the PO_4^{3-} and a soluble salt that contains the Ba^{2+} ion. Any one of the group 1A cations form a soluble salt with PO_4^{3-}. The most common ion, Na^+, will be used in this case. Ba^{2+} will form a soluble salt with the halide anions, NO_3^-, ClO_4^-, and $CH_3CO_2^-$. We will use $BaCl_2$.

Appendix A – Workbook Problems Solutions

Step 2: Write a balanced chemical equation.

$$Na_3PO_4 \ (aq) + BaCl_2 \ (aq) \rightarrow Ba_3(PO_4)_2 \ (s) + 3 \ NaCl \ (aq)$$

Step 3: Use solution stoichiometry to determine the molarity of each reactant needed (assume a 1 L solution of each reactant) knowing that you want to produce 2.5 g of $Ba_3(PO_4)_2$.

We can determine the number of moles of Na_3PO_4 and $BaCl_2$ from the amount of $Ba_3(PO_4)_2$ produced.

$$2.50 \text{ g } Ba_3(PO_4)_2 \times \frac{1 \text{ mol } Ba_3(PO_4)_2}{602 \text{ g } Ba_3(PO_4)_2} \times \frac{1 \text{ mol } Na_3PO_4}{1 \text{ mol } Ba_3(PO_4)_2} \times \frac{164 \text{ g } Na_3PO_4}{1 \text{ mol } Na_3PO_4} = 0.681 \text{ mol } Na_3PO_4$$

$$\frac{0.681 \text{ mol } Na_3PO_4}{1 \text{ L soln}} = 0.681 \text{ M } Na_3PO_4$$

$$2.50 \text{ g } Ba_3(PO_4)_2 \times \frac{1 \text{ mol } Ba_3(PO_4)_2}{602 \text{ g } Ba_3(PO_4)_2} \times \frac{1 \text{ mol } BaCl_2}{1 \text{ mol } Ba_3(PO_4)_2} \times \frac{208.23 \text{ g } BaCl_2}{1 \text{ mol } BaCl_2} = 0.865 \text{ mol } BaCl_2$$

$$\frac{0.865 \text{ mol } BaCl_2}{1 \text{ L soln}} = 0.865 \text{ M } BaCl_2$$

What key concept did you use? Solubility guidelines and stoichiometry (Law of Mass Conservation).

WP 4.4
Strategy: Identify the reactions as either a precipitation reaction or an acid–base neutralization reaction.

a. $Fe(NO_3)_3$ and Na_2S – neither compound is an acid or a base; therefore, this is a precipitation reaction.
b. NH_4Cl and $Hg_2(NO_3)_2$ – neither compound is an acid or a base; therefore, this is a precipitation reaction.
c. $Ba(OH)_2$ and H_2SO_4 – $Ba(OH)_2$ is a base, and H_2SO_4 is an acid; therefore, this is an acid–base neutralization reaction.

Step 1: Write the balanced molecular equation for each reaction.

If necessary, refer to the solubility rules in your textbook.

a. $2 \ Fe(NO_3)_3 \ (aq) + 3 \ Na_2S \ (aq) \rightarrow Fe_2S_3 \ (s) + 6 \ NaNO_3 \ (aq)$

b. $2 \ NH_4Cl \ (aq) + Hg_2(NO_3)_2 \ (aq) \rightarrow 2 \ NH_4NO_3 \ (aq) + Hg_2Cl_2 \ (s)$

c. $Ba(OH)_2 \ (aq) + H_2SO_4 \ (aq) \rightarrow BaSO_4 \ (s) + 2 \ H_2O \ (l)$

Step 2: Determine the presence of any strong electrolytes. Write the ionic equation, showing the strong electrolytes in terms of their free ions.

a. $Fe(NO_3)_3$, Na_2S and $NaNO_3$ are strong electrolytes.

$$2 \ Fe^{3+} \ (aq) + 6 \ NO_3^- \ (aq) + 6 \ Na^+ \ (aq) + 3 \ SO_4^{2-} \ (aq) \rightarrow Fe_2S_3 \ (s) + 6 \ Na^+ \ (aq) + 6 \ NO_3^- \ (aq)$$

Appendix A – Workbook Problems Solutions

b. NH_4Cl, $Hg_2(NO_3)_2$ and NH_4NO_3 are strong electrolytes.

$2\,NH_4^+\,(aq) + 2\,Cl^-\,(aq) + Hg_2^{2+}\,(aq) + 2\,NO_3^-\,(aq) \rightarrow 2\,NH_4^+\,(aq) + 2\,NO_3^-\,(aq) + Hg_2Cl_2\,(s)$

c. $Ba(OH)_2$ and H_2SO_4 are strong electrolytes.

$Ba^{2+}\,(aq) + 2\,OH^-\,(aq) + 2\,H^+\,(aq) + SO_4^{2-}\,(aq) \rightarrow BaSO_4\,(s) + H_2O\,(l)$

Step 3: Determine the presence of any spectator ions. Write the net ionic equation, dropping out any spectator ions that are present.

a. $2\,Fe^{2+}\,(aq) + 3\,S^{2-}\,(aq) \rightarrow Fe_2S_3\,(s)$

b. $Hg_2^{2+}\,(aq) + 2\,Cl^-\,(aq) \rightarrow Hg_2Cl_2\,(s)$

c. same as ionic equation.

Key Concepts: Solubilities, precipitation, and acid–base neutralization reactions

WP 4.5

Step 1: Identify the elements that have a fixed oxidation state, and assign that oxidation state to the element.

$KMnO_4$: elements with fixed oxidation states are K = +1 and O = –2.

HIO_4: elements with fixed oxidation states are H = +1 and O = –2.

Step 2: Determine the oxidation number of the remaining element, keeping in mind that the sum of all oxidation numbers must be equal to zero.

$KMnO_4$: (+1) + (Mn?) + 4(–2) = 0; (Mn?) = +7

HIO_4: (+1) + (I?) + 4(–2) = 0; (I?) = +7

Key Concept: Oxidation number

WP 4.6

Strategy: Determine the oxidation number of all species present.

SO_4^{2-}: Oxidation number of S = –6; oxidation number of O = –2. (Remember that the oxidation numbers of elements in an ion must add up to the charge on the ion.)

H^+: Oxidation number = +1

Al: Oxidation number = 0. (Remember that the oxidation number of an element in its elemental form = 0.)

Al^{3+}: Oxidation number = +3.

H_2SO_3: Oxidation number of H = +1; oxidation number of S = +4; oxidation number of O = −2.

H_2O: Oxidation number of H = +1; oxidation number of O = −2.

Step 1: Identify the species which have a change in oxidation number.

S: changes from a +6 (SO_4^{2-}) to a +4 (H_2SO_3).

Al: changes from 0 to +3.

Step 2: Based on the change in oxidation number, identify the species oxidized, the species reduced, and the oxidizing and reducing agents.

Species oxidized: Al (increase in oxidation number)

Species reduced: S (decrease in oxidation number)

Oxidizing agent: SO_4^{2-} (sulfur is present in the form of SO_4^{2-})

Reducing agent: Al

Key Concept: Oxidation–reduction

WP 4.7

Strategy: Remember that any element higher in the activity series will react with the ion of any element lower in the activity series. Also remember, metals above the H^+ ion in the activity series will displace the hydrogen ion from an acid to form H_2 gas.

Step 1: Predict the outcome of these reactions.

Sn (s) + NaCl (aq) → No reaction; Sn is below the Na^+ ion in the activity series. (Cl^- is a spectator ion).

Ca (s) + H_2S (aq) → CaS (s) + H_2 (g)

Fe (s) + $Pt(NO_3)_2$ (aq) → $Fe(NO_3)_2$ (aq) + Pt (s)

Key Concepts: Oxidation–reduction and the Activity Series

WP 4.8

Step 1: Write the unbalanced ionic equation.

SO_3^{2-} (aq) + CrO_4^{2-} (aq) → SO_4^{2-} (aq) + $Cr(OH)_3$ (s)

Step 2: Balance all atoms other than hydrogen and oxygen.

Same as unbalanced equation.

Appendix A – Workbook Problems Solutions

Step 3: Assign oxidation numbers to all atoms.

$$SO_3^{2-} + CrO_4^{2-} \rightarrow SO_4^{2-} + Cr(OH)_3$$

S in SO_3^{2-}: +4, −2; Cr in CrO_4^{2-}: +6, −2; S in SO_4^{2-}: +6, −2; Cr in $Cr(OH)_3$: +3, −2, +1

Step 4: Determine which atoms have changed oxidation number.

S: +4 → +6 (lost 2 e⁻); Cr: +6 → +3 (gained 3 e⁻)

Step 5: Determine the net increase and net decrease in oxidation number.

Net increase in oxidation number of oxidized atoms = 2; Net decrease in oxidation number of reduced atoms = 3.

Step 6: Multiply the species oxidized by the net decrease and the species reduced by the net increase.

$$3\ SO_3^{2-} + 2\ CrO_4^{2-} \rightarrow 3\ SO_4^{2-} + 2\ Cr(OH)_3$$

Step 7: Balance the oxygens by adding the appropriate number of H_2O's.

Reactant side of the equation has one less oxygen, so add 1 H_2O.

$$H_2O + 3\ SO_3^{2-} + 2\ CrO_4^{2-} \rightarrow 3\ SO_4^{2-} + 2\ Cr(OH)_3$$

Step 8: Balance the hydrogens by adding the appropriate number of H^+'s.

Reactant side of the equation has four less hydrogens, so add 4 H^+.

$$4\ H^+ + H_2O + 3\ SO_3^{2-} + 2\ CrO_4^{2-} \rightarrow 3\ SO_4^{2-} + 2\ Cr(OH)_3$$

Step 9: Make the solution basic by adding 1 OH^- for every H^+.

$$4\ OH^- + 4\ H^+ + H_2O + 3\ SO_3^{2-} + 2\ CrO_4^{2-} \rightarrow 3\ SO_4^{2-} + 2\ Cr(OH)_3 + 4\ OH^-$$

Step 10: If possible, combine OH^- and H^+ to form water.

The OH^- and H^+ on the reactant side of the equation can be combined to form H_2O.

$$4\ H_2O + H_2O + 3\ SO_3^{2-} + 2\ CrO_4^{2-} \rightarrow 3\ SO_4^{2-} + 2\ Cr(OH)_3 + 4\ OH^-$$

Step 11: If necessary, combine water molecules present on the same side of the equation and cancel water molecules on both sides of the equation.

Combining the water molecules on the reactant side gives

$$5\ H_2O + 3\ SO_3^{2-} + 2\ CrO_4^{2-} \rightarrow 3\ SO_4^{2-} + 2\ Cr(OH)_3 + 4\ OH^-$$

Step 12: Check your answer making sure atoms and charge are balanced.

Key Concept: Oxidation–reduction reactions, Law of Conservation of Mass, balancing redox equations

WP 4.9

Step 1: Write the unbalanced ionic equation:

$$CN^- + AsO_4^{3-} \rightarrow AsO_2^- + CNO^-$$

Step 2: Determine which species is being reduced and which species is being oxidized and write two unbalanced half reactions:

arsenate is being reduced, while cyanide is being oxidized.

$$CN^- \rightarrow CNO^-$$

$$AsO_4^{3-} \rightarrow AsO_2^-$$

Step 3: Balance each half reaction for atoms other than H and O.

same as above

Step 4: Add H_2O for any oxygens that are needed and H^+ for any hydrogens that are needed.

$$H_2O + CN^- \rightarrow CNO^- + 2 H^+$$

$$4 H^+ + AsO_4^{3-} \rightarrow AsO_2^- + 2H_2O$$

Step 5: Balance each reaction for charge

$$H_2O + CN^- \rightarrow CNO^- + 2 H^+ + 2 e^-$$

$$2 e^- + 4 H^+ + AsO_4^{3-} \rightarrow AsO_2^- + 2 H_2O$$

Step 6: Make the electron count the same in both reactions.

The oxidation process involves loss of two electrons, and the reduction process involves the gain of two electrons. Therefore, the electron count is the same in both reactions.

Step 7: Add the two half-reactions together, canceling anything that appears on both sides of the equation.

$$2 H^+ + AsO_4^{3-} + CN^- \rightarrow CNO^- + AsO_2^- + H_2O$$

Step 8: Make the solution basic by adding 1 OH^- for every H^+.

We need to add 1 OH^- for every H^+ to both sides of the equation.

$$2 OH^- + 2 H^+ + AsO_4^{3-} + CN^- \rightarrow CNO^- + AsO_2^- + H_2O + 2 OH^-$$

Step 9: If possible, combine OH^- and H^+ to form water.

The OH^- and H^+ on the reactant side of the equation can be combined to form H_2O.

Appendix A – Workbook Problems Solutions

$$2 H_2O + AsO_4^{3-} + CN^- \rightarrow CNO^- + AsO_2^- + H_2O + 2 OH^-$$

Step 10: If necessary, combine water molecules present on the same side of the equation and cancel water molecules on both sides of the equation.

Canceling the water on the product side gives

$$H_2O + AsO_4^{3-} + CN^- \rightarrow CNO^- + AsO_2^- + 2 OH^-$$

Step 11: Check your answer to make sure both atoms and charge are balanced.

Key Concept: Oxidation–reduction reactions, Law of Conservation of Mass, balancing redox equations

WP 4.10

Step 1: Determine the number of moles of I_3^- that reacted with the thiosulfate ion.

Calculate the number of moles of thiosulfate used in the titration.

$$\text{mol } S_2O_3^{2-} = 0.0500 \frac{\text{mol } S_2O_3^{2-}}{\text{L soln}} \times 0.04232 \text{ L soln} = 2.12 \times 10^{-3} \text{ mol } S_2O_3^{2-}$$

We know from the balanced equation that 2 mol of $S_2O_3^{2-}$ react with 1 mol of I_3^-. Therefore,

$$\text{mol } I_3^- = 2.12 \times 10^{-3} \text{ mol } S_2O_3^{2-} \times \frac{1 \text{ mol } I_3^-}{2 \text{ mol } S_2O_3^{2-}} = 1.06 \times 10^{-3} \text{ mol } I_3^-$$

Step 2: Determine the number of moles of OCl^- that reacted.

From the first chemical reaction given, we know that the mole ratio between I_3^- and OCl^- is 1:1. Therefore, 1.06×10^{-3} moles of OCl^- reacted.

Step 3: Determine the mass of NaOCl present.

$$\text{mass NaOCl} = 1.06 \times 10^{-3} \text{ mol OCl}^- \times \frac{1 \text{ mol NaOCl}}{1 \text{ mol OCl}^-} \times \frac{74.5 \text{ g NaOCl}}{1 \text{ mol NaOCl}} = 0.0790 \text{ g NaOCl}$$

Step 4: Determine the weight percent of NaOCl.

$$\% \text{ Weight} = \frac{0.0790 \text{ g NaOCl}}{1.500 \text{ g sample}} \times 100 = 5.26\% \text{ NaOCl}$$

Key Concept: Redox titration.

Appendix A – Workbook Problems Solutions

Chapter 5

WP 5.1

Step 1: Determine the value of m that will make λ the shortest.

λ is shortest when m is greatest. The largest value for m is ∞ (comparable to removing the electron from the atom). If $m = \infty$, the $1/m = 0$.

Step 2: Use the Balmer–Rydberg equation with $n = 4$ and solve for λ.

The Balmer–Rydberg equation becomes

$$\frac{1}{\lambda} = 1.097 \times 10^{-2} \text{ nm}^{-1} \left(\frac{1}{4^2}\right) = 6.856 \times 10^{-4} \text{ nm} \qquad \lambda = 1.459 \times 10^3 \text{ nm}$$

Step 3: Solve for λ using the Balmer–Rydberg equation and the values of m that make λ the longest.

Remember that $m > n$; therefore, $m > 4$. The smallest values for m if $m > 4$ is $m = 5$ and $m = 6$.

For $m = 5$:

$$\frac{1}{\lambda} = 1.097 \times 10^{-2} \text{ nm}^{-1} \left(\frac{1}{4^2} - \frac{1}{5^2}\right) = 2.468 \times 10^{-4} \text{ nm}^{-1} \qquad \lambda = 4.051 \times 10^3 \text{ nm}$$

For $m = 6$

$$\frac{1}{\lambda} = 1.097 \times 10^{-2} \text{ nm}^{-1} \left(\frac{1}{4^2} - \frac{1}{6^2}\right) = 3.809 \times 10^{-4} \text{ nm}^{-1} \qquad \lambda = 2.625 \times 10^3 \text{ nm}$$

Key Concept: Electromagnetic Radiation and Atomic Spectra

WP 5.2
Strategy: Use the equation for the energy of a photon. *Watch your units!*

$$E = \frac{(6.626 \times 10^{-34} \text{ J} \cdot \text{s})\left(3.00 \times 10^8 \, \frac{\text{m}}{\text{s}}\right)}{(4.051 \times 10^3 \text{ nm})\left(\frac{1 \text{ m}}{10^9 \text{ nm}}\right)} = 4.91 \times 10^{-20} \text{ J}$$

$$E = \frac{(6.626 \times 10^{-34} \text{ J} \cdot \text{s})\left(3.00 \times 10^8 \, \frac{\text{m}}{\text{s}}\right)}{(2.625 \times 10^3 \text{ nm})\left(\frac{1 \text{ m}}{10^9 \text{ nm}}\right)} = 7.57 \times 10^{-20} \text{ J}$$

Appendix A – Workbook Problems Solutions

$$E = \frac{(6.626 \times 10^{-34} \text{ J} \cdot \text{s})(3.00 \times 10^8 \frac{\text{m}}{\text{s}})}{(1.458 \times 10^3 \text{ nm})(\frac{1 \text{ m}}{10^9 \text{ nm}})} = 1.36 \times 10^{-19} \text{ J}$$

Key Concept: Particlelike properties of electromagnetic radiation

WP 5.3

Strategy: Determine the subshell associated with a value of $\ell = 3$.

When $\ell = 3$, the subshell designation is f.

Step 1: Identify the subshell with the value of n and the letter designation for $\ell = 3$.

The subshell is $4f$.

Key Concept: Wave mechanics and quantum numbers.

WP 5.4

Step 1: Determine the number of electrons in vanadium.

For V, $Z = 23$.

Step 2: Use the Aufbau principle to determine the ground–state electronic configuration.

$1s^2 2s^2 2p^6 3s^2 3p^6 4s^2 3d^3$

Step 3: Determine the noble gas in the previous row. Specify only those electrons in the unfilled subshells.

$[\text{Ar}]4s^2 3d^3$

Step 4: Draw the orbital–filling diagram.

[Ar] ↑↓   ↑ ↑ ↑ __ __
     4s   3d

Key Concept: Pauli Exclusion Principle, Hund's Rule, and the Aufbau Principle

WP 5.5

Strategy: Determine if the elements are s–block, p–block, or f–block elements.

Group 7A elements are p–block elements, and group 4B elements are d–block elements.

Step 1: Determine if other outer subshells need to be taken into consideration.

Appendix A – Workbook Problems Solutions

For both group 7A and group 4B elements, the ns subshell is filled.

Step 2: Determine the number of electrons in the outer block.

For group 7A, there are 5 np electrons. For group 4B, there are 2 $(n-1)d$ electrons.

The general electron configuration for group 7A is ns^2np^5. The general electron configuration for group 4B is $ns^2(n-1)d^2$.

Key Concept: Electron configuration and the periodic table.

Chapter 6

WP 6.1

Strategy Step: Consider the ground–state electron configuration for each element, and compare the Z_{eff} felt by the outer electrons.

  a. Na: [Ne] $3s^1$; Cs: [Kr] $6s^1$

  b. P: [Ne] $3s^23p^3$; S: [Ne] $3s^23p^4$

  c. B^{2+} [He] $2s^1$; C^{2+} [He] $2s^2$

Step 1: Determine the higher E_i based on considerations in the strategy.

  a. Na has the higher ionization energy. The $6s^1$ in Cs is farther away from the nucleus and feels a much weaker Z_{eff} than the $3s^1$ in Na.

  b. P has the highest ionization energy. The outer electron configuration for P shows a half–filled $3p$ subshell. This half–filled subshell has a greater stability; therefore, P is less likely to give up an electron.

  c. C^{2+} has the higher ionization energy. The $2s^2$ subshell is a filled subshell and, therefore, very stable. If B^{2+} loses an electron, it too will have a filled subshell.

Key Concepts: Ionization energies and electron configurations

WP 6.2

Step 1: Add the E_{i1} and E_{i2} for Mg to twice the E_{ea} for chlorine. Is the answer positive or negative?

$E = 738$ kJ/mol $+ 1451$ kJ/mol $+ 2(-348.6$ kJ/mol$) = 1492$ kJ/mol
(Note, the E_{ea} is multiplied by 2 since there are two Cl^- ions in $MgCl_2$.) This reaction is unfavorable.

Step 2: Construct a Born–Haber cycle for the formation of $MgCl_2$.

Mg $(s) \rightarrow$ Mg (g)	149.6 kJ/mol
$Cl_2 (g) \rightarrow 2$ Cl (g)	243.0 kJ/mol
Mg $(g) \rightarrow Mg^+(g) + e^-$	738.0 kJ/mol
$Mg^+ (g) \rightarrow Mg^{2+} (g) + e^-$	1451.0 kJ/mol
2 Cl $(g) + 2 e^- \rightarrow 2Cl^- (g)$	-697.2 kJ/mol
$Mg^{2+}(g) + 2Cl^- (g) \rightarrow MgCl_2 (s)$	-2526.0 kJ/mol

Appendix A – Workbook Problems Solutions

E = 149.6 kJ/mol + 243.0 kJ/mol + 738.0 kJ/mol + 1451.0 kJ/mol + (–697.2 kJ/mol) + (–2526 kJ/mol)
= –641.6 kJ/mol

Key Concept: The overall net energy process for the formation of an ionic compound is the sum of the individual steps. (You will learn in Chapter 8 that this is simply a statement of Hess's Law.)

WP 6.3

Strategy: Review the chemistry described on pages 104 – 107 of this book, and determine the outcome for each reaction.

a. $2 \text{ Na } (s) + 2 \text{ NH}_3 (l) \rightarrow 2 \text{ Na}^+ (soln) + \text{NH}_2^- (soln) + \text{H}_2 (g)$

b. $\text{Ca } (s) + 2 \text{ H}_2\text{O } (l) \rightarrow 2 \text{ M}^{2+} (aq) + 2 \text{ OH}^- (aq) + 2 \text{ H}_2 (g)$

c. $2 \text{ Al } (s) + 6 \text{ HBr } (g) \rightarrow 2 \text{ AlBr}_3 (s) + 3 \text{ H}_2 (g)$

d. $\text{Sr } (s) + \text{F}_2 (g) \rightarrow \text{SrF}_2 (s)$

e. $\text{Br}_2 (l) + \text{F}_2 (g) \rightarrow 2 \text{ BrF}$

Key Concepts: Main–group chemistry, redox reactions.

Chapter 7

WP 7.1

Step 1: Determine the total number of valence–shell electrons.

7 (from Br) + 21 (from 3 F) = 28

Step 2: Determine the connections.

In this case, Br is the central atom. (This leads to the most symmetrical structure.)

Step 3: Draw the bonds, and subtract the number of electrons used from the total number of valence electrons available.

$$\begin{array}{c} \text{F} \quad \quad \text{F} \\ \diagdown \quad \diagup \\ \text{Br} \\ | \\ \text{F} \end{array}$$

The three bonds use two electrons each for a total of six. That leaves 22 electrons.

Step 4: Complete the octets of the outer atoms.

$$\begin{array}{c} :\ddot{\text{F}} \quad \quad \ddot{\text{F}}: \\ \diagdown \quad \diagup \\ \text{Br} \\ | \\ :\ddot{\text{F}}: \end{array}$$

Step 5: Place any remaining electrons on the central atom.

Each F used six electrons to complete its octet for a total of 18 electrons. That leaves four electrons remaining.

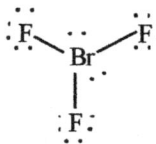

Key Concepts: Lewis theory and electron dot structures

WP 7.2
Step 1: Determine the formal charges (f.c.) for all atoms in each of the compounds.

Structure a
$$\text{f.c. (S)} = (6) - \tfrac{1}{2}(4) - 4 = 0$$
$$\text{f.c. (C)} = (4) - \tfrac{1}{2}(8) = 0$$
$$\text{f.c. (N)} = 5 - \tfrac{1}{2}(4) - 4 = -1$$

Structure b
$$\text{f.c. (S)} = 6 - \tfrac{1}{2}(6) - 2 = +1$$
$$\text{f.c. (C)} = 4 - \tfrac{1}{2}(8) = 0$$
$$\text{f.c. (N)} = 5 - \tfrac{1}{2}(2) - 6 = -2$$

Structure c
$$\text{f.c. (S)} = 6 - \tfrac{1}{2}(2) - 6 = -1$$
$$\text{f.c. (C)} = 4 - \tfrac{1}{2}(8) = 0$$
$$\text{f.c. (N)} = 5 - \tfrac{1}{2}(6) - 2 = 0$$

Step 2: Based on the formal charges of all of the elements in each compound as well as the electronegativities of the elements, determine the most likely structure.

The most likely structure is structure a. The electronegativities for structure a have the most electronegative element, N, with a –1 charge, while both C and S are neutral. In structure b, N has a –2 charge and S has a +1 charge. This structure can be ruled out because of the higher charge on nitrogen and greater charge separation. Structure c has the negative charge on S, which is a less electronegative element compared to N.

Key Concepts: Lewis theory and formal charges

WP 7.3
Strategy: Draw the electron–dot structure for each of the molecules. You will find the procedure for doing this in the example on pages 256 and 257 of your textbook and in WP 7.1 The electron– dot structures are

Appendix A – Workbook Problems Solutions

Step 1: Determine the number of charge clouds around the central atom. How many are used for bonding and how many are nonbonding?

For IF_5, there are six charge clouds around the iodine. Five clouds are used for bonding, and one cloud contains a lone pair of electrons.

For BrF_3, there are five charge clouds around bromine. Three clouds are used for bonding, and two clouds contain lone pairs of electrons.

Step 2: Based on your answer in step 2, determine the shape. Refer to Table 7.4 on pages 268–269 in your textbook if necessary.

IF_5: square pyramidal BrF_3: T–shaped

Key Concept: VSEPR theory

Chapter 8

WP 8.1
Step 1: Determine the ΔH for the reaction, taking into consideration that 1.81 moles of Si reacted.

$$-687 \frac{kJ}{mol\ Si} \times 1.81\ mol\ Si = -1.24 \times 10^3\ kJ$$

Step 2: Determine ΔE from the equation: $\Delta H = \Delta E + P\Delta V$. (Watch your units!)

$$P\Delta V = (0.5\ atm) \times (-22.4\ L) = -11.2\ atm \cdot L$$

$$-11.2\ atm \cdot L \times \frac{101,325\ Pa}{1\ atm} \times \frac{1 \frac{kg}{m \cdot s^2}}{1\ Pa} \times \frac{10^{-3}\ m^3}{1\ L} = -1.13 \times 10^3\ \frac{kg \cdot m^2}{s} = 1.13 \times 10^3\ J = -1.13\ kJ$$

$$\Delta E = (-1.24 \times 10^3\ kJ) - (-1.13\ kJ) = -1.23 \times 10^3\ kJ$$

Key Concept: Heat transfer at constant pressure: enthalpy

WP 8.2
Strategy: For an exothermic reaction, the amount of heat released by the reaction equals the amount of heat gained by calorimeter plus the amount of heat gained by solution. Determine the amount of heat gained by the solution by first calculating the mass of the solution.

$$100\ mL\ soln \times \frac{1.02\ g\ soln}{1\ mL\ soln} = 102\ g\ soln$$

$$102\ g\ soln \times 4.184 \frac{J}{g \cdot °C} \times (26.65°C - 23.35°C) = 1410\ J$$

Step 1: Determine the amount of heat gained by the calorimeter.

The amount of heat gained by the calorimeter is

$$25.0 \frac{J}{°C} \times (26.65°C - 23.35°C) = 82.5 J$$

Step 2: Determine the amount of heat released by the reaction.

The amount of heat released by the reaction is

1410 J + 82.5 J = 1492.5 J

This value should be reported as −1492.5 J since the reaction is exothermic.

Key Concepts: Calorimetry and specific heat.

WP 8.3
Strategy: Determine $\Delta H°$ for this reaction from the standard heats of formation.

Step 1: Determine the $\Delta H°$ using the $\Delta H_f°$ values in appendix B of your textbook.

	$\Delta H_f°$
PbS (s)	−100 kJ/mol
PbO (s)	−217.3 kJ/mol
SO$_2$ (g)	−296.8 kJ/mol

$$\Delta H = [(2 \text{ mol SO}_2 \times -296.8 \text{ kJ/mol SO}_2) + (2 \text{ mol PbO} \times -217.3 \text{ kJ/mol PbO})] - [(2 \text{ mol PbS} \times -100 \text{ kJ/mol PbS})]$$

$$\Delta H = -828 \text{ kJ}$$

Step 2: Knowing the ΔH for the reaction of 2 mol of PbS, calculate the ΔH for 2.50 g of PbS. (First convert grams of PbS to mol of Pbs.)

$$2.50 \text{ g PbS} \times \frac{1 \text{ mol PbS}}{239.2 \text{ g PbS}} = 1.05 \times 10^{-2} \text{ mol PbS}$$

$$\frac{-828 \text{ kJ}}{2 \text{ mol PbS}} \times 1.05 \times 10^{-2} \text{ mol PbS} = -4.35 \text{ kJ}$$

Key Concepts: Enthalpies of chemical change and heats of formation.

WP 8.4
1) 4 NH$_3$ (g) + 5 O$_2$ (g) → 4 NO (g) + 6 H$_2$O (g)
2) 2 NO (g) + O$_2$ (g) → 2 NO$_2$ (g)
3) 3 NO$_2$ (g) + H$_2$O (l) → 2HNO$_3$ (aq) + NO (g)

mol PbS

Step 1: For each step in the reaction, $\Delta H°$ can be calculated from $\Delta H_f°$ the values listed in Appendix B.

First reaction

	$\Delta H_f°$
NH$_3$ (g)	−46.1 kJ/mol

Appendix A – Workbook Problems Solutions

NO (g)   90.2 kJ/mol
H_2O (g)   −241.8 kJ/mol

$$\Delta H° = [(4 \text{ mol NO} \times 90.2 \text{ kJ/mol NO}) + (6 \text{ mol H}_2\text{O} \times -241.8 \text{ kJ/mol H}_2\text{O})] - [(4 \text{ mol NH}_3 \times -46.1 \text{ kJ/mol NH}_3)]$$
$$\Delta H° = -905.6 \text{ kJ}$$

Second reaction:

$\Delta H_f°$

NO (g)   90.2 kJ/mol
NO_2 (g)   33.2 kJ/mol

$$\Delta H° = [(2 \text{ mol NO}_2 \times 33.2 \text{ kJ/mol NO}_2) - (2 \text{ mol NO} \times 90.2 \text{ kJ/mol NO})] = -114 \text{ kJ}$$

Third reaction

$\Delta H_f°$

NO_2 (g)   33.2 kJ/mol
HNO_3 (aq)   −207.4 kJ/mol
H_2O (l)   −285.8 kJ/mol
NO (g)   90.2 kJ/mol

$$\Delta H° = [(2 \text{ mol HNO}_3 \times -207.4 \text{ kJ/mol HNO}_3) + (1 \text{ mol NO} \times 90.2 \text{ kJ/mol NO})] -$$
$$[(3 \text{ mol NO}_2 \times 33.2 \text{ kJ/mol NO}_2) + (1 \text{ mol H}_2\text{O} \times -285.8 \text{ kJ/mol H}_2\text{O})] = -138.4 \text{ kJ}$$

Step 2: Combine the individual reactions so that their sums will be the desired reaction.

4 NH_3 (g) + 2 H_2O (l) + 8 O_2 (g) → 4 HNO_3 (aq) + 6 H_2O (g)

We want to combine the reactions so that 4 mol of NH_3, 8 mol of O_2 and 2 mol H_2O appear on the reactant side and 4 mol of HNO_3 and 6 mol of H_2O (g) appear on the product side.

The first step in the process has 4 mol of NH_3 on the reactant side, so we will use that equation as is. The third step in the process has 2 mol of HNO_3 on the product side. Since we need 4 moles of HNO_3, we should multiply the coefficients in this equation, as well as the value for $\Delta H°$, by 2. Multiplying the third step in the process by 2 requires that we now multiply the second step in the process by 3 so that the NO_2 and NO will cancel out. Again, be sure to multiply the value of $\Delta H°$ by 3 as well.

4 NH_3 (g) + 5 O_2 (g) → 4~~NO~~ (g) + 6 H_2O (g) $\Delta H° = -905.6$ kJ

6~~NO_2~~ (g) + 2 H_2O (l) → 4 HNO_3 (aq) + 2~~NO~~ (g) $\Delta H° = -276.8$ kJ

6~~NO~~ (g) + 3 O_2 (g) → 6 NO_2 (g) $\Delta H° = -342$ kJ

---

4 NH_3 (g) + 2 H_2O (l) + 8 O_2 (g) → 4 HNO_3 (aq) + 6 H_2O (g) $\Delta H° = -1.524 \times 10^3$ kJ

Key Concepts: Hess's Law and heats of formation.

Appendix A – Workbook Problems Solutions

Chapter 9

WP 9.1

Strategy: Rearrange the ideal gas law to isolate the constants (the number of moles of gas and R).

$$nR = \left(\frac{PV}{T}\right)$$

Since nR remains constant that means that the relationship between the final temperature, pressure, and volume is the same as the relationship between the initial temperature, pressure, and volume. Write an equation showing this relationship.

$$nR = \left(\frac{PV}{T}\right)_{initial} = \left(\frac{PV}{T}\right)_{final}$$

Step 1: Rearrange the above equation to solve for the final volume.

$$V_{final} = \frac{T_{final} \times (PV)_{initial}}{T_{initial} \times P_{final}}$$

$$V_{final} = \frac{291.8 \text{ K} \times 754 \text{ mm Hg} \times 125 \text{ mL}}{296.6 \text{ K} \times 725 \text{ mm Hg}} = 128 \text{ mL}$$

Key Concept: Ideal Gas Law.

WP 9.2

Strategy: Remember that the reaction of a hydrocarbon with oxygen (combustion) produces carbon dioxide and water. With that in mind, write a balanced equation for the reaction.

$$C_3H_8 \text{ (g)} + 5\, O_2 \text{ (g)} \rightarrow 3\, CO_2 \text{ (g)} + 4\, H_2O \text{ (g)}$$

Calculate the number of moles of propane.

$$15.75 \text{ g } C_3H_8 \times \frac{1 \text{ mol } C_3H_8}{44.0 \text{ g } C_3H_8} = 0.358 \text{ mol } C_3H_8$$

Step 1: Using the balanced chemical equation, determine the number of moles of oxygen needed to react with the propane.

$$0.358 \text{ mol } C_3H_8 \times \frac{5 \text{ mol } O_2}{1 \text{ mol } C_3H_8} = 1.79 \text{ mol } O_2$$

Step 2: Rearrange the ideal gas law to determine the volume of oxygen needed.

$$V = \frac{nRT}{P}$$

$$V = \frac{1.79 \text{ mol} \times 0.08206 \frac{\text{L} \cdot \text{atm}}{\text{mol} \cdot \text{K}} \times 673.2 \text{ K}}{3.75 \text{ atm}} = 26.4 \text{ L}$$

Appendix A – Workbook Problems Solutions

Key Concepts: Stoichiometry and the Ideal Gas Law.

WP 9.3

Strategy: To begin, you need to think about the information you've been given and the information you're trying to find. You have the mass of the gas. You need the density of the gas. Determine what is needed to solve for the density.

To solve for density, you need to know the volume that the gas occupies. (Density = mass/volume)

Step 1: Use the ideal gas law to calculate the volume of the gas.

You can use the ideal gas law to solve for volume if you first convert grams of NO to moles of NO.

$$0.275 \text{ g NO} \times \frac{1 \text{ mol NO}}{30.0 \text{ g NO}} = 0.009\ 17 \text{ mol NO}$$

$$V = \frac{0.009\ 17 \text{ mol NO} \times 0.082\ 06 \frac{\text{L} \cdot \text{atm}}{\text{mol} \cdot \text{K}} \times 296.6 \text{ K}}{0.997 \text{ atm}} = 0.224 \text{ L}$$

Step 2: Determine the density of the gas.

$$d = \frac{0.275 \text{ g NO}}{0.224 \text{ L NO}} = 1.23 \text{ g/L}$$

Key Concepts: Stoichiometry and the Ideal Gas Law.

WP 9.4

Strategy: To solve this problem with the information given, you need to do a little thinking. First, write a mathematical equation for molar mass.

molar mass is the number of grams divided by the number of moles, n.

$$\text{molar mass} = \frac{g}{n}$$

Rearrange the above equation so that you are solving for n.

$$n = \frac{g}{\text{molar mass}}$$

Rearrange the ideal gas law so that you are solving for n.

$$n = \frac{PV}{RT}$$

Equate the last two equations.

Appendix A – Workbook Problems Solutions

$$\frac{g}{\text{Molar mass}} = \frac{PV}{RT}$$

Step 1: Rearrange the equation so that we are solving for the molar mass of the gas. (Keep in mind that density is grams/volume, $\frac{g}{V}$).

$$\text{Molar mass} = g \times \frac{RT}{PV}.$$

You can regroup the variables in this equation so that the density of a gas can be used.

$$\text{Molar mass} = \frac{g}{V} \times \frac{RT}{P} \quad \text{or} \quad \text{Molar mass} = d \times \frac{RT}{P}$$

Step 2: Solve for the molar mass of the gas.

$$\text{Molar mass} = 3.79 \frac{g}{L} \times \frac{0.082\,06 \frac{L \cdot atm}{mol \cdot K} \times 318.2\,K}{2.25\,atm} = 44.0 \frac{g}{mol}$$

Key Concepts: Stoichiometry and the Ideal Gas Law (plus a little bit of algebra and brain power)

WP 9.5

Step 1: Determine the partial pressure for each gas, using the ideal gas law.

You need to solve for the number of moles of each gas to use the Ideal Gas Law to solve for pressure.

$$\text{mol } H_2 = 13.9\,g\,H_2 \times \frac{1\,mol\,H_2}{2.0\,g\,H_2} = 6.95\,mol\,H_2$$

$$P = \frac{nRT}{V} = \frac{(6.95\,mol\,H_2)\left(0.0821 \frac{L \cdot atm}{mol \cdot K}\right)(423\,K)}{25.0\,L} = 9.65\,atm$$

$$\text{mol } N_2 = 64.8\,g\,N_2 \times \frac{1\,mol\,N_2}{28\,g\,N_2} = 2.31\,mol\,N_2$$

$$P = \frac{(2.31\,mol\,N_2)\left(0.0821 \frac{L \cdot atm}{mol \cdot K}\right)(423\,K)}{25.0\,L} = 3.21\,atm$$

$$\text{mol } NH_3 = 78.7\,g\,NH_3 \times \frac{1\,mol\,NH_3}{17\,g\,NH_3} = 4.63\,mol\,NH_3$$

$$P = \frac{(4.63\,mol\,NH_3)\left(0.0821 \frac{L \cdot atm}{mol \cdot K}\right)(423\,K)}{25.0\,L} = 6.43\,atm$$

Appendix A – Workbook Problems Solutions

Step 2: Sum the partial pressures to determine the total pressure.

P_{total} = 9.65 atm + 3.21 atm + 6.43 atm = 19.29 atm

Key Concept: Dalton's Law of Partial Pressures

Chapter 10

WP 10.1

Step 1: First, determine if any of the molecules listed are capable of hydrogen bonding.

$(CH_3)_2NH$ has a hydrogen bonded to nitrogen, which allows this molecule to undergo hydrogen bonding.

Step 2: Determine if any of the molecules listed are capable of ion–dipole or dipole–dipole interactions.

Ion–dipole forces can be ruled out, since neither of the two remaining molecules are ions. To determine the presence of dipole–dipole forces, you need to determine if a molecular dipole is present. This can be accomplished by first drawing the electron–dot structure for the molecules.

PCl_3 has a lone pair of electrons, which contributes substantially to a net molecular polarity. Therefore, this molecule is polar. PCl_5 has polar bonds that are symmetrical and, therefore, cancel each other out. This molecule is nonpolar.

Step 3: If any of the molecules are incapable of hydrogen bonding, ion–dipole interactions or dipole–dipole interactions, which type of intermolecular force will be present?

The intermolecular forces present in PCl_5 are London dispersion forces.

Step 4: Order the molecules from the weakest intermolecular force to the strongest intermolecular force.

$PCl_5 < PCl_3 < (CH_3)_2NH$

Key Concepts: Intermolecular forces and molecular polarity

WP 10.2

Strategy: Rearrange the equation $\Delta G = \Delta H - T\Delta S$ to solve for the temperature.

Remember, at a phase change the two phases (in this case, solid and liquid) are in equilibrium. Therefore, $\Delta G = 0$ and $\Delta H = T\Delta S$. To solve for temperature

Appendix A – Workbook Problems Solutions

$$T = \frac{\Delta H}{\Delta S} = \frac{5.94 \frac{\text{kJ}}{\text{mol}}}{\left(64.4 \frac{\text{J}}{\text{mol} \cdot \text{K}}\right)\left(\frac{1 \text{kJ}}{1000 \text{J}}\right)} = 92.2 \text{ K}$$

Key Concepts: Free energy change associated with phase changes

Chapter 11

WP 11.1

Step 1: You first need to calculate the moles of glucose.

The molecular mass of glucose is 180.2 g/mol.

$$17.84 \text{ g } C_6H_{12}O_6 \times \frac{1 \text{ mol } C_6H_{12}O_6}{180.2 \text{ g } C_6H_{12}O_6} = 0.099\ 00 \text{ mol } C_6H_{12}O_6$$

Step 2: Calculate the molality of the solution.

Molality is the number of moles of solute per kg of solvent.

$$m = \frac{0.099\ 00 \text{ mol } C_6H_{12}O_6}{0.250 \text{ kg H}_2\text{O}} = 0.396 \text{ m}$$

Step 3: Determine the mass percent.

To determine mass percent, you need to know the total mass of the solution.

Mass of solution = Mass of solute + Mass of solvent

Mass of solution = 17.84 g + 250 g = 268 g

$$\text{Mass \%} = \frac{17.84 \text{ g}}{268 \text{ g}} \times 100\% = 6.66\%$$

Step 4: Determine the molarity of the solution. (You need to use the density of the solution in this calculation.)

To determine the molarity of the solution, you need to know the volume of the solution. This can be calculated from the grams of solution and the density.

$$\text{Volume} = \frac{268 \text{ g}}{1.16 \frac{\text{g}}{\text{mL}}} = 231 \text{ mL} \qquad M = \frac{0.099\ 00 \text{ mol } C_6H_{12}O_6}{0.231 \text{ L soln}} = 0.428 \text{ M}$$

Key Concepts: Solutions and concentration units

WP 11.2

Step 1: If you have 100 g of solution, then you have 0.75 g of $CaCl_2$ and 99.25 g of H_2O. You first need to calculate the number of moles of $CaCl_2$ present.

Appendix A – Workbook Problems Solutions

$$\text{mol CaCl}_2 = 0.75 \text{ g CaCl}_2 \times \frac{1 \text{ mol CaCl}_2}{111.0 \text{ g CaCl}_2} = 6.8 \times 10^{-3} \text{ mol CaCl}_2$$

Step 2: You can calculate the molality of CaCl$_2$ from the moles of CaCl$_2$ and the mass of water. (Remember to use the total number of moles of ions.)

$$\text{mol ions} = 6.8 \times 10^{-3} \text{ mol CaCl}_2 \times \frac{1 \text{ mol Ca}^{2+}}{1 \text{ mol CaCl}_2} + 6.8 \times 10^{-3} \text{ mol CaCl}_2 \times \frac{2 \text{ mol Cl}^-}{1 \text{ mol CaCl}_2} = 2.0 \times 10^{-2} \text{ mol ions}$$

$$m = \frac{2.0 \times 10^{-2} \text{ mol ions}}{0.9925 \text{ kg H}_2\text{O}} = 2.0 \times 10^{-2} \, m$$

Step 3: Repeat steps 1 and 2 for citric acid.
If you have 100 g of solution, you have 0.75 g of citric acid and 99.25 g of water.

$$\text{mol C}_6\text{H}_8\text{O}_7 = 0.75 \text{ g C}_6\text{H}_8\text{O}_7 \times \frac{1 \text{ mol C}_6\text{H}_8\text{O}_7}{192 \text{ g C}_6\text{H}_8\text{O}_7} = 3.9 \times 10^{-3} \text{ mol C}_6\text{H}_8\text{O}_7$$

$$m = \frac{3.9 \times 10^{-3} \text{ mol C}_6\text{H}_8\text{O}_7}{0.9925 \text{ kg H}_2\text{O}} = 3.9 \times 10^{-3} \, m$$

Step 4: Compare the molality values for CaCl$_2$ and citric acid.

The CaCl$_2$ solution has the greater molality.

Key Concepts: Solutions and concentration units

WP 11.3

Strategy: Solve for X_{solv}, using Raoult's Law.

This solution consists of a nonvolatile solute in a volatile solvent. The form of Raoult's law that we will use is $P_{\text{soln}} = P_{\text{solv}} \times X_{\text{solv}}$. We are given both P_{soln} and P_{solv}. We can solve for X_{solv}.

$$X_{\text{solv}} = \frac{15.0 \text{ mm Hg}}{17.5 \text{ mm Hg}} = 0.857$$

Step 1: Solve for the number of moles of sucrose.

$$X_{\text{solv}} = \frac{\text{mol H}_2\text{O}}{\text{mol sucrose} + \text{mol H}_2\text{O}}$$

Rearranging this equation gives

Appendix A – Workbook Problems Solutions

$$X_{solv}(\text{mol sucrose} + \text{mol H}_2\text{O}) = \text{mol H}_2\text{O}$$

$$\text{mol sucrose} + \text{mol H}_2\text{O} = \frac{\text{mol H}_2\text{O}}{X_{solv}}$$

$$\text{mol sucrose} = \frac{\text{mol H}_2\text{O}}{X_{solv}} - \text{mol H}_2\text{O} = \frac{20.8}{0.857} - 20.8 = 3.47 \text{ mol sucrose}$$

$$\text{mol H}_2\text{O} = 375 \text{ g H}_2\text{O} \times \frac{1 \text{ mol H}_2\text{O}}{18.0 \text{ g H}_2\text{O}} = 20.8 \text{ mol H}_2\text{O}$$

Step 2: Calculate the mass of sucrose.

$$3.47 \text{ mol sucrose} \times \frac{342 \text{ g sucrose}}{1 \text{ mol sucrose}} = 1187 \text{ g sucrose}$$

Key Concept: Colligative properties – effects of vapor pressure on solutions

WP 11.4

Strategy: Solve for molality.

You can calculate ΔT_f from the freezing point of water and the freezing point of the solution.

fp solution = fp of water – ΔT_f

ΔT_f = fp. of water – fp of solution = 0°C – (–8.3°C) = 8.3°C

$\Delta T_f = K_f m$

$$m = \frac{8.3°C}{1.86°C/m} = 4.46 \, m$$

Step 1: Determine the number of moles of $(NH_4)_3PO_4$, taking into consideration that the molality is determined by the number of moles of solute particles in solution.

When $(NH_4)_3PO_4$ dissociates, it produces 3 mol of NH_4^+ and 1 mol PO_4^{3-} for a total of four moles of ions.

$$\frac{4.46 \text{ mol solute}}{1 \text{ kg solvent}} \times 0.500 \text{ kg solvent} = 2.23 \text{ mol ions}$$

$$2.23 \text{ mol ions} \times \frac{1 \text{ mol }(NH_3)_4PO_4}{4 \text{ mol ions}} = 0.558 \text{ mol ions}$$

Step 2: Determine the mass of $(NH_4)_3PO_4$.

$$0.553 \text{ mol }(NH_4)_3PO_4 \times \frac{149.0 \text{ g }(NH_4)_3PO_4}{1 \text{ mol }(NH_4)_3PO_4} = 83.0 \text{ g }(NH_4)_3PO_4$$

Key Concepts: Colligative properties and freezing point depression

Appendix A – Workbook Problems Solutions

WP 11.5

Strategy: Determine the molality of the solution.

This is accomplished by rearranging the equation for freezing–point depression. (The freezing point of benzene is 5.5°C.)

$$m = \frac{\Delta T_f}{K_f} \qquad \Delta T_f = \text{fp (benzene)} - \text{fp (solution)} = 5.5°C - 4.3°C = 1.2°C$$

$$m = \frac{1.2°C}{5.12°C/m} = 0.234\ m$$

Step 1: Calculate the actual number of moles in this solution.

This molality represents the number of moles of the unknown in 1 kg of benzene. The actual number of moles in this solution can be calculated using the mass of benzene in the solution.

$$\frac{0.234\ \text{mol unknown}}{1\ \text{kg benzene}} \times 0.025\ \text{kg benzene} = 0.005\ 85\ \text{mol unknown}$$

Step 2: Calculate the molar mass of the sample.

Knowing the mass and number of moles of the unknown, you can now calculate the molar mass of the sample.

$$\frac{2.50\ \text{g unknown}}{0.005\ 85\ \text{mol unknown}} = 427\ \text{g/mol}$$

Step 3: Determine the molecular formula.

To determine the molecular formula, you divide the molar mass of the unknown by the molar mass of the empirical formula.

$$\frac{427\ \text{g/mol}}{108\ \text{g/mol}} = 4$$

Multiplying the coefficients of the empirical formula by 4 gives $C_{24}H_{20}P_4$.

Key Concepts: Colligative properties and freezing–point depression

Chapter 12

WP 12.1

Step 1: Determine the order of the reaction with respect to NO.

In the first three solutions, the concentration of NO is changing while the concentration of H_2 remains constant. Therefore, we know that any changes that occur in the rate are a consequence of the change in concentration of NO. When the concentration of NO is doubled in the first two experiments, the rate increases by a factor of four. When the concentration of NO is tripled (experiments 1 and 3) the rate increases by a factor of nine. We know that $2^2 = 4$ and that $3^2 = 9$. Therefore, the rate of reaction with respect to NO depends on $[NO]^2$.

Step 2: Determine the order of the reaction with respect to H_2.

In comparing experiments 1, 4, and 5, the concentration of NO remains constant while the concentration of H_2 changes. Therefore, we know that any changes that occur in the rate are a consequence of the change in concentration of H_2. When the concentration of H_2 is doubled in experiments 1 and 4, the rate is doubled. When the concentration of H_2 is tripled in experiments 1 and 5, the rate is tripled. We know that $2^1 = 2$ and $3^1 = 3$. Therefore, the rate of reaction with respect to H_2 depends on $[H_2]$.

Step 3: Write the equation for the rate law.
We can now write the rate law.

Rate = $k[NO]^2[H_2]$

Step 4: Calculate the value of k.
We can calculate the value of k, using the data from any one of the five experiments. Using the data in experiment 4 gives

$$1.71 \times 10^{-5} \frac{mol}{L \cdot s} = k \left(0.15 \frac{mol}{L}\right)^2 \left(0.30 \frac{mol}{L}\right); \qquad k = \frac{1.71 \times 10^{-5} \frac{mol}{L \cdot s}}{6.75 \times 10^{-3} \frac{mol^3}{L^3}} = 2.53 \times 10^{-3} \frac{L^2}{mol^2 \cdot s}$$

Key Concept: Experimental determination of a rate law

WP 12.2
Step 1: Plot ln[A] versus time and 1/[A] versus time to determine the order of the reaction.

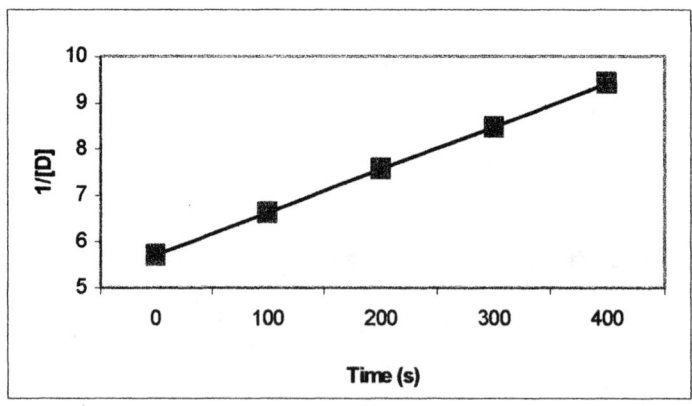

Appendix A – Workbook Problems Solutions

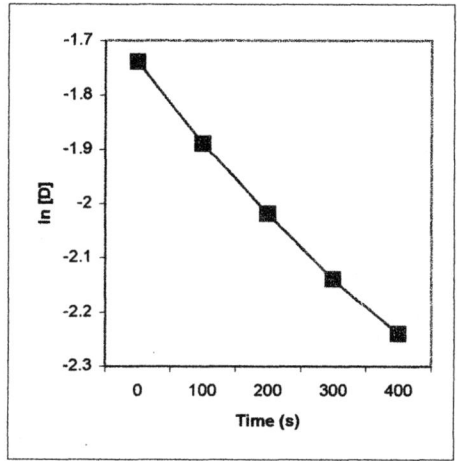

A close inspection of the two plots reveals that the plot of 1/[D] vs. time gives a better straight line; therefore, the reaction is 2nd order in [D]

Step 2: Use the integrated rate law of the appropriate order to determine k.

Second order rate law:

$$\frac{1}{[D]_t} = kt + \frac{1}{[D]_0} \qquad \frac{1}{(0.132)} = k(200\text{ s}) + \frac{1}{0.175} \qquad k = 9.31 \times 10^{-3}\text{ M}^{-1}\text{s}^{-1}$$

Step 3: Use the equation for half–life to determine the time required for 50% completion. (At 50% completion, half of the reactant has reacted.)

$$t_{\frac{1}{2}} = \frac{1}{k[D]_0} = \frac{1}{(9.31 \times 10^{-3}\text{ M}^{-1}\text{s}^{-1})(0.175\text{ M})} = 614\text{ s}$$

Step 4: Determine the concentration of D at 95% completion (5% of D remaining). Substitute the concentrations into the integrated rate law, and solve for t.

$$[D]_{95\%} = 0.175 - (0.175)(0.95) = 8.75 \times 10^{-3}$$

$$\frac{1}{8.75 \times 10^{-3}} = (9.31 \times 10^{-3})t + \frac{1}{0.175} = 1.17 \times 10^4\text{ s}$$

Key Concept: Integrated rate laws

Chapter 13

WP 13.1
Strategy: Write the equilibrium equation for K_c.

$$K_c = \frac{[PCl_3][Cl_2]}{[PCl_5]}$$

Step 1: Solve for K_c, using the molar concentrations given.

$$K_c = \frac{(1.26)(1.26)}{(2.75)} = 0.577$$

Step 2: Determine Δn.

$$\Delta n = 2 - 1 = 1$$

Step 3 Solve for K_p, using the equation that expresses the relationship between K_c and K_p.

$$K_p = (0.577)\left[\left(\frac{0.0821\ L\cdot atm}{mol\cdot K}\right)(500\ K)\right]^1 = 23.7$$

Key Concept: Equilibrium equations

WP 13.2
Step 1: Write the balanced equation for the reaction.

$$C\ (s) + H_2O\ (g) \rightleftarrows CO\ (g) + H_2\ (g)$$

Step 2: Make a table listing the initial concentration, the change in concentration and the equilibrium concentration. (Let x = the amount of substance that reacts.)

Principal Reaction	C (s) +	H$_2$O (g)	⇌	CO (g) +	H$_2$ (g)
Initial Concentration (M)		8.75 M		0	0
Change (M)		$-x$		$+x$	$+x$
Equilibrium Concentration (M)		$8.75 - x$		$+x$	$+x$

Step 3: From the balanced equation, write the equilibrium equation. Substitute the equilibrium concentrations in the equilibrium equation and solve for x. If necessary, use the quadratic equation.

$$K_c = 3.0\times 10^{-2} = \frac{[CO][H_2]}{[H_2O]} = \frac{(x)(x)}{(8.75-x)}$$

$$3.0\times 10^{-2}(8.75-x) = (x)^2$$

$$x^2 + (3.0\times 10^{-2})x - 0.263 = 0$$

$$x = \frac{-(3.0\times10^{-2}) \pm \sqrt{(3.0\times10^{-2}) - 4(-0.263)}}{2}$$

$x = 0.498$ or $x = -0.528$. The meaningful solution is $x = 0.498$. Substituting the value of x back into the equilibrium concentrations, we have

$[CO] = [H_2] = 0.498$ M

$[H_2O] = 8.75 - 0.498 = 8.25$ M

Key Concepts: Extent of reactions and calculating equilibrium concentrations.

Chapter 14

WP 14.1

Strategy: Use Table 14.1 (page 548 in your textbook) to determine which species is the stronger acid and which is the stronger base.

NH_4^+ and CO_3^{2-}: The two acids present in this reaction are NH_4^+ and HCO_3^-; the stronger acid is NH_4^+. The two bases present in this reaction are NH_3 and CO_3^{2-}; the stronger base is CO_3^{2-}.

HCN and $H_2PO_4^-$: The two acids present in this reaction are HCN and H_3PO_4; the stronger acid is H_3PO_4. The two bases present in this reaction are CN^- and $H_2PO_4^-$; the stronger base is CN^-.

Step 1: Write the reaction in the forward direction. (Remember that the stronger acid will react with the stronger base.)

$NH_4^+ (aq) + CO_3^{2-} (aq) \rightleftarrows NH_3 (aq) + HCO_3^- (aq)$

$H_3PO_4 (aq) + CN^- (aq) \rightleftarrows HCN (aq) + H_2PO_4^- (aq)$

Key Concept: Strengths of Brønsted–Lowry acids and bases

WP 14.2

Strategy: First calculate the $[H_3O^+]$, using the expression for the dissociation of water.

$1\times10^{-14} = [H_3O^+](2.35\times10^{-8})$

$[H_3O^+] = \dfrac{1\times10^{-14}}{2.35\times10^{-8}} = 4.26\times10^{-7}$

Step 1: Calculate the pH.

$pH = -\log(4.26\times10^{-7}) = 6.37$

Step 2: Determine if the solution is acidic, basic, or neutral.

Since pH < 7.0, the solution is acidic. However, in this case, we should describe the solution as *slightly* acidic, since the pH is not much less than 7.

Key Concepts: Relative values of [H$_3$O$^+$] and [OH$^-$] and the dissociation of water

WP 14.3
Step 1: Determine the concentration of [OH$^-$].

KOH is a strong base and exists in aqueous solution as K$^+$ and OH$^-$. Therefore, [OH$^-$] = initial concentration of KOH.

[OH$^-$] = 5.0 × 10^{-4} M

Step 2: Calculate the [H$_3$O$^+$].

The H$_3$O$^+$ concentration is calculated from the OH$^-$ concentration.

$$[H_3O^+] = \frac{K_w}{[OH^-]} = \frac{1.0 \times 10^{-14}}{5.0 \times 10^{-4}} = 2.0 \times 10^{-11} \text{ M}$$

Step 3: Calculate the pH.

pH = –log (2.0 × 10^{-11}) = 10.70

Key Concept: pH of strong acids and bases

WP 14.4
Step 1: Determine the species present initially.

HC$_3$H$_5$O$_3$ and H$_2$O

Step 2: Write the possible proton–transfer reactions.

HC$_3$H$_5$O$_3$ (aq) + H$_2$O (l) ⇌ H$_3$O$^+$ (aq) + C$_3$H$_5$O$_3^-$ (aq) $K_a = 1.4 \times 10^{-4}$

H$_2$O (l) + H$_2$O (l) ⇌ H$_3$O$^+$ (aq) + OH$^-$ (aq) $K_a = 1.0 \times 10^{-14}$

Step 3: Determine the principal reaction and write the equilibrium equation for that reaction.

$K_a \gg K_w$ The principle reaction is dissociation of HC$_3$H$_5$O$_3$

Step 4: Make a table showing the principal reaction and the initial and equilibrium concentrations.

Principal Reaction	HC$_3$H$_5$O$_3$ (aq)	⇌	H$_3$O$^+$ (aq)	+	C$_3$H$_5$O$_3^-$ (aq)
Initial Concentration	2.75		0		0
Change	–x		+x		+x
Eq. Concentration	2.75 – x		+x		+x

Step 5: Substitute the equilibrium concentrations into the equilibrium equation, and solve for x. (Since $K_a < 1.0 \times 10^{-3}$, assume x is negligible and that 2.75 – x ≅ 2.75.*)

---

*The initial concentration of HC$_3$H$_5$O$_3$ is known to the second decimal place (2.75). Usually x is considered negligible compared to the value of the initial concentration only if x < 0.01. In this case, x > .01. However, solving for x using the quadratic equation also gives x = 1.96 × 10^{-2}.

Appendix A – Workbook Problems Solutions

$$K_a = \frac{[H_3O^+][C_3H_5O_3^-]}{[HC_3H_5O_3]} = \frac{(x)(x)}{(2.75-x)}$$

$$1.4 \times 10^{-4} = \frac{x^2}{2.75} \qquad x = 1.96 \times 10^{-2}$$

Step 6: Calculate the big equilibrium concentrations.

$$[H_3O^+]_e = [C_3H_5O_3^-]_e = 1.96 \times 10^{-2} \text{ M}$$

$$[HC_3H_5O_3]_e = 2.75 - (1.96 \times 10^{-2}) = 2.73$$

Step 7: Calculate the pH.

$$pH = -\log(1.96 \times 10^{-2}) = 1.707$$

Key Concept: Dissociation of a weak acid

WP 14.5

Strategy: To determine the % dissociation, we need to first determine the concentration of the dissociated HA. This requires that we determine the equilibrium concentration of H_3O^+ and $C_6H_5COO^-$. We will use the procedure found on page 630 of your textbook.

Step 1: Determine the species present initially, write the possible proton–transfer reactions and determine the principal reaction.

The species present initially are C_6H_5COOH and H_2O (acid or base).

The possible proton–transfer reactions are

$$C_6H_5COOH\ (aq) + H_2O\ (l) \rightleftarrows H_3O^+\ (aq) + C_6H_5COO^-\ (aq) \qquad K_a = 6.5 \times 10^{-5}$$

$$H_2O\ (l) + H_2O\ (l) \rightleftarrows H_3O^+\ (aq) + OH^-\ (aq) \qquad\qquad\qquad K_w = 1.0 \times 10^{-14}$$

Since $K_a \gg K_w$, the principal reaction is dissociation of C_6H_5COOH.

Step 2: Make a table showing the principal reaction, initial concentration, change in concentration, and the equilibrium concentration.

Principal Reaction	$C_6H_5COOH\ (aq)$ $\rightleftarrows$	H_3O^+ +	$C_6H_5COO^-$
Initial Concentration	0.25	0	0
Change	$-x$	$+x$	$+x$
Equilibrium Concentration	$0.25 - x$	$+x$	$+x$

Step 3: Substitute the equilibrium concentrations into the equilibrium expression, and solve for x.

$$K_a = 6.5 \times 10^{-5} = \frac{[H_3O^+][C_6H_5COO^-]}{[C_6H_5COOH]} = \frac{(x)(x)}{(0.25-x)}$$

Appendix A – Workbook Problems Solutions

Given the value of K_a, assume that x is negligible compared with the initial concentration of the acid; therefore $0.25 - x \approx 0.25$. Using this value in the denominator, solve for x.

$$x^2 \approx (6.5 \times 10^{-5})(0.25)$$

$$x \approx 4.0 \times 10^{-3*}$$

Step 4: Calculate the equilibrium concentrations of the major species present.

The equilibrium concentrations of H_3O^+ and $C_6H_5COO^-$ are $[H_3O^+] = [C_6H_5COO^-] = 0.004$ M. This concentration also represents the amount of C_6H_5COOH that dissociated.

Step 5: Calculate the percent dissociation.

The percent dissociation can now be calculated.

$$\frac{0.004}{0.25} \times 100 = 1.6\%$$

Key Concept: Dissociation of a weak acid

WP 15.6

Step 1: Write the stepwise dissociation for arsenic acid.

The stepwise dissociation for arsenic acid is

$$H_3AsO_4\ (aq) + H_2O\ (l) \rightleftarrows H_3O^+\ (aq) + H_2AsO_4^-\ (aq)$$
$$H_2AsO_4^-\ (aq) + H_2O\ (l) \rightleftarrows H_3O^+\ (aq) + HAsO_4^{2-}\ (aq)$$
$$HAsO_4^{2-}\ (aq) + H_2O\ (l) \rightleftarrows H_3O^+\ (aq) + AsO_4^{3-}\ (aq)$$

Step 2: Determine the principal reaction.

Since $K_{a1} > K_{a2}$, K_{a3}, and K_w, we know that the principal reaction is the first dissociation step and that all of the H_3O^+ present is produced from this step. Therefore, we only need to consider the first dissociation step when calculating the pH of the solution.

Step 3: Make a table showing the principle reaction, initial concentration, change in concentration and equilibrium concentration of the reactant and products.

Principal Reaction	$H_3AsO_4\ (aq)$	$\rightleftarrows$	H_3O^+	+	$H_2AsO_4^-$
Initial Concentration	2.500×10^{-4}		0		0
Change	$-x$		$+x$		$+x$
Equilibrium Concentration	$(2.500 \times 10^{-4}) - x$		$+x$		$+x$

Step 4: Substitute the equilibrium concentrations into the equilibrium expression.

Assume that x is negligible; therefore, $(2.500 \times 10^{-4}) - x \approx 2.500 \times 10^{-4}$.

---

* The initial concentration of C_6H_5COOH is known to the second decimal place (0.25). Usually x is considered negligible compared to the value of the initial concentration only if $x < 0.01$. In this case, $x < 0.01$ and therefore, considered negligible. However, solving for x using the quadratic equation gives $x = 4.0 \times 10^{-3}$. Both values for x give an equilibrium concentration of 0.246 M for benzoic acid.

Appendix A – Workbook Problems Solutions

$$5.62 \times 10^{-3} = \frac{x^2}{2.500 \times 10^{-4}} \qquad x = 1.185 \times 10^{-3}$$

From the value of x, we know that our assumption is not valid.[*] Therefore, we must use the quadratic equation to solve for the equilibrium concentration of H_3O^+. Using $(2.500 \times 10^{-4}) - x$ and rearranging the equilibrium expression gives:

$$5.62 \times 10^{-3}[(2.500 \times 10^{-4}) - x] = x^2 \qquad x^2 + (5.62 \times 10^{-3})x - (1.405 \times 10^{-6})$$

$$x = \frac{-(5.62 \times 10^{-3}) \pm \sqrt{(5.62 \times 10^{-3})^2 - 4(-1.405 \times 10^{-6})}}{2}$$

$x = -5.5860 \times 10^{-3}$ or $x = 2.398 \times 10^{-4}$; The answer that makes chemical sense is 2.398×10^{-4} M. This value represents the equilibrium concentration of H_3O^+.

Step 5: Calculate the pH of the solution.

$$\text{pH} = -\log(2.398 \times 10^{-4}) = 3.62$$

Key Concept: Dissociation of a weak diprotic acid

WP 14.7
Strategy: Follow the procedure found on page 630 of your textbook.

Step 1: Determine the principle reaction.

$K_b > K_w$; therefore, the principal reaction is

$(CH_3)_3N\,(aq) + H_2O\,(l) \rightleftharpoons (CH_3)_3NH^+\,(aq) + OH^-\,(aq)$

Step 2: Construct a table with concentrations of the reactant and products.

Principal Reaction	$(CH_3)_3N\,(aq)$ $\rightleftharpoons$	$(CH_3)_3N^+$ +	OH^-
Initial Concentration	0.075	0	0
Change	$-x$	$+x$	$+x$
Equilibrium Concentration	$0.075 - x$	$+x$	$+x$

Step 3: Substitute the equilibrium concentrations into the equilibrium expression, and solve for x.

$$K_a = 6.5 \times 10^{-5} = \frac{[(CH_3)_3NH^+][OH^-]}{[(CH_3)_3N]} = \frac{(x)(x)}{0.075 - x}$$

Assume that x is negligible compared with the initial concentration of the base; therefore, $0.075 - x \approx x$. Using this value in the denominator, we can now solve for x. Remember: x represents the equilibrium concentration of the OH^- ion.

---

[*] A good rule of thumb is that x is negligible only if $K_a < 1.00 \times 10^{-3}$. (Remember, from Chapter 13, that the position of equilibrium lies to the left when $K_a < 1.00 \times 10^{-3}$).

$$6.5 \times 10^{-5} = \frac{x^2}{0.075} \qquad x = 0.0022$$

Step 4: Determine the equilibrium concentrations of the species present.

The equilibrium concentrations are $[(CH_3)_3N] = 0.075 - 0.0022 = 0.073$ M; $[OH^-] = [(CH_3)_3NH^+] = 0.0022$ M.

Step 5: Calculate the pH of the solution.

Knowing the equilibrium concentration of OH^-, we can calculate the H_3O^+ concentration from the K_w expression.

$$[H_3O^+] = \frac{1.0 \times 10^{-14}}{0.0022} = 4.5 \times 10^{-12}$$

$$pH = -\log(4.5 \times 10^{-12}) = 11.35$$

Key Concept: Dissociation of a weak base

WP 14.8

Strategy: From the preceding discussion, determine if KOCl produces an acidic, basic, or neutral aqueous solution, and write the hydrolysis reaction for this salt.

K^+ is an inert cation and will not react. However, OCl^- is the conjugate base of a weak acid and will react with water.

Step 1: Write the hydrolysis reaction for this salt.

$$OCl^- (aq) + H_2O (l) \rightleftarrows HOCl (aq) + OH^- (aq)$$

Step 2: Write an equilibrium expression for this reaction.

$$K_b = \frac{[HOCl][OH^-]}{[OCl^-]} \qquad K_b = \frac{1.0 \times 10^{-14}}{K_a(HOCl)} = \frac{1.0 \times 10^{-14}}{3.5 \times 10^{-8}} = 2.9 \times 10^{-7}$$

Step 3: Determine the principal reaction.

$K_b > K_w$; therefore, the principal reaction is hydrolysis of OCl^-.

Step 4: Construct a table.

Principal Reaction	$OCl^- (aq)$	$\rightleftarrows$	$HOCl (aq)$	+	OH^-
Initial Concentration	0.250		0		0
Change	$-x$		$+x$		$+x$
Equilibrium Concentration	$0.250 - x$		$+x$		$+x$

Step 5: Substitute the equilibrium concentrations into the equilibrium expression.

Assume that x is negligible; therefore, $0.250 - x \approx 0.250$.

Appendix A – Workbook Problems Solutions

$$2.9 \times 10^{-7} = \frac{(x)(x)}{0.250} \quad x = 2.7 \times 10^{-4}$$

Step 6: Calculate the pH of the solution.

The equilibrium concentration of OH^- is 2.7×10^{-4}. We need to determine the H_3O^+ concentration to calculate pH.

$$[H_3O^+] = \frac{1.0 \times 10^{-14}}{2.7 \times 10^{-4}} = 3.7 \times 10^{-11}$$

$pH = -\log(3.7 \times 10^{-11}) = 10.43$

Key Concept: Hydrolysis of a salt

Chapter 15

WP 15.1
Strategy: Use Table C.1 and C.3 in the appendix of your textbook to determine the formulas, K_a and K_b for benzoic acid and hydroxylamine.

$C_6H_5CO_2H \quad\quad K_a = 6.5 \times 10^{-5}$
$NH_2OH \quad\quad\quad K_b = 9.1 \times 10^{-9}$

Step 1: Write the net neutralization reaction by writing individual reactions for the acid, base, and water.

$C_6H_5CO_2H + H_2O \rightleftarrows H_3O^+ + C_6H_5CO_2^- \quad\quad K_a = 6.5 \times 10^{-5}$

$NH_2OH + H_2O \rightleftarrows NH_2OH_2^+ + OH^- \quad\quad K_b = 9.1 \times 10^{-9}$

$H_3O^+ + OH^- \rightleftarrows 2\,H_2O \quad\quad 1/K_w = 1.0 \times 10^{14}$

_____

$C_6H_5CO_2H + NH_2OH \rightleftarrows C_6H_5CO_2^- + NH_2OH_2^+$

Step 2: Calculate K_n based on the equilibrium constants for the individual reactions in step 1.

$K_n = K_a \times K_b \times (1/K_w) = (6.5 \times 10^{-5}) \times (9.1 \times 10^{-9}) \times (1.0 \times 10^{14}) = 59.2$

Step 3: Determine the position of equilibrium from the value of K_n.

$K_n > 1$; therefore, the position of equilibrium is to the right.

Step 4: Based on the values of K_a and K_b, predict the pH.

$$K_b(C_6H_5O_2^-) = \frac{K_w}{K_a} = \frac{1 \times 10^{-14}}{6.5 \times 10^{-5}} = 1.54 \times 10^{-10}$$

$$K_a(NH_2OH_2^+) = \frac{K_w}{K_b} = \frac{1 \times 10^{-14}}{9.1 \times 10^{-9}} = 1.10 \times 10^{-6}$$

Remember from Chapter 14 that the stronger the acid, the weaker the conjugate base and vice versa. The K_a for benzoic acid is greater than the K_b for hydroxylamine. Therefore, the strength of $C_6H_5O_2^-$ is less than the strength of $NH_2OH_2^+$. The calculations for K_a and K_b of the conjugate acid and conjugate base bear this out. Therefore, the pH of the solution will be less than 7.00.

Key Concepts: Neutralization reactions and the relative strengths of conjugate acids and bases

WP 15.2

Step 1: Determine the principal reaction and equilibrium concentrations.

The principal reaction and equilibrium concentration for this solution are:

Principal Reaction	HCOOH (aq) + H$_2$O (l)	⇌	H$_3$O$^+$ (aq) +	HCO$_2^-$ (aq)
Initial Concentration	0.45		0	0.55
Change	–x		+x	+x
Equilibrium Concentration	0.45 –x		+x	0.55 + x

Step 2: Using the equilibrium equation, solve for [H$_3$O$^+$].

If you solve the equilibrium equation for [H$_3$O$^+$], you obtain:

$$K_a = \frac{[H_3O^+][HCO_2^-]}{[HCOOH]}$$

$$[H_3O^+] = K_a \frac{[HCOOH]}{[HCO_2^-]}$$

K_a for formic acid equals 1.8×10^{-4}. Substituting the given information into the equilibrium expression gives

$$[H_3O^+] = (1.8 \times 10^{-4}) \frac{(0.45)}{(0.55)} = 1.5 \times 10^{-4}$$

Step 3: Solve for the pH.

pH = –log (1.5 x 10^{-4}) = 3.82

Step 4: Write the neutralization reaction that occurs on addition of HCl, and set up a table showing the number of moles present before and after the addition of HCl.

Neutralization Reaction	HCO$_2^-$ (aq) +	H$_3$O$^+$ (l)	⇌	HCOOH (aq) +	H$_2$O (l)
Before Reaction (mol)	0.55	0.10		0.45	
Change (mol)	–0.10	–0.10		+0.10	
After Reaction (mol)	0.45	0		0.55	

Step 5: Determine the concentrations of the buffer components after neutralization occurs.

Appendix A – Workbook Problems Solutions

Assuming that the volume of the solution does not change, the concentrations of the buffer components after neutralization are

$$[\text{HCO}_2^-] = \frac{0.45 \text{ mol}}{1.0 \text{ L}} = 0.45 \text{ M}$$

$$[\text{HCOOH}] = \frac{0.55 \text{ mol}}{1.0 \text{ L}} = 0.55 \text{ M}$$

Step 6: Substitute the concentrations of the buffer components into the equilibrium expression and calculate the pH.

$$[\text{H}_3\text{O}^+] = (1.8 \times 10^{-4}) \frac{0.55}{0.45} = 2.2 \times 10^{-4}$$

$$\text{pH} = -\log(2.2 \times 10^{-4}) = 3.66$$

Key Concept: Buffer solutions

WP 15.3

Strategy: Use the Henderson–Hasselbalch equation to calculate the [base]/[acid] ratio. This information provides the mole ratio of benzoate to benzoic acid.

$$\log \frac{[\text{benzoate}]}{[\text{benzoic acid}]} = \text{pH} - \text{pK}_a = 4.32 - 4.19 = 0.13$$

$$\frac{[\text{benzoate}]}{[\text{benzoic acid}]} = \text{antilog}(0.13) = 1.35$$

[benzoate] = 1.35[benzoic acid]

Step 1: Calculate the mass needed to prepare the buffer.

Sodium benzoate = $\text{NaC}_6\text{H}_5\text{CO}_2$ and benzoic acid = $\text{C}_6\text{H}_5\text{CO}_2\text{H}$.

The [benzoate] is 1.35 times greater than the [benzoic acid]. Therefore, you need 1 mol of $\text{C}_6\text{H}_5\text{CO}_2\text{H}$ for every 1.35 mol of $\text{NaC}_6\text{H}_5\text{CO}_2$. (The mol ratio for benzoate to sodium benzoate is 1:1. If you need 1.35 mol of $\text{C}_6\text{H}_5\text{CO}_2^-$, then you need 1.35 mol of $\text{NaC}_6\text{H}_5\text{CO}_2$.)

The mass of 1 mole of $\text{C}_6\text{H}_5\text{CO}_2\text{H}$ is

$$1 \text{ mol C}_6\text{H}_5\text{CO}_2\text{H} \times \frac{122 \text{ g C}_6\text{H}_5\text{CO}_2\text{H}}{1 \text{ mol C}_6\text{H}_5\text{CO}_2\text{H}} = 122 \text{ g C}_6\text{H}_5\text{CO}_2\text{H}$$

The mass of 1.35 moles of $\text{NaC}_6\text{H}_5\text{CO}_2$ is

$$1.35 \text{ mol NaC}_6\text{H}_5\text{CO}_2 \times \frac{144 \text{ g NaC}_6\text{H}_5\text{CO}_2}{1 \text{ mol NaC}_6\text{H}_5\text{CO}_2} = 194 \text{ g NaC}_6\text{H}_5\text{CO}_2$$

Of course, you can cut down the amount needed, as long as you keep the ratio of [benzoate] to [benzoic acid] = 1.35. If you decided to start with 0.01 moles of benzoic acid, then you would use

0.0135 mol of sodium benzoate. Therefore, you would prepare a solution using 0.122 g of benzoic acid and 0.194 g of sodium benzoate.

Key Concepts: Buffer solutions and the Henderson–Hasselbalch equation

WP 15.4

a) mL of acid required to reach the equivalence point:

Step 1: Write the neutralization reaction that occurs, and determine the number of mmoles of $(CH_3)_3N$ in the initial solution.

$$(CH_3)_3N\ (aq)\ +\ HCl\ (aq)\ \rightleftarrows\ (CH_3)_3NH^+\ (aq)\ +\ Cl^-\ (aq)$$

$$\text{mmol }(CH_3)_3N = 50.00\ \text{mL} \times 0.175 \frac{\text{mmol }(CH_3)_3N}{\text{mL }(CH_3)_3N} = 8.75\ \text{mmol }(CH_3)_3N$$

Step 2: Determine the mmol of HCl needed to react with the $(CH_3)_3N$ and the volume of 0.100 M HCl needed.

From the neutralization reaction, we know that the mol ratio of $(CH_3)_3N$:HCl is 1:1. Therefore,

mmol HCl = 8.75

$$\text{mL HCl} = \frac{8.75\ \text{mmol HCl}}{0.100\ \text{mmol HCl/mL HCl}} = 87.5\ \text{mL HCl}$$

b) Initial pH of the $(CH_3)_3N$ solution:

Step 1: Determine the initial pH of the solution, using the method learned in section 15.12 in your textbook.

Principal Reaction	$(CH_3)_3N\ (aq)$	+	$H_2O\ (l)$	$\rightleftarrows$	$(CH_3)_3NH^+\ (aq)$	+	$OH^-\ (aq)$
Initial Concentration	0.175 M				0		0
Change	$-x$				$+x$		$+x$
Equilibrium Concentration	0.175 M $- x$				$+x$		$+x$

$$K_b = \frac{[(CH_3)_3NH^+][OH^-]}{[(CH_3)N]}$$

From Table C.1 in the Appendix of the textbook, you will find $K_b = 6.5 \times 10^{-5}$.

$$6.5 \times 10^{-5} = \frac{(x)(x)}{(0.175-x)} \cong \frac{x^2}{0.175} \qquad x = [OH^-] = 3.37 \times 10^{-3}$$

$$[H_3O^+] = \frac{K_w}{[OH^-]} = \frac{1.0 \times 10^{-14}}{3.37 \times 10^{-3}} = 2.96 \times 10^{-12}; \qquad \text{pH} = -\log(2.96 \times 10^{-12}) = 11.53$$

Appendix A – Workbook Problems Solutions

c) pH of the solution after addition of 10.00 mL of acid:

Step 1: We need to use the Henderson–Hasselbalch equation to calculate the pH at this point. Before proceeding, do a little thinking. The Henderson–Hasselbalch equation has the form

$$pH = pK_a + \log \frac{[\text{base}]}{[\text{acid}]}$$

Keep in mind that the acid in this equation is the conjugate acid of $(CH_3)_3N$ and that the pK_a refers to the K_a of the conjugate acid of $(CH_3)_3N$. (Don't forget that the conjugate acid has one more H^+ than its conjugate base.) You are now armed with the information you need to proceed.

Step 2: Calculate the concentration of $(CH_3)_3N$ and its conjugate acid after the addition of 10.00 mL of HCl.

Appendix A – Workbook Problems Solutions

mmol (CH$_3$)$_3$N = 8.75 mmol – (0.100 M)(10.00 mL) = 7.75 mmol
                                    amount of HCl used that
                                      represents the amount of
                                      trimethylamine that reacted.

The total volume present = 50.00 mL (CH$_3$)$_3$N + 10.00 mL HCl added.

$$[(CH_3)_3N] = \frac{7.75 \text{ mmol } (CH_3)_3N}{60.00 \text{ mL soln}} = 0.129 \text{ M}$$

The conjugate acid of (CH$_3$)$_3$N is (CH$_3$)$_3$NH$^+$.

mmol (CH$_3$)$_3$NH$^+$ = mmol HCl used = (0.100 M)(10.00 mL) = 1.00 mmol

$$[(CH_3)_3NH^+] = \frac{1.00 \text{ mmol } (CH_3)_3NH^+}{60.00 \text{ mL}} = 0.0167 \text{ M}$$

Step 3: Determine the K_a and pK_a of the conjugate acid of (CH$_3$)$_3$N.

$$K_a(CH_3)_3NH^+ = \frac{K_w}{K_b(CH_3)_3N} = \frac{1.0 \times 10^{-14}}{6.5 \times 10^{-5}} = 1.5 \times 10^{-10}$$

pK_a = –log (1.5 × 10^{-10}) = 9.81

Step 4: Using the Henderson–Hasselbalch equation, calculate the pH of the solution.

$$pH = pK_a + \log\frac{[base]}{[acid]} = 9.81 + \log\frac{0.129}{0.0167} = 10.7$$

d) pH at the equivalence point:

Step 1: Once again, use the Henderson–Hasselbalch equation to calculate the pH. Remember, halfway to the equivalence point, the amount of the (CH$_3$)$_3$N that has reacted equals the amount of the conjugate acid that has been produced. That is, [base] = [acid].

Therefore, pH = pK_a = 9.81.

e) pH at the equivalence point:

Step 1: We need to do a little bit more thinking. At the equivalence point, all of the (CH$_3$)$_3$N present has now been converted to its conjugate acid. We now need to consider what is happening to the conjugate acid in an aqueous solution.

Step 2: Write the principle reaction that is occurring at the equivalence point.

(CH$_3$)$_3$NH$^+$ (aq) + H$_2$O (l) ⇌ (CH$_3$)$_3$N (aq) + H$_3$O$^+$ (aq)

Step 3: Calculate the concentration of the conjugate acid.

At the equivalence point, mmol of (CH$_3$)$_3$NH$^+$ = mmol of (CH$_3$)$_3$N reacted (= amount of (CH$_3$)$_3$N

A-55

present initially).

mmol $(CH_3)_3NH^+$ = 8.75

Total volume of soln = 50.00 mL $(CH_3)_3N$ + 87.5 mL HCl to reach eq point = 137.5 mL

$$[(CH_3)_3NH^+] = \frac{8.75 \text{ mmol}}{137.5 \text{ mL}} = 0.0636 \text{ M}$$

Step 4: Determine the pH of the solution, using the method learned in section 15.14 in your textbook.

K_a $(CH_3)_3NH^+$ = 1.5×10^{-10}

Principal Reaction	$(CH_3)_3NH^+$ (aq)	+	H_2O (l)	⇌	$(CH_3)_3N$ (aq)	+	H_3O^+ (aq)
Initial Concentration	0.064				0		0
Change	$-x$				$+x$		$+x$
Equilibrium Concentration	$0.0636 - x$				$+x$		$+x$

$$K_a = \frac{[H_3O^+][(CH_3)_3N]}{[(CH_3)_3NH^+]} \qquad 1.5 \times 10^{-10} = \frac{(x)(x)}{0.0636 - x} \cong \frac{x^2}{0.0636}$$

$x = [H_3O^+] = 3.1 \times 10^{-6}$ $\qquad$ pH = $-\log(3.14 \times 10^{-6})$ = 5.51

f) pH after the addition of 100.00 mL of 0.100 M HCl:
Step 1: Determine the number of mmoles of excess HCl.

mmol xs HCl = mmol HCl added − mmol HCl reacted = (0.100 M)(100.00 mL) − 8.75 mmol = 1.25

Step 2: Determine the concentration of excess HCl, and calculate the pH.

$$[HCl]_{xs} = \frac{1.25 \text{ mmol}}{150 \text{ mL}} = 8.33 \times 10^{-3} \text{ M} \qquad \text{pH} = -\log(8.33 \times 10^{-3}) = 2.08$$

Key Concepts: Stoichiometry, molarity, neutralization reactions, dissociation of a weak acid and a weak base, hydrolysis, buffer solutions

WP 15.5
Step 1: Write the solubility equilibrium expression for $Cd(OH)_2$.

The solubility equilibrium for $Cd(OH)_2$ is

$Cd(OH)_2$ (s) ⇌ Cd^{2+} (aq) + 2 OH^- (aq)

and the K_{sp} expression for this reaction is:

$K_{sp} = [Cd^{2+}][OH^-]^2$

Step 2: Determine the concentration of Cd^{2+} and OH^- and calculate the K_{sp}.

If $[Cd^{2+}] = 1.10 \times 10^{-5}$ then the $[OH^-]$ is 2 times that or 2.20×10^{-5}. Substituting these values into the equilibrium expression gives

$$K_{sp} = (1.10 \times 10^{-5})(2.20 \times 10^{-5})^2 = 5.32 \times 10^{-15}$$

Key Concept: Solubility equilibrium.

WP 15.6

Strategy: Write the solubility equilibrium for $Fe(OH)_3$ and the equilibrium expression. Using the method described in Chapter 13, calculate the molar solubility of $Fe(OH)_3$.

The solubility equilibrium for $Fe(OH)_3$ is

$$Fe(OH)_3\,(s) \rightleftarrows Fe^{3+}\,(aq) + 3\,OH^-\,(aq)$$

The equilibrium expression is

$$K_{sp} = [Fe^{3+}][OH^-]^3$$

Step 1: Let x be the number of mol/L of $Fe(OH)_3$ that dissolves. The saturated solution then contains x mol/L of Fe^{3+} and $3x$ mol/L of OH^-. Solve for K_{sp}.

Substituting this into the equilibrium expression gives

$$K_{sp} = 2.6 \times 10^{-39} = [Fe^{3+}][OH^-]^3 = (x)(3x)^3$$

Solving for x gives

$$2.6 \times 10^{-39} = 27x^4; \quad x = 9.9 \times 10^{-11}\,M$$

Key Concept: Solubility equilibrium

WP 15.7

Step 1: Write the solubility equation and solubility expression.

$$Sn(OH)_2\,(s) \rightleftarrows Sn^{2+}\,(aq) + 2\,OH^-\,(aq)$$

The equilibrium expression is

$$K_{sp} = [Sn^{2+}][OH^-]^2$$

Step 2: Using the pH, determine the $[OH^-]$ concentration.

$$[H_3O^+] = \text{antilog}\,(-9.35) = 4.47 \times 10^{-10}$$

$$[OH^-] = \frac{1 \times 10^{-14}}{4.47 \times 10^{-10}} = 2.24 \times 10^{-5}$$

Appendix A – Workbook Problems Solutions

Step 3: Let x be the number of mol/L of $Sn(OH)_2$ that dissolves. Construct a table showing the equilibrium concentrations.

Solubility Equilibrium	$Sn(OH)_2 (s)$ $\rightleftarrows$	$Sn^{2+} (aq)$ +	$2\, OH^- (aq)$
Initial Concentration		0	2.24×10^{-5}
Equilibrium Concentration		$+x$	$2.24 \times 10^{-5} + x$

Step 4: Calculate the solubility of $Sn(OH)_2$.

Given the very small value of K_{sp}, we can assume that x is negligible. Substituting the equilibrium concentrations into the K_{sp} expression gives

$$5.4 \times 10^{-27} = (x)(2.24 \times 10^{-5})^2 \quad x = 1.1 \times 10^{-17}$$

Key Concept: Solubility equilibrium and common ion effect

WP 15.8

Strategy: Write a metathesis reaction for the reaction between $Zn(NO_3)_2$ and Na_2CO_3, and apply the solubility rules in Chapter 4 to determine if a precipitate will form.

$$Zn(NO_3)_2 (aq) + Na_2CO_3 (aq) \rightarrow 2\, NaNO_3 (aq) + ZnCO_3 (s)$$

Step 1: Write the solubility equation and IP expression for the precipitate.

$$ZnCO_3 (s) \rightleftarrows Zn^{2+} (aq) + CO_3^{2-} (aq) \qquad K_{sp} = 1.2 \times 10^{-10}$$

The ion product expression is

$$IP = [Zn^{2+}][CO_3^{2-}]$$

Step 2: Calculate the IP and compare its value to the K_{sp}. Will a precipitate form?

Calculate the IP from the concentrations of Zn^{2+} and CO_3^{2-}. However, don't forget you mixed two solutions to give a total volume of 400 mL. Therefore, the concentrations of Zn^{2+} and CO_3^{2-} must be calculated based on the dilution that occurs on mixing the two solutions. Use the equation $M_i V_i = M_f V_f$.

$$[Zn^{2+}] = \frac{(0.75\, M)(150\, mL)}{400\, mL} = 0.28\, M$$

$$[CO_3^{2-}] = \frac{(1.50\, M)(250\, mL)}{400\, mL} = 0.938\, M$$

$$IP = (0.28)(0.938) = 0.26$$

$IP > K_{sp}$; Therefore, a precipitate will form.

Key Concepts: Position of equilibrium and precipitation of ionic compounds

Chapter 16

WP 16.1

Step 1: Write the net ionic equation for this reaction.

$$Ag^+ (aq) + Cl^- (aq) \rightarrow AgCl (s)$$

Step 2: Determine the information you will need to calculate ΔS_{total}

$$\Delta S_{total} = \Delta S_{system} - \frac{\Delta H_{rxn}}{T}$$

You need to calculate both ΔS_{system} and ΔH_{rxn}.

Step 3: Using appendix B.1 in your textbook, calculate ΔS_{system}.

$$\Delta S_{system} = [S°(AgCl)] - [S°(Ag^+) + S°(Cl^-)]$$

$$\Delta S_{system} = [96.2] - [72.7 + 56.5] = -33.0 \text{ J/K}$$

Step 4: Calculate ΔH_{rxn}

$$\Delta H°_{rxn} = [\Delta H°_f (AgCl)] - [\Delta H°_f (Ag^+) + \Delta H°_f (Cl^-)]$$

$$\Delta H°_{rxn} = [-127.1] - [105.6 + (-167.2)] = -65.5 \text{ kJ}$$

Step 5: Calculate ΔS_{total}.

When substituting the preceding values into the equation for ΔS_{total}, **watch your units!**

$$\Delta S_{total} = -33.0 \text{ J/K} - \frac{-6.55 \times 10^4 \text{ J}}{298 \text{ K}} = 187 \text{ J/K}$$

Key Concept: The Second Law of Thermodynamics

WP 16.2

Step 1: Determine $\Delta S°$.

$$\Delta H° = [(-1387.1) + (2 \times -92.3)] - [(2 \times -411.2) + (-814.0)] = 64.7 \text{ kJ}$$

Step 2: Determine $\Delta S°$.

$$\Delta S° = [(149.6) + (2 \times 186.8)] - [(2 \times 72.1) + (156.9)] = 222.1 \text{ J/K}$$

Step 3: Calculate $\Delta G°$.

When calculating $\Delta G°$, **watch your units!** (It is a common mistake for students to mix the kJ of $\Delta H°$ with the J of $\Delta S°$.) Since we are using standard enthalpies and entropies, $T = 298$ K.

Appendix A – Workbook Problems Solutions

$\Delta G° = 64.7 \text{ kJ} - (298)(0.2221 \text{ kJ/K}) = -1.49 \text{ kJ}$ (The reaction is spontaneous.)

Key Concept: Free–energy change of a system.

WP 16.3

Strategy: Determine $\Delta G°_{rxn}$ using appendix B.1 in your textbook.

$$\Delta G°_{rxn} = [\Delta G°_f(NH_3) + \Delta G°_f(H_2O)] - [\Delta G°_f(NH_4^+) + \Delta G°_f(OH^-)]$$

$$\Delta G°_{rxn} = [(1 \text{ mol } NH_3 \times -16.5 \text{ kJ/mol}) + (1 \text{ mol } H_2O \times -237.2 \text{ kJ/mol}] - $$
$$[(1 \text{ mol } NH_4^+ \times -79.4 \text{ kJ/mol}) + (1 \text{ mol } OH^- \times -157.3 \text{ kJ/mol})]$$

$$\Delta G°_{rxn} = -17.0 \text{ kJ}$$

Step 1: Based on the value of $\Delta G°_{rxn}$, determine if the reaction is spontaneous.

$\Delta G°_{rxn}$ has a negative value; therefore, the reaction is spontaneous.

Step 2: Determine the temperature at which the reaction will be spontaneous.

The equation needed to calculate the temperature of the spontaneous reaction is

$$T = \frac{\Delta H°}{\Delta S°}$$

Therefore, we need to calculate $\Delta H°$ and $\Delta S°$ for this reaction.

$$\Delta H° = [\Delta H°_f(NH_3) + \Delta H°_f(H_2O)] - [\Delta H°_f(NH_4^+) + \Delta H°_f(OH^-)]$$

$$\Delta H° = [(1 \text{ mol } NH_3 \times -46.1 \text{ kJ/mol}) + (1 \text{ mol } H_2O \times -285.8 \text{ kJ/mol})] -$$
$$[(1 \text{ mol } NH_4^+ \times -132.5 \text{ kJ/mol}) + (1 \text{ mol } OH^- \times -230.0 \text{ kJ/mol})]$$
$$\Delta H° = 30.6 \text{ kJ}$$

$$\Delta S° = [S°(NH_3) + S°(H_2O)] - [S°(NH_4^+) + S°(OH^-)]$$

$$\Delta S° = [(1 \text{ mol } NH_3 \times 192.3 \text{ J/mol·K}) + (1 \text{ mol } H_2O \times 69.9 \text{ J/mol·K})] -$$
$$[(1 \text{ mol } NH_4^+ \times 113 \text{ J/mol·K}) + (1 \text{ mol } OH^- \times -10.8 \text{ J/mol·K}]$$

$$\Delta S° = 160 \text{ J/K}$$

$$T = \frac{30.6 \text{ kJ}}{0.160 \text{ kJ/K}} = 191 \text{ K} = -82°\text{C}$$

Key Concept: Free–energy change and spontaneity

Chapter 17

WP 17.1

$Al\,(s)\,|\,Al^{3+}\,(aq)\,||\,Co^{2+}\,(aq)\,|\,Co\,(s)$

Strategy: Based on the shorthand notation above, determine which species is the cathode and which is the anode, and write the half–reaction that occurs at each electrode.

Remember, the shorthand notation for the anode is always written on the left of the salt–bridge symbol. Therefore, Al is the anode, and Co is the cathode.

Step 1: Write the half–reaction occurring at the anode. Remember, the reactant in each half–cell is written first, followed by the product.

Oxidation (loss of electrons) occurs at the anode. The half–reaction for this half–cell is

$Al\,(s)\;\rightarrow\;Al^{3+}\,(aq)\;+\;3\,e^-$

Step 2: Write the half–reaction occurring at the cathode.

Reduction (gain of electrons) occurs at the cathode. The half–reaction for this half–ell is

$Co^{2+}\,(aq)\;+\;2\,e^-\;\rightarrow\;Co\,(s)$

Step 3: Combine the two half–reactions to obtain the overall cell reaction. Don't forget that the number of electrons lost in an oxidation half–reaction must equal the number of electrons gained in the reduction half–reaction (Chapter 4).

For the number of electrons lost to equal the number of electrons gained, we must multiply the anode half–reaction by two and the cathode half–reaction by three. The two half–reactions would then be

$2\,Al\,(s)\;\rightarrow\;2\,Al^{3+}\,(aq)\;+\;6\,e^-$

$3\,Co^{2+}\,(aq)\;+\;6\,e^-\;\rightarrow\;3\,Co\,(s)$

Adding the two half–reactions together gives the overall equation:

$2\,Al\,(s)\;+\;3\,Co^{2+}\,(aq)\;\rightarrow\;2\,Al^{3+}\,(aq)\;+\;3\,Co\,(s)$

Step 5: Describe the cell indicated in the shorthand notation.

This cell would consist of a strip of aluminum as the anode dipping into an aqueous solution containing the Al^{3+} ion, such as $Al(NO_3)_3$. The cathode would consist of a strip of cobalt dipping into an aqueous solution of Co^{2+}, such as $CoSO_4$.

Key Concepts: Galvanic cells and oxidation–reduction reactions

WP 17.2

Strategy: Determine the position of the reducing agent relative to the position of the oxidizing agent, and based on your results, determine the spontaneity of the reactions.

In the first reaction, the reducing agent (the species undergoing oxidation) is Ag. Ag lies above the oxidizing agent in the table of standard reduction potentials. In the second reaction, the reducing agent is Br^-, which lies below the oxidizing agent, MnO_4^-.

A-61

Appendix A – Workbook Problems Solutions

Remember, to predict a spontaneous reaction, the reducing agent must lie below the oxidizing agent in the table of standard reduction potentials. For the first reaction, the reducing agent, Ag, lies above the oxidizing agent, Cu^{2+}. Therefore, the first reaction is nonspontaneous. For the second reaction, the reducing agent, Br^-, lies below the oxidizing agent, MnO_4^-. Therefore, the second reaction is spontaneous.

Key Concept: Use of the Table of Standard Reduction Potentials

WP 17.3

Strategy: Calculate the $E°_{cell}$ for the cell reaction and substitute the information given into the Nearnst equation and solve for [OH$^-$].

$$E°_{cell} = E°_{Ca \to Ca^{2+}} - E°_{ClO^- \to Cl^-}$$

$$E°_{cell} = 0.81 \text{ V} - (-2.87 \text{ V}) = 3.68 \text{ V}$$

$$E_{cell} = E°_{cell} - \frac{0.0592}{n} \log \frac{[Cl^-][Ca^{2+}][OH^-]^2}{[ClO^-]}$$

$$4.00 = 3.68 - \frac{0.0592}{2} \log[OH^-]^2$$

$$-10.81 = \log[OH^-]^2$$

Taking the antilog of each side gives

$$[OH^-]^2 = 1.55 \times 10^{-11} \qquad [OH^-] = 3.94 \times 10^{-6}$$

Step 1: Calculate the pH.

We first need to calculate the $[H_3O^+]$.

$$[H_3O^+] = \frac{1 \times 10^{-14}}{3.94 \times 10^{-6}} = 2.54 \times 10^{-9}$$

$$pH = -\log(2.54 \times 10^{-9}) = 8.60$$

Key Concept: Effect of the reaction mixture composition on the cell potential

WP 17.4

Strategy: First write the electrolysis reaction and consider the conversion process to calculate time.

$$Mg^{2+} + 2 e^- \to Mg (s)$$

Step 1: Treat the electrons as reactants in the chemical equation, and solve for the time required to produce 25 lbs of magnesium. Remember, the conversion factor of 96,500 C (or A·s) per mole of electrons.

Appendix A – Workbook Problems Solutions

$$25 \text{ lb Mg} \times \frac{1000 \text{ g}}{2.2056 \text{ lb}} \times \frac{1 \text{ mol Mg}}{24.3 \text{ g Mg}} \times \frac{2 \text{ mol e}^-}{1 \text{ mol Mg}} \times \frac{96{,}500 \text{ A} \cdot \text{s}}{1 \text{ mol e}^-} \times \frac{1}{20 \text{ A}} \times \frac{1 \text{ h}}{3600 \text{ s}} \times \frac{1 \text{ day}}{24 \text{ h}} = 52.1 \text{ days}$$

Key Concepts: Stoichiometry and the quantitative aspects of electrolysis

Chapter 18

There are no workbook problems in this chapter.

Chapter 19

WP 19.1

Step 1: Determine the number of moles of the compound that are present, using the freezing–point depression.

K_f (benzene) = 5.12°C·kg/mol

Using the equation, $\Delta T_f = K_f \times m$, we can solve for the number of moles of the boron compound.

$$6.37°\text{C} = \left(5.12° \text{ C} \cdot \text{kg/mol}\right) \frac{\text{mol compound}}{0.050 \text{ kg}}$$

$$\text{mol compound} = \frac{(6.37°\text{C})(0.050 \text{ kg})}{5.12° \text{ C} \cdot \text{kg/mol}} = 6.22 \times 10^{-2} \text{ mol}$$

Step 2: Calculate the molar mass from the number of moles and the mass of the compound.

$$\text{Molecular molar mass} = \frac{11.4 \text{ g}}{6.22 \times 10^{-2} \text{ mol}} = 183.2$$

Step 3: Determine the ratio of the molecular molar mass to the empirical molar mass. Multiply the empirical formula by the result.

$$\frac{183 \text{ g/mol}}{61.0 \text{ g/mol}} = 3$$

Molecular formula: $B_{15}H_{21}$

Key Concepts: Colligative properties, (freezing–point depression) and the Law of Definite Proportions

WP 19.2

Step 1: Determine the molecular formula.

The molecular formula is obtained by determining the ratio of the molecular molar mass to the empirical molar mass. The empirical formula is then multiplied by the result.

Appendix A – Workbook Problems Solutions

Empirical molar mass = 134.5 g/mol

$$\frac{269 \text{ g/mol}}{134.5 \text{ g/mol}} = 2$$

The molecular formula is Si_2Cl_6.

Step 2: Propose a structure.

To propose a structure, you need to take into consideration the descriptive chemistry of silicon. You know that silicon forms diamondlike structures and has relatively poor overlap of π orbitals.

From this information, you know that silicon can undergo sp^3 hybridization and will probably form structures much like sp^3 hybridized carbon. Therefore, a proposed structure for Si_2Cl_6 is to have 3 Cl's bonded to each Si and a Si – Si single bond.

$$\begin{array}{c} \text{Cl} \quad \text{Cl} \\ | \quad | \\ \text{Cl}-\text{Si}-\text{Si}-\text{Cl} \\ | \quad | \\ \text{Cl} \quad \text{Cl} \end{array}$$

Key Concepts: Law of Definite Proportion and the VSEPR theory

WP 19.3

Step 1: Write the balanced chemical equation.

$$NH_4NO_3 \,(s) \rightarrow N_2O \,(g) + 2\, H_2O \,(l)$$

Step 2: Calculate the pressure of N_2O, using the total pressure and the vapor pressure of water.

$$P_T = P_{H_2O} + P_{N_2O}$$

$$P_{N_2O} = 705 \text{ mm Hg} - 24 \text{ mm Hg} = 681 \text{ mm Hg}$$

Step 3: Calculate the number of moles of N_2O produced from 3.5 g NH_4NO_3.

$$3.5 \text{ g } NH_4NO_3 \times \frac{1 \text{ mol } NH_4NO_3}{80.0 \text{ g } NH_4NO_3} \times \frac{1 \text{ mol } N_2O}{1 \text{ mol } NH_4NO_3} = 4.38 \times 10^{-2} \text{ mol } N_2O$$

Step 4: Calculate the volume of N_2O, using the ideal gas law.

$$V = \frac{nRT}{P} = \frac{(4.38 \times 10^{-2})\left(0.0821 \frac{\text{L} \cdot \text{atm}}{\text{mol} \cdot \text{K}}\right)(295.15 \text{ K})}{681 \text{ mm Hg} \times \frac{1 \text{ atm}}{760 \text{ mm Hg}}} = 1.18 \text{ L}$$

Key Concepts: Law of conservation of mass, Dalton's law and the Ideal Gas Law

WP 19.4

Step 1: Write a balanced chemical equation for the two proposed reactions.

The unbalanced chemical equations for both possible reactions are

$PbS\ (s)\ +\ H_2O_2\ (aq)\ \rightarrow\ PbO_2\ (s)\ +\ SO_2\ (g)$

$PbS\ (s)\ +\ H_2O_2\ (aq)\ \rightarrow\ PbSO_4\ (s)$

Both reactions are redox reactions and must be balanced using the methods outlined in Chapter 4. In both reactions, the peroxide ion in H_2O_2 is being reduced to the oxide ion. From Appendix D in your textbook, you find that the half-reaction for the reduction of hydrogen peroxide is

$H_2O_2\ (aq)\ +\ 2\ H^+\ (aq)\ +\ 2\ e^-\ \rightarrow\ 2\ H_2O\ (l)$

For the reaction

$PbS\ (s)\ +\ H_2O_2\ (l)\ \rightarrow\ PbO_2\ (s)\ +\ SO_2\ (g)$

The two half-reactions are

$4\ H_2O\ (l)\ +\ PbS\ (s)\ \rightarrow\ PbO_2\ (s)\ +\ SO_2\ (g)\ +\ 8\ H^+\ +\ 8\ e^-$

$4\ H_2O_2\ (aq)\ +\ 8\ H^+\ (aq)\ +\ 8\ e^-\ \rightarrow\ 8\ H_2O\ (l)$

The overall balanced reaction is

$PbS\ (s)\ +\ 4\ H_2O_2\ (aq)\ \rightarrow\ PbO_2\ (s)\ +\ SO_2\ (g)\ +\ 4\ H_2O\ (l)$

For the reaction

$PbS\ (s)\ +\ H_2O_2\ (aq)\ \rightarrow\ PbSO_4\ (s)$

The two half-reactions are

$PbS\ (s)\ +\ 4\ H_2O\ (l)\ \rightarrow\ PbSO_4\ (s)\ +\ 8\ H^+\ +\ 8\ e^-$

$4\ H_2O_2\ (aq)\ +\ 8\ H^+\ (aq)\ +\ 8\ e^-\ \rightarrow\ 8\ H_2O\ (l)$

The overall balanced reaction is

$PbS\ (s)\ +\ 4\ H_2O_2\ (aq)\ \rightarrow\ PbSO_4\ (s)\ +\ 4\ H_2O\ (l)$

Step 2: Determine the $\Delta G°$ for each reaction.

For the reaction $PbS\ (s)\ +\ H_2O_2\ (aq)\ \rightarrow\ PbO_2\ (s)\ +\ SO_2\ (g)\ +\ 4\ H_2O\ (l)$

$\Delta G° = [(1\ mol\ PbO_2 \times -217.4\ kJ/mol\ PbO_2) + (1\ mol\ SO_2 \times -300.2\ kJ/mol\ SO_2) + (4\ mol\ H_2O \times -237.2\ kJ/mol\ H_2O)] - [(1\ mol\ PbS \times -98.7\ kJ/mol\ PbS) + (4\ mol\ H_2O_2 \times -120.4\ kJ/mol\ H_2O_2)]$

$\Delta G° = -886.1\ kJ$

For the reaction PbS (s) + 4 H$_2$O$_2$ (l) → PbSO$_4$ (s) + 4 H$_2$O (l)

$\Delta G° =$ [(1 mol PbSO$_4$ × −813.2 kJ/mol PbSO$_4$) + (4 mol H$_2$O × −237.2 kJ/mol H$_2$O)] − [(1 mol PbS × −98.7 kJ/mol PbS) + (4 mol H$_2$O$_2$ × −120.4 kJ/mol H$_2$O$_2$)]

$\Delta G° = -1,181.7$ kJ

Step 3: From your results, determine which reaction is more favorable.

The $\Delta G°$ for the reaction PbS (s) + 4 H$_2$O$_2$ (aq) → PbSO$_4$ (s) + 4 H$_2$O (l) is more negative than the $\Delta G°$ for the reaction PbS (s) + 4 H$_2$O$_2$ (aq) → PbO$_2$ (s) + SO$_2$ (g) + 4 H$_2$O (l). Therefore, the reaction PbS (s) + 4 H$_2$O$_2$ (l) → PbSO$_4$ (s) + 4 H$_2$O (aq) is more favored.

Key Concepts: Law of Conversation of Mass, the second law of thermodynamics, and Gibbs Free Energy

WP 19.5

Step 1: Write the balanced chemical equation recognizing that this is a redox reaction.

The two half-reactions are

6 H$_2$O (l) + I$_2$ (s) → 2 HIO$_3$ (aq) + 10 H$^+$ (aq) + 10 e$^-$

e$^-$ + H$^+$ (aq) + HNO$_3$ (aq) → NO$_2$ (g) + H$_2$O (l)

Electrons lost must equal electrons gained. Therefore,

6 H$_2$O (l) + I$_2$ (s) → 2 HIO$_3$ (aq) + 10 H$^+$ (aq) + 10 e$^-$

10 e$^-$ + 10 H$^+$ (aq) + 10 HNO$_3$ (aq) → 10 NO$_2$ (g) + 10 H$_2$O (l)

The overall balanced reaction is

I$_2$ (s) + 10 HNO$_3$ (aq) → 2 HIO$_3$ (aq) + 10 NO$_2$ (g) + 4 H$_2$O (l)

Step 2: Determine the number of moles of nitric acid needed from the mass of iodic acid and the balanced chemical equation.

$$15.0 \text{ g HIO}_3 \times \frac{1 \text{ mol HIO}_3}{175.9 \text{ g HIO}_3} \times \frac{10 \text{ mol HNO}_3}{2 \text{ mol HIO}_3} = 0.426 \text{ mol HNO}_3$$

Step 3: Determine the volume of nitric acid from the moles of nitric acid and the concentration of nitric acid.

$$0.426 \text{ mol HNO}_3 \times \frac{1 \text{ L}}{16.0 \text{ mol HNO}_3} = 2.66 \times 10^{-2} \text{ L} = 26.6 \text{ mL}$$

Key Concept: Law of Conservation of Mass

Chapter 20
There are no workbook problems in this chapter.

Chapter 21
There are no workbook problems in this chapter.

Chapter 22

WP 22.1
Step 1: Determine the amount of $^{99m}_{43}Tc$ (N) remaining after 2.50 hours (assume that $N_0 = 100$).

$$\ln\left(\frac{N}{N_o}\right) = -0.693\left(\frac{t}{t_{1/2}}\right)$$

$$\ln\left(\frac{N}{N_0}\right) = -0.693\left(\frac{2.50}{6.01}\right) = -0.288$$

$$\frac{N}{N_0} = \text{antiln}(-0.288) = 0.75$$

Step 2: Solve for the percent of $^{99m}_{43}Tc$ remaining

$N_0 = 100\%$ then $N = 75\%$.

Step 3: Determine the time needed for $^{99m}_{43}Tc$ to be reduced to 15%.

$$\ln\left(\frac{15}{100}\right) = -0.693\left(\frac{t}{6.01}\right)$$

$$-1.90 = -0.693\left(\frac{t}{6.01}\right)$$

$t = 16.5$ hours

Key Concept: Radioactive decay rates

WP 22.2
Step 1: First, calculate the total mass of the nucleons (95 n + 146 p)

Mass of 95 protons	(95)(1.007 28 amu)	= 95.691 60 amu
Mass of 146 neutrons	(146)(1.008 66 amu)	= 147.264 36 amu
Mass of 95 p + 146 n		= 242.955 96

Step 2: Determine the mass of the $^{241}_{95}Am$ nucleus by subtracting the mass of 95 e^- from the atomic mass.

Mass of $e^- = 5.486 \times 10^{-4}$ amu

Mass of $^{241}_{95}$Am = 241.056 823 amu − (95)(5.486×10^{-4} amu) = 241.004 71 amu

Step 3: Determine the mass defect.

Δm = 242.955 96 − 241.004 71 = 1.951 25 amu

Step 4: Convert the mass defect to grams per mole.

$$\frac{1.951\,25\,\text{amu}}{1\,\text{atom}} \times \frac{1.660\,54 \times 10^{-24}\,\text{g}}{1\,\text{amu}} \times \frac{6.022 \times 10^{23}\,\text{atoms}}{\text{mol}} = 1.951\,21\,\text{g/mol}$$

Step 5: Convert the mass defect into the binding energy.

$$\Delta E = 1.951\,21 \frac{\text{g}}{\text{mol}} \times \frac{10^{-3}\,\text{kg}}{1\,\text{g}} \times \left(3.00 \times 10^8 \frac{\text{m}}{\text{s}}\right)^2 = 1.76 \times 10^{14} \frac{\text{kg} \cdot \text{m}^2}{\text{s}^2 \cdot \text{mol}} = 1.76 \times 10^{14} \frac{\text{J}}{\text{mol}}$$

Step 6: Convert to MeV/nucleon.

$$1.76 \times 10^{14} \frac{\text{J}}{\text{mol}} \times \frac{1\,\text{mol}}{6.022 \times 10^{23}\,\text{nuclei}} \times \frac{1\,\text{MeV}}{1.60 \times 10^{-13}\,\text{J}} \times \frac{1\,\text{nucleus}}{241\,\text{nucleons}} = 7.56 \frac{\text{MeV}}{\text{nucleon}}$$

Key Concept: Energy changes during nuclear reactions

WP 22.3

Step 1: Calculate the mass change in the reaction.

$\Delta m = \Sigma$ mass of reactants − Σ mass of products

Δm = 241.0056 823 g/mol − [237.048 167 g/mol + 4.002 60 g/mol) = 6.056 × 10^{-3} g/mol

Step 2: Substitute the mass change into the Einstein equation, and solve for the energy.

$$\Delta E = 6.056 \times 10^{-3} \frac{\text{g}}{\text{mol}} \times \frac{10^{-3}\,\text{kg}}{\text{g}} \times \left(3.00 \times 10^8 \frac{\text{m}}{\text{s}}\right)^2 = 5.45 \times 10^{11} \frac{\text{kg} \cdot \text{m}^2}{\text{s}^2 \cdot \text{mol}} = 5.45 \times 10^{11} \frac{\text{J}}{\text{mol}}$$

Key Concepts: Nuclear fission and energy changes during nuclear reactions

Chapter 23

WP 23.1
Step 1: For each compound, determine the number of carbons in the chain based upon the parent name.

   a. 2–methyloctane: parent name – octane; has 8 carbons in the parent chain
   b. 3–ethylnonane: parent name – nonane; has 9 carbons in the parent chain
   c. isobutyldecane: parent name – decane; has 10 carbons in the parent chain

d. 5-ethyl-2,3-dimethylheptane: parent name – heptane; has seven carbons in the parent chain
e. 2,2,4-trimethylhexane: parent name – hexane; has six carbons in the parent chain

Step 2: Draw the parent chain and place the branches on the chain starting from either end. (Don't forget that carbon can have only four bonds. When adding branches, be sure to subtract the correct number of hydrogens.)

a.
$$\begin{array}{c} CH_3 \\ | \\ CH_3CHCH_2CH_2CH_2CH_2CH_2CH_3 \end{array}$$

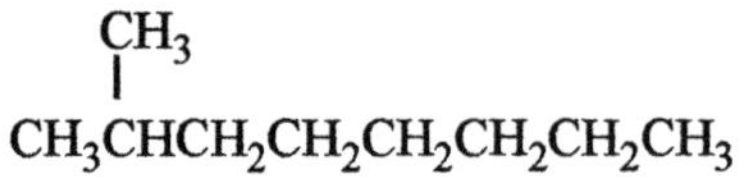

b.
$$\begin{array}{c} CH_3 \\ | \\ CHCH_3 \\ | \\ CH_3CH_2CHCH_2CH_2CH_2CH_2CH_3 \end{array}$$

c.
$$\begin{array}{c} H_3C CH_2CH_2CH_2CH_3 \\ \backslash | \\ CHCH_2CH \\ / | \\ H_3C CH_2CH_2CH_2CH_3 \end{array}$$

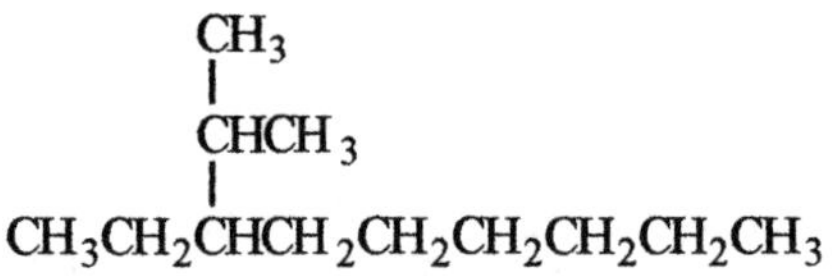

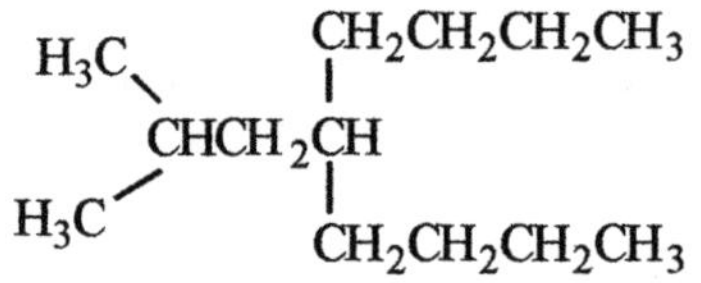

(Remember, removing the hydrogen from the end C on the branched-chain isobutane creates the isobutyl group. Also, C–C single bonds can rotate; therefore, the parent chain does not have to be all in one line.)

d.
$$\begin{array}{c} CH_3 CH_2CH_3 \\ | | \\ CH_3CHCHCH_2CHCH_2CH_3 \\ | \\ CH_3 \end{array}$$

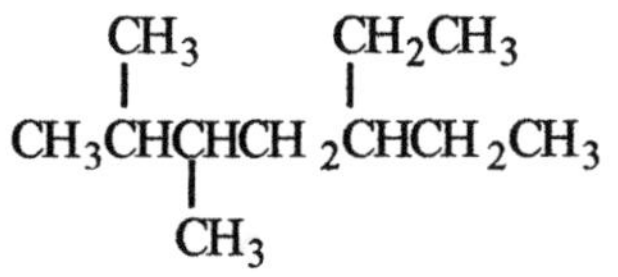

e.
$$\begin{array}{c} CH_3 \\ | \\ CH_3CCH_2CHCH_2CH_3 \\ | | \\ CH_3 CH_3 \end{array}$$

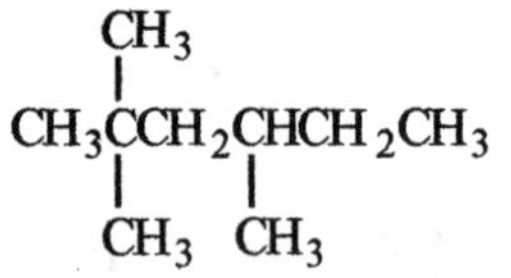

Key Concept: Naming alkanes

WP 23.2
Step 1: Use Table 23.1 on page 1000 in your textbook to identify the functional groups.

a. aldehyde
b. ether
c. ester
d. amide

Key Concept: Families of organic molecules.

Appendix A – Workbook Problems Solutions

WP 23.3

Step 1: Identify the parent chain in each molecule.

a. 3–octene: eight carbon chain with a double bond between the third and fourth carbon
b. 1–pentyne: five carbon chain with a triple bond between the first and second carbon
c. 1–pentene: five carbon chain with a double bond between the first and second carbon

Step 2: Draw the structure, placing the double or triple bond in the proper position. The positions of the branches are determined relative to the double or triple bond. Take into consideration any geometric isomers.

a. 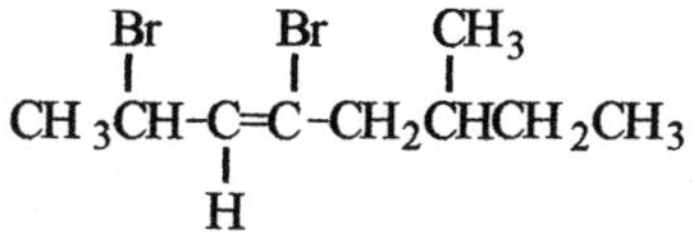 $CH_3CH-C=C-CH_2CHCH_2CH_3$ with Br, Br, H, and CH_3 substituents

b. $C{\equiv}C-CH-CHCH_3$ with Br, CH_3, and CH_3 substituents

c. $CH_3CH-CH_2-C=CH_3$ with CH_3 and CH_3 substituents

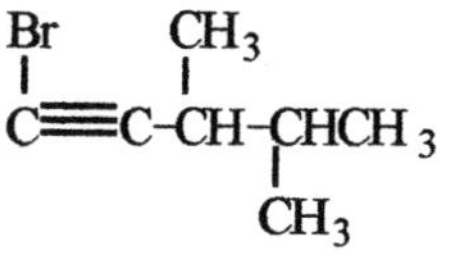

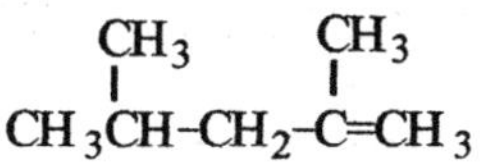

Key Concepts: Naming alkenes and alkynes

WP 23.4

Step 1: Draw either the ring or the parent chain placing the substituents in the correct position. For ring structures the actual placement of the substituents is not important, only that they are in the correct position relative to each other.

a. 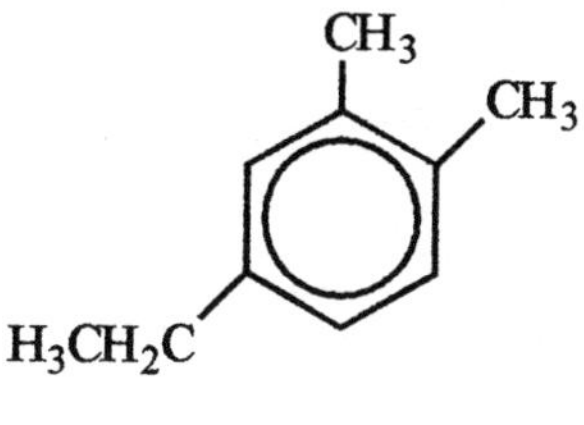 Benzene ring with CH_3, CH_3, and H_3CH_2C substituents

A-70

b. 4-bromochlorobenzene (Cl para to Br on benzene ring)

c. chlorobenzene

d. (2-pentyl)benzene — benzene ring with CH₃CHCH₂CH₂CH₃ substituent

Key Concept: Aromatic compounds and their reactions

 **PUTTING IT TOGETHER SOLUTIONS**

Chapter 3

We first need to write a balanced chemical equation for this problem. We will use the formula M_2S_3 for the metal sulfide since we know the metal has a +3 charge and the sulfide ion has a –2 charge (look at the periodic table). Our balanced equation is

$$M_2S_3 + 6\,HCl\,(aq) \rightarrow 2\,MCl_3 + 3\,H_2S\,(g)$$

We know that we used more HCl than we needed since we titrated the *excess* HCl with NaOH. We can determine the moles of HCl that reacted by calculating the difference between the original moles of HCl and the excess moles of HCl.

Original moles HCl = Reacted moles of HCl + Excess moles of HCl

$$\text{Original moles HCl} = 0.100\,L \times \frac{0.75\,\text{mol HCl}}{1\,L} = 0.075\,\text{mol HCl}$$

The excess moles of HCl are determined from the titration of NaOH with HCl, according to the reaction

$$HCl\,(aq) + NaOH\,(aq) \rightarrow NaCl\,(aq) + H_2O$$

$$\text{Excess moles HCl} = 0.0360\,L\,\text{NaOH} \times \frac{0.50\,M\,\text{NaOH}}{L} \times \frac{1\,\text{mol HCl}}{1\,\text{mol NaOH}} = 0.018\,\text{mol HCl}$$

Next, we calculate the number of moles of HCl that reacted with M_2S_3.

mol reacted HCl = 0.075 mol HCl – 0.018 mol HCl = 0.057 mol reacted HCl

We can now calculate the number of moles of M_2S_3 that reacted. (Refer back to the balanced equation.)

$$\text{mol}\,M_2S_3 = 0.057\,\text{mol HCl} \times \frac{1\,\text{mol}\,M_2S_3}{6\,\text{mol HCl}} = 0.0095\,\text{mol}\,M_2S_3$$

Knowing the number of moles of M_2S_3 and the number of grams of M_2S_3 that we started with, we can calculate the molar mass of M_2S_3.

$$\text{molar mass} = \frac{1.976\,g\,M_2S_3}{0.0095\,\text{mol}\,M_2S_3} = 208\,g/\text{mol}$$

Knowing the molar mass of M_2S_3 and the molar mass of sulfur, we can calculate the molar mass of M^{3+}.

Molar mass $M_2S_3 = (2 \times \text{molar mass}\,M^{3+}) + (3 \times \text{molar mass of}\,S^{2-})$

$2 \times \text{molar mass}\,M^{3+} = 208\,g/\text{mol} + (3 \times 32.0\,g/\text{mol}) = 112\,g/\text{mol}$

Appendix B – Putting It Together Solutions

$$\text{Molar mass } M^{3+} = \frac{112\,\text{g/mol}}{2} = 56.0\,\text{g/mol}$$

We have finally arrived at that moment when we can determine the identity of the metal. A quick study of the periodic table reveals that iron has a molar mass of 56.0 g/mol. We also know that iron can have a charge of +3.

We can determine the volume of H_2S (g) produced using the balanced chemical equation and knowing the density of H_2S (g) and the moles of HCl (aq).

$$\text{Volume of } H_2S \text{ produced} = 0.057\,\text{mol HCl} \times \frac{3\,\text{mol } H_2S}{6\,\text{mol HCl}} \times \frac{34.0\,\text{g } H_2S}{1\,\text{mol } H_2S} \cdot \frac{1\,\text{L } H_2S}{1.539\,\text{g } H_2S} = 0.630\,\text{L } H_2S = 630\,\text{mL } H_2S$$

Chapter 4

a. We know that our first reaction is between copper and nitric acid, that it is a redox reaction, and that the products are copper(II) nitrate and nitrogen dioxide gas. We can write an unbalanced chemical equation with this information and then use the rules for balancing redox reactions. (I prefer to use the half-reaction method.)

Cu (s) + HNO_3 (aq) → $Cu(NO_3)_2$ (aq) + NO_2 (g)

Net ionic equation:

Cu (s) + H^+ (aq) + NO_3^- (aq) → Cu^{2+} (aq) + NO_3^- (aq) + NO_2 (g)

Half-reactions:

Cu (s) → Cu^{2+} (aq)

H^+ (aq) + NO_3^- (aq) → NO_3^- (aq) + NO_2 (g)

The first half-reaction need be balanced only for charge.

Cu (s) → Cu^{2+} (aq) + 2 e^-

To balance the second half-reaction, we begin by balancing the number of N's present.

H^+ (aq) + 2 NO_3^- (aq) → NO_3^- (aq) + NO_2 (g)

We now balance the oxygens by adding water.

H^+ (aq) + 2 NO_3^- (aq) → NO_3^- (aq) + NO_2 (g) + H_2O

We now balance the hydrogens by adding H^+.

2 H^+ (aq) + 2 NO_3^- (aq) → NO_3^- (aq) + NO_2 (g) + H_2O

Finally, we balance for charge.

1 e^- + 2 H^+ (aq) + 2 NO_3^- (aq) → NO_3^- (aq) + NO_2 (g) + H_2O

We now need to make the electron count the same in both reactions.

Cu (s) → Cu^{2+} (aq) + 2 e$^-$

2(1 e$^-$ + 2 H$^+$ (aq) + 2 NO$_3^-$ (aq) → NO$_3^-$ (aq) + NO$_2$ (g) + H$_2$O)

Finally, we add the two half–reactions together and cancel anything that appears on both sides of the equation.

Cu (s) → Cu^{2+} (aq) + 2 e$^-$

2 e$^-$ + 4 H$^+$ (aq) + 4 NO$_3^-$ (aq) → 2 NO$_3^-$ (aq) + 2 NO$_2$ (g) + 2 H$_2$O

───

Cu (s) + 4 H$^+$ (aq) + 4 NO$_3^-$ (aq) → Cu^{2+} (aq) + 2 NO$_3^-$ (aq) + 2 NO$_2$ (g) + 2 H$_2$O

Once the Cu(NO$_3$)$_2$ is formed, it is reacted with NaOH. Using the solubility rules, we can write the molecular equation for this reaction.

Cu(NO$_3$)$_2$ (aq) + 2 NaOH (aq) → Cu(OH)$_2$ (s) + 2 NaNO$_3$ (aq)

The precipitate formed is Cu(OH)$_2$. This precipitate is heated and undergoes dehydration (the removal of water) forming an oxide. Since *matter cannot be created or destroyed,* we know that the oxide must be copper(II) oxide. (We know that the oxidation state of copper has not changed, since the description of the reaction indicates only the removal of water.)

2 Cu(OH)$_2$ (s) → CuO (s) + 2 H$_2$O

The next step involves reaction with sulfuric acid. Again, we use the solubility rules when writing the molecular equation.

CuO (s) + H$_2$SO$_4$ (aq) → CuSO$_4$ (aq) + H$_2$O

The final step in this cycle involves the reaction of CuSO$_4$ with zinc. Using the activity series, we know that zinc will replace copper in the CuSO$_4$ since Cu^{2+} is lower than zinc.

CuSO$_4$ (aq) + Zn (s) → Cu (s) + ZnSO$_4$ (aq)

b. We need to use the first balanced chemical equation to determine the amount of copper in the penny. We begin by determining the number of moles of HNO$_3$ used in the reaction.

$$\text{mol HNO}_3 = 0.057 \text{ L} \times \frac{7.5 \text{ mol HNO}_3}{\text{L soln}} = 0.428 \text{ mol HNO}_3$$

Knowing moles of HNO$_3$, we can now solve for grams of copper.

$$\text{g Cu} = 0.428 \text{ mol HNO}_3 \times \frac{1 \text{ mol Cu}}{4 \text{ mol HNO}_3} \times \frac{63.5 \text{ g Cu}}{1 \text{ mol Cu}} = 6.79 \text{ g Cu}$$

$$\% \text{ Cu} = \frac{6.79 \text{ g Cu}}{7.087 \text{ g sample}} \times 100 = 95.8\% \text{ Cu}$$

Appendix B – Putting It Together Solutions

Beginning in 1982, the U.S. Mint changed the composition of pennies from 95% copper and 5% zinc to 95% zinc and 5% copper.

c. We calculated that the mass of copper in the penny is 6.79 g. However, we recovered only 5.25 g of copper. First, let's calculate a percent recovery for copper.

$$\% \text{ recovery} = \frac{5.25 \text{ g Cu}}{6.79 \text{ g Cu}} \times 100 = 77.3\% \text{ recovery}$$

This percent recovery is not bad; however, it could be better. One possible source of error could easily be the loss of copper product during one of the many reactions. It is quite common to lose some of the copper (II) hydroxide when heating it to produce copper (II) oxide. This heating process takes a great deal of patience and should involve waving the test tube through the Bunsen burner. Many students are often impatient and hold the test tube directly over the flame, causing bumping to occur.

Chapter 5

a. We will use the half–reaction method to balance the redox reactions.

$$M^{3+} + Sn^{2+} \rightarrow M^{2+} + Sn^{4+}$$

For this reaction we need only to balance the charge in each half–reaction and then make the number of electrons lost equal the number of electrons gained.

$$2\,(M^{3+} + 1\,e^- \rightarrow M^{2+})$$

$$Sn^{2+} \rightarrow Sn^{4+} + 2\,e^-$$

Adding these two equations together gives: $2\,M^{3+} + Sn^{2+} \rightarrow 2\,M^{2+} + Sn^{4+}$

$$Sn^{2+} + HgCl_2 \rightarrow Sn^{4+} + Hg_2Cl_2 + 2\,Cl^-$$

The two half–reactions for this reaction are

$$Sn^{2+} \rightarrow Sn^{4+} + 2\,e^-$$

$$2\,HgCl_2 + 2\,e^- \rightarrow Hg_2Cl_2 + 2\,Cl^-$$

The number of electrons lost equals the number of electrons gained. The overall, balanced equation is

$$Sn^{2+} + 2\,HgCl_2 \rightarrow Sn^{4+} + Hg_2Cl_2 + 2\,Cl^-$$

$$M^{2+} + MnO_4^- \rightarrow M^{3+} + Mn^{2+}$$

The two half–reactions are

$$M^{2+} \rightarrow M^{3+} + 1\,e^-$$

$$MnO_4^- \rightarrow Mn^{2+}$$

We need to balance the oxygens in the second half–reaction.

$$MnO_4^- \rightarrow Mn^{2+} + 4\,H_2O$$

Now we balance the hydrogens.

$$8 H^+ + MnO_4^- \rightarrow Mn^{2+} + 4 H_2O$$

Finally, we balance for charge.

$$5 e^- + 8 H^+ + MnO_4^- \rightarrow Mn^{2+} + 4 H_2O$$

We need to make the number of electrons lost equal the number of electrons gained.

$$5 (M^{2+} \rightarrow M^{3+} + 1 e^-)$$

$$5 e^- + 8 H^+ + MnO_4^- \rightarrow Mn^{2+} + 4 H_2O$$

The overall reaction is

$$5 M^{2+} + 8 H^+ + MnO_4^- \rightarrow 5 M^{3+} + Mn^{2+} + 4 H_2O$$

b. $\text{mol KMnO}_4 = 0.0357 \text{ L soln} \times \dfrac{0.50 \text{ mol KMnO}_4}{1 \text{ L soln}} = 1.79 \times 10^{-2} \text{ mol KMnO}_4$

$\text{mol M} = 1.79 \times 10^{-2} \text{ mol KMnO}_4 \times \dfrac{5 \text{ mol M}}{1 \text{ mol KMnO}_4} = 8.93 \times 10^{-2} \text{ mol M}$

We calculate the molar mass of M to determine its identity.

$\text{Molar mass of M} = \dfrac{0.5 \text{ g M}}{8.93 \times 10^{-2} \text{ mol M}} = 56.0 \text{ g/mol}$

c. From the periodic table, we can identify M as Fe. The ground–state electron configuration of this atom is

$1s^2 \, 2s^2 \, 2p^6 \, 3s^2 \, 3p^6 \, 4s^2 \, 3d^6$

d. We can determine the wavelength from the ionization energy by first converting to units of joules and then using the equation for the energy of a photon.

$E = 7.0924 \text{ eV} \times \dfrac{1.602 \times 10^{-19} \text{ J}}{1 \text{ eV}} = 1.14 \times 10^{-18} \text{ J}$

$E = \dfrac{hc}{\lambda}$. Solving for λ gives $\lambda = \dfrac{hc}{E}$. Substituting the known values leads to

$\lambda = \dfrac{(6.626 \times 1^{-34} \text{ J} \cdot \text{s})(3.00 \times 10^8 \text{ m/s})}{1.14 \times 10^{-18} \text{ J}} = 1.75 \times 10^{-7} \text{ m}$

This wavelength would normally be reported in nanometers.

$\lambda = 175 \text{ nm}$

Chapter 7

Assume we have a 100 g sample. We then know that we have 40.9 g C, 54.5 g O and 4.5 g H. With this information, we can calculate the number of moles of each element present.

$$\text{mol C} = 40.9 \text{ g C} \times \frac{1 \text{ mol C}}{12.0 \text{ g C}} = 3.41 \text{ mol C}$$

$$\text{mol O} = 54.5 \text{ g O} \times \frac{1 \text{ mol O}}{16.0 \text{ g O}} = 3.41 \text{ mol O}$$

$$\text{mol H} = 4.5 \text{ g H} \times \frac{1 \text{ mol H}}{1.00 \text{ g H}} = 4.5 \text{ mol H}$$

Dividing by the smallest number of the three (3.4) gives 1 mol C, 1 mol O, and 1.3 mol H. To obtain the smallest whole-number ratio, we multiply each of these amounts by 3 to get 3 mol C, 3 mol O, and 4 mol H. The empirical formula is $C_3H_4O_3$. The empirical molar mass is 88.0 g/mol, the same as the molecular molar mass. Therefore, the empirical formula and the molecular formula are the same.

From the information given, we know that we have two C–C single bonds. Therefore, the three carbons in the molecular formula are bonded together. However, we also know that two of the carbons are sp^2-hybridized. If the three carbons are bonded together with only single bonds, this means that two of the carbons form double bonds to some element other than carbon. The element double-bonded to carbon cannot be hydrogen since hydrogen can have only two electrons surrounding it. Therefore, two of the carbons will each be double bonded to two oxygens. We also know that a majority of the hydrogens are bonded to a carbon. If the end carbon has a single bond to another carbon, it can form three bonds with hydrogen. We are now left with one oxygen and one hydrogen. The remaining oxygen is bonded to the other end carbon, with hydrogen bonded to that oxygen. (We can't have the reverse. If the hydrogen bonded to the carbon with the oxygen bonded to the hydrogen, four electrons would be surrounding the hydrogen.) The resulting structure is

```
      H    O   O
      |    ||  ||
  H—C—C—C—OH
      |
      H
```

Chapter 8

The unbalanced chemical equation for this reaction is:

$$NH_4NO_3 \text{ (s)} + Al \text{ (s)} \rightarrow N_2 \text{ (g)} + H_2O \text{ (g)} + Al_2O_3 \text{ (s)}$$

We can use the reported heats of formation in Appendix B to calculate the heat of reaction once we have balanced the equation. We can then determine the amount of heat evolved for the specific amount of NH_4NO_3.

We know from the information provided that this reaction is a redox reaction. (NH_4NO_3 is an oxidizing agent.) While it may be possible to balance this equation from inspection, let's use the rules for balancing redox reactions. Assume the reaction occurs in acidic conditions. The two half-reactions are

$NH_4NO_3 (s) \rightarrow N_2 (g)$

$2 Al (s) \rightarrow Al_2O_3 (s)$

Balance both equations for oxygens.

$NH_4NO_3 (s) \rightarrow N_2 (g) + 3 H_2O$

$2 Al (s) + 3 H_2O \rightarrow Al_2O_3 (s)$

Now balance the hydrogens.

$2 H^+ + NH_4NO_3 (s) \rightarrow N_2 (g) + 3 H_2O$

$2 Al (s) + 3 H_2O \rightarrow Al_2O_3 (s) + 6 H^+$

We now balance the electrons.

$2 e^- + 2 H^+ + NH_4NO_3 (s) \rightarrow N_2 (g) + 3 H_2O$

$2 Al (s) + 3 H_2O \rightarrow Al_2O_3 (s) + 6 H^+ + 6 e^-$

We now make the number of electrons lost equal the number of electrons gained by multiplying the top equation by 3.

$6 e^- + 6 H^+ + 3 NH_4NO_3 (s) \rightarrow 3 N_2 (g) + 9 H_2O$

$2 Al (s) + 3 H_2O \rightarrow Al_2O_3 (s) + 6 H^+ + 6 e^-$

We now add the two half-reactions to obtain the balanced equation.

$\cancel{6 e^-} + \cancel{6 H^+} + 3 NH_4NO_3 (s) \rightarrow 3 N_2 (g) + \overset{6}{\cancel{9}} H_2O$

$2 Al (s) + \cancel{3 H_2O} \rightarrow Al_2O_3 (s) + \cancel{6 H^+} + \cancel{6 e^-}$

───────────────────────────────────────

$3 NH_4NO_3 (s) + 2 Al (s) \rightarrow 3 N_2 (g) + 6 H_2O (g) + Al_2O_3 (s)$

We can now calculate the heat of reaction using the heats of formation found in Appendix B.

$\Delta H° =$
 [(1 mol Al_2O_3 × –1676 kJ/mol) + (6 mol H_2O × –241.8 kJ/mol)] – [(3 mol NH_4NO_3 × –365.6 kJ/mol)]
$\Delta H° = -2.03 \times 10^3$ kJ

This heat of reaction represents the heat evolved when 3 moles of NH_4NO_3 react. However, we know that we have 0.25 g of NH_4NO_3, which represents 3.13×10^{-3} mol of NH_4NO_3. The heat evolved will be

$$\frac{-2.03 \times 10^3 \text{ kJ}}{3 \text{ mol } NH_4NO_3} \times 3.13 \times 10^{-3} \text{ mol } NH_4NO_3 = -2.12 \text{ kJ}$$

Appendix B – Putting It Together Solutions

Chapter 9

Assume we have a 100 g sample. We then have 64.81 g C, 13.60 g H, and 21.59 g O. We can determine the empirical formula of this compound by calculating the moles of each substance.

$$64.81 \text{ g C} \times \frac{1 \text{ mol C}}{12.0 \text{ g C}} = 5.400 \text{ mol C}$$

$$13.60 \text{ g H} \times \frac{1 \text{ mol H}}{1.00 \text{ g H}} = 13.60 \text{ mol H}$$

$$21.59 \text{ g O} \times \frac{1 \text{ mol O}}{16.0 \text{ g O}} = 1.349 \text{ mol O}$$

Dividing by the smallest number of moles, we obtain the C:H:O mole ratio of 4.00:10.08:1.00. Therefore, the empirical formula is $C_4H_{10}O$. This empirical formula has a molar mass of 74.0 g/mol.

We can determine the molecular molar mass of the compound by using the Ideal Gas Law to calculate the number of moles of compound present in 2.57 g.

$$n = \frac{PV}{RT} = \frac{(0.420 \text{ atm})(2.00 \text{ L})}{\left(0.0821 \frac{\text{atm} \cdot \text{L}}{\text{mol} \cdot \text{K}}\right)(298.15 \text{ K})} = 0.0343 \text{ mol}$$

$$\text{Molar mass} = \frac{2.57 \text{ g}}{0.0343 \text{ mol}} = 74.9 \text{ g/mol}$$

The molecular molar mass and empirical molar mass are close enough in value to conclude that the empirical formula and the molecular formula are the same.

Chapter 10

We begin by drawing the electron–dot structure for SO_2. This molecule has a total of 18 valence electrons to distribute. Four electrons are used for bonding, and 12 electrons are used to complete the octets on the two oxygens. The remaining pair of electrons is placed on the sulfur. We need to create a double bond between one of the oxygens and the sulfur to complete the octet on sulfur. Either oxygen can be used; therefore, we have the following resonance structure:

The molecular shape is bent. The intermolecular forces present are dipole–dipole forces.

We can use the information found in Appendix B to calculate the heat of reaction for

$SO_2 (g) + 1/2 \, O_2 (g) \rightarrow SO_3 (g)$

$\Delta H° = [-395.7 \text{ kJ/mol}] - [-296.8 \text{ kJ/mol}] = -98.9 \text{ kJ/mol}$

Similarly, we can calculate the heat of reaction for

$SO_3 (g) + H_2O (l) \rightarrow H_2SO_4 (l)$

$\Delta H° = [-814.0 \text{ kJ/mol}] - [(-395.7 \text{ kJ/mol}) + (-285.8 \text{ kJ/mol})] = -132.5 \text{ kJ/mol}$

Chapter 11

We first need to determine the empirical formula from the combustion analysis. We can calculate the grams and moles of C and H from the masses of carbon dioxide and water produced.

$3.41 \text{ g } CO_2 \times \dfrac{1 \text{ mol } CO_2}{44.0 \text{ g } CO_2} \times \dfrac{1 \text{ mol C}}{1 \text{ mol } CO_2} = 7.76 \times 10^{-2} \text{ mol C}$ $7.76 \times 10^{-2} \text{ mol C} \times \dfrac{12.0 \text{ g C}}{1, \text{mol C}} = 0.930 \text{ g C}$

$1.40 \text{ g } H_2O \times \dfrac{1 \text{ mol } H_2O}{18.0 \text{ g } H_2O} \times \dfrac{2 \text{ mol H}}{1 \text{ mol } H_2O} = 0.155 \text{ mol H}$ $0.155 \text{ mol H} \times \dfrac{1.00 \text{ g H}}{1 \text{ mol H}} = 0.155 \text{ g H}$

We can calculate grams and moles of O from the grams of sample, carbon and hydrogen.

g sample = g C + g H + g O   g O = g sample − g C − g H = 1.50 g − 0.930 g − 0.155 g = 0.415 g

$0.415 \text{ g O} \times \dfrac{1 \text{ mol O}}{16.0 \text{ g O}} = 2.59 \times 10^{-2} \text{ mol O}$

Dividing by the smallest number of moles, we have a C:H:O mole ratio of 3:6:1. Therefore, our empirical formula is C_3H_6O. The molar mass for this formula is 58.0 g/mol.

We can determine the molecular molar mass from the osmotic pressure by first calculating the molarity of the solution we prepared.

$1.26 \text{ atm} = M \left(0.0821 \dfrac{L \cdot atm}{mol \cdot K} \right) (298.15 \text{ K})$ M = 0.0516 mol/L

mol of ethyl butyrate = $0.0516 \dfrac{mol}{L} \times 0.250 \text{ L} = 1.29 \times 10^{-2}$ mol

molecular molar mass = $\dfrac{1.50 \text{ g}}{1.29 \times 10^{-2} \text{ mol}} = 116.3 \text{ g/mol}$

The molecular molar mass is two times the empirical molar mass; therefore, the molecular formula is $C_6H_{12}O_2$.

Chapter 12

a. First step: molecularity = 1
Second step: molecularity = 2

b. Rate = $k[Ni(CO)_4]$

Appendix B – Putting It Together Solutions

c. From the preceding rate law, we know that this reaction is first–order in Ni(CO)$_4$. If the reaction is 80% complete, 20% of the initial amount of Ni(CO)$_4$ remains.

20% of Ni(CO)$_4$ = 0.20 × 0.25 M = 0.05 M Ni(CO)$_4$ remaining

The integrated rate law becomes

$$\ln\left(\frac{0.05}{0.25}\right) = -(9.9\times 10^{-3}\ s^{-1})t$$

$t = 162.6\ s = 2.71\ min$

d. Before calculating $\Delta H°$, we need to write a balanced chemical equation.

Ni (s) + 4 CO (g) → Ni(CO)$_4$ (g)

$\Delta H°$ = [1 mol Ni(CO)$_4$ × –602.91 kJ/mol] – [4 mol CO × –110.5 kJ/mol] = –160.9 kJ/mol

e. From the information given, we know we have a limiting reactant problem. Therefore, we need to solve for the number of moles of Ni (s) and CO (g).

mol Ni = 2.52 × 10^{-3}

$$n\ (CO) = \frac{PV}{RT} = \frac{0.95\ atm(0.730\ L)}{(0.0821\ L\cdot atm/mol\cdot K)(301.15\ K)} = 2.8\times 10^{-2}\ mol\ CO$$

The mole ratio for Ni to CO is 1:4. Therefore, we need 4 times as much CO as Ni. Multiplying the number of moles of Ni by 4, we obtain 1.02× 10^{-2} moles of CO needed to react with 2.55 × 10^{-2} moles of Ni. Therefore, CO is in excess and Ni is the limiting reactant. We can now calculate the grams of CO formed.

$$g\ Ni(CO)_4 = 2.52\times 10^{-3}\ mol\ Ni \times \frac{1\ mol\ Ni(CO)_4}{1\ mol\ Ni} \times \frac{170.7\ g\ Ni(CO)_4}{1\ mol\ Ni(CO)_4} = 0.430\ g\ Ni(CO)_4$$

Chapter 13

We can use Dalton's law to determine the pressure of all species present at equilibrium. However, it will be helpful to first set up a table showing the initial pressure, the changes that occur and the equilibrium pressure. Convert the units from mm Hg to atmosphere.

	2 NO (g) +	Br$_2$ (g) ⇌	2 NOBr (g)
Initial Pressure	0.129	0.0543	0
Change	– 2x	– x	+ 2x
Eq Pressure	0.129 – 2x	0.0543 – x	2x

The total pressure at equilibrium is equal to the sum of the pressures of each species at equilibrium. The total pressure at equilibrium is 0.1454 atm.

0.1454 = (0.129 – 2x) + (0.0543 – x) + 2x

x = 0.0379 atm

Appendix B – Putting It Together Solutions

At equilibrium

P(NO) = 0.129 atm − 2(0.0379 atm) = 0.0532 atm
P(Br$_2$) = 0.0543 atm − 0.0379 atm = 0.01643 atm
P(NOBr) = 2(0.0379 atm) = 0.0758 atm

$$K_p = \frac{(0.0758)^2}{(0.0532)^2(0.01643)} = 123.6$$

Chapter 14

To determine the amount of aniline used, we need to first know the theoretical yield expected in this reaction.

$$\% \text{ yield} = \frac{\text{Experimental yield}}{\text{Theoretical yield}} \times 100$$

$$\text{Theoretical yield} = \frac{200 \text{ g}}{90} \times 100 = 222 \text{ g}$$

We now calculate the number of moles of sulfanilic acid in 222 g of sulfanilic acid.

mol sulfanilic acid = 1.28 mol

The mole ratio of aniline to sulfanilic acid is 1:1. Therefore, we have 1.28 mol of aniline. The mass of aniline is 119 g. From this mass and the density, we calculate a volume of 117 mL.

Na(C$_6$H$_4$NH$_2$SO$_3$), sodium sulfanilate, contains the anion of a weak acid and the cation of a strong base. Therefore, the aqueous solution of this salt should be basic. The hydrolysis reaction for this salt is

C$_6$H$_4$NH$_2$SO$_3^-$ (aq) + H$_2$O (l) ⇌ HC$_6$H$_4$NH$_2$SO$_3$ (aq) + OH$^-$ (aq)

$$K_b = \frac{K_w}{K_a} = \frac{1 \times 10^{-14}}{5.9 \times 10^{-4}} = 1.69 \times 10^{-11}$$

We can calculate the pH as we do for any base dissociation problem.
Before setting up our table, we need to calculate the initial concentration of the sulfanilate ion.

$$\text{mol C}_6\text{H}_4\text{NH}_2\text{SO}_3^- = 3.75 \text{ g NaC}_6\text{H}_4\text{NH}_2\text{SO}_3\text{H} \times \frac{1 \text{ mol NaC}_6\text{H}_4\text{NH}_2\text{SO}_3\text{H}}{196 \text{ g NaC}_6\text{H}_4\text{NH}_2\text{SO}_3\text{H}} \times \frac{1 \text{ mol C}_6\text{H}_4\text{NH}_2\text{SO}_3^-}{1 \text{ mol NaC}_6\text{H}_4\text{NH}_2\text{SO}_3\text{H}}$$

$$= 1.91 \times 10^{-2} \text{ mol C}_6\text{H}_4\text{NH}_2\text{SO}_3^-$$

$$M = \frac{1.91 \times 10^{-2} \text{ mol C}_6\text{H}_4\text{NH}_2\text{SO}_3^-}{0.500 \text{ L}} = 3.83 \times 10^{-2} \text{ M}$$

We can now set up our table.

Appendix B – Putting It Together Solutions

Principal Reaction	$C_6H_4NH_2SO_3^-$ (aq) + H_2O (l) $\rightleftharpoons$ $HC_6H_4NH_2SO_3$ (aq) + OH^- (aq)		
Initial concentration	3.83×10^{-2}	0	0
Change	$-x$	$+x$	$+x$
Eq. concentration	$(3.83 \times 10^{-2}) - x$	$+x$	$+x$

We can assume that x is negligible, given the value of K_b. Therefore, $(3.83 \times 10^{-2}) - x \approx x$.

$$1.69 \times 10^{-11} = \frac{x^2}{3.83 \times 10^{-2}}$$

$$x^2 = 6.49 \times 10^{-13}$$

$$x = [OH^-]_e = 8.06 \times 10^{-7} \text{ M}$$

We can now calculate the $[H_3O^+]$.

$$K_w = [H_3O^+][OH^-]$$

$$[H_3O^+] = \frac{1 \times 10^{-14}}{8.06 \times 10^{-7}} = 1.24 \times 10^{-8}$$

$$pH = -\log(1.24 \times 10^{-8}) = 7.91$$

Chapter 15

a. $AlCl_3$ (aq) + H_3PO_4 (aq) $\rightarrow$ $AlPO_4$ (s) + 3 HCl (aq)

b. The amount of $AlPO_4$ produced is obtained by first determining the limiting reactant in this problem. Since the mole ratio of $AlCl_3$ to H_3PO_4 is 1:1, the reactant that is present in the least amount is the limiting reactant and will determine the amount of $AlPO_4$ produced.

$$\text{mol } AlCl_3 = 125 \text{ g } AlCl_3 \times \frac{1 \text{ mol } AlCl_3}{133.5 \text{ g } AlCl_3} = 0.936 \text{ mol } AlCl_3$$

$$\text{mol } H_3PO_4 = 2.50 \text{ L} \frac{0.75 \text{ mol } H_3PO_4}{\text{L}} = 1.875 \text{ mol } H_3PO_4$$

The limiting reactant is $AlCl_3$.

$$\text{g } AlPO_4 = 0.936 \text{ mol } AlCl_3 \times \frac{1 \text{ mol } AlPO_4}{1 \text{ mol } AlCl_3} \times \frac{122 \text{ g } AlPO_4}{1 \text{ mol } AlPO_4} = 114 \text{ g } AlPO_4$$

c. $AlPO_4$ (s) $\rightleftharpoons$ Al^{3+} (aq) + PO_4^{3-} (aq) $K_{sp} = 1.3 \times 10^{-20}$

At first glance, you may believe that to calculate the equilibrium concentrations of Al^{3+} and PO_4^{3-} you need to know the initial concentration of $AlPO_4$. However, the equilibrium concentrations are determined by the number of moles of Al^{3+} and PO_4^{3-} formed and the K_{sp}. Let x represent the number of moles of $AlPO_4$ that will dissolve. Therefore, the number of moles of Al^{3+} and PO_4^{3-} formed is also x. You now have

$K_{sp} = 1.3 \times 10^{-20} = (x)(x) = x^2;$ $x = 1.1 \times 10^{-10}$ M

d. The solubility of AlPO$_4$ will increase on the addition of HCl since the hydrogen ion can react with the phosphate ion to produce HPO$_4^{2-}$.

$H^+ (aq) + PO_4^{3-} (aq) \rightleftarrows HPO_4^{2-} (aq)$

$K_b(PO_4^{3-}) = \dfrac{K_w}{K_{a3}(H_3PO_4)} = \dfrac{1 \times 10^{-14}}{4.8 \times 10^{-13}} = 2.08 \times 10^{-2}$

Removal of the PO$_4^{3-}$ from solution will cause the reactant to shift right; thereby increasing the solubility of AlPO$_4$.

e. To determine if a precipitate will form, calculate Q and then compare this value to K_{sp}. (Remember that Q, the ion product, is calculated using the same formula for K_{sp}.

$Q = [Al^{3+}][PO_4^{3-}]$

Calculate the concentration of Al^{3+} and PO$_4^{3-}$ from the total volume of mixing the two solutions. Remember that when you dilute the solutions $M_f \times V_f = M_i \times V_i$. The total volume of the solution is 2.00 L.

$[Al^{3+}] = \dfrac{0.75 \text{ L} \times (5.00 \times 10^{-3} \text{ M})}{2.00 \text{ L}} = 1.88 \times 10^{-3}$ M

$[PO_4^{3-}] = \dfrac{1.25 \text{ L}(0.750 \text{ M})}{2.00 \text{ L}} = 0.469$ M

$Q = (1.88 \times 10^{-3})(0.469) = 8.82 \times 10^{-4}$

$Q \gg K_{sp}$; therefore, a precipitate will form. Given that the molar solubility of AlPO$_4$ is so small (1.1×10^{-10}), you can calculate the mass of AlPO$_4$ based on the number of moles of the limiting reactant, Al^{3+}.

mol Al^{3+} = 3.75 × 10^{-3}; mol PO$_4^{3-}$ = 0.938

g AlPO$_4$ = $3.75 \times 10^{-3} \times \dfrac{1 \text{ mol AlPO}_4}{1 \text{ mol Al}^{3+}} \times \dfrac{122 \text{ g AlPO}_4}{1 \text{ mol AlPO}_4} = 0.458$ g AlPO$_4$

Chapter 16

$\Delta G^\circ = -RT\ln K;$ $T = (273.15 + 37) = 310.15$ K

$-14{,}000 \text{ J} = -\left(8.314 \dfrac{}{\text{mol} \cdot \text{K}}\right)(310.15 \text{ K}) \ln K$

$\ln K = 5.43$ $K = 230$

$$K = \frac{[\text{Hb} \cdot \text{CO}][\text{O}_2]}{[\text{Hb} \cdot \text{O}_2][\text{CO}]} = 230$$

$$\frac{[\text{Hb} \cdot \text{CO}]}{[\text{Hb} \cdot \text{O}_2]} = 230$$

Chapter 17

$E°$ is determined by first writing the 2 half–reactions for the overall reaction and solving for $E° = E_{\text{anode}} + E_{\text{cathode}}$.

$\text{O}_2\,(g) + 4\,\text{H}^+\,(aq) + 4\,e^- \rightarrow 2\,\text{H}_2\text{O}\,(l)$ \qquad $E° = 1.23$ V

$2\,\text{Br}^-\,(aq) \rightarrow \text{Br}_2\,(l) + 2\,e^-$ \qquad $E° = -1.09$ V

$E° = 1.23$ V $+ (-1.09$ V$) = 0.14$

The reaction is spontaneous since $E°$ is positive.

To determine the value of E after addition of the buffer, we need to first determine the [H$^+$] and then substitute this value into the Nernst equation.

$K_a\,(\text{HCHO}_2) = 1.8 \times 10^{-4}$; \qquad p$K_a = 3.74$

$$\text{pH} = \text{p}K_a + \log\frac{[\text{CHO}_2^-]}{[\text{HCHO}_2]} = 3.74 + \log\frac{0.25}{0.35} = 3.59$$

$$[\text{H}^+] = \text{antilog}(-3.59) = 2.57 \times 10^{-4}$$

We now have the information needed to substitute into the Nernst equation.

$$E = E° - \frac{0.0592}{n}\log\frac{1}{[\text{Br}^-][\text{H}^+]}$$

After balancing the number of electrons in the two half–reactions, $n = 4$. The equation then becomes

$$E = 0.14\text{ V} - \frac{0.0592}{4}\log\left[\frac{1}{(2.57\times 10^{-4})}\right]^4 = -0.0725\text{ V}$$

The reaction is not spontaneous under these non-standard state conditions.

Chapter 18

By now, you should recognize that this problem is a limiting reactant problem from the information given. Before we can calculate the volume of hydrogen, we need to know the number of moles of hydrogen. We can determine the number of moles of hydrogen produced by identifying the limiting reactant. However, before we can begin any stoichiometric reaction, we need to write a balanced chemical equation.

Appendix B – Putting It Together Solutions

The unbalanced chemical equation for the reaction that is described is

NaH (s) + SO$_2$ (l) → Na$_2$S$_2$O$_4$ (s) + H$_2$ (g)

First, balance all elements other than hydrogen and oxygen.

2 NaH (s) + 2 SO$_2$ (l) → Na$_2$S$_2$O$_4$ (s) + H$_2$ (g)

We now have a balanced equation and do not need to use either the oxidation–number method or half–reaction method.

Determining the limiting reactant is very straightforward, given that we have a 1:1 mole ratio of NaH to SO$_2$. Therefore, the limiting reactant is the reactant containing the smaller number of moles.

$$\text{mol NaH} = 32.83 \text{ g NaH} \times \frac{1 \text{ mol NaH}}{24.0 \text{ g NaH}} = 1.367 \text{ mol NaH}$$

$$\text{mass SO}_2 = 125.25 \text{ mL} \times \frac{1.434 \text{ g SO}_2}{1 \text{ mL}} = 180.00 \text{ g SO}_2$$

$$\text{mol SO}_2 = 180.00 \text{ g SO}_2 \times \frac{1 \text{ mol SO}_2}{64.0 \text{ g SO}_2} = 2.8125 \text{ mol SO}_2$$

The limiting reactant is NaH. We can now calculate moles of H$_2$.

$$\text{mol H}_2 = 1.367 \text{ mol NaH} \times \frac{1 \text{ mol H}_2}{2 \text{ mol NaH}} = 0.6835 \text{ mol H}_2$$

Knowing moles of hydrogen, we can calculate the volume of hydrogen, using the ideal gas law.

$$V = \frac{nRT}{P} = \frac{0.6835 \text{ mol H}_2 \left(0.0821 \frac{\text{L} \cdot \text{atm}}{\text{mol} \cdot \text{K}}\right)(298.15 \text{ K})}{1 \text{ atm}} = 16.73 \text{ L}$$

Chapter 19

As a first attempt to obtain the balanced equation, write the reactants and products as stated in the problem.

SO$_2$ (g) + 2 H$_2$S (aq) → 3 S (s) + 2 H$_2$O (l)

To determine the volume of H$_2$S needed, first calculate the moles of SO$_2$ (g). Assume that all of the sulfur in one ton of coal is converted to SO$_2$ (g).

$$\text{mol SO}_2 = 0.035 \text{ tons S} \times \frac{2000 \text{ lb}}{1 \text{ ton}} \times \frac{453.59 \text{ g}}{1 \text{ lb}} \times \frac{1 \text{ mol S}}{32.0 \text{ g S}} \times \frac{1 \text{ mol SO}_2}{1 \text{ mol S}} = 992.2 \text{ mol SO}_2$$

From the balanced equation, we know that 2 mol of H$_2$S react with 1 mol of SO$_2$. Therefore, the moles of H$_2$S (aq) needed is 1.98×10^3.

Now use the ideal gas law to calculate the volume of H$_2$S.

Appendix B – Putting It Together Solutions

$$V = \frac{\left(0.0821 \frac{L \cdot atm}{K \cdot mol}\right)(1.98 \times 10^3 \text{ mol})(300 \text{ K})}{\left(740 \text{ mm Hg} \times \frac{1 \text{ atm}}{760 \text{ mm Hg}}\right)} = 5.01 \times 10^4 \text{ L}$$

Calculate the amount of sulfur produced from the moles of SO_2.

$$g \text{ S} = 992.2 \text{ mol } SO_2 \times \frac{3 \text{ mol S}}{1 \text{ mol } SO_2} \times \frac{32.0 \text{ g S}}{1 \text{ mol S}} = 9.53 \times 10^4 \text{ g S}$$

Chapter 20

First, write the balanced net ionic equation for the reaction. Given that this is a redox reaction involving oxygen, use the method of half–reactions.

$Cr_2O_7^{2-}$ (aq) → Cr^{2+} (aq)

Fe^{2+} (aq) → Fe^{3+} (aq)

The first half–reaction is balanced by first balancing the chromium and then balancing the oxygen and hydrogens by the addition of H_2O and H^+. The result is

8 e^- + 14 H^+ (aq) + $Cr_2O_7^{2-}$ (aq) → 2 Cr^{2+} (aq) + 7 H_2O (l)

The second half–reaction is balanced by balancing the charge

Fe^{2+} (aq) → Fe^{3+} (aq) + e^-

Multiply the second half–reaction by 8 so that the number of electrons lost equals the number of electrons gained.

8 e^- + 14 H^+ (aq) + $Cr_2O_7^{2-}$ (aq) → 2 Cr^{2+} (aq) + 7 H_2O (l)

8 Fe^{2+} (aq) → 8 Fe^{3+} (aq) + 8 e^-

Adding the two half–reactions gives the following balanced net ionic equation:

14 H^+ (aq) + $Cr_2O_7^{2-}$ (aq) + 8 Fe^{2+} (aq) → 2 Cr^{2+} (aq) + 7 H_2O (l) + 8 Fe^{3+} (aq)

With this balanced reaction and the data provided, determine the mass of iron in the ore.

$$g \text{ Fe} = 0.02205 \text{ L} \times \frac{0.105 \text{ mol } Cr_2O_7^{2-}}{1 \text{ L}} \times \frac{8 \text{ mol } Fe^{2+}}{1 \text{ mol } Cr_2O_7^{2-}} \times \frac{1 \text{ mol Fe}}{1 \text{ mol } Fe^{2+}} \times \frac{55.8 \text{ g Fe}}{1 \text{ mol Fe}} = 1.03 \text{ g Fe}$$

$$\% \text{ Fe} = \frac{1.03 \text{ g Fe}}{1.213 \text{ g sample}} \times 100 = 84.9\%$$

Appendix B – Putting It Together Solutions

Chapter 21

Write the balanced, net ionic equations for the reaction of Fe^{2+} with MnO_4^- and the reaction of Fe^{2+} with $Cr_2O_7^{2-}$ (see Chapter 20 Putting It Together problem).

The two half–reactions for the reaction of Fe^{2+} with MnO_4^- in acidic solution are

MnO_4^- (aq) → Mn^{2+} (aq)

Fe^{2+} (aq) → Fe^{3+} (aq)

For the first half–reaction, balance the oxygens by the addition of water and H^+.

$5\,e^- + 8\,H^+$ (aq) + MnO_4^- (aq) → Mn^{2+} (aq) + $4\,H_2O$ (l)

For the second half–reaction, balance the charge.

Fe^{2+} (aq) → Fe^{3+} (aq) + e^-

Now multiply the second half–reaction by 5 so that the number of electrons lost equals the number of electrons gained.

$5\,e^- + 8\,H^+$ (aq) + MnO_4^- (aq) → Mn^{2+} (aq) + $4\,H_2O$ (l)

$5\,Fe^{2+}$ (aq) → $5\,Fe^{3+}$ (aq) + $5\,e^-$

Adding the two half–reactions together gives

$8\,H^+$ (aq) + MnO_4^- (aq) + $5\,Fe^{2+}$ (aq) → Mn^{2+} (aq) + $4\,H_2O$ (l) + $5\,Fe^{3+}$ (aq)

The balanced, net ionic equation for the reaction of Fe^{2+} with $Cr_2O_7^{2-}$ is

$14\,H^+$ (aq) + $Cr_2O_7^{2-}$ (aq) + $8\,Fe^{2+}$ (aq) → $2\,Cr^{2+}$ (aq) + $7\,H_2O$ (l) + $8\,Fe^{3+}$ (aq)

From the molarity and volume of $FeSO_4$, we know that we have 4.0×10^{-3} moles of Fe^{2+}. We also know that

Moles of Fe^{2+} consumed in the first reaction + Moles of Fe^{2+} consumed in the second reaction = Total moles of Fe^{2+} = 4.0×10^{-3} moles Fe^{2+}

By determining the number of moles of Fe^{2+} consumed in the second reaction, we can determine the moles of Fe^{2+} consumed in the first reaction, which will allow us to determine the moles of manganese present.

For the reaction

$14\,H^+$ (aq) + $Cr_2O_7^{2-}$ (aq) + $8\,Fe^{2+}$ (aq) → $2\,Cr^{2+}$ (aq) + $7\,H_2O$ (l) + $8\,Fe^{3+}$ (aq)

$$\text{mol } Fe^{2+} = 0.0224\,L \times \frac{0.0100 \text{ mol } Cr_2O_7^{2-}}{1\,L} \times \frac{8 \text{ mol } Fe^{2+}}{1 \text{ mol } Cr_2O_7^{2-}} = 1.79 \times 10^{-3} \text{ mol } Fe^{2+}$$

The moles of Fe^{2+} consumed in the reaction

Appendix B – Putting It Together Solutions

$$8\,H^+\,(aq) + MnO_4^-\,(aq) + 5\,Fe^{2+}\,(aq) \rightarrow Mn^{2+}\,(aq) + 4\,H_2O\,(l) + 5\,Fe^{3+}\,(aq)$$

Mole Fe^{2+} = $(4.00 \times 10^{-3}\,\text{mol}\,Fe^{2+}) - (1.79 \times 10^{-3}\,\text{mol}\,Fe^{2+}) = 2.21 \times 10^{-3}$ Moles Fe^{2+} consumed in the first reaction

We can now calculate the mass of manganese in the steel sample.

$$g\,Mn = (2.21 \times 10^{-3}\,\text{mol}\,Fe^{2+}) \times \frac{1\,\text{mol}\,MnO_4^-}{5\,\text{mol}\,Fe^{2+}} \times \frac{1\,\text{mol}\,Mn}{1\,\text{mol}\,MnO_4^-} \times \frac{54.9\,\text{g}\,Mn}{1\,\text{mol}\,Mn} = 0.0243\,\text{g}\,Mn$$

$$\%\,Mn = \frac{0.0243\,\text{g}\,Mn}{0.450\,\text{g}\,\text{steel}} \times 100 = 5.4\%$$

Chapter 22

The formula for sodium perchlorate is $NaClO_4$. The amount of chlorine in 37.8 mg in $NaClO_4$ is

$$37.8\,\text{mg}\,NaClO_4 \times \frac{1\,\text{mmol}\,NaClO_4}{122.5\,\text{mg}\,NaClO_4} \times \frac{1\,\text{mmol}\,Cl}{1\,\text{mmol}\,NaClO_4} \times \frac{35.5\,\text{mg}\,Cl}{1\,\text{mmol}\,Cl} = 11.0\,\text{mg}\,Cl$$

Of the 11.0 mg Cl present, 31.0% is ^{36}Cl or 3.41 mg.

From the information provided, we can calculate k.

$$t_{1/2} = (3.0 \times 10^5\,\text{yr}) \times \left(\frac{365\,d}{1\,\text{yr}}\right) \times \left(\frac{24\,h}{1\,d}\right) \left(\frac{3600\,s}{1\,h}\right) = 9.46 \times 10^{12}\,s$$

$$k = \frac{0.693}{9.46 \times 10^{12}\,\text{sec}} = 7.32 \times 10^{-14}\,s^{-1}$$

We know that the decay rate = $k \times N$, where N is the number of radioactive nuclei. We can calculate N from the 3.41 mg ^{36}Cl.

$$N = (3.41 \times 10^{-3}\,g\,^{36}Cl) \times \frac{1\,\text{mol}\,^{36}Cl}{36.0\,g\,^{36}Cl} \times \frac{6.022 \times 10^{23}\,\text{nuclei}\,^{36}Cl}{1\,\text{mol}\,^{36}Cl} = 5.70 \times 10^{19}\,\text{nuclei}$$

We can now calculate the decay rate:

Rate = $(7.32 \times 10^{-14}\,s^{-1}) \times 5.70 \times 10^{19}$ nuclei = 4.17×10^6 nuclei/s or 4.17×10^6 disintegrations/s.

Chapter 23

Let's assume we have 100 g of sample. We then have:

71.89 g C; 5.991 mol C
12.13 g H; 12.13 mol H
15.98 g O; 0.9988 mol O

The empirical formula is $C_{5.991}H_{12.13}O_{0.9988}$. Dividing by the smallest subscript gives $C_6H_{12}O$. We know that when an ester undergoes hydrolysis, it falls apart into its constituent alcohol and carboxylic acid. We also know that an ester has the carboxylate group COO^-. Therefore, we know that we need to at least double the value of the subscripts in the empirical formula so that we have two oxygens. This gives us $C_{12}H_{24}O_2$, which has a molar mass of 200 g/mol. Subtracting the molar mass of the $CH_3CH_2O^-$ fragment, which adds the H^+ from water (45 g/mol), leaves us with a molar mass of 155 g/mol. Adding 17 g/mol to this value (to compensate for the addition of the OH^- group to the acid) gives us a molar mass of 172 g/mol. Therefore, the molecular formula for ethyl caprate is $C_{12}H_{24}O_2$.

Chapter 24

Using the Henderson–Haselbach equation, we can calculate the ratio of [AMP—OH^-] to [AMP—O^{2-}].

$$pH = pK_a + \log \frac{[AMP-O^{2-}]}{[AMP-OH^-]}$$

$$7.40 = 7.21 + \log \frac{[AMP-O^{2-}]}{[AMP-OH^-]}$$

$$\log \frac{[AMP-O^{2-}]}{[AMP-OH^-]} = 7.40 - 7.21 = 0.19$$

$$\frac{[AMP-O^{2-}]}{[AMP-OH^-]} = 1.55; \quad \frac{[AMP-O^{2-}]}{[AMP-OH^-]} = 0.646$$

SELF–TEST SOLUTIONS

Chapter 1

True–False

1. F. The symbol for the element cobalt is Co.

2. F. Sodium belongs in the group referred to as the alkali metals.

3. T

4. F. Semimetals do have properties somewhere between those of metals and nonmetals, but they are poor conductors of electricity.

5. T

6. F. The unit most commonly used in the laboratory to measure volume is either the liter (dm^3) or milliliter (cm^3).

7. T

8. T

9. T

10. T

Multiple Choice

1. d
2. b
3. a
4. c
5. b
6. d
7. d
8. b
9. c
10. d
11. d
12. b
13. a
14. b
15. a

Matching

Hypothesis—e
Theory—h
Element—f

Appendix C – Self Test Solutions

Main group elements—j
Matter—i
Volume—a
Precision—d
Accuracy—c
Dimensional Analysis—g
Intensive property—b

Fill–in–the–Blank

1. Sc
2. fourth period, 4B
3. germanium; semimetal
4. noble gases
5. pascal
6. the amount of matter in an object; kilogram
7. number and unit label
8. 1×10^{-9}
9. 100
10. 0.001
11. one additional estimated digit
12. either of the original numbers
13. intensive; the mass of an object to its volume
14. experimentation, hypothesis, theory
15. violently; alkaline; alkali
16. the study of the composition, properties, and transformations of matter and of chemical laws that are responsible for the changes that take place in nature.
17. Chemical laws
18. International System of Units (SI); seven
19. groups; periods
20. properties

Problems

1. $3.89 \; \mu g \times \dfrac{1 \; g}{1 \times 10^6 \; \mu g} \times \dfrac{1000 \; mg}{1 \; g} = 3.89 \times 10^{-3} \; mg$

   $3.89 \; \mu g \times \dfrac{1 \; g}{1 \times 10^6 \; \mu g} \times \dfrac{1 \times 10^9 \; ng}{1 \; g} = 3.89 \times 10^3 \; ng$

2. $\dfrac{7.984}{3.2} = 2.5$ (The least amount of significant figures, two, is found in the denominator.)

   $7.53 \times 23.945 \; 62 = 180$ (The least amount of significant figures, three, is found in the number 7.53.)

   $3.268 + 4 = 7$ (There are no digits to the right of the decimal point in the number 4.)

   $34.21 - 0.039 = 34.17$ (There are only two digits to the right of the decimal point in 34.21.)

3. 2.942×10^6; the first digit removed is 5, which is followed by more nonzero digits; therefore, you round up one digit.

   2.9×10^6; the first digit removed is less than 5; therefore, all of the digits to the right of the 9 are

Appendix C – Self Test Solutions

simply dropped.

4. The conversion factor needed to solve this problem is 1 in. = 2.54 cm

$$35.67 \text{ cm} \times \frac{1 \text{ in.}}{2.54 \text{ cm}} = 14.04 \text{ in.}$$

5. $°F = \left(\frac{9}{5} \times 78.5\right) + 32 = 173 \ °F$

   $K = 78.5 + 273.15 = 351.6 \text{ K}$

6. The conversion factor needed to solve this problem is 1 km = 0.6214 mi.

$$700 \text{ km} \times \frac{0.6214 \text{ mi}}{1 \text{ km}} = 435 \text{ mi}$$

$$103 \ \frac{\text{km}}{\text{hr}} \times \frac{0.6214 \text{ mi}}{1 \text{ km}} = 64.0 \ \frac{\text{mi}}{\text{hr}}$$

7. Volume of metal cylinder = 59.72 mL − 25.34 mL = 34.38 mL

   Density $= \frac{53.487 \text{ g}}{34.38 \text{ mL}} = 1.556 \text{ g/mL}$

8. The conversion factors needed to solve this problem are
   1 lb = 453.6 g; 1 qt = 946 mL; 1 gal = 4 qt

   This problem can be solved in a single step.

$$1.556 \ \frac{\text{g}}{\text{mL}} \times \frac{1 \text{ lb}}{453.6 \text{ g}} \times \frac{946 \text{ mL}}{1 \text{ qt}} \times \frac{4 \text{ qt}}{1 \text{ gal}} = 12.99 \ \frac{\text{lb}}{\text{gal}}$$

9. mass = $2.33 \ \frac{\text{g}}{\text{mL}} \times 5.78 \text{ mL} = 13.5 \text{ g}$

10. Neighbors vertical to As are P (phosphorus) and Sb (antimony). Neighbors horizontal to As are Ge (germanium) and Se (selenium).

11. O (oxygen), S (sulfur), Se (selenium), Te (tellurium), Po (polonium). This group is referred to as the chalcogens.

12. 93 million miles = 9.3×10^7 mi. However, the speed of light is expressed in terms of m/s. Once we convert miles to meters, we can calculate time in seconds and convert to minutes.

$$9.3 \times 10^7 \text{ mi} \times \frac{1.6093 \text{ km}}{1 \text{ mi}} \times \frac{1000 \text{ m}}{1 \text{ km}} \times \frac{1 \text{ s}}{3.00 \times 10^8 \text{ m}} \times \frac{1 \text{ min}}{60 \text{ s}} = 8.3 \text{ min}$$

13. The volume of a cylinder is calculated with the formula $\pi r^2 h$. Therefore, the volume of the titanium cylinder is

$$\text{Volume} = \pi \left(\frac{2.75}{2}\right)^2 5.05 = 30.0 \text{ cm}^3$$

Appendix C – Self Test Solutions

$$\text{Mass} = \text{Density} \times \text{Volume} \qquad \text{Mass} = 4.55 \frac{g}{cm^3} \times 30.0 \text{ cm}^3 = 137 \text{ g}$$

14. Density is a temperature dependent property because most substances change in volume when heated or cooled.

15. a. If the density of a substance is less than the density of water, then the substance floats on water. Therefore, olive oil will float on water. b. If the density of a substance is greater than the density of water, then the substance sinks in water. 1,1,1-trichloroethane will sink in water.

Challenge Problem

On a mercury thermometer the liquid range of gallium is 2373.22°C (the difference in temperature between the melting and boiling points). On the gallium thermometer the liquid range of gallium is 500°C. The size adjustments for these two scales are

$$1°C \times \frac{500° \text{Ga}}{2373.22°C} = 0.211° \text{Ga} \qquad 1° \text{Ga} \times \frac{273.22°C}{500° \text{Ga}} = 4.75°C$$

Therefore, 1°C is 4.75 times greater than 1°Ga, and 1°Ga is 0.211 times smaller than 1°C.

The melting point of gallium is higher by 29.78° on the Celsius scale, so we would first do the size adjustment followed by the zero point adjustment when converting from °Ga to °C.

$$°C = (4.75 \times ° \text{Ga}) + 29.78$$

We can now convert the melting point of calcium from °Ga to °C.

$$°C = (4.75 \times 17° \text{Ga}) + 29.78 = 837.28°C$$

We can determine the accuracy of the thermometer by comparing the difference between the converted and reported value to the reported value.

$$\frac{(839° - 837.28°)}{839°} \times 100 = 0.21\%$$

Chapter 2

True–False

1. T

2. F. In a chemical reaction, the atoms only rearrange the way they are combined; the atoms themselves are not changed.

3. F. Cathode rays consist of tiny, negatively charged particles.

4. F. The nucleus is composed of two kinds of particles: protons and neutrons.

Appendix C – Self Test Solutions

5. T

6. T

7. F. A cation is a positively charged particle resulting from the loss of an electron.

8. T

9. F. The correct name for FeS is iron(II) sulfide.

10. T

Multiple Choice

1. d
2. a
3. d
4. c
5. c
6. b
7. b
8. c
9. b
10. d

Fill–in–the–Blank

1. small whole–number multiples of each other

2. 7000 times more massive than an electron and have a positive charge twice the magnitude of the charge on the electron.

3. protons

4. protons

5. neutrons

6. atomic mass

7. nonmetal atoms

8. cation and anion

9. acid

10. subscript

11. atoms

12. compounds; chemical reaction

13. chemical bonds

14. Ionic solids; ions

Appendix C – Self Test Solutions

15. nucleus

Matching

Al_2O_3—i
Cations—j
FeO—o
Anions—l
NH_4Cl p
Bases—h
$CaCO_3$—e
Isotopes—b
SF_4—g
Protons—n
TiF_3—q
Neutrons—f
PCl_3—c
Electrons—d
$BaSO_3$—m
HgO—a
$Cr(OH)_3$—k

Problems

1. First compound: $N:O$ ratio $= \dfrac{0.681 \text{ g N}}{0.778 \text{ g O}} = 0.875$

    Second compound: $N:O$ ratio $= \dfrac{0.560 \text{ g N}}{1.28 \text{ g O}} = 0.438$

    $\dfrac{N:O \text{ ratio in 1st compound}}{N:O \text{ ratio in 2nd compound}} = \dfrac{0.875}{0.438} = 2.0$

2. Ammonia: $N:H = \dfrac{8.75 \text{ g N}}{3.75 \text{ g H}} = 2.33$

    Hydrazine: $N:H = \dfrac{11.48 \text{ g N}}{3.28 \text{ g H}} = 3.50$

    $\dfrac{N:H \text{ ratio in ammonia}}{N:H \text{ ratio in hydrazine}} = \dfrac{2.33}{3.50} = 0.666 = \dfrac{2}{3}$

3. Average atomic mass $= (0.4782 \times 150.9 \text{ amu}) + (0.5218 \times 152.9 \text{ amu}) = 151.9 \text{ amu}$

4. Average atomic mass
    $= (203.973 \text{ amu} \times 0.0148) + (205.9745 \text{ amu} \times 0.236) + (206.9759 \text{ amu} \times 0.226) + (207.9766 \text{ amu} \times 0.523) = 207 \text{ amu}$

5. a) p = 8, n = 10, e = 8
    b) p = 26, n = 31, e = 24
    c) p = 79, n = 118, e = 79
    d) p = 53, n = 78, e = 54

6. a) Al^{3+} b) Se^{2-}

Appendix C – Self Test Solutions

7. a) NO_3^- b) ClO_4^- c) HSO_4^- d) HS^-

8. a) K^+ b) Ca^{2+} c) Mg^{2+} d) Li^+

9. $150 \text{ atoms} \times \dfrac{55.85 \text{ amu}}{1 \text{ atom}} \times \dfrac{1.6605 \times 10^{-24} \text{ g}}{1 \text{ amu}} = 1.39 \times 10^{-20} \text{ g}$

10. $\left(2.107 \times 10^{-22} \dfrac{\text{g}}{\text{atom}}\right) \times \left(6.02 \times 10^{23} \text{ atoms}\right) = 126.841 \text{ g}$

   This answer is very close to the atomic mass of iodine.

11. The chemical law needed to work this problem is the Law of Conservation of Mass. According to this law, "Matter cannot be created nor destroyed, only changed from one form to another." In other words, the total mass of the reactants must equal the mass of the products. In this chemical reaction, the mass of the reactants is the mass of the iron and the nitric acid. The mass of the nitric acid is calculated from the volume and density.

   $25.00 \text{ mL} \times 1.40 \dfrac{\text{g}}{\text{mL}} = 35.0 \text{ g}$

   Mass of reactants = 15.51 g Fe + 35.0 g HNO_3 = 50.51 g reactants

   The mass of products also must equal 50.51 g. Knowing the mass of products, we can now calculate the mass of H_2 gas and determine the volume of H_2 from its density.

   50.51 g products = mass of H_2 + 49.958 g $Fe(NO_3)_2$; mass of H_2 = 0.55 g H_2

   $\text{Volume of } H_2 = 0.55 \text{ g } H_2 \times \dfrac{1 \text{ L}}{0.0899 \text{ g } H_2} = 6.1 \text{ L } H_2$

12. a. lithium fluoride   b. gallium chloride   c. calcium nitride

    d. vanadium(V) oxide   e. scandium(III) nitrate   f. manganese(VII) oxide

    g. iron(III) bromide   h. potassium iodate   i. potassium dichromate

    j. tin(IV) nitrite   k. dinitrogen pentoxide   l. hydrogen bromide

    m. chlorous acid   n. hydrobromic acid   o. sulfurous acid

    p. dinitrogen oxide

13. $C_3H_5N_3O_9$

14. a. RbI   b. Mg_3P_2 c. $CoCl_2$ d. Cu_2O

    e. $(NH_4)_2S_2O_3$ f. CrN   g. HCN (g)   h. SO_3

    i. HF (aq)   j. CaH_2 k. $HBrO_3$ l. $(Na)_2SO_4$

    m. $NaC_2H_3O_2$ n. $KMnO_4$ o. HClO   p. $KHCO_3$

Appendix C – Self Test Solutions

Challenge Problem

Enflurane is a molecular compound since it consists of all nonmetals. The structural formula for this molecule is:

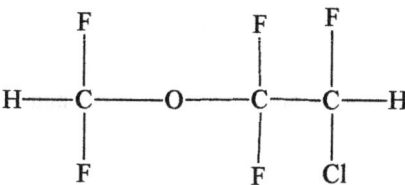

The total mass of each element in this compound is calculated by taking the atomic mass found underneath the element on the periodic table (units = amu) and multiplying the number of atoms indicated in the formula.

mass C = 3 × 12.0 amu = 36.0 amu

mass H = 2 × 1.0 amu = 2.0 amu

mass O = 1 × 16.0 amu = 16.0 amu

mass F = 5 × 19.0 amu = 95.0 amu

mass Cl = 1 × 35.5 amu = 35.5 amu

mass $C_3H_2ClOF_5$ = 36.0 amu + 2.0 amu + 16.0 amu + 95.0 amu + 35.5 = 184.5 amu

% Mass of each element = $\dfrac{\text{total mass of element}}{\text{mass } C_3H_2ClOF_5} \times 100$

% Mass C = $\dfrac{36.0 \text{ a.m.u.}}{184.5 \text{ a.m.u.}} \times 100 = 19.5\%$

% Mass H = $\dfrac{2.0 \text{ a.m.u.}}{184.5 \text{ a.m.u.}} \times 100 = 1.1\%$

% Mass O = $\dfrac{16.0 \text{ a.m.u.}}{184.5 \text{ a.m.u.}} \times 100 = 8.67\%$

% Mass Cl = $\dfrac{35.5 \text{ amu}}{184.5 \text{ amu}} \times 100 = 19.2\%$

% Mass F = $\dfrac{95.0 \text{ amu}}{184.5 \text{ amu}} \times 100 = 51.5\%$

Chapter 3

True–False

1. F. A balanced chemical equation obeys the law of mass conservation.

2. F. When balancing a chemical equation, only the coefficients may be changed; the subscripts of the formula units must stay the same.

3. T

4. F. Chemical symbols represent the behavior of atoms and molecules on both the microscopic and macroscopic level.

5. T

6. T

7. F. The limiting reactant determines the extent to which a chemical reaction takes place.

8. F. When diluting a solution, the number of moles of solute stays the same; only the total volume of solution changes.

9. T

10. F. The molecular formula of a compound may be the same as the empirical formula.

Matching

Combustion reaction—c
Microscopic level—f
Macroscopic level—m
Formula mass—e
Molecular mass—i
Molar mass—k
Excess reactant—a
Limiting reactant—j
Molarity—l
Titration—d
Percent composition—g
Empirical formula—b
Molecular formula—h

Fill–in–the–Blank
1. balanced
2. Avogadro's number
3. stoichiometry
4. molarity
5. volume; number of moles
6. limiting reactant
7. Titration
8. percent composition
9. carbon dioxide; water
10. moles and grams

Appendix C – Self Test Solutions

Problems

1. a. $Mg_3N_2 + 8\,HCl \rightarrow 3\,MgCl_2 + 2\,NH_4Cl$

   b. $CaCO_3 + 2\,HNO_3 \rightarrow CO_2 + H_2O + Ca(NO_3)_2$

   c. $2\,C_8H_{18} + 25\,O_2 \rightarrow 16\,CO_2 + 18\,H_2O$: Remember that combustion refers to the reaction with oxygen and that hydrocarbons produce CO_2 and water when they undergo combustion. To balance this equation, first begin by balancing the C's and H's as shown below.

   $C_8H_{18} + O_2 \rightarrow 8\,CO_2 + 9\,H_2O$

   Notice that this gives you a total of 25 O's on the product side. Because the oxygen on the reactant side is present in the form of O_2, you would need 25/2 as the coefficient. This would give the following balanced equation:

   $C_8H_{18} + 25/2\,O_2 \rightarrow 8\,CO_2 + 9\,H_2O$

   However, coefficients are not allowed to be fractions. If we multiply all of the coefficients in the equation by 2, we will remove the fraction and end up with the smallest whole–number ratio of molecules.

   $2\,C_8H_{18} + 25\,O_2 \rightarrow 16\,CO_2 + 18\,H_2O$

   d. $2\,Al + 3\,H_2SO_4 \rightarrow Al_2(SO_4)_3 + 3\,H_2$

   e. $2\,AgNO_3 + Na_2SO_4 \rightarrow Ag_2SO_4 + 2\,NaNO_3$

2. First, write the balanced chemical equation:

   $Ca_3P_2 + 6\,H_2O \rightarrow 2\,PH_3 + 3\,Ca(OH)_2$

   Next, convert grams of Ca_3P_2 to moles of Ca_3P_2

   $15.4\ \text{g}\ Ca_3P_2 \times \dfrac{1\ \text{mol}\ Ca_3P_2}{182.3\ \text{g}\ Ca_3P_2} = 0.0845\ \text{mol}\ Ca_3P_2$

   Knowing moles of Ca_3P_2, we can convert to moles of PH_3 and then grams of PH_3.

   $0.0845\ \text{mol}\ Ca_3P_2 \times \dfrac{2\ \text{mol}\ PH_3}{1\ \text{mol}\ Ca_3P_2} \times \dfrac{34\ \text{g}\ PH_3}{1\ \text{mol}\ PH_3} = 5.75\ \text{g}\ PH_3$

3. First, check that the chemical equation given is balanced. Next, convert grams of thioacetamide to moles.

   $5.75\ \text{g}\ CH_3CSNH_2 \times \dfrac{1\ \text{mol}\ CH_3CSNH_2}{75.0\ \text{g}\ CH_3CSNH_2} = 0.0767\ \text{mol}$

   Knowing moles of thioacetamide, we can now calculate grams of water and grams of hydrogen sulfide.

   $0.0767\ \text{mol}\ CH_3CSNH_2 \times \dfrac{1\ \text{mol}\ H_2O}{1\ \text{mol}\ CH_3CSNH_2} \times \dfrac{18.0\ \text{g}\ H_2O}{1\ \text{mol}\ H_2O} = 1.38\ \text{g}\ H_2O$

   $0.0767\ \text{mol}\ CH_3CSNH_2 \times \dfrac{1\ \text{mol}\ H_2S}{1\ \text{mol}\ CH_3CSNH_2} \times \dfrac{34.0\ \text{g}\ H_2S}{1\ \text{mol}\ H_2S} = 2.61\ \text{g}\ H_2S$

4. First, check to see if the chemical equation is balanced. It is not! Balance the equation.

$2 \text{ NaHSO}_3 + \text{Na}_2\text{CO}_3 \rightarrow 2 \text{ Na}_2\text{SO}_3 + \text{CO}_2 + \text{H}_2\text{O}$

Next, calculate the theoretical yield of sodium sulfite by converting grams of sodium hydrogen sulfite to moles of sodium hydrogen sulfite. Next, convert moles of sodium hydrogen sulfite to moles of sodium sulfite and then to grams of sodium sulfite.

$$75.8 \text{ g NaHSO}_3 \times \frac{1 \text{ mol NaHSO}_3}{104.0 \text{ g NaHSO}_3} = 0.729 \text{ mol NaHSO}_3$$

$$0.729 \text{ mol NaHSO}_3 \times \frac{2 \text{ mol Na}_2\text{SO}_3}{2 \text{ mol NaHSO}_3} \times \frac{126.0 \text{ g Na}_2\text{SO}_3}{1 \text{ mol Na}_2\text{SO}_3} = 91.8 \text{ g Na}_2\text{SO}_3$$

Knowing the percent yield and theoretical yield, we can now calculate the actual yield.

$$64\% = 100\% \times \frac{\text{actual yield}}{91.8 \text{ g Na}_2\text{SO}_3} \qquad \text{Actual yield} = 59 \text{ g Na}_2\text{SO}_3$$

5. The chemical equation for this problem is

$\text{HNO}_3 + \text{NH}_3 \rightarrow \text{NH}_4\text{NO}_3$

This problem is a limiting reactant problem. First, we need to determine how many moles of HNO_3 and NH_3 are present.

$$156.5 \text{ g HNO}_3 \times \frac{1 \text{ mol HNO}_3}{63.0 \text{ g HNO}_3} = 2.48 \text{ mol HNO}_3$$

$$275.0 \text{ g NH}_3 \times \frac{1 \text{ mol NH}_3}{17.0 \text{ g NH}_3} = 16.2 \text{ mol NH}_3$$

The required mole ratio is 1 mol of nitric acid for every 1 mol of ammonia. Therefore, the number of moles of ammonia needed to react with nitric acid is 2.48. Since we have more than enough moles of ammonia to react with nitric acid, ammonia is the excess reactant and nitric acid is the limiting reactant.

Moles of nitric acid consumed: 2.48 mol HNO_3
Moles of ammonia consumed: 2.48 mol NH_3
Moles of ammonia not consumed: 16.2 mol NH_3 − 2.48 mol NH_3 = 13.7 mol NH_3

The amount of NH_4NO_3 produced is determined by the 2.48 mol of HNO_3

$$2.48 \text{ mol HNO}_3 \times \frac{1 \text{ mol NH}_4\text{NO}_3}{1 \text{ mol HNO}_3} \times \frac{80.0 \text{ g NH}_4\text{NO}_3}{1 \text{ mol NH}_4\text{NO}_3} = 198 \text{ g NH}_4\text{NO}_3$$

6. First, calculate the number of moles of NaOH.

$$0.500 \text{ L} \times \frac{3.75 \text{ mol NaOH}}{1 \text{ L}} = 1.88 \text{ mol NaOH}$$

Next, determine how many grams of NaOH are in 1.88 mol NaOH.

$$1.88 \text{ mol NaOH} \times \frac{40.0 \text{ g NaOH}}{1 \text{ mol NaOH}} = 75.2 \text{ g NaOH}$$

Appendix C – Self Test Solutions

To prepare this solution, you would dissolve 75.2 g of NaOH in water and dilute to a final volume of 500 mL.

7. To prepare this solution, we need to know the volume of the concentrated phosphoric acid. We will use the equation

$$V_i = \frac{M_f \times V_f}{M_i}.$$

Substituting in the given information leads to

$$V_i = \frac{0.125 \text{ M} \times 2.50 \text{ L}}{15.0 \text{ M}} = 0.0208 \text{ L} = 20.8 \text{ mL}$$

To prepare this solution, **carefully** add 20.8 mL to ≈ 1 L of water and dilute to 2.5 L. **CAUTION**: Always add acid to water, not water to acid!

8. First, calculate the number of moles of zinc.

$$2.45 \text{ g Zn} \times \frac{1 \text{ mol Zn}}{65.4 \text{ g Zn}} = 0.0375 \text{ mol Zn}$$

The Zn:HCl mole ratio is 1:2. Therefore, the moles of HCl = 0.0750. The volume of HCl needed for this reaction can now be calculated.

$$\frac{0.0750 \text{ mol HCl}}{0.150 \text{ mol HCl / L}} = 0.500 \text{ L} = 500 \text{ mL}$$

9. The balanced equation for this reaction is

$$Pb(NO_3)_2 + K_2CrO_4 \rightarrow PbCrO_4 + 2 \text{ KNO}_3$$

This problem is a limiting reactant problem. First, calculate the number of moles of $Pb(NO_3)_2$ and K_2CrO_4.

$$0.050 \text{ L} \times \frac{0.025 \text{ mol Pb(NO}_3)_2}{1 \text{ L}} = 1.25 \times 10^{-3} \text{ mol Pb(NO}_3)_2$$

$$0.050 \text{ L} \times \frac{0.10 \text{ mol K}_2 \text{ CrO}_4}{1 \text{ L}} = 5.0 \times 10^{-3} \text{ mol K}_2 \text{ CrO}_4$$

The $Pb(NO_3)_2$:K_2CrO_4 ratio is 1:1. Therefore, 1.25×10^{-3} moles of K_2CrO_4 are needed to react with 1.25×10^{-3} moles of $Pb(NO_3)_2$. There is more than enough K_2CrO_4 for this reaction. Therefore, the excess reactant is K_2CrO_4, and the limiting reactant is $Pb(NO_3)_2$.

Moles of $Pb(NO_3)_2$ consumed: 1.25×10^{-3} mol $Pb(NO_3)_2$
Moles of K_2CrO_4 consumed: 1.25×10^{-3} mol K_2CrO_4
Moles of K_2CrO_4 not consumed: $(5.0 \times 10^{-3}$ mol $K_2CrO_4) - (1.25 \times 10^{-3}$ mol $K_2CrO_4)$
$= 3.75 \times 10^{-3}$ mol K_2CrO_4

Grams of $PbCrO_4$ produced can now be calculated.

Appendix C – Self Test Solutions

$$1.25 \times 10^{-3} \text{ mol Pb(NO}_3)_2 \times \frac{1 \text{ mol PbCrO}_4}{1 \text{ mol Pb(NO}_3)_2} \times \frac{323.2 \text{ g PbCrO}_4}{1 \text{ mol PbCrO}_4} = 0.404 \text{ g PbCrO}_4$$

10. The C:H:O mole ratio is 9:8:4. Convert this mole ratio into a mass ratio by assuming there is a 1 mole sample present.

$$1 \text{ mol C}_9\text{H}_8\text{O}_4 \times \frac{9 \text{ mol C}}{1 \text{ mol C}_9\text{H}_8\text{O}_4} \times \frac{12.0 \text{ g C}}{1 \text{ mol C}} = 108 \text{ g C}$$

$$1 \text{ mol C}_9\text{H}_8\text{O}_4 \times \frac{8 \text{ mol H}}{1 \text{ mol C}_9\text{H}_8\text{O}_4} \times \frac{1.00 \text{ g H}}{1 \text{ mol H}} = 8.00 \text{ g H}$$

$$1 \text{ mol C}_9\text{H}_8\text{O}_4 \times \frac{4 \text{ mol O}}{1 \text{ mol C}_9\text{H}_8\text{O}_4} \times \frac{16.0 \text{ g O}}{1 \text{ mol O}} = 64.0 \text{ g O}$$

To determine the percent composition, divide the mass of each element present by the total mass of the compound and multiply by 100.

$$\%C = \frac{108 \text{ g C}}{180 \text{ g C}_9\text{H}_8\text{O}_4} \times 100\% = 60.0\%$$

$$\%H = \frac{8.00 \text{ g H}}{180 \text{ g C}_9\text{H}_8\text{O}_4} \times 100\% = 4.44\%$$

$$\%O = \frac{64.0 \text{ g O}}{180 \text{ g C}_9\text{H}_8\text{O}_4} \times 100\% = 35.6\%$$

11. Assuming that you have a 100 g sample gives 38.7 g C, 9.7 g H, and 51.6 g O. Convert these masses to number of moles.

$$38.7 \text{ g C} \times \frac{1 \text{ mol C}}{12.0 \text{ g C}} = 3.22 \text{ mol C}$$

$$9.7 \text{ g H} \times \frac{1 \text{ mol H}}{1.0 \text{ g H}} = 9.7 \text{ mol H}$$

$$51.6 \text{ g O} \times \frac{1 \text{ mol O}}{16.0 \text{ g O}} = 3.22 \text{ mol O}$$

Knowing the relative number of moles, find the ratio by dividing the larger number by the smaller number.

$$\frac{9.7}{3.22} = 3; \text{ the C:H:O ratio of 1:3:1 gives the empirical formula CH}_3\text{O}.$$

To determine the molecular formula, you must find the multiplier.

$$\frac{62 \text{ g/mol}}{31 \text{ g/mol}} = 2; \text{ therefore, the molecular formula is C}_2\text{H}_6\text{O}_2.$$

12. Because you have been given information about both reactants in this problem, you must first check to see if one of the reactants is in excess. Begin by writing the balanced chemical equation for the

Appendix C – Self Test Solutions

reaction.

$$Ag_2CO_3 + 2\,HNO_3 \rightarrow CO_2 + 2\,AgNO_3 + H_2O$$

Next, determine the number of moles of both reactants.

$$9.85\,g\,Ag_2CO_3 \times \frac{1\,mol\,Ag_2CO_3}{275.8\,g\,Ag_2CO_3} = 0.0357\,mol\,Ag_2CO_3$$

$$0.250\,L\,HNO_3 \times \frac{0.250\,mol\,HNO_3}{1\,L} = 0.0625\,mol\,HNO_3$$

The Ag_2CO_3:HNO_3 mole ratio is 1:2; therefore you need 0.0714 moles of HNO_3 (2 × 0.0357). You have only 0.0625 moles of HNO_3, so it is the limiting reactant and will determine the number of grams of silver nitrate produced.

$$0.0625\,mol\,HNO_3 \times \frac{2\,mol\,AgNO_3}{2\,mol\,HNO_3} \times \frac{169.9\,g\,AgNO_3}{1\,mol\,AgNO_3} = 10.6\,g\,AgNO_3$$

13. Begin by finding the molar amounts of C and H in CO_2 and H_2O.

$$25.56\,g\,CO_2 \times \frac{1\,mol\,CO_2}{44.0\,g\,CO_2} \times \frac{1\,mol\,C}{1\,mol\,CO_2} = 0.581\,mol\,C$$

$$10.46\,g\,H_2O \times \frac{1\,mol\,H_2O}{18.0\,g\,H_2O} \times \frac{2\,mol\,H}{1\,mol\,H_2O} = 1.16\,mol\,H$$

Next, carry out mole–to–gram conversions to find the number of grams of C and H in the original sample.

$$0.581\,mol\,C \times \frac{12\,g\,C}{1\,mol\,C} = 6.97\,g\,C$$

$$1.16\,mol\,H \times \frac{1.0\,g\,H}{1\,mol\,H} = 1.16\,g\,H$$

Subtract the masses of C and H from the mass of the starting sample to determine the mass of O.

$$12.78\,g - 6.97\,g - 1.16\,g = 4.65\,g$$

Convert the mass of oxygen to moles of oxygen.

$$4.65\,g\,O \times \frac{1\,mol\,O}{16.0\,g\,O} = 0.291\,mol\,O$$

Find the ratio of the numbers of moles by dividing the larger numbers of moles by the smaller number of moles.

$$\frac{1.16}{0.291} = 4 \qquad \frac{0.581}{0.291} = 2$$

This gives a mole ratio of C:H:O of 2:4:1. Therefore, the empirical formula is C_2H_4O. To determine the molecular formula, you must determine the multiplier, $\dfrac{88.0 \text{ g/mol}}{44.0 \text{ g/mol}} = 2$; therefore, the molecular formula for this compound is $C_4H_8O_2$.

14. The formula for calcium phosphite is $Ca_3(PO_3)_2$. The Ca:P:O mole ratio is 3:2:6. Begin by converting this mole ratio into a mass ratio by assuming 1 mole of sample is present.

$$1 \text{ mol } Ca_3(PO_3)_2 \times \dfrac{3 \text{ mol Ca}}{1 \text{ mol } Ca_3(PO_3)_2} \times \dfrac{40.1 \text{ g Ca}}{1 \text{ mol Ca}} = 120.3 \text{ g Ca}$$

$$1 \text{ mol } Ca_3(PO_3)_2 \times \dfrac{2 \text{ mol P}}{1 \text{ mol } Ca_3(PO_3)_2} \times \dfrac{31.0 \text{ g P}}{1 \text{ mol P}} = 62.0 \text{ g P}$$

$$1 \text{ mol } Ca_3(PO_3)_2 \times \dfrac{6 \text{ mol O}}{1 \text{ mol } Ca_3(PO_3)_2} \times \dfrac{16.0 \text{ g O}}{1 \text{ mol O}} = 96.0 \text{ g O}$$

The percent composition is determined by dividing the mass of each element present by the total mass of the compound and multiplying by 100%.

$$\% \text{ Ca} = \dfrac{120.3 \text{ g Ca}}{278.3 \text{ g } Ca_3(PO_3)_2} \times 100\% = 43.2\%$$

$$\% \text{ P} = \dfrac{62.0 \text{ g P}}{278.3 \text{ g } Ca_3(PO_3)_2} \times 100\% = 22.3\%$$

$$\% \text{ O} = \dfrac{96.0 \text{ g O}}{278.3 \text{ g } Ca_3(PO_3)_2} \times 100\% = 34.5\%$$

15. Assuming a 100 g sample gives us 74.2 g C, 9.3 g H, and 16.5 g O. Convert these masses to numbers of moles.

$$74.2 \text{ g C} \times \dfrac{1 \text{ mol C}}{12 \text{ g C}} = 6.18 \text{ mol C}$$

$$9.3 \text{ g H} \times \dfrac{1 \text{ mol H}}{1.0 \text{ g H}} = 9.3 \text{ mol H}$$

$$16.5 \text{ g O} \times \dfrac{1 \text{ mol O}}{16.0 \text{ g O}} = 1.03 \text{ mol O}$$

Knowing the relative numbers of moles, find the ratio by dividing the two larger numbers by the smaller number.

$$\dfrac{6.18}{1.03} = 6 \qquad \dfrac{9.3}{1.03} = 9$$

The C:H:O ratio of 6:9:1 gives an empirical equation of C_6H_9O. To determine the molecular formula, find the multiplier.

$$\dfrac{194 \text{ g/mol}}{97 \text{ g/mol}} = 2; \text{ therefore, the molecular formula is } C_{12}H_{18}O_2.$$

Appendix C – Self Test Solutions

Challenge Problem

We solve this problem in much the same way we solved the "Putting It Together" problem. The only difference between the two is that this time we know the identity of all our reactants and we have more chemical reactions involved in the process. We can determine the original number of moles of Br_2 present from the information we have regarding $KBrO_3$ and the balanced equation in which $KBrO_3$ is converted to Br_2.

$$\text{Original mol } Br_2 = 0.025 \text{ L } KBrO_3 \times \frac{0.15 \text{ mol } KBrO_3}{L} \times \frac{3 \text{ mol } Br_2}{1 \text{ mol } KBrO_3} = 0.011 \text{ mol } Br_2$$

We can determine the number of moles of Br_2 in excess by determining the number of moles of I_2 that reacted with $Na_2S_2O_3$. (Remember that I_2 is generated by reacting KI with the excess Br_2 in the reaction.)

$$\text{mol } I_2 = 0.0325 \text{ L } Na_2S_2O_3 \times \frac{0.25 \text{ mol } Na_2S_2O_3}{L} \times \frac{1 \text{ mol } I_2}{2 \text{ mol } Na_2S_2O_3} = 4.06 \times 10^{-3} \text{ mol } I_2$$

$$\text{mol } Br_2 = 4.06 \times 10^{-3} \text{ mol } I_2 \times \frac{1 \text{ mol } Br_2}{1 \text{ mol } I_2} = 4.06 \times 10^{-3} \text{ mol } Br_2 \text{ (Remember that this value represents}$$
the number of excess moles of Br_2.)

Moles Br_2 reacted = original moles Br_2 – excess moles Br_2 = $0.011 - (4.06 \times 10^{-3}) = 6.94 \times 10^{-3}$ moles Br_2 reacted

Grams of ascorbic acid can now be calculated.

$$g \, C_6H_8O_6 = (6.94 \times 10^{-3} \text{ mol } Br_2) \times \frac{1 \text{ mol } C_6H_8O_6}{1 \text{ mol } Br_2} \times \frac{176 \text{ g } C_6H_8O_6}{1 \text{ mol } C_6H_8O_6} = 1.22 \text{ g } C_6H_8O_6$$

Chapter 4

True–False
1. T
2. F. H_3PO_4 incompletely dissociates in water and is therefore a weak acid.
3. T
4. F. $BaCl_2$ is a soluble salt: there is no reaction.
5. F. The driving force for an acid–base neutralization reaction is the formation of water.
6. T
7. F. The reducing agent is the species that causes reduction to occur and is therefore oxidized.
8. F. The oxidation number of Br in $NaBrO_3$ is +5. (Na = +1, 3O = –6: +1 + Br + (–6) = 0; Br = +5).
9. T
10. F. Hg is below H_2 in the activity series.

Multiple Choice
1. d
2. d
3. d
4. a
5. a
6. b

Appendix C – Self Test Solutions

7. c
8. a
9. a
10. d

Matching
Electrolyte—d
Strong electrolyte—f
Spectator ions—h
Solubility—a
Oxidation—g
Reduction—j
Oxidizing agent—i
Reducing agent—c
Acid—b
Base—e

Fill–in–the–Blank
1. activity series
2. neutralization reaction; salt and water
3. redox titration
4. redox reactions
5. oxidation number
6. weak electrolytes
7. oxidation number
8. half reactions
9. half–reaction
10. H^+ or H_3O^+ ion

Problems

1. $2\ NaCl\ (aq)\ +\ Pb(NO_3)_2\ (aq)\ \rightarrow\ PbCl_2\ (s)\ +\ 2\ NaNO_3\ (aq)$ Molecular equation

   $2\ Na^+\ (aq) + 2\ Cl^-\ (aq) + Pb^{2+}\ (aq) + 2\ NO_3^-\ (aq) \rightarrow\ PbCl_2\ (s) + 2\ Na^+\ (aq) + 2\ NO_3^-\ (aq)$ Ionic equation

   $2\ Cl^-\ (aq)\ +\ Pb^{2+}\ (aq)\ \rightarrow\ PbCl_2\ (s)$ Net ionic equation

2. $HClO_4\ (aq)\ +\ KOH\ (aq)\ \rightarrow\ KClO_4(aq)\ +\ H_2O\ (l)$ Molecular equation

   $H^+\ (aq) + ClO_4^-\ (aq) + K^+\ (aq) + OH^-\ (aq) \rightarrow\ K^+\ (aq) + ClO_4^-\ (aq) + H_2O\ (l)$ Ionic equation

   $H^+\ (aq) + OH^-\ (aq) \rightarrow\ H_2O\ (l)$ Net ionic equation

3. $K_2CO_3\ (aq)\ +\ CaCl_2\ (aq)\ \rightarrow\ 2\ KCl\ (aq)\ +\ CaCO_3\ (s)$ molecular equation

   $2\ K^+\ (aq) + CO_3^{2-}\ (aq) + Ca^{2+}\ (aq) + 2\ Cl^-\ (aq) \rightarrow\ 2\ K^+\ (aq) + 2\ Cl^-\ (aq) + CaCO_3\ (s)$ Ionic equation

   $CO_3^{2-}\ (aq) + Ca^{2+}\ (aq) \rightarrow\ CaCO_3\ (s)$ Net ionic equation

4. HCO_3^-: $(H^+)(C^?)(O^{2-})$ $1(+1) + (?) + 3(-2) = -1$ net charge
   $? = -1 - (+1) - 3(-2) = +4$

5. $MgSO_4$: $(Mg^{2+})(S^{?})(O^{2-})$ $(+2) + (?) + 4(-2) = 0$ net charge
   $? = 0 - (+2) - 4(-2) = +6$

6. SF_6: $(S^{?})(F^-)$ $(?) + 6(-1) = 0$ net charge
   $? = 0 - 6(-1) = +6$

7. $Mg \rightarrow Mg^{2+}$ oxidized, reducing agent; $O_2 \rightarrow O^{2-}$ reduced, oxidizing agent

8. Cr^{6+} $(Cr_2O_7^{2-}) \rightarrow Cr^{3+}$ reduced, oxidizing agent; $Sn^{2+} \rightarrow Sn^{4+}$ oxidized, reducing agent

9. S^{2-} (FeS) $\rightarrow S^{6+}$ (SO_4^{2-}) oxidized, reducing agent; N^{5+} $(NO_3^-) \rightarrow N^{2+}$ (NO) reduced, oxidizing agent

10. C^{2-} $(C_2H_4) \rightarrow C^{4+}$ (CO_2) oxidized, reducing agent; $O_2 \rightarrow O^{2-}$ (CO_2, H_2O) reduced, oxidizing agent

11. $Na(s) + H_2O \rightarrow NaOH(aq) + H_2(g)$

12. $Fe(s) + H_2O(g) \rightarrow Fe(OH)_2(s) + H_2(g)$

13. $Mg(s) + H_2O \rightarrow$ no reaction

14. $Cu(s) + Zn(SO_4)(aq) \rightarrow$ no reaction

15. $Sn^{2+} + MnO_4^- \longrightarrow Sn^{4+} + Mn^{2+}$
    ↑   ↑ ↑   ↑   ↑
    +2   +7 −2   +4   +2

Sn: $+2 \rightarrow +4$; Mn: $+7 \rightarrow +2$ net increase in oxidation number of oxidized atoms = 2; net decrease in oxidation number of reduced atoms = 5. Multiply net increase by 5 and the net decrease by 2.

$5 Sn^{2+} + 2 MnO_4^- \rightarrow 5 Sn^{4+} + 2 Mn^{2+}$

Product side of the equation has eight less oxygens, so add 8 H_2O; Reactant side has 16 less hydrogens, so add 16 H^+.

$5 Sn^{2+} + 2 MnO_4^- + 16 H^+ \rightarrow 5 Sn^{4+} + 2 Mn^{2+} + 8 H_2O$

16. $S_2O_3^{2-} + Cl_2 \rightarrow SO_4^{2-} + Cl^-$

$S_2O_3^{2-} + Cl_2 \rightarrow 2 SO_4^{2-} + 2 Cl^-$
↑ ↑   ↑   ↑ ↑   ↑
+2 −2   0   +6 −2   −1

S: $+2 \rightarrow +6$; Cl: $0 \rightarrow -1$; net increase in oxidation number of oxidized atoms = 8; net decrease in oxidation number of reduced atoms = 2. Multiply the net increase by 1 and the net decrease by 4.

$S_2O_3^{2-} + 4 Cl_2 \rightarrow 2 SO_4^{2-} + 8 Cl^-$

Reactant side of the equation has five less oxygens, so add 5 H_2O's. Product side of the equation has 10 less hydrogens, so add 10 H^+'s.

$5 H_2O + S_2O_3^{2-} + 4 Cl_2 \rightarrow 2 SO_4^{2-} + 8 Cl^- + 10 H^+$

Appendix C – Self Test Solutions

17. $MnO_4^- + C_2O_4^{2-} \rightarrow MnO_2 + CO_3^{2-}$

Remember to first balance all atoms other than hydrogen and oxygen and assign oxidation numbers.

$MnO_4^- + C_2O_4^{2-} \rightarrow MnO_2 + 2\, CO_3^{2-}$
+7 –2   +3 –2   +4 –2   +4 –2

Mn: +7 → +4; C: +3 → +4; net increase in oxidation number of oxidized atoms = 2 (each carbon atom has a change of +1 in oxidation number; however, there are two carbon atoms, so the net increase = +2); net decrease in oxidation number of reduced atoms = 3. Multiply the net increase by 3 and the net decrease by 2.

$2\, MnO_4^- + 3\, C_2O_4^{2-} \rightarrow 2\, MnO_2 + 6\, CO_3^{2-}$

The reactant side has two less oxygens, so add 2 H_2O's. Product side of the equation has four less H^+ so add 4 H^+'s.

$2\, H_2O + 2\, MnO_4^- + 3\, C_2O_4^{2-} \rightarrow 2\, MnO_2 + 6\, CO_3^{2-} + 4\, H^+$

Add 4 OH^- to both sides of the equation to neutralize the H^+.

$4\, OH^- + 2\, H_2O + 2\, MnO_4^- + 3\, C_2O_4^{2-} \rightarrow 2\, MnO_2 + 6\, CO_3^{2-} + 4\, H^+ + 4\, OH^-$

Combine the hydrogens and oxygens on the product side to make 4 H_2O, and cancel the two water molecules from the reactant side.

$4\, OH^- + 2\, MnO_4^- + 3\, C_2O_4^{2-} \rightarrow 2\, MnO_2 + 6\, CO_3^{2-} + 2\, H_2O$

18. Sulfur is oxidized and chromium is reduced, so the two half–reactions are

$HSO_3^- \rightarrow SO_4^{2-}$

$Cr_2O_7^{2-} \rightarrow Cr^{3+}$

Balance for elements other than O and H.

$HSO_3^- \rightarrow SO_4^{2-}$

$Cr_2O_7^{2-} \rightarrow 2\, Cr^{3+}$

Add H_2O for any oxygens that are needed and H^+ for any hydrogens that are needed.

$HSO_3^- + H_2O \rightarrow SO_4^{2-} + 3\, H^+$

$Cr_2O_7^{2-} + 14\, H^+ \rightarrow 2\, Cr^{3+} + 7\, H_2O$

Balance each reaction for charge.

$HSO_3^- + H_2O \rightarrow SO_4^{2-} + 3\, H^+ + 2\, e^-$

$6\, e^- + Cr_2O_7^{2-} + 14\, H^+ \rightarrow 2\, Cr^{3+} + 7\, H_2O$

Appendix C – Self Test Solutions

Make the electron count the same in both reactions.

$$3(HSO_3^- + H_2O \rightarrow SO_4^{2-} + 3H^+ + 2e^-)$$

$$6e^- + Cr_2O_7^{2-} + 14H^+ \rightarrow 2Cr^{3+} + 7H_2O$$

Add the two half-reactions together, canceling anything that appears on both sides of the equation.

$$3HSO_3^- + Cr_2O_7^{2-} + 5H^+ \rightarrow 3SO_4^{2-} + 2Cr^{3+} + 4H_2O$$

19. Lead is oxidized and Cl is reduced, so the two half-reactions are

$$Pb(OH)_3^- \rightarrow PbO_2$$

$$OCl^- \rightarrow Cl^-$$

Atoms other than O and H are already balanced. Add H_2O for any oxygens that are needed and H^+ for any hydrogens that are needed.

$$Pb(OH)_3^- \rightarrow PbO_2 + H_2O + H^+$$

$$2H^+ + OCl^- \rightarrow Cl^- + H_2O$$

Balance each reaction for charge.

$$Pb(OH)_3^- \rightarrow PbO_2 + H_2O + H^+ + 2e^-$$

$$2e^- + 2H^+ + OCl^- \rightarrow Cl^- + H_2O$$

Electron count is the same in both reactions. Add the two half reactions, canceling anything that appears on both sides of the equation.

$$Pb(OH)_3^- + H^+ + OCl^- \rightarrow PbO_2 + 2H_2O + Cl^-$$

Add one OH^- to both sides of the equation to neutralize the H^+ on the reactant side.

$$Pb(OH)_3^- + H_2O + OCl^- \rightarrow PbO_2 + 2H_2O + Cl^- + OH^-$$

Notice that water is on both the reactant and the product side of the equation. The final net reaction is

$$Pb(OH)_3^- + OCl^- \rightarrow PbO_2 + H_2O + Cl^- + OH^-$$

20. $Br_2 \rightarrow Br^- + BrO_3^-$

The bromine is both oxidized and reduced. The two half-reactions are

$$Br_2 \rightarrow Br^-$$

$$Br_2 \rightarrow BrO_3^-$$

Balance the atoms other than hydrogen and oxygen.

$Br_2 \rightarrow 2\ Br^-$

$Br_2 \rightarrow 2\ BrO_3^-$

Add H_2O for oxygens that are needed and H^+ for hydrogens that are needed.

$Br_2 \rightarrow 2\ Br^-$

$6\ H_2O + Br_2 \rightarrow 2\ BrO_3^- + 12\ H^+$

Balance each reaction for charge.

$2\ e^- + Br_2 \rightarrow 2\ Br^-$

$6\ H_2O + Br_2 \rightarrow 2\ BrO_3^- + 12\ H^+ + 10\ e^-$

Make the electron count the same in both reactions.
$5\ (2\ e^- + Br_2 \rightarrow 2\ Br^-)$

$6\ H_2O + Br_2 \rightarrow 2\ BrO_3^- + 12\ H^+ + 10\ e^-$

Add the two half–reactions, canceling anything that is common on both sides of the equation.

$6\ H_2O + 6\ Br_2 \rightarrow 10\ Br^- + 2\ BrO_3^- + 12\ H^+$

Add 12 OH^-'s to both sides of the equation to neutralize the H^+.

$12\ OH^- + 6\ H_2O + 6\ Br_2 \rightarrow 10\ Br^- + 2\ BrO_3^- + 12\ H^+ + 12\ OH^-$

Combine the H^+ and OH^- on the product side of the equation to form H_2O and cancel any water molecules that are common to both sides of the equation.

$12\ OH^- + 6\ Br_2 \rightarrow 10\ Br^- + 2\ BrO_3^- + 6\ H_2O$

Reduce the coefficients to their smallest whole–number ratio.

$6\ OH^- + 3\ Br_2 \rightarrow 5\ Br^- + BrO_3^- + 3\ H_2O$

21. $3\ HSO_3^{2-} + Cr_2O_7^{2-} + 5\ H^+ \rightarrow 3\ SO_4^{2-} + 2\ Cr^{3+} + 4\ H_2O$

$$\frac{0.3143\ \text{mol}\ Na_2SO_3}{1\ L} \times \frac{1\ \text{mol}\ SO_3^{2-}}{1\ \text{mol}\ Na_2SO_3} \times 0.02500\ L = 7.858 \times 10^{-3}\ \text{mol}\ SO_3^{2-}$$

$$7.858 \times 10^{-3}\ \text{mol}\ SO_3^{2-} \times \frac{1\ \text{mol}\ Cr_2O_7^{2-}}{3\ \text{mol}\ SO_3^{2-}} = 2.619 \times 10^{-3}\ \text{mol}\ Cr_2O_7^{2-}$$

$$2.619 \times 10^{-3}\ \text{mol}\ Cr_2O_7^{2-} \times \frac{1\ \text{mol}\ K_2Cr_2O_7}{1\ \text{mol}\ Cr_2O_7^{2-}} \times \frac{1}{0.02842\ L} = 0.09216\ \frac{\text{mol}\ K_2Cr_2O_7}{L}$$

22. Using either the oxidation–number or half–reaction method will produce the following balanced equation:

Appendix C – Self Test Solutions

$$16 \text{ H}^+ + 2 \text{ Cr}_2\text{O}_7^{2-} + \text{C}_2\text{H}_5\text{OH} \rightarrow 4 \text{ Cr}^{3+} + 2 \text{ CO}_2 + 11 \text{ H}_2\text{O}$$

$$\text{mol K}_2\text{Cr}_2\text{O}_7 = \frac{5.000 \times 10^{-3} \text{ mol K}_2\text{Cr}_2\text{O}_7}{1 \text{ L soln}} \times 0.006522 \text{ L soln} = 3.26 \times 10^{-5} \text{ mol K}_2\text{Cr}_2\text{O}_7$$

$$\text{mol C}_2\text{H}_5\text{OH} = 3.26 \times 10^{-5} \text{ mol K}_2\text{Cr}_2\text{O}_7 \times \frac{1 \text{ mol C}_2\text{H}_5\text{OH}}{2 \text{ mol K}_2\text{Cr}_2\text{O}_7} = 1.63 \times 10^{-5} \text{ mol C}_2\text{H}_5\text{OH}$$

Knowing moles, we can now calculate grams of $\text{C}_2\text{H}_5\text{OH}$ and determine the percent weight of alcohol in the blood.

$$\text{g C}_2\text{H}_5\text{OH} = 1.63 \times 10^{-5} \text{ mol C}_2\text{H}_5\text{OH} \times \frac{46.0 \text{ g C}_2\text{H}_5\text{OH}}{1 \text{ mol C}_2\text{H}_5\text{OH}} = 7.50 \times 10^{-4} \text{ g C}_2\text{H}_5\text{OH}$$

$$\% \text{ Weight} = \frac{7.50 \times 10^{-4} \text{ g C}_2\text{H}_5\text{OH}}{0.50 \text{ g sample}} \times 100 = 0.15\%; \text{ yes, this person is legally drunk.}$$

23. a. $\text{Na}_2\text{S} (aq) + \text{Zn(NO}_3)_2 (aq) \rightarrow 2 \text{ NaNO}_3 (aq) + \text{ZnS} (s)$ molecular equation

$2 \text{ Na}^+ (aq) + \text{S}^{2-} (aq) + \text{Zn}^{2+} (aq) + 2 \text{ NO}_3^- (aq) \rightarrow 2 \text{ Na}^+ (aq) + 2 \text{ NO}_3^- (aq) + \text{ZnS} (s)$ ionic equation

Spectator ions: Na^+ and NO_3^-

$\text{Zn}^{2+} (aq) + \text{S}^{2-} (aq) \rightarrow \text{ZnS}$ net ionic equation

b. We were given information about both reactants. Therefore, we must determine which reactant is the limiting reactant. Begin by determining the number of moles of each reactant present.

$$\text{mol Na}_2\text{S} = 7.5 \text{ g Na}_2\text{S} \times \frac{1 \text{ mol Na}_2\text{S}}{78.0 \text{ g Na}_2\text{S}} = 9.6 \times 10^{-2}$$

$$\text{mol Zn(NO}_3)_2 = 6.8 \text{ g Zn(NO}_3)_2 \times \frac{1 \text{ mol Zn(NO}_3)_2}{189.4 \text{ g Zn(NO}_3)_2} = 3.6 \times 10^{-2} \text{ mol Zn(NO}_3)_2$$

The required mole ratio is 1 mol of Na_2S for every 1 mol $\text{Zn(NO}_3)_2$. Therefore, the limiting reactant is $\text{Zn(NO}_3)_2$. The mass of ZnS produced is

$$\text{g ZnS} = 3.6 \times 10^{-2} \text{ mol Zn(NO}_3)_2 \times \frac{1 \text{ mol ZnS}}{1 \text{ mol Zn(NO}_3)_2} \times \frac{97.4 \text{ g ZnS}}{1 \text{ mol ZnS}} = 3.5 \text{ g ZnS}$$

c. The amount of leftover reactant is

xs $\text{Na}_2\text{S} = 0.096$ mol Na_2S available – 0.036 mol Na_2S reacted = 0.060 mol Na_2S leftover

$$\text{g Na}_2\text{S} = 6.0 \times 10^{-2} \text{ mol Na}_2\text{S} \times \frac{78.0 \text{ g Na}_2\text{S}}{1 \text{ mol Na}_2\text{S}} = 4.7 \text{ g Na}_2\text{S}$$

d. $\% \text{ Yield} = \dfrac{2.53 \text{ g ZnS}}{3.5 \text{ g ZnS}} \times 100 = 72\%$

24. $Mg\ (s) + CuSO_4\ (aq) \rightarrow MgSO_4\ (aq) + Cu\ (s)$ Molecular equation

$Mg\ (s) + Cu^{2+}\ (aq) + SO_4^{2-}\ (aq) \rightarrow Mg^{2+}\ (aq) + SO_4^{2-}\ (aq) + Cu\ (s)$ Ionic equation

$Mg\ (s) + Cu^{2+}\ (aq) \rightarrow Mg^{2+}\ (aq) + Cu\ (s)$ Net ionic equation

We were given information about both reactants. Therefore, we must determine which reactant is the limiting reactant. Begin by determining the number of moles of each reactant present.

$$\text{mol Mg} = 4.8\ g\ Mg \times \frac{1\ mol\ Mg}{24.3\ g\ Mg} = 0.20\ mol\ Mg$$

$$\text{mol CuSO}_4 = \frac{0.75\ mol\ CuSO_4}{1\ L\ soln} \times 0.250\ L\ soln = 0.19\ mol\ CuSO_4$$

The required mole ratio is 1 mol of Mg for every 1 mol of $CuSO_4$. Therefore, the limiting reactant is $CuSO_4$. The mass of copper produced is

$$g\ CuSO_4 = 0.19\ mol\ CuSO_4 \times \frac{1\ mol\ Cu}{1\ mol\ CuSO_4} \times \frac{63.5\ g\ Cu}{1\ mol\ Cu} = 12\ g\ Cu$$

The amount of leftover reactant is

xs Mg = 0.20 mol Mg available − 0.19 mol Mg reacted = 0.01 mol Mg left over

$$g\ Mg\ \text{left over} = 0.01\ mol\ Mg \times \frac{24.3\ g\ Mg}{1\ mol\ Mg} = 0.24\ g\ Mg$$

$$\%\ \text{Yield} = \frac{9.2\ g\ Cu}{12\ g\ Cu} \times 100 = 77\%$$

25. We begin this problem by determining the number of moles of I_3^- produced in the reaction between Cu^{2+} and I^-. We can determine the number of moles of I_3^- from the number of moles of $Na_2S_2O_3$ used in the titration.

$$\text{mol } S_2O_3^{2-} = 0.2500 \frac{\text{mol } S_2O_3^{2-}}{L\ soln} \times 0.020\ 80\ l\ soln = 0.005\ 200\ \text{mol } S_2O_3^{2-}$$

From the balanced equation given in Workbook Problem 4.10 on page 84, we know that 2 mol of $S_2O_3^{2-}$ react with 1 mol of I_3^-. Therefore, the number of moles of I_3^- produced in the reaction between Cu^{2+} and I^- is

$$\text{mol } I_3^- = 0.005\ 200\ \text{mol } S_2O_3^{2-} \times \frac{1\ mol\ I_3^-}{2\ mol\ S_2O_3^{2-}} = 0.002\ 600\ \text{mol } I_3^-$$

Knowing the moles of I_3^- produced, we can now calculate the number of moles and the mass of Cu^{2+} that reacted.

Appendix C – Self Test Solutions

$$\text{mol Cu}^{2+} = 0.002\,600 \text{ mol } I_3^- \times \frac{2 \text{ mol Cu}^{2+}}{1 \text{ mol } I_3^-} = 0.00520 \text{ mol Cu}^{2+}$$

$$\text{g Cu}^{2+} = 0.00520 \text{ mol Cu}^{2+} \times \frac{63.5 \text{ g Cu}^{2+}}{1 \text{ mol Cu}^{2+}} = 0.3302 \text{ g Cu}^{2+}$$

$$\%\text{Cu} = \frac{0.3302 \text{ g Cu}}{0.500 \text{ g sample}} \times 100 = 66.0\%$$

Challenge Problems

1. The precipitate that forms on the addition of hydrochloric acid to test tube A could be silver chloride and/or lead(II) chloride. However, when you add hydrogen sulfide to test tube B, no precipitate forms. If Pb^{2+} was present in the original solution, you should have obtained the PbS precipitate on addition of the hydrogen sulfide to test tube B. You can also rule out the presence of Ni^{2+} and Cu^{2+} since you did not obtain a precipitate in test tube B. The formation of a precipitate on the addition of sodium carbonate indicates the presence of Ba^{2+} in the original solution. Therefore, the ions present in the original solution are Ag^+ and Ba^{2+}.

2. The precipitate that was formed is $BaSO_4$. The net ionic equation for the formation of this precipitate is

$$Ba^{2+}(aq) + SO_4^{2-}(aq) \rightarrow BaSO_4$$

To start, we know that the mass of the solid is equal to the mass of Na_2SO_4 and K_2SO_4 present. (All mathematical equations are numbered in an attempt to help you follow the solution strategy.)

Mass of solid = Mass of Na_2SO_4 + Mass of K_2SO_4 (1)

We also know that the mass of $BaSO_4$ is determined by the amount of sulfate ion in Na_2SO_4 and K_2SO_4.

Mass of $BaSO_4$ = Mass of $BaSO_4$ from Na_2SO_4 + Mass of $BaSO_4$ from K_2SO_4 (2)

We can write stoichiometric equations that show the relationship between the mass of $BaSO_4$ to either the mass of Na_2SO_4 or the mass of K_2SO_4.

$$\text{Mass of Ba}_2SO_4 \text{ from Na}_2SO_4 = \text{Mass Na}_2SO_4 \times \frac{1 \text{ mol Na}_2SO_4}{142 \text{ g Na}_2SO_4} \times \frac{1 \text{ mol Ba}_2SO_4}{1 \text{ mol Na}_2SO_4} \times \frac{233.3 \text{ g Ba}_2SO_4}{1 \text{ mol Ba}_2SO_4}$$

After cancelation, this equation reduces to

$$\text{Mass of Ba}_2SO_4 \text{ from Na}_2SO_4 = \text{Mass of Na}_2SO_4 \times \frac{233.3 \text{ g Ba}_2SO_4}{142 \text{ g Na}_2SO_4} \quad (3)$$

Using the same approach for the mass of Ba_2SO_4 from K_2SO_4, we find that

$$\text{Mass of BaSO}_4 \text{ from K}_2SO_4 = \text{Mass of K}_2SO_4 \times \frac{233.3 \text{ g BaSO}_4}{174.2 \text{ g K}_2SO_4} \quad (4)$$

Appendix C – Self Test Solutions

$\dfrac{233.3\text{ g BaSO}_4}{142\text{ g Na}_2\text{SO}_4}$ and $\dfrac{233.3\text{ g BaSO}_4}{174.2\text{ g K}_2\text{SO}_4}$ simply represent factors that describe the relationships between the mass of $BaSO_4$ to Na_2SO_4 and K_2SO_4. In subsequent equations, we will drop the units from these fractions.

We can show the mass of Na_2SO_4 in terms of the mass of the solid and the mass of K_2SO_4. Rearranging the first equation **(1)** we wrote gives:

Mass Na_2SO_4 = mass of solid – mass of K_2SO_4 = 0.500 g – mass of K_2SO_4 **(5)**

Substituting equation **5** into equation **3** gives:

Mass of Ba_2SO_4 from $Na_2SO_4 = (0.500\text{ g} - \text{mass }K_2SO_4) \times \dfrac{233.3}{142}$.

Using equation **2**, we can now write a mathematical equation showing the relationship between the mass of $BaSO_4$ and the mass of K_2SO_4.

Mass of $BaSO_4 = \left((0.500\text{ g} - \text{mass }K_2SO_4) \times \dfrac{233.3}{142}\right) + \left(\text{mass }K_2SO_4 \times \dfrac{233.3}{174.2}\right)$

Mass of $BaSO_4 = ((0.500\text{ g} - \text{mass }K_2SO_4) \times 1.64) + (\text{mass }K_2SO_4 \times 1.34)$

$0.716\text{ g} = 0.820 - 1.64(\text{mass }K_2SO_4) + 1.34(\text{mass }K_2SO_4)$

$0.716\text{ g} - 0.820 = -0.30(\text{mass }K_2SO_4)$

$-0.104 = -0.30(\text{mass }K_2SO_4)$

Mass of K_2SO_4 = 0.347 g

We can solve for the mass of Na_2SO_4 by substituting the mass of K_2SO_4 into equation **5**.

Mass of Na_2SO_4 = 0.500 g – 0.347 g = 0.153 g

Chapter 5

True–False

1. F. Frequency and wavelength are inversely related.
2. T
3. F. The energy of the photons depends only on the frequency.
4. T
5. F. Quantum mechanics concentrates on the wave properties of an electron.
6. T
7. F. The principal quantum number defines the size and energy level of the electron.
8. F. The energy levels of various orbitals in the hydrogen atom depend only on the value of n; in multielectron atoms, the energy levels of various orbitals depends on the value of both n and ℓ

Appendix C – Self Test Solutions

9. F. The increase in atomic radius down a group is due to the fact that successively larger valence-shell orbitals are occupied.
10. F. Elements in each group have similar valence electron configurations.

Multiple Choice
1. b
2. d
3. d
4. c
5. b
6. c
7. c
8. b
9. c

Matching
Photoelectric effect—h
Einstein—e
de Broglie—g
Schrödinger—b
Heisenberg—a
Pauli—d
Hund—i
Aufbau principle—f
Mendeleev—c

Fill-in the Blank
1. electromagnetic spectrum
2. own unique
3. blackbody radiation
4. dual wave/particle
5. position in three dimensional space
6. shell
7. the energy difference between the higher and lower-energy orbitals
8. distance of the electron from the nucleus
9. effective nuclear charge
10. electron configuration
11. ground state configuration
12. chemical properties
13. similar valence-shell electron configurations
14. total energy of the atom.
15. decreases

Problems

1. $\lambda \times \nu = c$; $\lambda = \dfrac{c}{\nu}$; $\lambda = \dfrac{3.00 \times 10^8 \text{ m/s}}{1.58 \times 10^{13} \text{ s}^{-1}} = 1.90 \times 10^{-5} \text{ m} = 1.90 \times 10^4 \text{ nm}$

   $E = h\nu$; $E = (6.626 \times 10^{-34} \text{ J} \cdot \text{s}) \times (1.58 \times 10^{13} \text{ s}^{-1}) = 1.05 \times 10^{-20} \text{ J}$

2. $\lambda \times \nu = c$; $\nu = \dfrac{3.00 \times 10^8 \text{ m/s}}{6.708 \times 10^{-7} \text{ m}} = 4.472 \times 10^{14} \text{ s}^{-1}$

   The color of this radiation is orange-red.

Appendix C – Self Test Solutions

3. $E = \dfrac{hc}{\lambda}$; $E = \dfrac{(6.626 \times 10^{-34} \text{ J} \cdot \text{s}) \times (3.00 \times 10^{8} \text{ m/s})}{1.30 \times 10^{-7} \text{ m}} = 1.53 \times 10^{-18}$ J

This represents the energy of one photon at a wavelength of 130 nm. The energy of 1 mole of photons is calculated in the following manner.

$\left(1.53 \times 10^{-18} \dfrac{\text{J}}{\text{photon}}\right) \times \left(6.022 \times 10^{23} \dfrac{\text{photon}}{\text{mol}}\right) = 9.21 \times 10^{5} \dfrac{\text{J}}{\text{mol}} = 921 \dfrac{\text{kJ}}{\text{mol}}$

4. $\dfrac{1}{\lambda} = 1.097 \times 10^{-2} \text{ nm}^{-1} \left(\dfrac{1}{2^{2}} - \dfrac{1}{6^{2}}\right) = 2.438 \times 10^{-3} \text{ nm}^{-1}$; $\lambda = 410.2$ nm

$E = \dfrac{(6.626 \times 10^{-34} \text{ J} \cdot \text{s}) \times (3.00 \times 10^{8} \text{ m/s})}{4.102 \times 10^{-7} \text{ m}} = 4.85 \times 10^{-19}$ J

5. For a 4p orbital the allowed quantum numbers are $n = 4$; $\ell = 1$; and m_ℓ can be +1, 0 or –1.

6. 3p

7. $n = 5$, $\ell = 4$, m_ℓ can be +4, +3, +2, +1, 0, –1, –2, –3, or –4.

8. The first line in the Paschen series results from the electron falling from the $n = 4$ to the $n = 3$ shell.

$\dfrac{1}{\lambda} = 1.097 \times 10^{-2} \text{ nm}^{-1} \left(\dfrac{1}{3^{2}} - \dfrac{1}{4^{2}}\right) = 5.332 \times 10^{-4} \text{ nm}^{-1}$; $\lambda = 1.875 \times 10^{3}$ nm

$E = \dfrac{(6.626 \times 10^{-34} \text{ J} \cdot \text{s})(3.00 \times 10^{8} \text{ m/s})}{1.875 \times 10^{-6} \text{ m}} = 1.060 \times 10^{-19}$ J

This represents the energy of one photon at a wavelength of 1.875×10^{3} nm. We must now calculate the energy of one mole of photons.

$1.060 \times 10^{-19} \dfrac{\text{J}}{\text{photon}} \times \dfrac{6.022 \times 10^{23} \text{ photons}}{1 \text{ mol}} \times \dfrac{1 \text{ kJ}}{1000 \text{ J}} = 63.84$ kJ / mol

9. $E = \dfrac{(6.626 \times 10^{-34} \text{ J} \cdot \text{s}) \times (3.00 \times 10^{8} \text{ m/s})}{9.73 \times 10^{-8} \text{ m}} = 2.04 \times 10^{-18}$ J

We know that the hydrogen atom is in the ground state, therefore, m = 1. Use the Balmer–Rydberg equation to solve for n.

$\dfrac{1}{97.3 \text{ nm}} = 1.097 \times 10^{-2} \text{ nm}^{-1} \left(\dfrac{1}{1} - \dfrac{1}{n^{2}}\right)$; $n = 4$

10. S: $1s^2\ 2s^2\ 2p^6\ 3s^2\ 3p^4$ [Ne] ↑↓ ↑↓ ↑ ↑ ; 2 unpaired e⁻
                                                3s   3p

   Sr: $1s^2\ 2s^2\ 2p^6\ 3s^2\ 3p^6\ 4s^2\ 3d^{10}\ 4p^6\ 5s^2$ [Kr] ↑↓; 0 unpaired e⁻
                                                                              5s

   Pb: $1s^2\ 2s^2\ 2p^6\ 3s^2\ 3p^6\ 4s^2\ 3d^{10}\ 4p^6\ 5s^2 4d^{10}\ 5p^6\ 6s^2\ 4f^{14}\ 5d^{10}\ 6p^2$

Appendix C – Self Test Solutions

[Xe] ↑↓ ↑↓ ↑↓ ↑↓ ↑↓ ↑↓ ↑↓   ↑↓ ↑↓ ↑↓ ↑↓ ↑↓   ↑↓   ↑ ↑ __   2 unpaired e⁻
           4f   5d   6s   6p

Ni: $1s^2\ 2s^2\ 2p^6\ 3s^2\ 3p^6\ 4s^2\ 3d^8$ [Ar] ↑↓   ↑↓ ↑↓ ↑↓ ↑ ↑ ; 2 unpaired e⁻
                                                   4s   3d

11. The general valence–shell electron configuration for group VA is $ns^2\ np^3$. These elements are located in the p block.

12. $[Ar]4s^2 3d^7$: Co
    $[Xe]6s^2 4f^{14} 5d^{10} 6p^4$: Po
    $[Kr]5s^2 4d^{10} 5p^2$: Sn

13. Ba > Pb > Ir

14. Check your answer with Figure 5.17, page 172 in your textbook.

Challenge Problem

1. The first shell in any atom or ion is the 1s shell and it can only hold two electrons. The second shell is the $n = 2$ shell. If $n = 2$, $\ell = 0$, 1 giving the 2s and 2p subshells. The 2s subshell holds two electrons and the 2p subshell holds six electrons giving a total of eight electrons for the second subshell. The third shell is the $n = 3$ shell. If $n = 3$, $\ell = 0$, 1, 2 giving the 3s, 3p, and 3d subshells. The 3s subshell holds two electrons, the 3p subshell holds six electrons, and the 3d subshell can hold up to 10 electrons giving a total of 18 electrons in the third subshell. However, the 3+ ion has only 13 electrons in the third shell. Since the 3d subshell is the highest–energy subshell in the third shell, it has five electrons instead of the maximum number of 10.

   The electron configuration for the ion is: $1s^2\ 2s^2\ 2p^6\ 3s^2\ 3p^6\ 3d^5$

   At first glance, it would appear that we would just add three more electrons to the 3d subshell to write the electron configuration of the atom. Not so! Remember the note given with this problem. The 4s electrons are lost before the 3d electrons. However, know from Figure 5.9, page 174 in the text, that the 4s orbital is lower in energy than the 3d orbital. When removing electrons from the orbitals, we always remove the valence electrons (highest value of n) first. In this case, the valence electrons are the 4s electrons. (The "building up" of the periodic table and the removal of electrons are not the reverse of each other. When building up the table, we added a proton for every electron added. However, we remove only electrons when forming ions.) Therefore, two of the three electrons removed to create the 3+ ion were in the 4s orbital. The remaining electron was in the 3d orbital.

   The electron configuration for the atom is: $1s^2\ 2s^2\ 2p^6\ 3s^2\ 3p^6\ 3d^6\ 4s^2$

   From Figure 5.17, page 185 in your textbook, we can now identify the element as Fe.

2. Let's begin this discussion by writing the electron configuration for Al and all of the ions given.

   Al: $1s^2\ 2s^2\ 2p^6\ 3s^2\ 3p^1$

   Al⁺: $1s^2\ 2s^2\ 2p^6\ 3s^2$

   Al²⁺: $1s^2\ 2s^2\ 2p^6\ 3s^1$

   Al³⁺: $1s^2\ 2s^2\ 2p^6$

Appendix C – Self Test Solutions

Al^{4+}: $1s^2\ 2s^2\ 2p^5$

Recall that a filled or half–filled subshell has an added stability. If you compare the electron configuration of Al to Al^+, you notice that loss of an electron leads to an electron configuration in which all of the subshells are full. Also, the $3s^1$ electron is highly shielded from the nucleus by the $n = 1$ and $n = 2$ electrons. Therefore, the $3s^1$ electron is lost rather easily. Loss of the second electron to form Al^{2+} is more difficult since this electron feels a higher Z_{eff}. (Loss of an electron leads to a higher Z_{eff} because there will be less shielding of the nucleus.) Loss of the third electron requires even more energy than loss of the second electron because the remaining $2s^1$ electron feels an even greater Z_{eff}. The E_4 is much greater than E_3. However, close inspection of the Al^{3+} ion reveals that this ion has the same electron configuration as Ne. This electron configuration has a closed shell, which is very stable. As a consequence, Al^{3+} is a stable ion and reluctant to give up any additional electrons.

Chapter 6

True–False

1. T
2. F. The E_i increases across a period because Z_{eff} increases and the radius of the atoms shrink.
3. T
4. F. The tendency of the atom to accept an electron increases as the value of E_{ea} becomes more negative.
5. F. An element with a low E_i can transfer an electron to an atom with a negative E_{ea}.
6. T
7. T
8. F. The alkaline earth metals are less reactive than the alkali metals because the E_i (alkaline earth metals) > E_i (alkali metals).
9. F. The halogens are powerful oxidizing agents.
10. T

Multiple Choice

1. c
2. d
3. d
4. b
5. d
6. d
7. a
8. b
9. c
10. c

Matching

Ionization energy—c
Electron affinity—a
Born–Haber cycle—e
Lattice energy—d
Octet rule—b

Appendix C – Self Test Solutions

Fill–in–the–Blank
1. ionization energy
2. reactivity
3. less; E_i (alkaline earth metals) > E_i (alkali metals)
4. decreasing cation size
5. $-\frac{1}{2}$
6. a metal hydroxide, $M(OH)_2$
7. oxidizing agents
8. lighter, more reactive halogen
9. oxidizing agents
10. the availability of a *d* subshell

Problems

1. Ge < Si < P; ionization energies increase across a period (P > Si) and decrease down a group (Si > Ge).

2. In: The fourth electron is a valence electron in Sn but a core electron in In.

3. Fluorine has a higher Z_{eff} than oxygen; neon is a noble gas and has a filled valence shell, making it very difficult to gain an electron.

4. Use a Born–Haber cycle. However, this time the net energy change is known, and we need to solve for the heat of sublimation of aluminum.

Al (s) → Al (g)	? kJ/mol
Al (g) → Al$^+$ (g) + e$^-$	578 kJ/mol
Al$^+$ (g) → Al^{2+} (g) + e$^-$	1,817 kJ/mol
Al^{2+} (g) → Al^{3+} (g) + e	2,745 kJ/mol
Br$_2$ (l) → Br$_2$ (g)	30.9 kJ/mol
3/2 Br$_2$ (g) → 3 Br (g)	3/2(193 kJ/mol)
3 Br (g) → 3 Br$^-$ (g)	−3(324.8 kJ/mol)
Al^{3+} (g) + 3 Br$^-$ (g) → AlBr$_3$ (s)	−5,361 kJ/mol
Overall net energy change	−527.2 kJ/mol

   −527.2 = ? + 578 kJ/mol + 1,817 kJ/mol + 2,745 kJ/mol + 30.9 kJ/mol + 3/2(193 kJ/mol) + [−3(324.8 kJ/mol)] − 5,361 kJ/mol

   ? = 347.8 kJ/mol

5. a. Li (s) + H$_2$O → LiOH (aq) + H$_2$ (g)
   Half–reactions: 2 (Li (s) → Li$^+$ (aq) + e$^-$)
   2 H$_2$O + 2 e$^-$ → 2 OH$^-$ (aq) + H$_2$ (g)
   Balanced reaction: 2 Li (s) + 2 H$_2$O → 2 Li$^+$ (aq) + 2 OH$^-$ (aq) + H$_2$ (g)
   Species oxidized: Li (s)
   Species reduced: H$_2$O

   b. Cl$_2$ (g) + I$_2$ (s) + H$_2$O → HCl (aq) + HIO$_3$ (aq)
   Half reactions: (Cl$_2$ (g) + 2 H$^+$ + 2 e$^-$ → 2 HCl (aq))5
   I$_2$ (s) + 6 H$_2$O → 2 HIO$_3$ (aq) + 10 H$^+$ + 10 e$^-$
   Balanced reaction: 5 Cl$_2$ (g) + I$_2$ (s) + 6 H$_2$O → 10 HCl (aq) + 2 HIO$_3$ (aq)

Appendix C – Self Test Solutions

Species oxidized: I_2 (s)
Species reduced: Cl_2 (g)

6. a. Ca (s) + Br_2 (l) → 2 $CaBr_2$ (s)
   b. 2 Al (s) + N_2 (g) → 2 AlN (s)
   c. K (s) + O_2 (g) → KO_2 (s)
   d. 2 Na (s) + H_2 (g) → 2 NaH (s)
   e. H_2 (g) + I_2 (g) → 2 HI (g)

7. a. $AlCl_3$; Al^{3+} has the higher charge
   b. $AlCl_3$; Al^{3+} is the smaller cation
   c. $SnCl_2$; both Sn^{2+} and Cl^- have the smaller size

8. ClF_3: oxidation number Cl = +3; oxidation number F = –1
   IF_7: oxidation number I = +7; oxidation number F = –1
   ICl_5: oxidation number I = +5; oxidation number Cl = –1
   IBr: oxidation number I = +1; oxidation number Br = –1

9. We know we need to work this problem as a limiting reagent problem because we were given data for both reactants. First, we need to calculate the number of moles of each reactant present and determine which reactant is the limiting reactant.

$$4.50 \text{ g BeF}_2 \times \frac{1 \text{ mol BeF}_2}{47.0 \text{ g BeF}_2} = 0.0957 \text{ mol BeF}_2$$

$$8.23 \text{ g Mg} \times \frac{1 \text{ mol Mg}}{24.3 \text{ g Mg}} = 0.339 \text{ mol Mg}$$

The BeF_2:Mg mole ratio is 1:1. Therefore, BeF_2 is the limiting reagent since we need 0.339 mol BeF_2 to react with 0.339 mol Mg. We can now calculate the amount of Be produced in this reaction.

$$0.0957 \text{ mol BeF}_2 \times \frac{1 \text{ mol Be}}{1 \text{ mol BeF}_2} \times \frac{9.0 \text{ g Be}}{1 \text{ mol Be}} = 0.861 \text{ g Be}$$

10. Cl_2 (g) + 3 F_2 (g) → 2 ClF_3 (g); Br_2 (l) + 5 F_2 (g) → 2 BrF_5 (g)

Challenge Problem

The chemical equation for the reaction of the alkaline earth bromide with silver nitrate is

MBr_2 + 2 $AgNO_3$ → $M(NO_3)_2$ + 2 AgBr

a. We can calculate the mass of the bromide from the mass of silver bromide produced.

$$\text{mass of Br}^- = 7.05 \text{ g AgBr} \times \frac{1 \text{ mol AgBr}}{187.8 \text{ g AgBr}} \times \frac{1 \text{ mol MBr}_2}{2 \text{ mol AgBr}} \times \frac{2 \text{ mol Br}^-}{1 \text{ mol MBr}_2} \times \frac{79.9 \text{ g Br}^-}{1 \text{ mol Br}^-} = 3.00 \text{ g Br}^-$$

$$\text{Mass \% of Br}^- \text{ in MBr}_2 = \frac{3.00 \text{ g Br}^-}{3.75 \text{ g MBr}_2} \times 100 = 80.0\%$$

Appendix C – Self Test Solutions

b. We can determine the identity of the metal by first determining the molar mass of the metal bromide and subtracting the molar mass of bromine from the molar mass of the metal bromide.

$$\text{mol MBr}_2 = 7.05 \text{ g AgBr} \times \frac{1 \text{ mol AgBr}}{187..8 \text{ g AgBr}} \times \frac{1 \text{ mol MBr}_2}{2 \text{ mol AgBr}} = 1.88 \times 10^{-2} \text{ mol MBr}_2$$

$$\text{Molar mass MBr}_2 = \frac{3.75 \text{ g MBr}_2}{1.89 \times 10^{-2} \text{ mol MBr}_2} = 200 \text{ g/mol}$$

Molar mass MBr_2 = Molar mass M + 2 Molar mass Br

Molar mass of M = 200 g/mol – 160 g/mol = 40 g/mol

From the periodic table, we know that the alkaline earth metal with a molar mass of 40 g/mol is calcium.

c. For the reaction of calcium with bromine

Ca (s) + Br_2 (l) → $CaBr_2$ (s)

The chemical equation for the reaction of $CaBr_2$ with $AgNO_3$ was given earlier.

d. The mole ratio for the reaction between calcium and bromine is 1:1. For every mole of calcium we need 1 mole of Br_2. Let''s begin by calculating the number of moles of calcium present.

$$\text{mol Ca} = 2.25 \text{ g Ca} \times \frac{1 \text{ mol Ca}}{40.0 \text{ g Ca}} = 5.63 \times 10^{-2} \text{ mol Ca}$$

We need to have 5.63×10^{-2} mol of Br_2 as well. We now calculate the number of moles of Br_2 present.

$$\text{mol Br}_2 = 7.5 \times 10^{22} \text{ molecules Br}_2 \times \frac{1 \text{ mol Br}_2}{6.022 \times 10^{23} \text{ molecules}} = 0.13 \text{ mol Br}_2$$

Br_2 is in excess, and calcium is the limiting reactant. The mass of the unreacted Br_2 can now be calculated.

mol Br_2 = mol Br_2 reacted + mol Br_2 unreacted

0.13 mol = 0.056 mol + mol Br_2 unreacted

mol Br_2 unreacted = 0.13 mol – 0.056 mol = 0.074 mol unreacted Br_2

$$\text{g Br}_2 \text{ unreacted} = 0.074 \text{ mol Br}_2 \times \frac{160 \text{ g Br}_2}{1 \text{ mol Br}_2} = 12 \text{ g Br}_2$$

Chapter 7

True–False
1. T
2. F. The formation of a covalent bond leads to lower energy; therefore, energy is released when a covalent bond is formed.

Appendix C – Self Test Solutions

3. F. The bond order refers to the number of electron pairs between atoms.
4. F. The pairing of dots in an electron–dot structure does not correspond to the pairing of electrons in the electron configuration.
5. T
6. F. Some elements are able to expand their octets because they have unfilled d orbitals; other elements will have less than an octet.
7. F. Polar covalent bonds do not involve the complete transfer of electrons.
8. T
9. T
10. T

Fill–in–the–Blank
1. electron–dot structure
2. it is surrounded by four pairs of electrons or has no more electrons to share
3. more than one pair of electrons
4. elements beyond the second row
5. resonance hybrid
6. similar electronegativities
7. overlaps a singly occupied valence orbital on another atom.
8. sp^2
9. additive and subtractive
10. the number of atomic orbitals combined.

Matching
Bond length—i
Bond dissociation energy—c
Coordinate covalent bond—d
Polar covalent bond—j
Electronegativity—a
VSEPR model—b
hybrid orbitals—k
σ bond—e
π bond—l
Molecular orbital—f
Paramagnetic—h
Diamagnetic g

Problems

1. a) CrO_4^{2-}; number of valence electrons = 32; number of bonding electrons = 8; number of electrons used for O's octet = 24

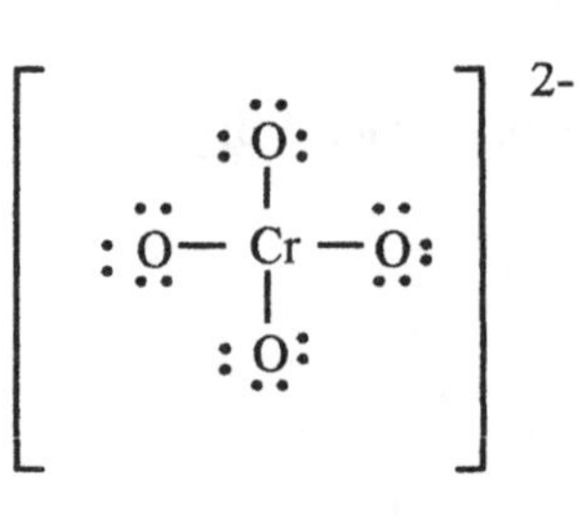

$$\left[\begin{array}{c} :\ddot{O}: \\ | \\ :\ddot{O} - Cr - \ddot{O}: \\ | \\ :\ddot{O}: \end{array} \right]^{2-}$$

b) IF_6^+; number of valence electrons = 48; number of bonding electrons = 12; number of electrons used for F's octet = 36

Appendix C – Self Test Solutions

$$\left[\begin{array}{c} :\ddot{F}: \\ :\ddot{F}\diagdown \mid \diagup \ddot{F}: \\ I \\ :\ddot{F}\diagup \mid \diagdown \ddot{F}: \\ :\ddot{F}: \end{array} \right]^{+}$$

c) ClF_3; number of valence electrons = 28; number of bonding electrons = 6; number of electrons used for F's octet = 18; number of of electrons left over = 4

$$\begin{array}{c} :\ddot{F}: \\ \mid \\ :\ddot{C}l - \ddot{F}: \\ \mid \\ :\ddot{F}: \end{array}$$

d) H_2F^+; number of valence electrons = 8; number of bonding electrons = 4; number of electrons left over = 4

$$\left[H \diagup \overset{\cdot\cdot}{\underset{\cdot\cdot}{F}} \diagdown H \right]^{+}$$

e) PF_4^-; number of valence electrons = 34; number of bonding electrons = 8; number of electrons used for F's octet = 24; number of electrons left over = 2

$$\left[\begin{array}{c} :\ddot{F}: \\ \mid \ddot{F}: \\ :P \diagup \\ \mid \ddot{F}: \\ :\ddot{F}: \end{array} \right]^{-}$$

f) XeF_4; number of valence electrons = 36; number of bonding electrons = 8; number of electrons used for F's octet = 24; number of electrons leftover = 4

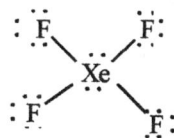

g) BF_3; number of valence electrons = 24. number of bonding electrons = 6; number of electrons used for F's octet = 18. (Remember that B is one of the few elements that will have less than an octet around it.)

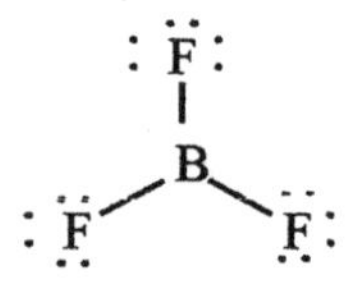

Appendix C – Self Test Solutions

[Lewis structure of BF₃ shown]

2. SO₃; number of valence electrons = 24; number of bonding electrons = 6; number of electrons used for O's octet = 18

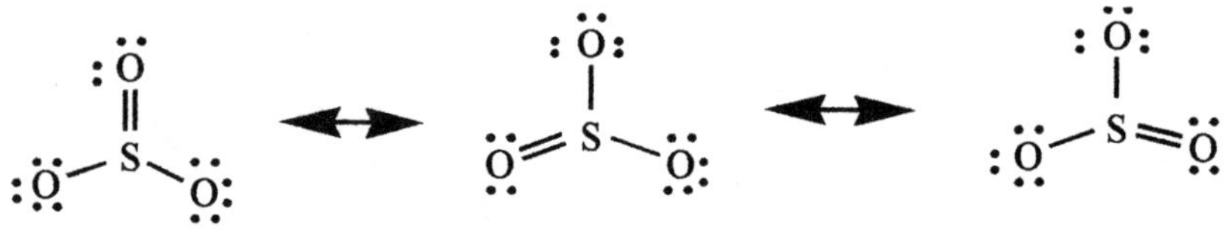

[Three resonance structures of SO₃ shown]

3. C–H; EN (C) = 2.5, EN (H) = 2.1; ΔEN = 0.4, nonpolar
   Na–Cl; EN (Na) = 0.9; EN (Cl) = 3.0; ΔEN = 2.1, ionic
   C–N; EN (C) = 2.5; EN (N) = 3.0; ΔEN = 0.5; polar covalent
   O–H; EN (O) = 3.5; EN (H) = 2.1; ΔEN = 1.4; polar covalent
   F–F; EN (F) = 4.0; ΔEN = 0; nonpolar

4. S: 6 valence electrons; 8 bonding electrons; Formal charge = 6 – 4 = 2
   double bonded O: 6 valence electrons; 4 bonding electrons; 4 nonbonding electrons; Formal charge = 6 – 2 – 4 = 0
   single bonded O: 6 valence electrons; 2 bonding electrons; 6 nonbonding electrons; Formal charge = 6 – 1 – 6 = –1.

5. Following the rules for drawing electron–dot structures on pages 256 and 257 in your textbook, we find that three resonance structures exist for this molecule.

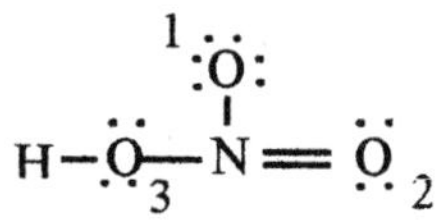

[Three resonance structures shown: Structure a, Structure b, Structure c]

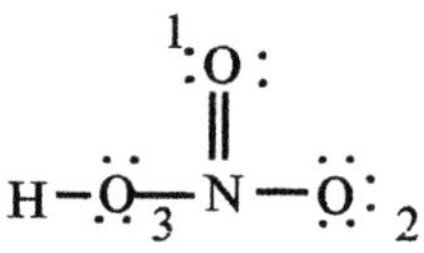

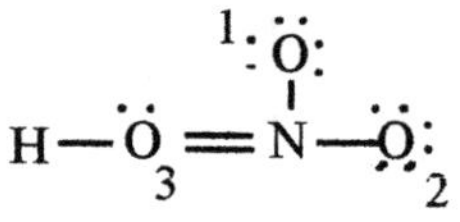

We now calculate formal charges for each structure.

Structure a
formal charge (N) = 5 – 1/2(8) = +1
formal charge (O)₁ = 6 – 6 – 1/2(2) = –1
formal charge (O)₂ = 6 – 4 – 1/2(4) = 0
formal charge (O)₃ = 6 – 4 – 1/2(4) = 0

Structure b
formal charge (N) = 5 – 1/2(8) = +1
formal charge (O)₁ = 6 – 4 – 1/2(4) = 0
formal charge (O)₂ = 6 – 6 – 1/2(2) = –1
formal charge (O)₃ = 6 – 4 – 1/2(2) = 0

Structure c
formal charge (N) = 5 – 1/2(8) = +1

Appendix C – Self Test Solutions

formal charge $(O)_1 = 6 - 6 - 1/2(2) = -1$
formal charge $(O)_2 = 6 - 6 - 1/2(2) = -1$
formal charge $(O)_3 = 6 - 2 - 1/2(6) = +1$

Structures a and b are the most stable structures. Structure c puts a formal charge of +1 on oxygen, which is very electronegative, and therefore makes the structure more unstable, compared with structures a and b, in which the oxygens carry a formal charge of either –1 or 0.

6. Assume that nitrogen is the central atom, and follow the rules for drawing an electron–dot structure. Fluorine is the more electronegative atom and, therefore, less likely to share a lone pair of electrons to form a double bond. Formal charge calculations show this to be true.

:F̈—N̈=Ö: :Ö—N̈=F̈:

Structure a   Structure b

Structure a is favored.

Structure a:
formal charge (F) = 7 – 6 – 1/2(1) = 0
formal charge (N) = 5 – 2 – 1/2(6) = 0
formal charge (O) = 6 – 4 – 1/2(4) = 0

Structure b
formal charge (F) = 7 – 4 – 1/2(4) = +1
formal charge (N) = 5 – 2 – 1/2(6) = 0
formal charge (O) = 6 – 6 – 1/2(2) = –1

7. CrO_4^{2-}: number of charge clouds = 4; number of bonds = 4; number of lone pairs = 0; shape = tetrahedral

   IF_6^+: number of charge clouds = 6, number of bonds = 6; number of lone pairs = 0; shape = octahedral

   ClF_3: number of charge clouds = 5; number of bonds = 3; number of lone pairs = 2; shape = T–shaped

   H_2F^+: number of charge clouds = 4; number of bonds = 2; number of lone pairs = 2; shape = bent

   PF_4: number of charge clouds = 5; number of bonds = 4; number of lone pairs = 1; shape = seesaw

   XeF_4: number of charge clouds = 6; number of bonds = 4; number of lone pairs = 2; shape = square planar

   BF_3: number of charge clouds = 3; number of bonds = 3; shape = trigonal planar

8. The carbon that is bonded to three H atoms and one C atom is surrounded by four charge clouds; therefore, the geometry around this carbon is tetrahedral. The other carbon atom is surrounded by three charge clouds; therefore, the geometry around this carbon is trigonal planar.

9. The carbon atom is surrounded by three charge clouds; therefore, its hybridization is sp^2. The phosphorous and nitrogen atoms are surrounded by four charge clouds; therefore, their hybridization is sp^3. (Don't forget the lone pair on nitrogen.)

10. N: bond order = 1/2(8 – 2) = 3; O: bond order = 1/2(8 – 4) = 2; F: bond order = 1/2(8 – 6) = 1

Appendix C – Self Test Solutions

Challenge Problem

Assume 100 grams of sample. We then have 24.3 g C, 71.6 g Cl, and 4.1 g H. Knowing the mass of each element, we can calculate the number of moles of each element.

$$\text{mol C} = 24.3 \text{ g C} \times \frac{1 \text{ mol C}}{12.0 \text{ g C}} = 2.03 \text{ mol C}$$

$$\text{mol Cl} = 71.6 \text{ g Cl} \times \frac{1 \text{ mol Cl}}{35.5 \text{ g Cl}} = 2.02 \text{ mol Cl}$$

$$\text{mol H} = 4.1 \text{ g H} \times \frac{1 \text{ mol H}}{1.00 \text{ g H}} = 4.1 \text{ mol H}$$

These molar amounts lead to the empirical equation CH_2Cl. The structure for this empirical formula is

```
       Cl
       |
   H — C
       |
       H
```

This structure is incomplete in that carbon is surrounded by three bonds instead of four. Chlorine is much more electronegative than carbon and, therefore, would not share a lone pair of electrons to form a double bond. Multiplying the subscripts in the empirical formula by 2 leads to the simple molecular formula of $C_2H_4Cl_2$. Two possible structures for this formula are:

```
     H   Cl              Cl  Cl
     |   |               |   |
 H — C — C — H       H — C — C — H
     |   |               |   |
     H   Cl              H   H
```

Chapter 8

True–False
1. F. $\Delta E = 0$ when the system is isolated from the surroundings.
2. T
3. T
4. F. If a system expands doing PV work, $w = -P\Delta V$.
5. F. For reactions carried out at constant pressure, the energy change is due to both heat transfer and PV work.
6. T
7. F. A reaction can absorb energy and still be spontaneous if the entropy of the reaction is positive and $T\Delta S > \Delta H$.
8. F. Specific heat is intensive.
9. F. The most stable forms of all elements in their standard state have $\Delta H_f^\circ = 0$.
10. F. ΔG is negative for a spontaneous process.

Matching
Energy—j
Temperature—l

Appendix C – Self Test Solutions

System—b
State function—f
Work—k
Enthalpy—c
Heat of fusion—m
Sublimation—a
Heat capacity—e
Hess's Law—i
Heat of combustion—g
Spontaneous process—d
Entropy—h

Fill–in–the–Blank
1. potential energy; the storage medium
2. heat
3. the First Law of Thermodynamics
4. into the system from the surroundings
5. surroundings; system; positive, $-P\Delta V$
6. 1 atm pressure of each gas, 298.15 K, and 1 M concentration for solutions.
7. heats of reaction
8. to the surroundings; system; negative
9. standard heat of formation
10. enthalpy changes for the corresponding bond–breaking reactions
11. Gibbs free energy

Problems

1. Since we know the gas absorbs 15 kJ of heat, we know that $q = +15$ kJ. To calculate work, we use the equation $w = -P\Delta V$, where $P = 5.00$ atm and $\Delta V = (350$ mL $- 750$ mL$) = -400$ mL.

   $w = -[5.00 \text{ atm} \times (-0.400 \text{ L})] = 2.00$ L·atm

   $$2.00 \text{ L·atm} \times 101 \frac{\text{J}}{\text{L·atm}} = 202 \text{ J}$$

   Work is being done on the system since the change in volume is due to a contraction of the system. Knowing q and w allows us to calculate ΔE.

   For the system, $\Delta E = q + w$; $\Delta E = 15$ kJ $+ 0.202$ kJ $= 15$ kJ.

   For the surroundings, $\Delta E = -15$ kJ

2. $\Delta H = 575$ J; $\Delta E = \Delta H - P\Delta V = 575$ J $- (200$ J$) = 375$ J

3. $\Delta H = -2862.7$ kJ for the reaction of 10 mol N_2O. 3.98 g N_2O is equivalent to 0.0905 mol N_2O.

   $$\text{Heat evolved} = 0.0905 \text{ mol} \times \frac{-2862.7 \text{ kJ}}{10 \text{ mol}} = -25.9 \text{ kJ}$$

4. From Table 8.1 in your textbook, we find that the specific heat of iron is 0.450 J/g·°C.

   $$0.450 \frac{\text{J}}{\text{g·°C}} = \frac{\text{heat}}{78.0 \text{ g} \times 15.0\text{°C}}; \quad \text{heat} = 527 \text{ J}$$

Appendix C – Self Test Solutions

5. In this problem, the heat of the reaction is determined from the temperature change of the known quantity of solution in the calorimeter. The heat absorbed by the reaction is equal to the heat released by the solution or $q_{rxn} = -q_{soln}$. We can solve for the final temperature by using the equation

$$q_{rxn} = -q_{soln} = -[(\text{sp. heat}) \times (\text{mass of soln}) \times (\Delta T)]$$

7.75 g of NH_4NO_3 is equivalent to 0.0969 mol NH_4NO_3; therefore $q_{rxn} = 2.50$ kJ

$$2{,}500 \text{ J} = -4.18 \frac{\text{J}}{\text{g} \cdot ^\circ\text{C}} \times 117.75 \text{ g} \times \Delta T; \quad \Delta T = -5.08 ^\circ\text{C}$$

$$\Delta T = T_f - T_i; \quad -5.08 \,^\circ\text{C} = T_f - 25.00 \,^\circ\text{C}; \quad T_f = 19.92 ^\circ\text{C}$$

6. From the overall reaction, we know that we want to have 2 mol of Al and 1 mol of Fe_2O_3 on the reactant side and 1 mol of Al_2O_3 and 2 mol of Fe on the product side. We can use the first reaction as given, since that reaction has 2 mol of Al on the reactant side. We need to reverse the second reaction that is given, since we need the 1 mol of Fe_2O_3 on the reactant side. Remember, if you reverse the reaction, you must reverse the sign of ΔH°.

$$\begin{array}{ll}
2\,Al\,(s) + \tfrac{3}{2}O_2\,(g) \rightarrow Al_2O_3\,(s) & \Delta H^\circ = -1676 \text{ kJ} \\
Fe_2O_3\,(s) \rightarrow 2\,Fe\,(s) + \tfrac{3}{2}O_2\,(g) & \Delta H^\circ = +824.2 \text{ kJ} \\
\hline
2\,Al\,(s) + Fe_2O_3\,(s) \rightarrow Al_2O_3\,(s) + 2\,Fe\,(s) & \Delta H^\circ = -852 \text{ kJ}
\end{array}$$

Note that the $\tfrac{3}{2} O_2$ molecules on the reactant side of the first reaction cancel out with the $\tfrac{3}{2} O_2$ molecules on the product side of the second reaction.

7. $SO_3\,(g) + H_2O\,(l) \rightarrow H_2SO_4\,(l)$

$$\Delta H^\circ = [(-814.0 \text{ kJ})] - [(-395.7 \text{ kJ}) + (-285.8 \text{ kJ})] = -132.5 \text{ kJ}$$

$2\,KClO_3\,(s) \rightarrow 2\,KCl\,(s) + 3\,O_2\,(g)$

$$\Delta H^\circ = [(2 \times -436.7 \text{ kJ})] - [(2 \times -397.7 \text{ kJ})] = -78.0 \text{ kJ}$$

8. $CH_3CH{=}CH_2 + HCl \rightarrow CH_3CHClCH_3$

Reactant: break 1 H–Cl bond (432 kJ) and 1 C=C bond (635 kJ); Products: form 1 C–H bond (410 kJ) 1 C–Cl bond (330 kJ), and 1 C–C bond (350 kJ)

$$\Delta H^\circ = [D(H{-}Cl) + D(C{=}C)] - [D(C{-}H) + D(C{-}Cl) + D(C{-}C)] = [432 + 635] - [410 + 330 + 350]$$
$$= -23 \text{ kJ}$$

9. $\Delta H^\circ = [(3 \times -393.5 \text{ kJ}) + (4 \times -285.8 \text{ kJ})] - [(-105)] = -2218.7 \text{ kJ}$

-2218.7 kJ represents the heat released when 1 mol of C_3H_8 reacts.

$$\frac{-2218.7 \text{ kJ}}{1 \text{ mol } C_3H_8} \times \frac{1 \text{ mol } C_3H_8}{44.0 \text{ g } C_3H_8} = -50.4 \frac{\text{kJ}}{\text{g } C_3H_8}$$

10. $CaO\,(s) + 2\,NH_4Cl\,(s) \rightarrow 2\,NH_3\,(g) + CaCl_2\,(s)$ $\quad\quad \Delta S = $ positive

Appendix C – Self Test Solutions

1 mole	2 moles	2 moles	1 mole
solid	solid	gas	solid

$$BaCl_2\ (aq)\ +\ Na_2SO_4\ (aq)\ \rightarrow\ BaSO_4\ (s)\ +\ 2\ NaCl\ (aq) \qquad \Delta S = \text{negative}$$

3 moles ions — 3 moles ions — 1 mole solid — 4 moles ions

11. From the overall reaction, we know that we need 2 mol of graphite and 3 mol of hydrogen on the reactant side of the equation and 1 mol of ethane on the product side of the equation. Therefore, we need to multiply the first equation by 2, and the second equation by 3. We need to reverse the third equation and multiply it by $\frac{1}{2}$. Remember, whatever we do to the coefficients, we also do to the value of $\Delta H°$. If we reverse an equation, we change the sign of $\Delta H°$.

2 C (graphite) + ~~2 O$_2$ (g)~~ → ~~2 CO$_2$ (g)~~ $\qquad \Delta H° = -787$ kJ

3 H$_2$ (g) + ~~$\frac{3}{2}$ O$_2$ (g)~~ → ~~3 H$_2$O (l)~~ $\qquad \Delta H° = -857.4$ kJ

~~2 CO$_2$ (g)~~ + ~~3 H$_2$O (l)~~ → C$_2$H$_6$ (g) + ~~7/2 O$_2$ (g)~~ $\qquad \Delta H° = 1559.8$ kJ

───

2 C (graphite) + H$_2$ (g) → C$_2$H$_6$ (g) $\qquad \Delta H° = -84.6$ kJ

12. Balanced overall reaction: N$_2$H$_4$ (l) + 2 H$_2$O$_2$ (l) → N$_2$ (g) + 4 H$_2$O (l)

From the above overall reaction, we know that we need 1 mol of hydrazine and 2 mol of hydrogen peroxide on the reactant side of the equation and that we need 1 mol of nitrogen and 4 mol of water on the product side. We can use the first equation as given. The third equation should be reversed and multiplied by 2, and the second equation should be multiplied by 2. Remember, whatever we do to the coefficients we also do to the value of $\Delta H°$. If we reverse an equation, we change the sign of $\Delta H°$.

N$_2$H$_4$ (l) + ~~O$_2$ (g)~~ → N$_2$ (g) + 2 H$_2$O (l) $\qquad \Delta H° = -621.6$ kJ

2 H$_2$O$_2$ (l) → ~~2 H$_2$ (g)~~ + ~~2 O$_2$ (g)~~ $\qquad \Delta H° = 375.6$ kJ

~~2 H$_2$ (g)~~ + O$_2$ (g) → 2 H$_2$O (l) $\qquad \Delta H° = -571.6$ kJ

───

N$_2$H$_4$ (l) + 2 H$_2$O$_2$ (l) → N$_2$ (g) + 4 H$_2$O (l) $\qquad \Delta H° = -817.6$ kJ

13. $\Delta G° = \Delta H° - T\Delta S°$; when using this equation, be careful with your units. $\Delta H°$ is reported in kJ/mol while $\Delta S°$ is reported in J/mol·K.

$\Delta G° = 32.9$ kJ/mol $- [298.15\text{ K} \times 0.2265\text{ J/mol·K}] = -34.6$ kJ/mol

$\Delta G°$ is negative; therefore the reaction is spontaneous.

14. When $\Delta G° = 0$, the reaction changes from a nonspontaneous reaction to a spontaneous reaction. Therefore, we can calculate the temperature by using the equation

$$T = \frac{\Delta H°}{\Delta S°}; \quad T = \frac{180.4 \text{ kJ/mol}}{0.4214 \text{ kJ/mol·K}} = 428.1\text{ K} = 154.9°\text{C}$$

15. The overall balanced equation for this reaction is

LiOH (aq) + HCl (aq) → LiCl (aq) + H$_2$O (l)

Appendix C – Self Test Solutions

We know that we need 1 mol of aqueous LiOH and 1 mol of aqueous HCl on the reactant side and 1 mol of aqueous LiCl and 1 mol of liquid water on the product side. Let's begin by reversing the third and fourth equations provided.

LiOH (aq) → LiOH (s) $\Delta H° = 19.2$ kJ

HCl (aq) → HCl (g) $\Delta H° = 77.0$ kJ

We will use the fifth equation as is:

LiCl (s) → LiCl (aq) $\Delta H° = -36.0$ kJ

We now have our two reactants and one of our products. We can use the information in Appendix B in your textbook to obtain the equation for our remaining product:

H$_2$ (g) + 1/2 O$_2$ (g) → H$_2$O (l) $\Delta H° = -285.8$ kJ

While these equations provide us with our reactants and products, they also provide us with substances that we do not need. We need to cancel out the presence of LiOH (s), HCl (g), LiCl (s), H$_2$ (g), and O$_2$ (g). For that we need to look at the other equations provided and think about information we can obtain from Appendix B in your textbook. If we reverse the first equation provided in the problem, we can cancel out LiOH (s).

LiOH (s) → Li (s) + 1/2 O$_2$ (g) + 1/2 H$_2$ (g) $\Delta H° = 487.0$ kJ

We also can cancel out the O$_2$ (g) and 1/2 of the H$_2$ (g). Using the second equation as is allows us to cancel out the LiCl (s). However, we need to multiply the coefficients and, therefore, the $\Delta H°$, by 1/2.

Li (s) + 1/2 Cl$_2$ (g) → LiCl (s) $\Delta H° = -407.5$ kJ

(Notice that we have now introduced another substance we don't need, Cl$_2$. Let's not worry about this for now.) We need to cancel the HCl (g) and can do so by using the heat of formation for HCl (g) found in Appendix B in your textbook. However, we need to reverse the reaction.

HCl (g) → 1/2 H$_2$ (g) + 1/2 Cl$_2$ (g) $\Delta H° = 92.3$ kJ

We not only cancel the HCl (g) with this equation, we also cancel the remaining 1/2 H$_2$ (g) and the 1/2 Cl$_2$ (g). We can now add all of the equations together to obtain our overall reaction.

LiOH (aq) → ~~LiOH (s)~~ $\Delta H° = 19.2$ kJ

HCl (aq) → ~~HCl (g)~~ $\Delta H° = 77.0$ kJ

~~LiCl (s)~~ → LiCl (aq) $\Delta H° = -36.0$ kJ

~~H$_2$ (g) + 1/2 O$_2$ (g)~~ → H$_2$O (l) $\Delta H° = -285.8$ kJ

~~LiOH (s) → Li (s) + 1/2 O$_2$ (g) + 1/2 H$_2$ (g)~~ $\Delta H° = 487.0$ kJ

~~Li (s) + 1/2 Cl$_2$ (g) → LiCl (s)~~ $\Delta H° = -407.5$ kJ

~~HCl (g) → 1/2 H$_2$ (g) + 1/2 Cl$_2$ (g)~~ $\Delta H° = 92.3$ kJ

Appendix C – Self Test Solutions

$$\text{LiOH (aq) + HCl (aq)} \rightarrow \text{LiCl (aq) + H}_2\text{O (l)} \qquad \Delta H° = -53.8 \text{ kJ}$$

Challenge Problem

The $\Delta H°$ given for this reaction is for the production of 2 mol of ammonia. Given both the density of NH_3 and the volume produced, we can calculate the grams and moles of NH_3 produced.

$$\text{mol NH}_3 = 0.75 \text{ L NH}_3 \times \frac{0.696 \text{ g NH}_3}{1 \text{ L NH}_3} \times \frac{1 \text{ mol NH}_3}{17.0 \text{ g NH}_3} = 0.031 \text{ mol NH}_3$$

Knowing the number of moles of NH_3 produced, we can now calculate the amount of heat evolved.

$$\frac{-91.8 \text{ kJ}}{2 \text{ mol NH}_3} \times 0.031 \text{ mol NH}_3 = -1.4 \text{ kJ}$$

To determine the amount of heat needed for this reaction to occur using the molar heat capacity of nitrogen, we first need to calculate the number of moles of nitrogen used in the production of 0.031 mol of NH_3.

$$0.031 \text{ mol NH}_3 \times \frac{1 \text{ mol N}_2}{2 \text{ mol NH}_3} = 0.0155 \text{ mol N}_2$$

We can now determine the heat needed for this reaction with the temperature change of 375°C.

$$29.12 \frac{\text{J}}{\text{mol} \cdot °\text{C}} \times 0.0155 \text{ mol} \times 375°\text{C} = 169 \text{ J}$$

Chapter 9

Multiple Choice
1. c
2. b
3. a
4. c
5. c
6. a
7. b
8. b
9. c

Matching
Gas—j
Pressure—k
Atmospheric pressure—l
Boyle's law—h
Charles' law—m
Avogadro's law—i
Ideal-gas law—b
Dalton's law—f
Partial pressures—c
Kinetic Molecular Theory—e
Graham's law—a

Appendix C – Self Test Solutions

Diffusion—g
Effusion—d

Fill-in-the-Blank
1. only 0.10% of the volume of a gas is occupied by the molecules
2. higher than
3. 273.15 K and 1 atm
4. the total molar amount of gas present, temperature, volume, and the partial pressure of gases in a mixture
5. the number of moles of a component of a gas mixture divided by the total number of moles in the mixture
6. absolute temperature
7. rates of effusion
8. increases; decrease
9. smaller
10. chlorofluorocarbons

Problems

1. a. If the level of mercury in the arm connected to the gas is lower, then $P_{gas} = P_{atm} + P_{Hg}$; $P_{gas} = 784$ mm Hg + 38 mm Hg = 822 mm Hg.
   b. If the level of mercury in the arm connected to the gas is higher, then $P_{gas} + P_{Hg} = P_{atm}$; $P_{gas} = 747$ mm Hg − 23 mm Hg = 724 mm Hg

2. The relationship between the heights of columns of fluids in a manometer can be stated as

$$h_b = h_a \times \frac{d_a}{d_b}$$

In the above equation, $h_a = P_{Hg}$ and $h_b =$ the height of the silicon oil column.

a. $140 \text{ mm oil} \times \dfrac{1.30 \text{ g/mL oil}}{13.6 \text{ g/mL Hg}} = 13.4 \text{ mm Hg}$; $P_{gas} = P_{atm} + P_{Hg} = 760 \text{ mm Hg} + 13.4 \text{ mm Hg} = 773 \text{ mm Hg}$

b. $175 \text{ mm oil} \times \dfrac{1.30 \text{ g/mL oil}}{13.6 \text{ g/mL Hg}} = 16.7 \text{ mm Hg}$; $P_{gas} + P_{Hg} = P_{atm}$; $760 \text{ mm Hg} - 16.7 \text{ mm Hg} = 743$ mm Hg

3. $0.40 \text{ g O}_2 \times \dfrac{1 \text{ mol O}_2}{32.0 \text{ g O}_2} = 0.0125 \text{ mol O}_2$;

$$P = \dfrac{0.0125 \text{ mol O}_2 \times 0.082\,06 \dfrac{\text{L·atm}}{\text{mol·K}} \times 296.6 \text{ K}}{0.500 \text{ L}} = 0.61 \text{ atm}$$

4. $n = \dfrac{1 \text{ atm} \times 0.250 \text{ L}}{0.082\,06 \dfrac{\text{L·atm}}{\text{mol·K}} \times 273 \text{ K}} = 0.0112 \text{ mol}$; $0.0112 \text{ mol NO}_2 \times \dfrac{46.0 \text{ g NO}_2}{1 \text{ mol NO}_2} = 0.515 \text{ g NO}_2$

5. a. Assume a 100 g sample, and convert 82.8 g C and 17.2 g H to moles.

$82.8 \text{ g C} \times \dfrac{1 \text{ mol C}}{12.0 \text{ g C}} = 6.90 \text{ mol C}$; $17.2 \text{ g H} \times \dfrac{1 \text{ mol H}}{1.01 \text{ g H}} = 17.0 \text{ mol H}$

Appendix C – Self Test Solutions

This gives rise to a 6.9:17.0 C to H mole ratio.

$$\frac{17.0}{6.9} = 2.5$$

For every mole of carbon there are 2.5 mol of hydrogen or a 1:2.5 mole ratio. However, formulas must consist of whole numbers, so we need to multiply both numbers in the ratio by 2 which gives a 2:5 mole ratio and the empirical formula C_2H_5.

b. To solve for molar mass, we need to find the number of moles of gas, since we already know the mass of the gas.

$$n = \frac{0.996 \text{ atm} \times 0.350 \text{ L}}{0.082\ 06 \frac{\text{L} \cdot \text{atm}}{\text{mol} \cdot \text{K}} \times 296 \text{ K}} = 0.0144 \text{ mol}; \quad \frac{0.835 \text{ g}}{0.0144 \text{ mol}} = 58.0 \text{ g/mol}$$

c. We find the molecular formula by finding the multiplier (dividing the molar mass of the compound by the molar mass of C_2H_5).

$$\frac{58.1 \text{ g/mL}}{29.0 \text{ g/mL}} = 2; \text{ the molecular formula is } C_4H_{10}.$$

6. $3.45 \text{ g H}_2\text{S} \times \frac{1 \text{ mol H}_2\text{S}}{34.0 \text{ g H}_2\text{S}} = 0.101 \text{ mol H}_2\text{S}$

$$V = \frac{0.101 \text{ mol H}_2\text{S} \times 0.082\ 06 \frac{\text{L} \cdot \text{atm}}{\text{mol} \cdot \text{K}} \times 298 \text{ K}}{1.01 \text{ atm}} = 2.45 \text{ L}$$

$$d = \frac{3.45 \text{ g}}{2.45 \text{ L}} = 1.41 \text{ g/L}$$

7. molar mass $= 1.49 \frac{\text{g}}{\text{L}} \times \frac{0.082\ 06 \frac{\text{L} \cdot \text{atm}}{\text{mol} \cdot \text{K}} \times 291 \text{ K}}{0.980 \text{ atm}} = 36.3 \text{ g/mol}$

8. First, find the number of moles of O_2, then convert to the number of moles of NO.

$$n = \frac{0.855 \text{ atm} \times 75 \text{ L}}{0.082\ 06 \frac{\text{L} \cdot \text{atm}}{\text{mol} \cdot \text{K}} \times 373 \text{ K}} = 2.10 \text{ mol}; \quad 2.10 \text{ mol O}_2 \times \frac{4 \text{ mol NO}}{5 \text{ mol O}_2} = 1.68 \text{ mol NO}$$

We can now calculate the number of liters of NO.

$$V = \frac{nRT}{P} = \frac{1.68 \text{ mol NO} \times 0.082\ 06 \frac{\text{L} \cdot \text{atm}}{\text{mol} \cdot \text{K}} \times 773 \text{ K}}{0.967 \text{ atm}} = 110 \text{ L}$$

9. To solve this problem, we must first calculate the number of moles of NH_3 and O_2 and convert to the number of moles of NO and H_2O. Once we know the <u>total</u> number of moles of product, we can then convert to the total pressure.

$$n = \frac{0.951 \text{ atm} \times 0.175 \text{ L}}{0.082\,06 \frac{\text{L}\cdot\text{atm}}{\text{mol}\cdot\text{K}} \times 303 \text{ K}} = 6.69 \times 10^{-3} \text{ mol O}_2$$

$$n = \frac{0.807 \text{ atm} \times 0.275 \text{ L}}{0.082\,06 \frac{\text{L}\cdot\text{atm}}{\text{mol}\cdot\text{K}} \times 323 \text{ K}} = 8.37 \times 10^{-3} \text{ mol NH}_3$$

From the balanced equation, we find that oxygen is the limiting reactant.

$$6.69 \times 10^{-3} \text{ mol O}_2 \times \frac{4 \text{ mol NH}_3}{5 \text{ mol O}_2} = 5.35 \times 10^{-3} \text{ mol NH}_3$$

We need to calculate the number of moles of NO and H_2O based on the number of moles of O_2.

$$6.69 \times 10^{-3} \text{ mol O}_2 \times \frac{4 \text{ mol NO}}{5 \text{ mol O}_2} = 5.35 \times 10^{-3} \text{ mol NO};$$

$$6.69 \times 10^{-3} \text{ mol O}_2 \times \frac{6 \text{ mol H}_2\text{O}}{5 \text{ mol O}_2} = 8.03 \times 10^{-3} \text{ mol H}_2\text{O}$$

total number of moles produced = $(5.35 \times 10^{-3}) + (8.03 \times 10^{-3}) = (1.34 \times 10^{-2})$

mol NH_3 remaining = $(8.37 \times 10^{-3}) - (5.35 \times 10^{-3}) = 3.02 \times 10^{-3}$

total number of moles present = $(1.34 \times 10^{-2}) + (3.02 \times 10^{-3}) = 1.64 \times 10^{-2}$ mol

We can now solve for the total pressure of the products by using the total number of moles present along with the information given.

$$P = \frac{1.64 \times 10^{-2} \text{ mol} \times 0.082\,06 \frac{\text{L}\cdot\text{atm}}{\text{mol}\cdot\text{K}} \times 473 \text{ K}}{0.500 \text{ L}} = 1.27 \text{ atm}$$

$$1.27 \text{ atm} \times \frac{760 \text{ mm Hg}}{1 \text{ atm}} = 965 \text{ mm Hg}$$

10. The balanced equation is

$$C_3H_8 \text{ (g)} + 5 O_2 \text{ (g)} \rightarrow 3 CO_2 \text{ (g)} + 4 H_2O \text{ (g)}$$

$$8.92 \text{ g C}_3\text{H}_8 \times \frac{1 \text{ mol C}_3\text{H}_8}{44.0 \text{ g C}_3\text{H}_8} = 0.203 \text{ mol C}_3\text{H}_8$$

$$0.203 \text{ mol C}_3\text{H}_8 \times \frac{3 \text{ mol CO}_2}{1 \text{ mol C}_3\text{H}_8} = 0.609 \text{ mol CO}_2$$

Appendix C – Self Test Solutions

$$V = \frac{0.609 \text{ mol} \times 0.082\,06 \,\frac{\text{L} \cdot \text{atm}}{\text{mol} \cdot \text{K}} \times 273 \text{ K}}{1 \text{ atm}} = 13.6 \text{ L}$$

11. The number of moles and pressure of the gas is being held constant, so we can use the equation

$$\frac{nR}{P} = \left(\frac{V}{T}\right)_{\text{initial}} = \left(\frac{V}{T}\right)_{\text{final}} \qquad \frac{575 \text{ mL}}{296 \text{ K}} = \frac{350 \text{ mL}}{T_{\text{final}}}; \quad T_{\text{final}} = \frac{350 \text{ mL} \times 296 \text{ K}}{575 \text{ mL}} = 180 \text{ K}$$

$180 - 273.15 = -93°\text{C}$

12. $X_{\text{CH}_4} = \dfrac{0.68 \text{ atm}}{2.0 \text{ atm}} = 0.34$; $X_{\text{C}_3\text{H}_8} = \dfrac{1.05 \text{ atm}}{2.0 \text{ atm}} = 0.52$; $X_{\text{C}_4\text{H}_{10}} = \dfrac{0.27 \text{ atm}}{2.0 \text{ atm}} = 0.14$

13. a. Since the total pressure depends on the total molar amount of gas present, we can write

    $P_{\text{total}} = (n_1 + n_2 + n_3)\left(\dfrac{RT}{V}\right)$. We can determine the total number of moles of gas present by solving for $(n_1 + n_2 + n_3)$.

    $$(n_1 + n_2 + n_3) = \frac{1.25 \text{ atm} \times 15.0 \text{ L}}{\left(0.08206 \dfrac{\text{atm} \cdot \text{L}}{\text{mol} \cdot \text{K}}\right) 301 \text{ K}} = 0.759 \text{ mol}$$

    b. To determine the mole fraction of each gas, we need to know the number of moles of each gas present.

    $4.0 \text{ g O}_2 \times \dfrac{1 \text{ mol O}_2}{32.0 \text{ g O}_2} = 0.125 \text{ mol O}_2$; $X_{\text{O}_2} = \dfrac{0.125 \text{ mol O}_2}{0.759 \text{ total mol}} = 0.165$

    $3.0 \text{ g CO}_2 \times \dfrac{1 \text{ mol CO}_2}{44.0 \text{ g CO}_2} = 0.068 \text{ mol CO}_2$; $X_{\text{CO}_2} = \dfrac{0.068 \text{ mol CO}_2}{0.759 \text{ total mol}} = 0.090$

    $0.759 \text{ total mol} - 0.125 \text{ mol O}_2 - 0.068 \text{ mol CO}_2 = 0.566 \text{ mol N}_2$

    $X_{\text{N}_2} = \dfrac{0.566 \text{ mol N}_2}{0.759 \text{ total mol}} = 0.746$

    c. $P_{\text{O}_2} = 0.165 \times 1.25 \text{ atm} = 0.206 \text{ atm}$; $P_{\text{CO}_2} = 0.090 \times 1.25 \text{ atm} = 0.113 \text{ atm}$

    $P_{\text{N}_2} = 0.746 \times 1.25 \text{ atm} = 0.932 \text{ atm}$

    d. $0.566 \text{ mol N}_2 \times \dfrac{28.0 \text{ g N}_2}{1 \text{ mol N}_2} = 12.3 \text{ g N}_2$

14. Gas pressure is created when the gas particles collide with the walls of the container. If the volume of the container is decreased while the temperature and number of moles of gas is constant, there will be more collisions between the gas particles and the walls of the container, causing an increase in the pressure.

Appendix C – Self Test Solutions

15. $\dfrac{\text{Rate of effusion of He}}{\text{Rate of effusion of Ar}} = \sqrt{\dfrac{39.9 \text{ g/mL}}{4.00 \text{ g/mL}}} = 3.16$

Challenge Problem

To determine the molecular formula, we need to first determine the empirical formula of the compound. The data provided allows us to calculate the mole ratios of the elements in the compound. We can then determine both the empirical and molecular formulas.

We will begin by calculating the percent nitrogen. From Dalton's law, we know that

$P_T = P_{N_2} + P_{H_2O}$; $P_{N_2} = 746.0$ mm Hg $- 22.110$ mm Hg $= 724.0$ mm Hg

We now know the pressure, volume, and temperature of the N_2 and can calculate the number of moles of N_2 present, using the ideal gas law.

$$n = \dfrac{PV}{RT} = \dfrac{\left(\dfrac{724.0 \text{ mm Hg}}{760 \text{ mm Hg}}\right)(0.01890 \text{ L})}{\left(0.0821 \dfrac{\text{atm} \cdot \text{L}}{\text{mol} \cdot \text{K}}\right)(296.95 \text{ K})} = 7.385 \times 10^{-4} \text{ mol N}_2$$

We have 2 mol of N for every mol of N_2 produced. Therefore,

$7.385 \times 10^{-4} \text{ mol N}_2 \times \dfrac{2 \text{ mol N}}{1 \text{ mol N}_2} \times \dfrac{14.0 \text{ g N}}{1 \text{ mol N}} = 2.068 \times 10^{-2} \text{ g N}$

$\% \text{N} = \dfrac{0.02068 \text{ g N}}{0.2394 \text{ g sample}} \times 100 = 8.638 \% \text{ N}$

We will now calculate the mg of C and H from the combustion data. (The molar mass of a compound can carry the units of g/mol or mg/mmol.)

$17.57 \text{ mg CO}_2 \times \dfrac{1 \text{ mmol CO}_2}{44.00 \text{ mg CO}_2} \times \dfrac{1 \text{ mmol C}}{1 \text{ mmol CO}_2} \times \dfrac{12.00 \text{ mg C}}{1 \text{ mmol C}} = 4.792 \text{ mg C}$

$4.319 \text{ mg H}_2\text{O} \times \dfrac{1 \text{ mmol H}_2\text{O}}{18.00 \text{ mg H}_2\text{O}} \times \dfrac{2 \text{ mmol H}}{1 \text{ mmol H}_2\text{O}} \times \dfrac{1.00 \text{ g H}}{1 \text{ mmol H}} = 0.4799 \text{ mg H}$

We can calculate the mass of N present from the mass percent we calculated at the beginning.

mg N = 0.08638 × 6.478 mg = 0.0.5596 mg N

Finally, we can calculate the mass of O.

mg sample = mg C + mg H + mg N + mg O

mg O = 6.478 mg − 4.792 mg C − 0.4799 mg H − 0.5596 mg N = 0.6465 mg O

$\text{mmol C} = 4.797 \text{ mg C} \times \dfrac{1 \text{ mmol C}}{12.01 \text{ mg C}} = 0.3994 \text{ mol C}$

Appendix C – Self Test Solutions

$$\text{mmol H} = 0.4799 \text{ mg H} \times \frac{1 \text{ mmol H}}{1.00 \text{ mg H}} = 0.4799 \text{ mmol H}$$

$$\text{mmol N} = 0.5596 \text{ mg N} \times \frac{1 \text{ mmol N}}{14.0 \text{ mg N}} = 0.03997 \text{ mmol N}$$

$$\text{mmol O} = 0.6465 \text{ mg O} \times \frac{1 \text{ mmol O}}{16.00 \text{ mg O}} = 0.04041 \text{ mmol O}$$

Dividing by the smallest number of mmoles gives the following ratio of C:H:N:O – 9.99:12.00:1.00:1.01. Therefore, the empirical formula is $C_{10}H_{12}NO$, which has a molar mass of 162 g/mol. The ratio of the empirical molar mass to the molecular molar mass is 162 g/mol:324 g/mol or 1:2. Therefore, the molecular formula is $C_{20}H_{24}N_2O_2$.

Chapter 10

True–False
1. F. The polarity of a molecule is due to the net sum of the individual bond polarities and lone–pair contributions. If that net sum is 0, then it is possible to have a molecule with polar bonds be nonpolar.
2. T
3. F. Smaller molecules and lighter atoms with fewer electrons are relatively nonpolarizable.
4. T
5. F. Strong intermolecular forces result in a higher surface tension, which leads to a decrease in surface area.
6. F. For solid → liquid, there is an increase in both ΔH and ΔS.
7. T
8. F. Constructive interference leads to an increase in the intensity of the wave.
9. F. CO_2 is a linear molecule with double bonds between the C and O's; SiO_2 has four single bonds between Si and four O's in a covalent network structure.
10. F. A supercritical fluid is neither a liquid nor solid.

Multiple Choice
1. b
2. c
3. d
4. d
5. a
6. c
7. b
8. b
9. a
10. d

Fill–in–the–Blank
1. instantaneous dipole; induces a temporary dipole
2. London dispersion force
3. surface tension
4. gas → liquid, gas → solid; liquid → solid
5. all intermolecular forces must be overcome to change a liquid to a gas
6. an increase in temperature leads to an increase in the kinetic energy of the molecules and allows them to escape the liquid
7. amorphous solid

8. hexagonal closest–packing
9. sp^3; tetrahedral
10. the pressure of a gas at the critical point is so high and the molecules are so close together that it is hard to distinguish between the gas and a liquid (that the temperature of a liquid at the critical point is so high that it is hard to distinguish between the liquid and the gas)

Problems
1. To determine the polarity of the molecules, you must first determine the Lewis structures of the molecules.

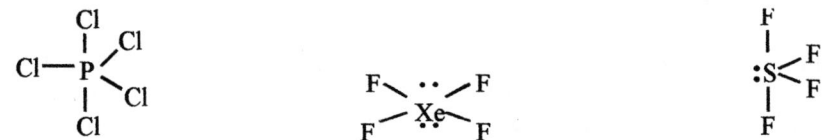

a. PCl_5 – The bonds in this molecule are polar; however, the net sum of the bond polarities is equal to zero. Therefore, the molecule is nonpolar.

b. XeF_4 – The bonds in this molecule are also polar, plus there are two lone pairs present. However, once again, the net sum of the bond polarities is equal to zero; also the dipoles created by the lone pairs cancel because they are separated by 180°. Therefore, the molecule is nonpolar.

c. SF_4 – The bond polarities in this molecule do not cancel; and this molecule is therefore, polar.

2. The intermolecular forces in N_2H_4 are stronger than the intermolecular forces in C_2H_4. The forces present in N_2H_4 are hydrogen bonds, while the forces present in C_2H_4 are London dispersive forces. (C_2H_4 is a nonpolar molecule.)

3. $T = \dfrac{\Delta H_{vap}}{\Delta S_{vap}}$; $T = \dfrac{98{,}000 \text{ J/mol}}{84.8 \text{ J/K} \cdot \text{mol}} = 1156 \text{ K} = 883°\text{C}$

4.
$$\ln P_2 = \ln P_1 + \dfrac{\Delta H_{vap}}{R}\left(\dfrac{1}{T_1} - \dfrac{1}{T_2}\right)$$

$$\ln P_2 = \ln(760 \text{ mm Hg}) + \dfrac{56{,}800 \dfrac{\text{J}}{\text{mol}}}{8.3145 \dfrac{\text{J}}{\text{mol} \cdot \text{K}}}\left(\dfrac{1}{461.4 \text{ K}} - \dfrac{1}{423.2 \text{ K}}\right) = 2.30;\ P_2 = 200 \text{ mm Hg}$$

5. a. $172 \text{ pm} = 2(350 \text{ pm}) \times \sin\theta$; $\sin\theta = \dfrac{172 \text{ pm}}{(2 \times 350 \text{ pm})} = 0.246;\ \theta = 14.2°$

b. $\sin\theta = \dfrac{172 \text{ pm}}{2 \times 975 \text{ pm}} = 0.0882;\ \theta = 5.06°$

6. $d = \dfrac{n\lambda}{2\sin\theta}$

$d = \dfrac{175 \text{ pm}}{2 \times 0.259} = 338 \text{ pm}$

Appendix C – Self Test Solutions

$$d = \frac{175 \text{ pm}}{2 \times 0.423} = 207 \text{ pm}$$

$$d = \frac{175 \text{ pm}}{2 \times 0.574} = 153 \text{ pm}$$

7. Volume of cube = d^3 = $(3.89 \times 10^{-8} \text{ cm})^3$ = 5.89×10^{-23} cm^3;

$$\text{Density of unit cell} = \frac{\text{Mass of unit cell}}{\text{Volume of unit cell}}$$

$$\text{Mass of unit cell} = \left(12.02 \frac{\text{g}}{\text{cm}^3}\right) \times \left(5.89 \times 10^{-23} \text{ cm}^3\right) = 7.08 \times 10^{-22} \text{ g}$$

$$\text{Mass of unit cell} = \text{number of atoms} \left(\frac{\text{molar mass of Pd}}{6.022 \times 10^{23}}\right)$$

$$\text{Number of atoms} = \frac{(7.08 \times 10^{-22} \text{ g}) \times (6.022 \times 10^{23} \text{ atoms/mol})}{106.42 \text{ g/mol}} = 4 \text{ atoms}$$

Each unit cell has four atoms. To determine what type of cubic unit cell this is, we must first count the number of atoms in a simple cube. Each cube has an atom at the corner, but each of these atoms is shared by eight cubes. Therefore, the total number of corner atoms = 1/8 × 8 = 1. This leaves three atoms unaccounted for. If the unit cell were body–centered, we would have only one more atom for a total of two. Since we know we have four atoms, we can discount this type of cell. For a face–centered cell, there are six faces, each with one atom that is shared by two faces. This gives a total of 1/2 × 6 = 3 atoms. If we add these three face atoms to the one corner atom, we have a total of four atoms per unit cell. Therefore, the type of unit cell is face–centered.

8. Diagonal of the cell = 4r; from the Pythagorean theorem, we know that $d^2 + d^2 = (4r)^2$; $2d^2 = 16r^2$; $2(352 \text{ pm})^2 = 16r^2$; r = 124 pm

Challenge Problem

To solve this problem, we need to set up two Clasius–Clapeyron equations, one for each liquid.

Liquid *Q*

$$\ln \frac{150}{x} = \left(\frac{3.50 \times 10^4 \frac{\text{J}}{\text{mol}}}{8.314 \frac{\text{J}}{\text{mol} \cdot \text{K}}}\right) \left(\frac{1}{T_2} - \frac{1}{313.15}\right)$$

Liquid *M*

$$\ln \frac{300}{x} = \left(\frac{2.20 \times 10^4 \frac{\text{J}}{\text{mol}}}{8.314 \frac{\text{J}}{\text{mol} \cdot \text{K}}}\right) \left(\frac{1}{T_2} - \frac{1}{313.15}\right)$$

We now have two equations with two unknowns. We can solve for the temperature by solving these equations simultaneously. Let's begin this process by simplifying both equations. Remember the rules for logarithms. Taking the logarithm of a quotient is the same as subtracting two logarithms. Let's also simplify the quantity $-\Delta H_{vap}/R$.

$$\ln 150 - \ln x = \left(4.21 \times 10^3\right)\left(\frac{1}{T_2} - \frac{1}{313.15}\right)$$

$$\ln 300 - \ln x = \left(2.65 \times 10^3\right)\left(\frac{1}{T_2} - \frac{1}{313.15}\right)$$

Let's simplify further by calculating the ln 150 and the ln 300. We also will distribute the value of $-\Delta H_{vap}/R$ over the quantity $\left(\frac{1}{T_2} - \frac{1}{313.15}\right)$.

$$-\ln x = \frac{4.21 \times 10^3}{T_2} - 13.44 - 5.01 = \frac{4.21 \times 10^3}{T_2} - 18.45$$

$$-\ln x = \frac{2.65 \times 10^3}{T_2} - 8.46 - 5.70 = \frac{2.65 \times 10^3}{T_2} - 14.16$$

We can now solve for T_2. We need to subtract the second equation from the first equation.

$$-\ln x = \frac{4.21 \times 10^3}{T_2} - 18.45$$

$$-\ln x = \frac{2.65 \times 10^3}{T_2} - 14.16$$

$$0 = \frac{1.56 \times 10^3}{T_2} - 4.29$$

We can solve for T_2 by rearranging this equation.

$$4.29 = \frac{1.56 \times 10^3}{T_2}; \quad T_2 = \frac{1.56 \times 10^3}{4.29} = 364 \text{ K}; \quad T_2 = 91°\text{C}$$

Chapter 11

True–False
1. T
2. F. "Like dissolves like." Polar solutes can be dissolved in polar solvents.
3. F. Disorder increases when a solution is formed; therefore, the entropy of solution is positive.
4. T
5. F. The solution described is a saturated solution.

Appendix C – Self Test Solutions

6. T
7. F. Solutions with a nonvolatile solute will have a lower vapor pressure than the pure solvent.
8. F. The vapor pressure of the solution will be less than that predicted by Raoult's law.
9. T
10. F. The colligative property considered to be the most accurate for determining molar mass is osmotic pressure.

Multiple Choice

1. c
2. d
3. a
4. c
5. d
6. d
7. d
8. a
9. b
10. c

Fill-in-the-Blank

1. the minor component; the major component
2. positive; due to the increase in molecular randomness upon dissolution.
3. temperature; volume; temperature
4. less
5. the amount of dissolved solute; chemical identity of the solute
6. smaller
7. solvated; hydrated
8. ΔS_{fusion} is greater for a solution than pure solvent and $T_f = \dfrac{\Delta H_f}{\Delta S_f}$
9. the migration of solvent and other small molecules through a semipermeable membrane
10. because measurements are made at the temperature specified in the equation

Matching

Colloids—c
Supersaturated solution—e
Solubility—a
Semipermeable membrane—f
Osmotic pressure—b
Fractional distillation—d

Problems

1. a. CCl_4 — nonpolar substance; London dispersion forces; Br_2 — nonpolar substance; London dispersion forces; will form a solution
   b. CH_3OH — polar substance; hydrogen bonding; Br_2 — nonpolar substance; London dispersion forces; will not form a solution
   c. KCl — ionic substance; ionic attractions; NH_3 — polar substance; hydrogen bonding; will form a solution
   d. NH_3 — polar substance; hydrogen bonding; H_2O — polar substance; hydrogen bonding; will form a solution

2. Assume 100 g of solution. We now have 9.99 g of NaOH and 90.01 g of water. To calculate both molality and molarity, we need to convert grams of NaOH to moles of NaOH.

Appendix C – Self Test Solutions

$$9.99 \text{ g NaOH} \times \frac{1 \text{ mol NaOH}}{40.0 \text{ g NaOH}} = 0.250 \text{ mol NaOH}$$

$$\frac{0.250 \text{ mol NaOH}}{0.09001 \text{ kg}} = 2.78 \ m$$

To calculate molarity, we also need to know the volume of the solution which we can obtain from the mass and density of the solution.

$$\frac{100 \text{ g soln}}{1.109 \text{ g soln}/\text{mL}} = 90.2 \text{ mL} \qquad \frac{0.250 \text{ mol NaOH}}{0.0902 \text{ L soln}} = 2.77 \text{ M}$$

3. Assume that you have 1 L of solution; therefore, you also have 0.838 moles of acetic acid. To calculate molality, you need to know the number of kg of solvent. This can be calculated from the grams of solution and grams of solute. (g soln = g solvent + g solute)

$$0.838 \text{ mol CH}_3\text{COOH} \times \frac{60.0 \text{ g CH}_3\text{COOH}}{1 \text{ mole CH}_3\text{COOH}} = 50.3 \text{ g CH}_3\text{COOH}$$

$$1000 \text{ mL soln} \times \frac{1.0055 \text{ g soln}}{1 \text{ mL soln}} = 1,005.5 \text{ g soln}; \quad 1,005.5 \text{ g soln} - 50.3 \text{ g solute} = 955.2 \text{ g solvent}$$

$$\frac{0.838 \text{ mol CH}_3\text{COOH}}{0.9552 \text{ kg H}_2\text{O}} = 0.877 \ m$$

To calculate mass %, simply divide the grams of acetic acid by the grams of solution.

$$\frac{50.3 \text{ g CH}_3\text{COOH}}{1,005.5 \text{ g soln}} \times 100 = 5.00\%$$

4. By assuming we have 1 kg of solvent, we know that we have 8.0 moles of NH_3. We need to know the grams of NH_3 for both calculations.

$$8.0 \text{ mol NH}_3 \times \frac{17.0 \text{ g NH}_3}{1 \text{ mol NH}_3} = 136 \text{ g NH}_3$$

We also need to know the mass of the solution for both calculations.

$136 \text{ g NH}_3 + 1000 \text{ g H}_2\text{O} = 1136 \text{ g soln}$

To calculate molarity, determine the volume of the solution from the mass and density of the solution.

$$\frac{1,136 \text{ g soln}}{0.950 \text{ g soln}/\text{mL}} = 1,200 \text{ mL soln} \qquad \frac{8 \text{ mol NH}_3}{1.2 \text{ L soln}} = 6.67 \text{ M}$$

To calculate mass %, we divide the mass of NH_3 by the mass of the solution.

$$\frac{136 \text{ g NH}_3}{1,136 \text{ g soln}} \times 100 = 12.0\%$$

Appendix C – Self Test Solutions

5. From the definition for ppb, we know that this solution contains 0.0325 mg of Hg in a 1 L sample. The number of grams of Hg in a 50 mL sample is

$$\frac{0.0325 \text{ mg Hg}}{1 \text{ L sample}} \times 0.050 \text{ L} \times \frac{1 \times 10^{-3} \text{ g}}{1 \text{ mg}} = 1.6 \times 10^{-6} \text{ g Hg}$$

To calculate molarity, convert grams of Hg to moles.

$$1.63 \times 10^{-6} \text{ g Hg} \times \frac{1 \text{ mol Hg}}{200.6 \text{ g Hg}} = 8.0 \times 10^{-9} \text{ mol Hg}; \qquad \frac{8.0 \times 10^{-9} \text{ mol Hg}}{0.050 \text{ L}} = 1.6 \times 10^{-7} \text{ M}$$

6. Solubility $= 3.34 \times 10^{-3} \dfrac{\text{mol}}{\text{L} \cdot \text{atm}} \times \left(852 \text{ mm Hg} \times \dfrac{1 \text{ atm}}{760 \text{ mm Hg}} \right) = 3.74 \times 10^{-3} \dfrac{\text{mol}}{\text{L}}$

7. Use the first set of data to calculate the Henry's law constant.

$$5.85 \times 10^{-4} \frac{\text{mol}}{\text{L}} = k \times \left(725 \text{ mm Hg} \times \frac{1 \text{ atm}}{760 \text{ mm Hg}} \right) \qquad k = 6.84 \times 10^{-4} \frac{\text{mol}}{\text{L} \cdot \text{atm}}$$

$$\text{Solubility} = 6.84 \times 10^{-4} \frac{\text{mol}}{\text{L} \cdot \text{atm}} \left(\frac{725 \text{ mm Hg}}{760 \frac{\text{mm Hg}}{\text{atm}}} \right) = 6.53 \times 10^{-4} \frac{\text{mol}}{\text{L}}$$

8. To calculate the vapor pressure of the solution, we need to know the mole fraction of water. To calculate the mole fraction of water, we need to know the moles of water and the total number of moles of particles in the solution.

$$72.1 \text{ g H}_2\text{O} \times \frac{1 \text{ mol H}_2\text{O}}{18.0 \text{ g H}_2\text{O}} = 4.01 \text{ mol H}_2\text{O}$$

$$15.8 \text{ g NaCl} \times \frac{1 \text{ mol NaCl}}{58.5 \text{ g NaCl}} = 0.270 \text{ mol NaCl}$$

Because we are using an ionic solid, we must take into account the number of moles of particles, not just the number of moles of formula units of NaCl. There are 0.270 mol of Na^+ and 0.270 mol of Cl^- in 0.270 mol of NaCl. This gives 0.540 mol of ions in the solution. This amount will be used to calculate the total number of moles of particles present in the solution.

$$X_{H_2O} = \frac{4.01 \text{ mol H}_2\text{O}}{4.55 \text{ mol particles}} = 0.881 \qquad P_{\text{soln}} = 0.881 \times 42.175 \text{ mm Hg} = 37.2 \text{ mm Hg}$$

9. $8.0 \text{ g C}_6\text{H}_6 \times \dfrac{1 \text{ mol C}_6\text{H}_6}{78.0 \text{ g C}_6\text{H}_6} = 0.103 \text{ mol C}_6\text{H}_6;$

$15.0 \text{ g C}_6\text{H}_5\text{CH}_3 \times \dfrac{1 \text{ mol C}_6\text{H}_5\text{CH}_3}{92.0 \text{ g C}_6\text{H}_5\text{CH}_3} = 0.163 \text{ mol C}_6\text{H}_5\text{CH}_3$

$$X_{C_6H_5CH_3} = \frac{0.163 \text{ mol } C_6H_5CH_3}{0.266 \text{ mol particles}} = 0.613; \qquad X_{C_6H_6} = \frac{0.103 \text{ mol } C_6H_6}{0.266 \text{ mol particles}} = 0.387$$

$$P_{soln} = (0.387 \times 93.4 \text{ mm Hg}) + (0.613 \times 26.9 \text{ mm Hg}) = 52.6 \text{ mm Hg}$$

10. To calculate the molar mass of the unknown, we need to know both the mass and number of moles of unknown. The mass of the unknown was given. We can find the number of moles of unknown by determining the mole fraction of heptane and the unknown. Remember that the mole fraction of heptane is the number of moles of heptane divided by the number of moles of heptane and the unknown,

$$X_{heptane} = \frac{\text{mol heptane}}{\text{mol heptane} + \text{mol unknown}}$$

Start by calculating the moles of heptane.

$$250 \text{ g } C_7H_{16} \times \frac{1 \text{ mol } C_7H_{16}}{100.0 \text{ g } C_7H_{16}} = 2.50 \text{ mol } C_7H_{16}$$

Let x be the number of moles of unknown. The mole fraction for heptane and the unknown are

$$X_{C_7H_{16}} = \frac{2.50 \text{ mol } C_7H_{16}}{2.50 \text{ mol } C_7H_{16} + x \text{ mol unk}} \qquad X_{unk} = \frac{x \text{ mol unk}}{2.50 \text{ mol } C_7H_{16} + x \text{ mol unk}}$$

Substituting these mole fractions into the equation for the vapor pressure of a solution gives

$$639 \text{ mm Hg} = \left(\frac{2.50 \text{ mol } C_7H_{16}}{2.50 \text{ mol } C_7H_{16} + x \text{ mol unk}}\right)(791 \text{ mm Hg}) + \left(\frac{x \text{ mol unk}}{2.50 \text{ mol } C_7H_{16} + x \text{ mol unk}}\right)(352 \text{ mm Hg})$$

$$639 \text{ mm Hg} = \frac{1}{2.50 \text{ mol } C_7H_{16} + x \text{ mol unk}}[(2.50 \text{ mol } C_7H_{16} \times 791 \text{ mm Hg}) + (x \text{ mol unk} \times 352 \text{ mm Hg})]$$

For simplification, let's drop the units out of the calculations, keeping in mind that the units for i are moles of unknown.

$$639(2.50 + x) = (2.50 \times 791) + (x \times 352);$$

$$(1.60 \times 10^3) + 639x = (1.98 \times 10^3) + 352x; \qquad 287x = 380; \quad x = 1.32$$

Now that we know the number of moles of unknown, we can calculate the molar mass of the unknown.

$$\frac{150 \text{ g unk}}{1.32 \text{ mol unk}} = 114 \text{ g/mol}$$

11. To calculate both the boiling point and freezing point of a solution, we need to know the molality of the solution. Remember, when calculating the molality of an ionic substance, we need to use the number of moles of solute particles, not formula units.

Appendix C – Self Test Solutions

$$10.0 \text{ g CaCl}_2 \times \frac{1 \text{ mol CaCl}_2}{111.1 \text{ g CaCl}_2} = 9.00 \times 10^{-2} \text{ mol CaCl}_2$$

There are three moles of ions for every one mole of $CaCl_2$, so the number of moles of solute particles is 0.270. The molality of the solution is

$$\frac{0.270 \text{ mol ions}}{0.090 \text{ kg H}_2\text{O}} = 3.00 \ m$$

To calculate the boiling point and freezing point of the solution, we need to calculate ΔT_b and ΔT_f. K_b for water = 0.51°C/m and K_f for water = 1.86°C/m.

$\Delta T_b = (0.51° \text{ C}/m)(3.00 \ m) = 1.53° \text{ C}$; $\Delta T_f = (1.86° \text{ C}/m)(3.00 \ m) = 5.58° \text{ C}$

bp soln = 100.00°C + 1.53°C = 101.53°C   fp soln = 0.00°C – 5.58°C = –5.58°C

12. To calculate molar mass, we again need to find the number of moles of solute. This time, we will do that by finding the molality of the solution from the change in freezing point. From the definition of molality, we can then determine the exact number of moles in the solution.

$\Delta T_f = 5.5°\text{C} - 3.5°\text{C} = 2.0°\text{C}$; $K_f = 5.12°\text{C}/m$; $m = \dfrac{2.0°\text{C}}{5.12°\text{C}/m} = 0.391 \ m$

$\dfrac{0.391 \text{ mol solute}}{1 \text{ kg benzene}} \times 0.050 \text{ kg benzene} = 0.0196 \text{ mol solute}$ $\dfrac{2.50 \text{ g solute}}{0.0196 \text{ mol solute}} = 128 \text{ g/mol}$

13. $2.5 \text{ g ethanol} \times \dfrac{1 \text{ mol ethanol}}{46.0 \text{ g ethanol}} = 0.0543 \text{ mol ethanol}$ $\dfrac{0.0543 \text{ mol ethanol}}{0.010 \text{ kg water}} = 5.43 \ m$

$K_f = 1.86°\text{C}/m$ $\Delta T_f = (1.86° \text{ C}/m)(5.43 \ m) = 10.1° \text{ C}$;

fp soln = 0.0° C – 10.1° C = –10.1°C

Since ΔH_{fusion} for the solvent is the same as ΔH_{fusion} for the solution, we now have all the information we need to calculate ΔS_{fusion} from the equation $\Delta H_{fusion} = T\Delta S_{fusion}$.

$$\frac{6.01 \text{ kJ/mol}}{(273 - 10.1)} \times \frac{1000 \text{ J}}{1 \text{ kJ}} = 22.9 \text{ J/mol} \cdot \text{K}$$

14. $8.0 \text{ g C}_6\text{H}_{12}\text{O}_6 \times \dfrac{1 \text{ mol C}_6\text{H}_{12}\text{O}_6}{180.0 \text{ g C}_6\text{H}_{12}\text{O}_6} = 0.044 \text{ mol C}_6\text{H}_{12}\text{O}_6$; $\dfrac{0.044 \text{ mol C}_6\text{H}_{12}\text{O}_6}{1 \text{ L soln}} = 0.044 \text{ M}$

$$\Pi = 0.044 \frac{\text{mol}}{\text{L}} \times 0.08206 \frac{\text{L} \cdot \text{atm}}{\text{mol} \cdot \text{K}} \times 288 \text{ K} = 1.04 \text{ atm}$$

15. $\left(16.4 \text{ mm Hg} \times \dfrac{1 \text{ atm}}{760 \text{ mm Hg}}\right) = \text{M} \times 0.08206 \dfrac{\text{atm} \cdot \text{L}}{\text{mol} \cdot \text{K}} \times 288 \text{ K}$; $\text{M} = 9.13 \times 10^{-4}$

Since we have 1 L of solution, we know that we have 9.13×10^{-4} moles of the tripeptide. We can now calculate the molar mass.

$$\frac{0.250 \text{ g}}{9.13 \times 10^{-4} \text{ mol}} = 274 \text{ g/mol}$$

Challenge Problem

The reaction for the metal with excess hydrochloric acid can be written as

M (s) + x HCl (aq) → MCl$_x$ (s) + x/2 H$_2$ (g)

We can calculate the value of x/2 (moles of H$_2$) using the ideal gas law and the data provided.

$$x/2 = n = \frac{PV}{RT} = \frac{\left(755 \text{ mm Hg} \times \frac{1 \text{ atm}}{760 \text{ mm Hg}}\right)(0.2586 \text{ L})}{\left(0.0821 \frac{\text{L} \cdot \text{atm}}{\text{mol} \cdot \text{K}}\right)(296.15 \text{ K})} = 1.06 \times 10^{-2}$$

We know from our balanced equation that x represents the number of moles of chloride ions present in MCl$_x$. If x/2 = 1.06 × 10^{-2}, then x = 2.12 × 10^{-2}.

Using the freezing point of the solution, we can calculate the number of moles of ions present. (Remember, when calculating freezing point depressions for ionic substances, we must take into consideration the number of moles of ions present. K_f for water = 1.86°C·kg/molal.)

$$2.37°\text{C} = \left(1.86 \frac{°\text{C} \cdot \text{kg}}{\text{molal}}\right) \times \frac{\text{mol of ions}}{0.025 \text{ kg H}_2\text{O}}$$

mol MCl$_x$ = 3.18 × 10^{-2}

Knowing the total moles of ions and the moles of Cl$^-$, we can calculate the moles of M^{x+} since the total moles of ions = moles M^{x+} + moles Cl$^-$.

moles M^{x+} = (3.18 × 10^{-2}) − (2.12 × 10^{-2}) = 1.06 × 10^{-2}

Knowing the mass of the metal and the number of moles of metal, we can calculate the molar mass.

$$\frac{1.455 \text{ g}}{1.06 \times 10^{-2} \text{ mol}} = 137.3 \text{ g/mol}$$

This molar mass is the molar mass of barium.

Chapter 12

True–False
1. T
2. F. The reaction rate usually decreases as the reaction runs out of reactants.
3. F. The exponents m and n can be determined only from experiments.
4. F. Doubling the concentration of A will cause the rate to increase by a factor of 4 for a second order reaction.

Appendix C – Self Test Solutions

5. T
6. T
7. F. Molecularity is the number of molecules only on the reactant side of an elementary reaction.
8. F. The balanced equation for an overall reaction only describes the stoichiometry of the reaction.
9. F. Only if the reaction mechanism is consistent with the observed rate law for the reaction.
10. F. Industrial processes use heterogeneous catalysts.

Multiple Choice

1. c
2. b
3. c
4. b
5. c
6. a
7. d
8. a
9. d
10. a

Matching
Kinetics—e
Integrated rate law—h
Half–life—f
Reaction mechanism—g
Elementary step—a
Molecularity—j
Bimolecular reaction—k
Potential energy profile—b
Transition state—c
Catalyst—d
Homogeneous catalyst—i

Fill–in–the–Blank
1. the coefficients in the balanced equation
2. how a change in concentration can affect the rate
3. experimentally
4. doubling; eight
5. first–order
6. individual molecular event
7. the molecularity of the reaction
8. slowest step in the reaction mechanism
9. 10°C
10. orientation; transition state

Problems

1. $\dfrac{\Delta[C]}{\Delta t}$ can be calculated by first plotting the data and then determining the slope of the right triangle created from $\Delta[C]$ and Δt.

Appendix C – Self Test Solutions

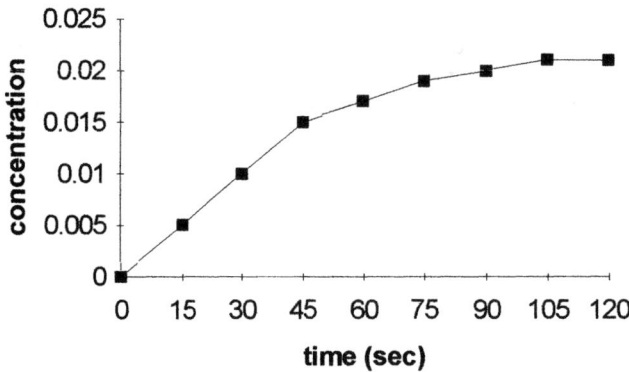

$$\frac{\Delta[C]}{\Delta t} = \frac{0.019 - 0.015}{75 - 45} = 1.3 \times 10^{-4} \text{ M/s}$$

2. Rate of appearance of $H_2 = \left[1.47 \times 10^{-3} \text{ mol PH}_3 / (L \cdot s)\right] \times \frac{6 \text{ mol H}_2}{4 \text{ mol PH}_3} = 2.20 \times 10^{-3} \text{ mol H}_2 / (L \cdot s)$

3. In the first three experiments, the concentration of NO changes while that of Cl_2 remains constant. Therefore, the change in rate for these three experiments is dependent only on the change in concentration of NO. When the concentration of NO is doubled from 0.025 M to 0.050 M, the rate increases by a factor of 4. When the concentration of NO is tripled from 0.025 M to 0.075 M, the rate increases by a factor of 9. We know that $2^2 = 4$ and that $3^2 = 9$. Therefore, the rate of the reaction with respect to NO depends on $[NO]^2$. In the first, fourth, and fifth experiments, the concentration of NO remains constant while that of the Cl_2 is changed. Therefore, the change in rate for these three experiments is dependent only on the change in concentration of Cl_2. When the concentration of Cl_2 is doubled from 0.025 M to 0.050 M, the rate increases by a factor of 2. When the concentration of Cl_2 is tripled from 0.025 M to 0.075 M, the rate increases by a factor of 3. We know that $2^1 = 2$ and that $3^1 = 3$. Therefore, the rate of reaction with respect to Cl_2 depends on $[Cl_2]$. The overall rate law is: rate = $k[NO]^2[Cl_2]$. The overall order of the reaction is third order. k can be calculated from any one of the five experiments.

$$117.2 \text{ mol}/(L \cdot s) = k(0.025 \text{ mol}/L)^2 (0.075 \text{ mol}/L) ;$$
$$k = \frac{117.2 \text{ mol}/(L \cdot s)}{4.69 \times 10^{-5} \text{ mol}^3/L^3} = 2.50 \times 10^6 \text{ L}^2/(\text{mol}^2 \cdot s)$$

4. In the first three experiments, the concentration of B is held constant, while the concentration of A is changing. Therefore, the change in rate for these three experiments is dependent only on the change in concentration of A. When the concentration of A is doubled, the rate increases by a factor of 8. When the concentration of A is tripled, the rate increases by a factor of 27. We know that $2^3 = 8$ and that $3^3 = 27$. Therefore, the rate of reaction with respect to A is proportional to $[A]^3$. In the first, fourth, and fifth experiments, the concentration of A is held constant, while the concentration of B is changing. Therefore, the rate of reaction in these three experiments is dependent only on the concentration of B. When the concentration of B is doubled, the rate remains the same. When the concentration of B is tripled, the rate remains the same. (This is equivalent to multiplying the rate by 1.) We know that $2^0 = 1$ and that $3^0 = 1$. Therefore, the rate of reaction with respect to B is proportional to $[B]^0$ or 1. The overall rate law is rate = $k[A]^3$. The overall order of the reaction is third order. k can be calculated from any one of the five experiments.

Appendix C – Self Test Solutions

$$1.56 \times 10^{-2} \text{ mol}/(\text{L} \cdot \text{s}) = k(0.150 \text{ mol}/\text{L})^3$$

$$k = \frac{1.56 \times 10^{-2} \text{ mol}/(\text{L} \cdot \text{s})}{(3.38 \times 10^{-3} \text{ mol}^3/\text{L}^3)} = 4.62 \text{ L}^2/(\text{mol}^2 \cdot \text{s})$$

$$\text{Rate} = 4.62 \text{ L}^2/(\text{mol}^2 \cdot \text{s}) \times (0.639 \text{ mol}/\text{L})^3 = 1.21 \text{ mol}/(\text{L} \cdot \text{s})$$

5. When the reaction has reached 35% completion, then 0.500×0.35 or 0.175 M has reacted. This means that the concentration of cyclopropane after 35% completion is $0.500 - 0.175$ or 0.325 M. We can now use the integrated rate law for a first–order reaction to calculate t.

$$\ln\left(\frac{0.325 \text{ M}}{0.500 \text{ M}}\right) = -(1.16 \times 10^{-6} \text{ s}^{-1})t \qquad t = \frac{-0.4308}{-1.16 \times 10^{-6} \text{ s}^{-1}} = 3.71 \times 10^5 \text{ s}$$

When the reaction has reached 70% completion, the concentration of cyclopropane is 0.150 M.

$$\ln\left(\frac{0.150}{0.500}\right) = -(1.16 \times 10^{-6} \text{ s}^{-1})t \qquad t = \frac{-0.1204}{-1.16 \times 10^{-6} \text{ s}^{-1}} = 1.04 \times 10^6 \text{ s}$$

6. The half–life can be determined from the plotted data by finding the time when the concentration of PH_3 is 0.225 M.

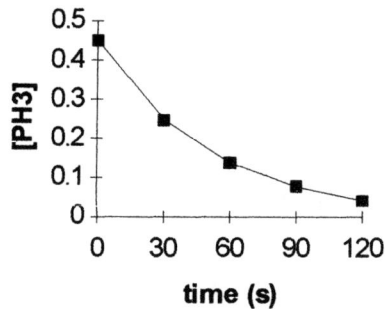

The estimated half–life from the plot is 35 s. k can be calculated by using the integrated rate law for a first–order reaction.

$$\ln\frac{0.248}{0.450} = -k(30.0 \text{ s}); \qquad k = -\left(\frac{-0.596}{30.0 \text{ s}}\right) = 1.99 \times 10^{-2} \text{ s}^{-1}$$

The calculated half–life is

$$t_{\frac{1}{2}} = \frac{0.693}{1.99 \times 10^{-2} \text{ s}^{-1}} = 34.8 \text{ s}$$

7. $\dfrac{1}{[O_3]} = \left[1.40 \times 10^{-2} \text{ L}/(\text{mol} \cdot \text{s})\right]\left(24 \text{ h} \times \dfrac{3600 \text{ s}}{1 \text{ h}}\right) + \dfrac{1}{2.75 \text{ M}} = 1.21 \times 10^3 \text{ L}/\text{mol}$

$[O_3] = 8.26 \times 10^{-4}$ M

8. We can determine the order of the reaction by plotting ln[HI] versus t. If we obtain a straight line, we know we have a first–order reaction. If we don't obtain a straight line, then we need to plot 1/[HI] versus t. If we obtain a straight line, we know we have a second–order reaction.

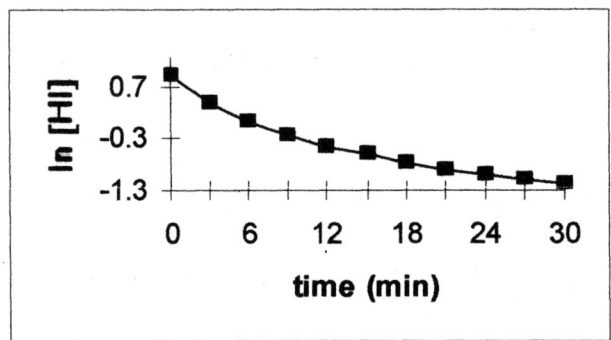

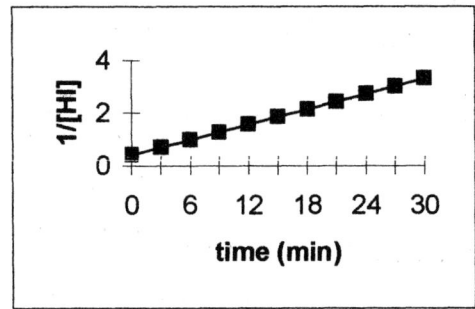

From the two plots, we can see that the reaction is second–order. We can use the integrated rate law for a second order reaction to calculate the value of k and the half–life.

$$\frac{1}{1.45 \text{ M}} = k(3 \text{ min}) + \frac{1}{2.50 \text{ M}} \; ; \; k = \frac{\dfrac{1}{1.45 \text{ M}} - \dfrac{1}{2.50 \text{ M}}}{3 \text{ min}} = 9.66 \times 10^{-2} \text{ L}/(\text{mol} \cdot \text{min})$$

$$t_{\frac{1}{2}} = \frac{1}{[9.66 \times 10^{-2} \text{ L}/(\text{mol} \cdot \text{min})]2.50 \text{ mol}/\text{L}} = 4.14 \text{ min}$$

9. Both steps are bimolecular. Rate = $k[NO_2][F_2]$.

10. Overall reaction: $2 H_2O_2 \rightarrow 2 H_2O + O_2$; OH· and HO_2 are reaction intermediates. The first step is unimolecular. The second and third steps are bimolecular. The first step is the slowest step.

11. Plot the data to prove that we can use the Arrhenius equation to calculate E_a.

Appendix C – Self Test Solutions

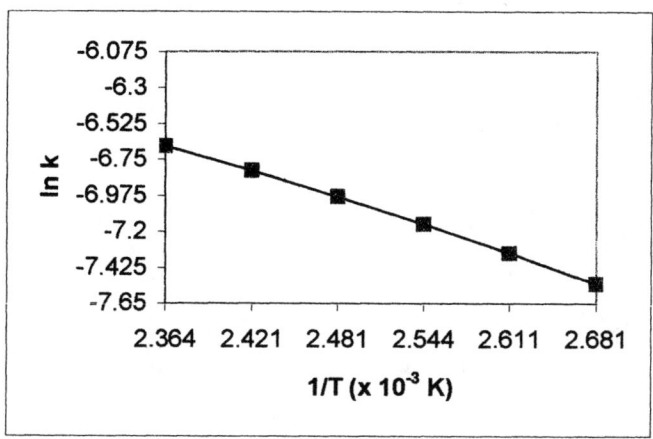

From the plot, we can see that we have a straight line. To obtain the slope, we calculate $\frac{\Delta y}{\Delta x}$ for the two farthest points. (Don't forget that you are plotting ln k versus $1/T$ in Kelvin.)

$$\text{Slope} = \frac{-6.66 - (-7.53)}{(2.36 \times 10^{-3}) - (2.68 \times 10^{-3})} = -2.7 \times 10^3$$

Remember that the slope $= \frac{-E_a}{R}$; therefore

$$E_a = (2.7 \times 10^3 \text{ K})(8.314 \text{ J/mol} \cdot \text{K}) = 2.2 \times 10^4 \text{ J/mol}$$

12. $\ln(1.67 \times 10^{-4}) = \left(\frac{-1.118 \times 10^5 \text{ J/mol}}{8.314 \text{ J/mol} \cdot \text{K}}\right)\left(\frac{1}{341 \text{K}}\right) + \ln A$; $\ln A = 30.7$; $A = 2 \times 10^{13}$

13. $\ln\left(\frac{2.8 \times 10^{-3}}{9.3 \times 10^{-6}}\right) = \left(\frac{-1.0025 \times 10^5 \text{ J/mol}}{8.314 \text{ J/(mol} \cdot \text{K)}}\right)\left(\frac{1}{T_2} - \frac{1}{350 \text{K}}\right)$

$5.71 = -1.206 \times 10^4 \text{ K}\left(\frac{1}{T_2} - \frac{1}{350 \text{K}}\right)$ $-4.74 \times 10^{-4} \text{ K}^{-1} = \frac{1}{T_2} - \frac{1}{350 \text{ K}}$

$\frac{1}{T_2} = 2.38 \times 10^{-3} \text{ K}^{-1}$ $T_2 = 420 \text{ K}$

14.

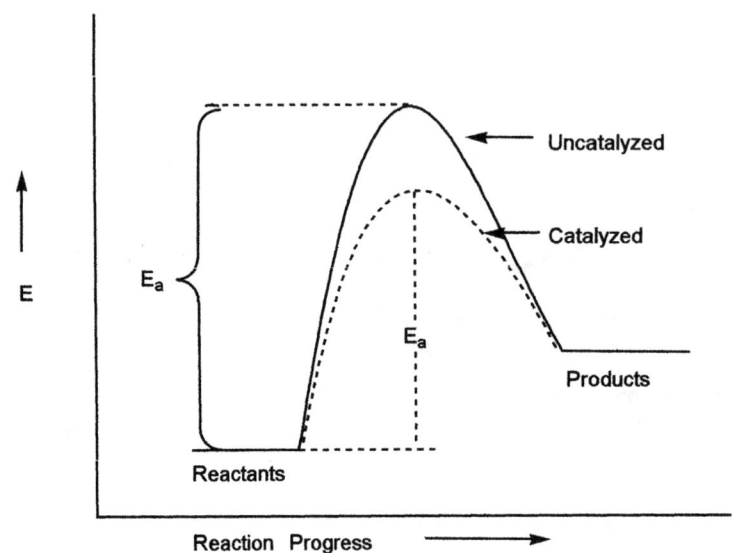

Challenge Problem

To calculate the initial rate, we need to know a rate law, rate constant, and the initial concentrations of C_2H_5I and/or OH^-. From the information given, we know that rate = $k[C_2H_5I][OH^-]$ since the doubling of the concentration of each reactant while the other is held constant results in a doubling of the rate.

The rate constant, k, is calculated using the Arrhenius equation, $k = Ae^{-E_a/RT}$.

$k = 2.10 \times 10^{11} \text{ M}^{-1}\text{s}^{-1} e^{-(8.68 \times 10^4 \text{ J/mol})/(8.314 \text{ J/mol·K})(308.15\text{K})} = 4.06 \times 10^{-4} \text{ M}^{-1}\text{s}^{-1}$.

We now need to calculate the initial concentrations of OH^- and C_2H_5I, keeping in mind that the final volume of the solution is 500 mL.

mol OH^- = mol KOH = 8.46×10^{-3}

mol C_2H_5I = 8.839×10^{-3}

$[OH^-] = \dfrac{8.46 \times 10^{-3} \text{ mol OH}^-}{0.500 \text{ L}} = 1.692 \times 10^{-2}$

$[C_2H_5I] = \dfrac{8.839 \times 10^{-3} \text{ mol C}_2\text{H}_5\text{I}}{0.500 \text{ L}} = 1.768 \times 10^{-3}$

Rate = $(4.06 \times 10^{-4} \text{ M}^{-1}\text{s}^{-1})(1.692 \times 10^{-2} \text{ M})(1.768 \times 10^{-3} \text{ M}) = 1.21 \times 10^{-7} \text{ M·s}^{-1}$

Appendix C – Self Test Solutions

Chapter 13

True–False
1. F. The concentrations in the reaction quotient expression are not necessarily equilibrium values.
2. T
3. F. If $Q_c > K_c$, the reaction proceeds from right to left.
4. F. The reaction will shift to the left.
5. T
6. F. When an inert gas is added to a reaction mixture, there is no change in the molar concentrations of reactants and products and, therefore, no effect on the composition of the equilibrium mixture.
7. T
8. F. The rates of the forward and reverse reactions increase by the same factor. Therefore, the addition of a catalyst does not affect the composition of the equilibrium mixture.
9. T
10. T

Multiple Choice
1. d
2. b
3. b
4. c
5. c
6. c
7. d
8. c
9. a
10. b

Matching
Chemical equilibrium—f
Equilibrium constant, K_c—g
Homogeneous equilibria—e
Heterogeneous equilibria—c
Reaction quotient, Q_c—h
Reversible reaction—d
Dynamic state—a
Le Châtelier's principle—b

Fill–in–the–Blank
1. decreases; increases; level off at constant equilibrium values
2. reciprocal of the original equilibrium expression.
3. not included
4. hardly proceeds
5. in the direction that consumes the added reactant or product
6. in the direction of greater number of moles of gas
7. ignored
8. no effect; there is no change in the molar concentrations of reactants or products
9. increases
10. k_f and k_r

Problems

1. a. $K_c = \dfrac{[NH_3]^2[H_2O]^4}{[H_2]^7[NO_2]^2}$ b. $K_c = \dfrac{1}{[Cl_2]^3}$ c. $K_c = \dfrac{[I_2][HBr]^2}{[Br_2][HI]^2}$

   d. $K_c = [CO_2][H_2O]$

2. $K_p = \dfrac{(p_{NO})^2}{(p_{N_2})(p_{O_2})}$ $K_p = \dfrac{(0.050)^2}{(0.25)(0.198)} = 5.1 \times 10^{-2}$

3. $K_c = \dfrac{K_p}{(RT)^{\Delta n}}$; $\Delta n = 0$; $K_c = K_p = 5.1 \times 10^{-2}$

4. a. mainly products b. mainly products c. mainly reactants

5. To determine if a system is at equilibrium, you must first solve for the value of Q_c and then compare it to the value of K_c.

   $Q_c = \dfrac{[PCl_3][Cl_2]}{[PCl_5]}$ $Q_c = \dfrac{(0.75)(0.75)}{(1.25 \times 10^{-3})} = 4.5 \times 10^2$

   Since $Q_c \gg K_c$, the system is not at equilibrium. The reaction would need to shift from right to left to achieve equilibrium.

6. To solve this problem, we need to use the approach outlined in Figure 13.6 in your text and on page 208 of this chapter.
   a. The balanced equation is given.
   b. The equilibrium concentrations of N_2O and O_2 can be determined by allowing x to be the concentration of N_2O that reacts on going to the equilibrium state. Since the mole ratios are 2:3:4, if x mol/L of N_2O reacts, then $1.5x$ mol/L of O_2 reacts and $2x$ mol/L of NO_2 are produced.

   This can be summarized in the following table.

Principal Reaction	2 N_2O (g)	+	3 O_2 (g)	⇌	4 NO_2 (g)
Initial Concentration (M)	0.0292		0.0713		0
Change (M)	$-2x$		$-3x$		$+4x$
Eq Concentration (M)	$0.0292 - 2x$		$0.0713 - 3x$		$+4x$

   We know that the equilibrium concentration of NO_2 is 0.0284 M and can, therefore, solve for the value of x.

   $4x = 0.0284$ $x = 7.1 \times 10^{-3}$

   The value of x can be substituted back into the expressions for the equilibrium concentrations of N_2O and O_2.

   $[N_2O] = 0.0292 - (2 \times 0.0071) = 0.0150$; $[O_2] = 0.0713 - (3 \times 0.0071) = 0.0500$

   c. The value of K_c can now be calculated by substituting the equilibrium concentrations of all species into the equilibrium expression.

Appendix C – Self Test Solutions

$$K_c = \frac{[NO_2]^4}{[N_2O]^2[O_2]^3} \; ; \qquad K_c = \frac{(0.0284)^4}{(1.50\times10^{-2})^2(5.00\times10^{-2})^3} = 23.1$$

7. To solve this problem, we need to use the approach outlined in Figure 13.6 in your textbook and on page 208 of this chapter.
   a. The balanced equation is given.
   b. The initial concentrations of SO_2 and NO_2 are 0.0650 M. Let x be the concentration of SO_2 that reacts on going to the equilibrium state. Since the mole ratios are 1:1:1:1, if x mol/L of SO_2 reacts, then x mol/L of NO_2 also reacts and x mol/L of SO_3 and NO are produced. This can be summarized in the following table:

Principal Reaction	SO_2 (g)	+ NO_2 (g)	⇌	SO_3 (g)	+	NO (g)
Initial Concentration (M)	0.0650	0.0650		0		0
Change	$-x$	$-x$		$+x$		$+x$
Eq Concentration (M)	$0.0650-x$	$0.0650-x$		$+x$		$+x$

   c. The equilibrium expression for the reaction is $K_c = \dfrac{[NO][SO_3]}{[SO_2][NO_2]}$. The equilibrium concentrations from the table can be substituted into this expression, and we can solve for x.

   $$85.0 = \frac{(x)(x)}{(0.0650-x)(0.0650-x)} = \left(\frac{x}{0.0650-x}\right)^2$$

   Taking the square root of both sides gives

   $$\pm 9.22 = \frac{x}{0.0650-x}$$

   Solving for x, we obtain two solutions. The equation with the positive square root of 85 gives

   $$+9.22(0.0650-x) = x \qquad +0.599 = x + 9.22x \qquad x = \frac{0.599}{10.22} = 5.86\times10^{-2}$$

   The equation with the negative square root of 85 gives

   $$-9.22(0.0650-x) = x; \qquad -0.599 = x - 9.22x \qquad x = \frac{-0.599}{-8.22} = 7.29\times10^{-2}$$

   Because the initial concentrations of SO_2 and NO_2 are 0.0650, x cannot exceed 0.0650. Therefore, $x = 7.29 \times 10^{-2}$ should be discarded as chemically unreasonable and $x = 5.86 \times 10^{-2}$ is chosen as the solution. The equilibrium concentrations are

   $[SO_2] = [NO_2] = 0.0650 - (5.86\times10^{-2}) = 0.0064$ M; $\quad [SO_3] = [NO] = 0.0586$ M

8. a. reaction shifts to the right
   b. shifts left (K_c decreases)
   c. shifts to the right (greater number of moles of gas)
   d. no effect
   e. shifts to the left

9. $K_c = \dfrac{k_f}{k_r}$; $\qquad K_c = \dfrac{7.5 \times 10^{-7}}{3.2 \times 10^{-2}} = 2.3 \times 10^{-5}$

10. To solve this problem, we need to use the approach outlined in Figure 13.6 in your text and on page 208 in this chapter.
    a. The balanced equation is given.

    b. The initial concentration of $HCONH_2$ is 0.250 M. Let x be the concentration of $HCONH_2$ that reacts on going to the equilibrium state. Since the mole ratios are 1:1:1, if x mol/L of $HCONH_2$ reacts, then x mol/L of NH_3 and CO are produced. This can be summarized in the following table:

Principal Reaction	$HCONH_2$ (g) ⇌	NH_3 (g) +	CO (g)
Initial Concentration	0.250	0	0
Change	$-x$	$+x$	$+x$
Eq Concentration (M)	$0.250 - x$	$+x$	$+x$

The equilibrium expression for the reaction is $K_c = \dfrac{[NH_3][CO]}{[HCONH_2]}$. The equilibrium concentrations from the table can be substituted into this expression

$$4.84 = \dfrac{(x)(x)}{(0.250-x)} = \dfrac{x^2}{0.250-x}$$

To solve for x, we need to rearrange this expression and use the quadratic equation.

$4.84(0.250-x) = x^2 \qquad x^2 + 4.84x - 1.21 = 0;$

$$x = \dfrac{-4.84 \pm \sqrt{(4.84)^2 - 4(-1.21)}}{2} \qquad x = -5.08 \text{ or } 0.238$$

The mathematical solution that makes chemical sense is 0.238.

c. The equilibrium concentrations can now be calculated.

$[HCONH_2] = 0.250 - 0.238 = 0.012$ M $\qquad [NH_3] = [CO] = 0.238$ M

11. The position of equilibrium is to the left, given the very small value of K_c. We can calculate the value of Q_c from the concentrations given.

$$Q_c = \dfrac{[H_2]^4[CS_2]}{[H_2S]^2[CH_4]} = \dfrac{(0.300)^4(0.075)}{(3.00)^2(1.50)} = 4.50 \times 10^{-5}$$

$Q_c > K_c$; therefore, the reaction mixture is not at equilibrium. The reaction will proceed from right to left to reach equilibrium.

Appendix C – Self Test Solutions

Challenge Problem

The equilibrium expression is:

$$K_c = \frac{[N_2O][O_2]}{[NO_2][NO]}$$

The initial concentrations (mol/L) for each species are

$[NO_2] = 0.05$ M; $[NO] = 0.075$ M; $[N_2O] = 0.0375$ M; $[O_2] = 0.0625$ M

The mole ratio of NO_2 to NO is 1:1. Therefore, the amount of NO_2 that reacts equals the amount of NO that reacts. Define this amount as x. We can summarize the initial and equilibrium concentrations in the following table:

Principal Reaction	NO_2 (g) +	NO (g) ⇌	N_2O (g) +	O_2 (g)
Initial Concentration (M)	0.05	0.075	0.025	0.0625
Change (M)	$-x$	$-x$	$+x$	$+x$
Eq. Concentration (M)	$0.05 - x$	$0.075 - x$	$0.025 + x$	$0.0625 + x$

$$0.914 = \frac{(0.0375 + x)(0.0625 + x)}{(0.05 - x)(0.075 - x)}$$

$$0.914 = \frac{(0.00234 + 0.0375x + 0.0625x + x^2)}{(0.00375 - 0.075x - 0.05x + x^2)}$$

$$0.00343 - 0.114x + x^2 = 0.00234 + 0.100x + x^2$$

Rearranging the above equation leads to

$0.00109 = 0.00509x;$ $x = 0.00509$

Substituting the value of x into the equilibrium concentration leads to

$[NO_2]_e = 0.05 - 0.00509 = 0.0449$ M
$[NO]_e = 0.07 - 0.00509 = 0.0699$ M
$[N_2O]_e = 0.025 + 0.00509 = 0.0426$ M
$[O_2]_e = 0.0625 + 0.00509 = 0.0676$ M

Chapter 14

True–False
1. F. One of the limitations of the Arrhenius theory is that it does not account for the basicity of substances that do not contain OH^-.
2. T
3. F. A conjugate base has one less proton than its acid.
4. F. In every acid–base reaction, the proton is transferred to the stronger base.
5. F. The strength of an acid and the strength of its conjugate base are inversely related (the stronger the acid, the weaker its conjugate base).
6. T
7. F. The principal reaction for a diprotic acid is the dissociation of H_2A.

8. F. A salt solution prepared by reacting a strong acid with a weak base is acidic; therefore, the pH is less than 7.
9. F. The strength of a binary acid increases down a group in the periodic table.
10. F. Acid strength increases with increasing electronegativity of Y.

Multiple Choice
1. b
2. c
3. d
4. d
5. d
6. b
7. b
8. a
9. d
10. d

Matching
Ion–product for water – d
Conjugate acid–base pair—e
Base–dissociation equilibrium—f
Polyprotic acids—a
Hydronium ion—c
pH scale—g
Acid–base indicator—h
Lewis acid—i
Brønsted–Lowry base—b

Fill–in–the Blank
1. far to the left
2. acid dissociation constant, K_a
3. dissociation of water
4. acid–base properties
5. acid strength; base strength
6. strength and polarity
7. polarity
8. increasing oxidation number
9. vacant valence orbitals; accept a pair of electrons
10. Electrically neutral

Appendix C – Self Test Solutions

Problems

1. We first need to solve for the concentration of the H_3O^+ ion.

   $[H_3O^+]$ = antilog (–9.87) = 1.3×10^{-10} M

   $[OH^-] = \dfrac{1.0 \times 10^{-14}}{1.3 \times 10^{-10}} = 7.7 \times 10^{-5}$ M

2. $Ba(OH)_2$ is a strong base and completely dissociates in water according to the reaction

   $Ba(OH)_2\ (aq) \rightarrow Ba^{2+}\ (aq) + 2\ OH^-\ (aq)$.

   We know that pH = log$[H_3O^+]$. To determine $[H_3O^+]$, we need to determine $[OH^-]$. We can determine the $[OH^-]$ by first calculating the number of moles of $Ba(OH)_2$ present. Once we know the moles of $Ba(OH)_2$ present, we can calculate the moles of OH^- and the molarity of OH^-.

   0.583 g $Ba(OH)_2 \times \dfrac{1\text{ mol } Ba(OH)_2}{171.3\text{ g } Ba(OH)_2} = 3.40 \times 10^{-3}$ mol $Ba(OH)_2$

   3.40×10^{-3} mol $Ba(OH)_2 \times \dfrac{2\text{ mol }OH^-}{1\text{ mol } Ba(OH)_2} = 6.81 \times 10^{-3}$ mol OH^-;

   $\dfrac{6.81 \times 10^{-3}\text{ mol }OH^-}{0.125\text{ L}} = 5.45 \times 10^{-2}$ M

   $[H_3O^+] = \dfrac{1.0 \times 10^{-14}}{5.45 \times 10^{-2}} = 1.83 \times 10^{-13}$; pH = –log($1.83 \times 10^{-13}$) = 12.737

3. $HSO_4^-\ (aq) + CH_3CO_2^-\ (aq) \rightleftarrows SO_4^{2-}\ (aq) + CH_3COOH$

   $HNO_2\ (aq) + NH_3\ (aq) \rightleftarrows NO_2^-\ (aq) + NH_4^+\ (aq)$

4. $0.51\% = \dfrac{[\text{acid}]\text{ dissociated}}{1.50 \times 10^{-3}} \times 100$ [acid] dissociated = 7.65×10^{-6}

   The [acid] dissociated represents the $[H_3O^+]$ and $[A^-]$ at equilibrium. Therefore, we can calculate the pH using this concentration.

   pH = –log(7.65×10^{-6}) = 5.116

   The equilibrium concentration of veronal is 1.49×10^{-3} (calculated from the initial and dissociated concentrations). We can now calculate K_a.

   $K_a = \dfrac{(7.65 \times 10^{-6})(7.65 \times 10^{-6})}{(1.49 \times 10^{-3})} = 3.93 \times 10^{-8}$

5. Knowing the pH, we can calculate the $[H_3O^+]$ at equilibrium.

   $[H_3O^+]$ = antilog (–5.89) = 1.29×10^{-6}.

   This concentration also represents the amount of histidine that dissociated and can be used to calculate the equilibrium concentration of histidine.

$(2.50 \times 10^{-3}) - (1.29 \times 10^{-6}) = 2.50 \times 10^{-3}$

Since $[H_3O^+] = [A^-]$ at equilibrium, we can now calculate K_a.

$$K_a = \frac{(1.29 \times 10^{-6})(1.29 \times 10^{-6})}{(2.50 \times 10^{-3})} = 6.65 \times 10^{-10}$$

6. To solve this problem, use the procedure found on page 628 in your textbook.

Steps 1–5: $K_b > K_w$

Principal Reaction	C_6H_5N (aq) + H_2O (l) ⇌	$C_6H_5NH^+$ (aq) +	OH^- (aq)
Initial Concentration	0.150	0	0
Change	$-x$	$+x$	$+x$
Equilibrium Concentration	$0.150 - x$	$+x$	$+x$

Substituting these values into the equilibrium expression gives

$$K_a = 1.8 \times 10^{-9} = \frac{[C_6H_5NH^+][OH^-]}{[C_6H_5N]} = \frac{(x)(x)}{(0.150-x)}$$

Assume that $0.150 - x \approx 0.150$. The above expression then simplifies to

$x^2 = (1.8 \times 10^{-9})(0.150); \quad x = 1.64 \times 10^{-5}$ M

Step 6: $[OH^-] = x = 1.64 \times 10^{-5}$

Step 7: $[H_3O^+] = \dfrac{1.0 \times 10^{-14}}{1.64 \times 10^{-5}} = 6.1 \times 10^{-10}$ M

Step 8: pH = $-\log (6.1 \times 10^{-10}) = 9.21$

7. To solve this problem, use the procedure found on page 628 in your textbook.

Steps 1–5: $K_{a1} \gg K_w$

Principal Reaction	$H_2C_2O_4$ (aq) + H_2O (l) ⇌	H_3O^+ (aq) +	$HC_2O_4^-$ (aq)
Initial Concentration	0.025	0	0
Change	$-x$	$+x$	$+x$
Equilibrium Concentration	$0.025 - x$	$+x$	$+x$

Substituting these values into the equilibrium expression gives:

$$K_{a1} = 5.9 \times 10^{-2} = \frac{[H_3O^+][HC_2O_4^-]}{[H_2C_2O_4]} = \frac{(x)(x)}{(0.025-x)}$$

Due to the value of K_{a1}, we cannot assume that the value of x is negligible. Therefore, we must use the quadratic equation to solve for x. Rearranging the above expression gives:

Appendix C – Self Test Solutions

$$(1.48 \times 10^{-3}) - (5.9 \times 10^{-2})x = x^2$$

$$x^2 + (5.9 \times 10^{-2})x - (1.48 \times 10^{-3}) = 0$$

$$x = \frac{-(5.9 \times 10^{-2}) \pm \sqrt{(5.9 \times 10^{-2})^2 - 4(-1.48 \times 10^{-3})}}{2}$$

$$x = 1.9 \times 10^{-2} \quad \text{or} \quad x = -7.8 \times 10^{-2}$$

The solution that makes chemical sense is $x = 1.9 \times 10^{-2}$

Step 6: $[H_2C_2O_4] = 0.0250 - 0.019 = 0.006$ M; $[H_3O^+] = [HC_2O_4^-] = x = 0.019$

Step 7: To calculate the $[C_2O_4^{2-}]$, we use the second dissociation step of $H_2C_2O_4$.

$$HC_2O_4^- \,(aq) + H_2O \,(l) \rightleftarrows H_3O^+ \,(aq) + C_2O_4^{2-} \,(aq)$$

$$K_{a2} = 6.4 \times 10^{-5} = \frac{[H_3O^+][C_2O_4^{2-}]}{[HC_2O_4^-]}$$

Because $K_{a1} \gg K_{a2}$, the $[H_3O^+]$ and $[HC_2O_4^-]$ at equilibrium in the second dissociation step are equal to 0.019. (x, the $[C_2O_4^{2-}]$ in this dissociation step, is negligible).

$$6.5 \times 10^{-5} = \frac{0.019x}{0.019}$$

$$x = 6.5 \times 10^{-5}$$

$$[C_2O_4^{2-}] = x = 6.5 \times 10^{-5} \text{ M}$$

The $[OH^-]$ is calculated using the concentration of H_3O^+ calculated in the first dissociation step.

$$[OH^-] = \frac{1.0 \times 10^{-14}}{0.019} = 5.26 \times 10^{-13} \text{ M}$$

Step 8: pH = $-\log(0.019) = 1.72$

8. Let the formula HA represent the formula for saccharin. To solve this problem use the procedure outlined in this chapter and found on page 628 in your textbook.

Steps 1–5: $K_a > K_w$

Principal Reaction	HA (aq) + H$_2$O (l) ⇄	H$_3$O$^+$ (aq) +	A$^-$ (aq)
Initial Concentration	0.750	0	0
Change	$-x$	$+x$	$+x$
Equilibrium Concentration	$0.750 - x$	$+x$	$+x$

Substituting these values into the equilibrium expression gives:

$$K_a = 2.1 \times 10^{-12} = \frac{[H_3O^+][A^-]}{[HA]} = \frac{(x)(x)}{(0.750 - x)}$$

Assume that $0.750 - x \approx 0.750$. The above expression then simplifies to:

$x^2 = (2.1 \times 10^{-12})(0.750); \qquad x = 1.25 \times 10^{-6}$

Step 6: $[H_3O^+] = 1.25 \times 10^{-6}$

Step 7: $pH = -\log(1.25 \times 10^{-6}) = 5.903$

9. Let the formula B represent the formula for benzylamine. To solve this problem, use the procedure found on page 628 in your textbook.

Steps 1–5: $K_b > K_w$

Principal Reaction	B (aq) + H$_2$O (l)	⇌	BH$^+$ (aq) +	OH$^-$ (aq)
Initial Concentration	0.150		0	0
Change	$-x$		$+x$	$+x$
Equilibrium Concentration	$0.150 - x$		$+x$	$+x$

Substituting these values into the equilibrium expression gives

$$K_b = 2.14 \times 10^{-5} = \frac{[BH^+][OH^-]}{[B]} = \frac{(x)(x)}{(0.150-x)}$$

Assume that $0.150 - x \approx 0.150$. The above expression then simplifies to:

$x^2 = (2.14 \times 10^{-5})(0.150) \qquad x = 1.79 \times 10^{-3}$

Step 6: $[OH^-] = 1.79 \times 10^{-3}$

Step 7: $[H_3O^+] = \dfrac{1.0 \times 10^{-14}}{1.79 \times 10^{-3}} = 5.59 \times 10^{-12}$

Step 8: $pH = -\log(5.59 \times 10^{-12}) = 11.253$

10. a. Formic acid: $K_b = \dfrac{1.0 \times 10^{-14}}{1.8 \times 10^{-4}} = 5.6 \times 10^{-11}$

   Ascorbic acid: $K_b = \dfrac{1.0 \times 10^{-14}}{8.0 \times 10^{-5}} = 1.2 \times 10^{-10}$

   Hypochlorous acid: $K_b = \dfrac{1.0 \times 10^{-14}}{3.5 \times 10^{-8}} = 2.8 \times 10^{-7}$

   b. Hydrazine: $K_a = \dfrac{1.0 \times 10^{-14}}{8.9 \times 10^{-7}} = 1.1 \times 10^{-8}$

   Aniline: $K_a = \dfrac{1.0 \times 10^{-14}}{4.3 \times 10^{-10}} = 2.3 \times 10^{-5}$

   Methylamine: $K_a = \dfrac{1.0 \times 10^{-14}}{3.7 \times 10^{-4}} = 2.7 \times 10^{-11}$

Appendix C – Self Test Solutions

11. K_b for OCl⁻ was calculated in problem 10 and is 2.8×10^{-7}. K_a for NH_4^+ is $\dfrac{1.0 \times 10^{-14}}{1.8 \times 10^{-5}} = 5.6 \times 10^{-10}$.

    K_b (OCl⁻) > K_a (NH_4^+). Therefore, the solution is basic.

12. KCN is a salt derived from a weak acid (HCN) and a strong base (KOH). Therefore, we can predict that this solution will be basic. The hydrolysis reaction for this salt is:

    CN⁻ (aq) + H₂O (l) ⇌ HCN (aq) + OH⁻ (aq)

    The K_b for this reaction can be calculated from K_w and K_a for HCN.

    $$K_b = \dfrac{1.0 \times 10^{-14}}{4.9 \times 10^{-10}} = 2.04 \times 10^{-5}$$

    To determine the HCN equilibrium concentration and the pH of the solution, we use the procedure found on page 628 in your text. However, we need to first determine the initial concentration of CN⁻.

    $$2.75 \text{ g KCN} \times \dfrac{1 \text{ mol KCN}}{65.1 \text{ g KCN}} \times \dfrac{1 \text{ mol CN}^-}{1 \text{ mol KCN}} = 0.0422 \text{ mol CN}^-$$

    $$\dfrac{0.0422 \text{ mol CN}^-}{0.250 \text{ L}} = 0.169 \text{ M}$$

    Steps 1–5: $K_b > K_w$

Principal Reaction	CN⁻ (aq) + H₂O (l) ⇌	HCN (aq) +	OH⁻ (aq)
Initial Concentration	0.169	0	0
Change	$-x$	$+x$	$+x$
Equilibrium Concentration	$0.169 - x$	$+x$	$+x$

    Substituting these values into the equilibrium expression gives

    $$K_b = 2.04 \times 10^{-5} = \dfrac{[\text{HCN}][\text{OH}^-]}{[\text{CN}^-]} = \dfrac{(x)(x)}{(0.169 - x)}$$

    Assume that $0.169 - x \approx 0.169$. The above expression then simplifies to:

    $x^2 = (2.04 \times 10^{-5})(0.169)$ $\qquad x = 1.86 \times 10^{-3}$

    Step 6: $x = [\text{HCN}] = [\text{OH}^-] = 1.86 \times 10^{-3}$

    step 7: $[H_3O^+] = \dfrac{1.0 \times 10^{-14}}{1.86 \times 10^{-3}} = 5.38 \times 10^{-12}$

    Step 8: pH = $-\log(5.38 \times 10^{-12}) = 11.266$

13. a. stronger acid: HI – For binary acids, acid strength increases down a group due to the increase in the size of the atom.

b. Stronger acid: HBr—For binary acids, acid strength increases across a period due to the increase in the electronegativity of the nonmetal atom.

c. Stronger acid: H_3PO_4—For oxoacids, acid strength increases with increasing electronegativity on Y.

d. Stronger acid: H_3PO_4—For oxoacids, acid strength increases with increasing oxidation number (more oxygen atoms) on Y.

14. Lewis acid: Co^{3+} Lewis base: CN^-

Challenge Problem

Assume 100 g of solution. We now have 50 g of H_3PO_4. The volume of the solution can be calculated from the mass of the solution and the density of the solution.

mL of solution = 75.0 mL

mol H_3PO_4 = 0.511 mol (Calculated knowing the mass of H_3PO_4 and the molar mass of H_3PO_4.)

$$M = \frac{0.511 \text{ mol } H_3PO_4}{0.0750 \text{ L}} = 6.80 \text{ M}$$

1. Write the stepwise dissociation for phosphoric acid.

$H_3PO_4 \, (aq) + H_2O \, (l) \rightleftarrows H_3O^+ \, (aq) + H_2PO_4^- \, (aq)$ $K_{a1} = 7.5 \times 10^{-3}$

$H_2PO_4^- \, (aq) + H_2O \, (l) \rightleftarrows H_3O^+ \, (aq) + HPO_4^{2-} \, (aq)$ $K_{a2} = 6.2 \times 10^{-8}$

$HPO_4^{2-} \, (aq) + H_2O \, (l) \rightleftarrows H_3O^+ \, (aq) + PO_4^{3-} \, (aq)$ $K_{a3} = 4.8 \times 10^{-13}$

2. Determine the principal reaction.

Since $K_{a1} > K_{a2}, K_{a3}$, we know that the principal reaction is the first dissociation step and that all of the H_3O^+ present is produced from this step. Therefore, we need only to consider the first dissociation step when calculating the pH of the solution.

3. Make a table showing the principal reaction, initial concentration, change in concentration, and equilibrium concentration of the reactant and products.

Principal Reaction	$H_3PO_4 \, (aq)$	$\rightleftarrows$	H_3O^+	+ $H_2PO_4^-$
Initial Concentration	6.80		0	0
Change	$-x$		$+x$	$+x$
Eq Concentration	$6.80 - x$		$+x$	$+x$

4. Substitute the equilibrium concentrations into the equilibrium expression. Since $K_{a1} > 1.0 \times 10^{-3}$, you cannot assume that x is negligible.

Appendix C – Self Test Solutions

$$7.5 \times 10^{-3} = \frac{x^2}{6.80 - x}$$

$$x^2 + (7.5 \times 10^{-3})x - (5.1 \times 10^{-2}) = 0$$

$$x = \frac{-(7.5 \times 10^{-3}) \pm \sqrt{(7.5 \times 10^{-3})^2 + 4(5.1 \times 10^{-2})}}{2}$$

$x = -0.230$ or $x = 0.222$. The answer that makes chemical sense is 0.222 M. This value represents the equilibrium concentration of H_3O^+ and $H_2PO_4^-$.

5. Calculate the pH from this value.

pH = –log (0.222) = 0.653

6. Equilibrium concentrations are:

$[H_3PO_4]_e$ = 6.80 M – 0.222 M = 6.58 M
$[H_3O^+]_e = [H_2PO_4^-]_e$ = 0.222 M

We now repeat the process, using the chemical equation for $H_2PO_4^-$ and the equilibrium concentration calculated above.

Principal Reaction	$H_2PO_4^-$ (aq)	⇌	H_3O^+ (aq)	+	HPO_4^{2-} (aq)
Initial Concentration	0.222		0.222		0
Change	–x		+x		+x
Eq Concentration	0.222 – x		0.22 + x		+x

The value of K_{a2} is small enough so that x is negligible; therefore $0.222 - x \approx 0.222$ and $0.22 + x \approx 0.222$.

$$6.2 \times 10^{-8} = \frac{(0.222)x}{0.222}$$

$x = 6.2 \times 10^{-8}$

This value represents the equilibrium concentration of $[H_2PO_4^-]$

$[HPO_4^{2-}]_e = 6.2 \times 10^{-8}$ M

Finally, we will once again go through the process to calculate the equilibrium concentration of PO_4^{2-}.

Principal Reaction	HPO_4^{2-} (aq)	⇌	H_3O^+ (aq)	+	PO_4^{3-} (aq)
Initial concentration	6.2×10^{-8}		0.222		0
Change	–x		+x		+x
Eq. concentration	$(6.2 \times 10^{-8}) - x$		0.222 + x		+x

The value of K_{a3} is small enough so that x is negligible; therefore $(6.2 \times 10^{-8}) - x \approx 6.2 \times 10^{-8}$.

$$4.8 \times 10^{-13} = \frac{(0.222)x}{6.2 \times 10^{-8}}$$

$$x = 1.34 \times 10^{-19}$$

This value of x represents the equilibrium concentration of PO_4^{3-}.

$$[PO_4^{3-}]_e = 1.34 \times 10^{-19}$$

We now need to calculate the concentration of OH^-. Given that practically all of the H_3O^+ is produced in the first dissociation step, we will calculate OH^- based on the concentration of H_3O^+ calculated in this step.

$$K_w = [H_3O^+][OH^-]$$

$$1.0 \times 10^{-14} = 0.222[OH^-]$$

$$[OH^-] = \frac{1.0 \times 10^{-14}}{0.222} = 4.50 \times 10^{-14}$$

Chapter 15

True/False
1. T
2. F. When a weak acid (HClO) reacts with a strong base (NaOH) the pH of the solution is greater than 7.00.
3. F. Adding the conjugate base of a weak acid to a solution of a weak acid increases the concentration of a product in an equilibrium reaction. Therefore, the reaction shifts to the left.
4. T. ($pK_a = 3.35$ for HNO_2)
5. F. The best indicator to use is phenolphthalein. (See Figure 16.7 on page 682 in your textbook.)
6. F. The solubility of FeS increases with decreasing pH since FeS contains the basic anion S^{2-}. Increasing the amount of H_3O^+ in a solution removes the S^{2-} from solution, shifting the solubility equilibrium to the right. This is illustrated by the following equations:
   $FeS\,(s) \rightleftarrows Fe^{2+}\,(aq) + S^{2-}\,(aq)$ $S^{2-}\,(aq) + H_3O^+\,(aq) \rightleftarrows HS^-\,(aq) + OH^-\,(aq)$
7. T
8. T. (I.P. $> K_{sp}$)
9. T

Fill-in-the-Blank
1. proton transfer; strong acid; weak base
2. weak acid; weak base; strong acids; strong bases
3. common-ion effect
4. weak acid and its conjugate base; weak base and its conjugate acid; pH
5. buffer capacity
6. pK_a
7. equivalence point
8. solubility
9. K_f
10. ion product (IP)

Problems

Appendix C – Self Test Solutions

1. a. HNO$_3$ and KOH: This is a reaction between a strong acid and a strong base. The net ionic equation is H$_3$O$^+$ (aq) + OH$^-$ (aq) → 2 H$_2$O (l); pH = 7.

   b. HNO$_3$ and NH$_2$NH$_2$: NH$_2$NH$_2$ is a weak base. The net ionic equation is:

   H$_3$O$^+$ (aq) + NH$_2$NH$_2$ (aq) → H$_2$O (l) + NH$_2$NH$_3^+$ (aq); pH < 7.00.

   c. Benzoic acid (C$_6$H$_5$COOH) and KOH: Benzoic acid is a weak acid. The net ionic equation is C$_6$H$_5$COOH + OH$^-$ (aq) → C$_6$H$_5$CO$_2^-$ (aq) + H$_2$O (l); pH > 7.00.

   d. Benzoic acid and pyridine (C$_5$H$_5$N): Benzoic acid is a weak acid and pyridine is a weak base. The net ionic equation is C$_6$H$_5$COOH (aq) + C$_5$H$_5$N (aq) ⇌ C$_6$H$_5$CO$_2^-$ (aq) + C$_5$H$_5$NH$^+$ (aq). The pH is determined by comparing the K_a of benzoic acid to the K_b of pyridine. $K_a > K_b$; therefore, the solution is acidic.

2. Because the salt N$_2$H$_5$Cl is 100% dissociated, the species present initially are N$_2$H$_4$, N$_2$H$_5^+$, Cl$^-$, and H$_2$O. N$_2$H$_4$ is the strongest base present ($K_b \gg K_w$). This gives the following table:

Principal Reaction	N$_2$H$_4$ (aq) + H$_2$O (l) ⇌	N$_2$H$_5^+$ (aq) +	OH$^-$ (aq)
Initial Concentration (M)	0.15	0.10	0
Change (M)	$-x$	$+x$	$+x$
Eq Concentration (M)	$0.15 - x$	$0.10 + x$	$+x$

The common ion in this problem is N$_2$H$_5^+$. The equilibrium equation for the principal reaction is

$$K_b = 8.9 \times 10^{-7} = \frac{[N_2H_5^+][OH^-]}{[N_2H_4]} = \frac{(0.10+x)(x)}{(0.15-x)} \approx \frac{(0.10)(x)}{(0.15)}$$

x is assumed to be negligible because K_b is so small and the equilibrium is shifted to the left due to the common–ion effect.

$$x = [OH^-] = \frac{(8.9 \times 10^{-7})(0.15)}{(0.10)} = 1.3 \times 10^{-6}$$

Note that the assumption concerning the size of x is justified. Since we now know the [OH$^-$], we can calculate the [H$_3$O$^+$].

$$[H_3O^+] = \frac{1.0 \times 10^{-14}}{1.3 \times 10^{-6}} = 7.7 \times 10^{-9}$$

pH = –log (7.7 × 10^{-9}) = 8.11

3. The principal reaction and equilibrium concentrations for this solution are

Principal Reaction	HClO (aq) + H$_2$O (l) ⇌	H$_3$O$^+$ (aq) +	ClO$^-$ (aq)
Initial Concentration (M)	0.50	0	0.35
Change (M)	$-x$	$+x$	$+x$
Equilibrium Concentration (M)	$0.50 - x$	$+x$	$0.35 + x$

If we solve the equilibrium equation for H$_3$O$^+$, we obtain

$$K_a = \frac{[H_3O^+][ClO^-]}{[HClO]}$$

$$[H_3O^+] = K_a \frac{[HClO]}{[ClO^-]}$$

K_a for HClO is 3.5×10^{-8}. Substituting the given information into the equilibrium expression gives

$$[H_3O^+] = (3.5 \times 10^{-8})\frac{(0.50)}{(0.35)} = 5.0 \times 10^{-8}$$

pH = $-\log(5.0 \times 10^{-8})$ = 7.30

To determine the pH of the solution after addition of NaOH, we must take into account the neutralization reaction that takes place.

Neutralization Reaction	HClO (aq) +	OH$^-$ (aq)	$\rightleftharpoons$ H$_2$O (l) +	ClO$^-$ (aq)
Before Reaction (mol)	0.50	0.10		0.35
Change (mol)	-0.10	-0.10		+0.10
After Reaction (mol)	0.40	0		0.45

Substituting these values into the expression for [H$_3$O$^+$], we can then calculate the pH.

$$[H_3O^+] = (3.5 \times 10^{-8})\frac{(0.40)}{(0.45)} = 3.1 \times 10^{-8}$$

pH = $-\log(3.1 \times 10^{-8})$ = 7.51

To determine the pH of the solution after addition of HCl, we must take into account the neutralization reaction that takes place.

Neutralization Reaction	ClO$^-$ (aq) +	H$_3$O$^+$ (aq)	$\rightleftharpoons$ H$_2$O (l) +	HClO (aq)
Before reaction (mol)	0.35	0.10		0.50
Change (mol)	-0.10	-0.10		+0.10
After Reaction (mol)	0.25	0		0.60

Substituting these values into the expression for [H$_3$O$^+$], we can then calculate the pH.

$$[H_3O^+] = (3.5 \times 10^{-8})\frac{(0.60)}{(0.25)} = 8.4 \times 10^{-8}$$

pH = $-\log(8.4 \times 10^{-8})$ = 7.08

4. We can solve this problem by rearranging the Henderson–Hasselbalch equation

$$\log \frac{[\text{base}]}{[\text{acid}]} = \text{pH} - pK_a$$

Appendix C – Self Test Solutions

The pK_a for lactic acid is $-\log(1.4 \times 10^{-4}) = 3.85$

Substituting the pH and pK_a values gives

$$\log \frac{[\text{lactate}]}{[\text{lactic acid}]} = 4.25 - 3.85 = 0.40$$

Taking the antilog of both sides gives

$$\frac{[\text{lactate}]}{[\text{lactic acid}]} = 2.51$$

Notice that this gives the ratio of the concentration of the lactate ion to lactic acid. To determine the ratio of the concentration of lactic acid to the lactate ion, we simply take the reciprocal of 2.51.

$$\frac{[\text{lactic acid}]}{[\text{lactate}]} = 0.40$$

5. Before we can determine the change in pH that occurs, we need to calculate the pH of the buffer solution. Using the Henderson–Hasselbalch equation, we get

$$pH = pK_a + \log\frac{[HCO_2^-]}{[HCOOH]} \qquad pH = 3.74 + \log\frac{0.10}{0.25} = 3.34$$

To determine the pH of the solution after addition of HCl, we must take into account the neutralization reaction that takes place.

Neutralization Reaction	HCO_2^- (aq)	+ H_3O^+ (aq)	→	HCOOH (aq)	+ H_2O (l)
Before Reaction (mol)	0.10	0.150		0.25	
Change (mol)	–0.10	–0.10		–0.10	
After Reaction (mol)	0	0.050		0.35	

Note that the limiting reactant in this problem is the HCO_2^- ion. The HCl present swamps this buffer out, and the pH is calculated from the $[H_3O^+]$ after the neutralization reaction.

$pH = -\log(0.050) = 1.30$

The buffer capacity of this solution is much less than the buffer capacity of the 0.45 M HCOOH/0.55 M HCO_2^-, which experienced only a 0.16 change in pH.

6. a. 0.0 mL of NaOH: $[H_3O^+] = 0.0500$ M; pH = 1.301

   b. 10.0 mL of NaOH: mmol H_3O^+ = mmol H_3O^+ initial – mmol OH^- added
      mmol H_3O^+ = (50.0 mL)(0.0500 mmol/mL) – (10.0 mL)(0.1000 mmol/mL) = 1.50 mmol

$$[H_3O^+]\text{ after neutralization} = \frac{(1.50\text{ mmol})}{(60.0\text{ mL})} = 0.0250 \qquad pH = -\log(0.0250) = 1.602$$

c. 25.0 mL of NaOH: At this point, the number of mmol of H_3O^+ = the number of mmol of OH^-, which is the equivalence point. Since this is a strong acid–strong base titration, the pH = 7.000.

d. 30.0 mL of NaOH: mmol of excess OH^- = mmol of OH^- added – mmol of HCl initial
mmol of excess OH^- = (30.0 mL)(0.1000 mmol/mL) – (50.0 mL)(0.0500 mmol/mL) = 0.5000 mmol

$$[OH^-] \text{ after neutralization} = \frac{0.50 \text{ mmol}}{80 \text{ mL}} = 6.25 \times 10^{-3}$$

$$[H_3O^+] = \frac{1.0 \times 10^{-14}}{6.25 \times 10^{-3}} = 1.6 \times 10^{-12} \qquad pH = -\log(1.6 \times 10^{-12}) = 11.80$$

7. a. 0.0 mL HCl: pH is calculated in the same manner as a weak base.

Principal Reaction	CN^- (aq) + H_2O (l) ⇌	HCN (aq) +	OH^- (aq)
Initial concentration (M)	0.0500	0	0
Change (M)	–x	+x	+x
Equilibrium Concentration (M)	0.0500 – x	+x	+x

$$K_b = \frac{K_w}{K_a} = \frac{1.0 \times 10^{-14}}{4.9 \times 10^{-10}} = 2.04 \times 10^{-5} = \frac{[HCN][OH^-]}{[CN^-]} = \frac{(x)(x)}{(0.0500-x)} \approx \frac{(x)^2}{(0.0500)}$$

$$x = [OH^-] = 1.01 \times 10^{-3} \text{ M} \qquad [H_3O^+] = \frac{1.0 \times 10^{-14}}{1.01 \times 10^{-3}} = 9.9 \times 10^{-12}$$

$$pH = -\log(9.9 \times 10^{-12}) = 11.00$$

b. 10.0 mL HCl: mmol HCN after neutralization = mmol of HCl added
mmol HCN after neutralization = (10.0 mL)(0.1000 mmol/mL) = 1.00 mmol HCN

$$[HCN] = \frac{1.00 \text{ mmol}}{60.0 \text{ mL}} = 1.67 \times 10^{-2}$$

mmol CN^- after neutralization = mmol CN^- initial – mmol HCN
mmol CN^- after neutralization = (50.0 mL)(0.0500 mmol/mL) – 1.00 mmol = 1.50 mmol CN^-

$$[CN^-] = \frac{1.50 \text{ mmol}}{60.0 \text{ mL}} = 2.50 \times 10^{-2}$$

We can calculate the pH from the Henderson–Hasselbalch equation, using the pK_a for HCN.

$$pH = pK_a + \log\frac{[CN^-]}{[HCN]} \qquad pH = 9.31 + \log\frac{(2.50 \times 10^{-2})}{(1.67 \times 10^{-2})} = 9.49$$

c. 25.0 mL HCl: mmol HCN after neutralization = (25.0 mL)(0.1000 mmol/mL) = 2.50 mmol

$$[HCN] = \frac{2.50 \text{ mmol}}{75.0 \text{ mL}} = 3.33 \times 10^{-2}$$

mmol CN^- after neutralization = mmol CN^- initial – mmol HCN

Appendix C – Self Test Solutions

mmol CN⁻ after neutralization = 2.50 – 2.50 = 0

The pH is determined from the equilibrium expression for HCN.

$$K_a = 4.9 \times 10^{-10} = \frac{(x)^2}{3.33 \times 10^{-2}} \qquad x = [H_3O^+] = 4.04 \times 10^{-6}$$

pH = –log (4.04 × 10⁻⁶) = 5.394

d. 30.0 mL HCl: mmol H_3O^+ added = (30.0 mL)(0.1000 mmol/mL) = 3.00 mmol
mmol HCN present = 2.50 mmol
mmol H_3O^+ present = mmol H_3O^+ added – mmol CN⁻ present
= 3.00 mmol – 2.50 mmol = 0.50 mmol

$$[H_3O^+] = \frac{0.50 \text{ mmol}}{80.0 \text{ mL}} = 6.3 \times 10^{-3} \qquad pH = -\log(6.3 \times 10^{-3}) = 2.20$$

8. The solubility equilibrium for $Mg(OH)_2$ is

$$Mg(OH)_2 \text{ (s)} \rightleftharpoons Mg^{2+} \text{ (aq)} + 2\ OH^- \text{ (aq)}$$

If 1.12×10^{-4} mol is the amount of $Mg(OH)_2$ that dissolves in 1.0 L of solution then the $[Mg^{2+}] = 1.12 \times 10^{-4}$ M and the $[OH^-] = 2(1.12 \times 10^{-4}) = 2.24 \times 10^{-4}$ M. Substituting these values into the equilibrium expression gives

$$K_{sp} = [Mg^{2+}][OH^-]^2 = (1.12 \times 10^{-4})(2.24 \times 10^{-4})^2 = 5.62 \times 10^{-12}$$

9.

Solubility Equilibrium	$Cu_3(PO_4)_2$ (s) ⇌	3 Cu^{2+} (aq) +	2 PO_4^{3-} (aq)
Equilibrium Concentration (M)		+3 x	+2 x

$$K_{sp} = 1.4 \times 10^{-37} = [Cu^{2+}]^3[PO_4^{3-}]^2 = (3x)^3(2x)^2 = 108x^5$$
$$x = 1.7 \times 10^{-8}$$

10.

Solubility Equilibrium	$BaSO_4$ (s) ⇌	Ba^{2+} (aq) +	SO_4^{2-} (aq)
Equilibrium Concentration (M)		+ x	0.50 + x

$$K_{sp} = 1.1 \times 10^{-10} = [Ba^{2+}][SO_4^{2-}] = (x)(0.50 + x) \approx (x)(0.50)$$
$$x = 2.2 \times 10^{-10}$$

11. a. CuBr: Not more soluble. Br⁻ is the conjugate base of a very strong acid and is very unreactive.
   b. Ag_2S: More soluble. S^{2-} is the conjugate base of a weak acid.
   c. AgCN: More soluble. CN⁻ is the conjugate base of a very weak acid.
   d. MgF_2: More soluble. F⁻ is the conjugate base of a weak acid.

12. Because K_f for $Cu(NH_3)_4^{2+}$ is large (5.6×10^{11}), nearly all the Cu^{2+} from $Cu(NO_3)_2$ will be converted to $Cu(NH_3)_4^{2+}$.

$$Cu^{2+}(aq) + 4\,NH_3(aq) \rightleftarrows Cu(NH_3)_4^{2+}$$

We will use the procedure found in Example 16.11 (page 694 in your textbook) to calculate $[Cu^{2+}]$ and $[Cu(NH_3)_4^{2+}]$. Conversion of 0.25 mol/L of Cu^{2+} to $Cu(NH_3)_4^{2+}$ consumes 1.00 mol/L of NH_3 (due to a 1:4 mole ratio of Cu^{2+} to NH_3). Assuming 100% conversion to $Cu(NH_3)_4^{2+}$ gives the following concentrations:

$[Cu^{2+}] = 0\,M$

$[Cu(NH_3)_4^{2+}] = 0.25\,M$

$[NH_3] = 5.0 - 1.00 = 4.0\,M$

After conversion of Cu^{2+} to $Cu(NH_3)_4^{2+}$, assume that a small amount of back-reaction occurs, producing Cu^{2+}.

$$Cu(NH_3)_4^{2+} \rightleftarrows Cu^{2+}(aq) + 4\,NH_3(aq)$$

Dissociation of x mol/L of $Cu(NH_3)_4^{2+}$ produces x mol/L of Cu^{2+} and $4x$ mol/L of NH_3. The equilibrium concentrations are

$[Cu(NH_3)_4^{2+}] = 0.25 - x$

$[Cu^{2+}] = x$

$[NH_3] = 4.0 + 4x$

The following table summarizes this reasoning under the balanced equation

Principal Reaction	$Cu^{2+}(aq)$	+	$4\,NH_3(aq)$	$\rightleftarrows$	$Cu(NH_3)_4^{2+}(aq)$
Initial Concentration (M)	0.25		5.0		0
After 100% Reaction (M)	0		4.0		0.25
Equilibrium Concentration (M)	x		$4.0 + 4x$		$0.25 - x$

Substituting the equilibrium concentrations into the expression for K_f and making the approximation that x is negligible compared with 0.25 gives

$$K_f = 5.6 \times 10^{11} = \frac{[Cu(NH_3)_4^{2+}]}{[Cu^{2+}][NH_3]^4} = \frac{(0.25 - x)}{(x)(4.0 + x)^4} \approx \frac{(0.25)}{(x)(4.0)^4}$$

$$[Cu^{2+}] = x = \frac{(0.25)}{(5.6 \times 10^{11})(4.0)^4} = 1.74 \times 10^{-15}\,M$$

$[Cu(NH_3)_4] = 0.25 - (1.74 \times 10^{-15}\,M = 0.25\,M$

Appendix C – Self Test Solutions

13. The equation for this reaction is

    $$3\,CaCl_2\,(aq) + 2\,Na_3PO_4\,(aq) \rightarrow Ca_3(PO_4)_2\,(s) + 6\,NaCl$$

    To determine if a precipitate will form, we must calculate the ion product and compare its value to K_{sp}. (Remember, the concentrations we use for calculating ion product must be determined using $M_iV_i = M_fV_f$. On mixing, both solutions are diluted by ½.) The ion–product expression is

    $$IP = [Ca^{2+}]^3[PO_4^{3-}]^2$$

    $[Ca^{2+}] = 0.875\ M \qquad [PO_4^{3-}] = 1.25\ M$

    $$IP = (0.875)^3(1.25)^2 = 1.05$$

    K_{sp} for $Ca_3(PO_4)_2 = 2.1 \times 10^{-33}$; IP $\gg K_{sp}$; therefore, a precipitate will form.

14. $K_{spa}\,(CuS) = 6 \times 10^{-16}$; $K_{spa}\,(FeS) = 6 \times 10^2$

    $$Q_c = \frac{[Cu^{2+}][H_2S]^2}{[H_3O^+]^2} = \frac{(0.005)(0.10)^2}{(0.30)^2} = 5.6 \times 10^{-4}$$

    Since $K_{spa}\,(CuS) < Q_c < K_{spa}\,(FeS)$, Cu^{2+} will selectively precipitate out of solution.

15. a. The presence of ppt A indicates that group I ions are present. Ppt B contains group III cations since a precipitate formed on addition of H_2S and NH_3. The white precipitate formed after addition of $(NH_4)_2CO_3$ to soln b indicates that cations from group IV are present. The fleeting violet color in the flame test of soln b indicates that K^+ is present.

    b. Since ppt A dissolved after addition of hot water, it appears that Pb^{2+} is present. This is confirmed by addition of K_2CrO_4. The resulting yellow precipitate is $PbCrO_4$. However, we cannot rule out the possibility of the presence of Ag^+. On addition of NH_3, no visible change occurred. However, it is possible that AgCl dissolved. We can test for the presence of Ag^+ ion by the addition of HNO_3. HNO_3 will react with NH_3 according to the following neutralization equilibrium:

    $$H_3O^+\,(aq) + NH_3\,(aq) \rightleftarrows NH_4^+\,(aq) + H_2O\,(l) \qquad K_n = 1.8 \times 10^9$$

    Because K_n is greater than K_f for $Ag(NH_3)_2^+$, the neutralization reaction will predominate, removing NH_3 from the formation equilibrium. This causes Ag^+ to be available for the solubility equilibrium and AgCl to precipitate out.

Challenge Problem

For PbS

$$PbS\,(s) \rightleftarrows Pb^{2+}\,(aq) + S^{2-}\,(aq) \qquad K_{sp}$$

In the presence of an acid, the S^{2-} ion will react with the H^+ ion to form HS^-. The HS^- will react with H^+ to form $H_2S\,(aq)$. The chemical equations for these two reactions are

$$H^+\,(aq) + S^{2-}\,(aq) \rightleftarrows HS^-\,(aq) \qquad K' = 1/K_{a2}$$

$$H^+ (aq) + HS^- (aq) \rightleftharpoons H_2S (aq) \qquad K'' = 1/K_{a1}$$

(Note that the first equation is the reverse of the second dissociation step for H_2S and that the second equation is the reverse of the first dissociation step for H_2S. You learned in chapter 13 that you use the reciprocal of the equilibrium constant when using the reverse reaction.)

Adding the three chemical equations together, we have:

$$PbS (s) + 2 H^+ (aq) \rightleftharpoons Pb^{2+} (aq) + H_2S (aq) \qquad K_{spa} = \frac{K_{sp}}{K_{a2} \cdot K_{a1}} = \frac{[Pb^{2+}][H_2S]}{[H^+]^2}$$

$$K_{spa} = 3.0 \times 10^{-7}$$

To determine the molar solubility of PbS, we need to determine the concentration of Pb^{2+} at equilibrium. We work this problem as we work any equilibrium problem, keeping in mind that we know the initial concentration of the H^+ from the information we have regarding the buffer.

$[HCHO_2] = 0.25$ M and $[CHO_2^-] = 0.35$ M; $K_a = 1.8 \times 10^{-4}$

$$pH = pK_a + \log\frac{[CH_2O^-]}{[HCH_2O]} = 3.74 + \log\frac{0.35}{0.25} = 3.89$$

$[H^+] = 1.30 \times 10^{-4}$

Principal Reaction	PbS (s) +	2 H^+ (aq)	$\rightleftharpoons$	Pb^{2+} (aq)	+	H_2S (aq)
Initial concentration		1.30×10^{-4}		0		0
Change		$-2x$		$+x$		$+x$
Equilibrium Conc.		$(1.30 \times 10^{-4}) - 2x$		$+x$		$+x$

Substituting the equilibrium concentrations into the equilibrium expression, we have

$$3.0 \times 10^{-7} = \frac{(x)(x)}{[(1.3 \times 10^{-4}) - 2x]^2}$$

The value of K_{spa} is small enough that x can be considered negligible.

$$3.0 \times 10^{-7} = \frac{(x)^2}{1.3 \times 10^{-4}}$$

$x = [Pb^+] = 7.1 \times 10^{-8}$

Appendix C – Self Test Solutions

Chapter 16

True–False
1. F. The spontaneity of a reaction is determined by thermodynamics; the speed of the reaction is determined by kinetics. A reaction can be spontaneous and very slow.
2. F. Some spontaneous reactions move to a state of higher potential energy by absorbing heat from the surroundings.
3. F. Enthalpy alone does not account for the direction of spontaneous change. ΔS_{total} determines the spontaneity of a reaction.
4. T
5. T
6. F. When a molecule breaks in two, there is an increase in the disorder of the system, which results in an increase of the entropy of the system.
7. F. The first law of thermodynamics does not indicate the spontaneity of a process. It only keeps track of the energy flow between the system and surroundings.
8. T
9. T
10. F. If the reaction proceeds spontaneously in the forward direction, the free energy decreases since a spontaneous reaction has a negative ΔG.

Matching
Spontaneous process—d
Entropy—f
Third law of thermodynamics—g
Standard molar entropy—c
First law of thermodynamics—a
Second law of thermodynamics—e
Standard free energy of formation—b

Fill–in–the–Blank
1. equilibrium
2. disorder or randomness
3. the number of ways the state can be achieved
4. temperature
5. increases
6. sign
7. entropy of the system plus surroundings
8. decreases
9. ΔG_f°
10. < 0; > 0

Problems
1. Keep in mind that the probabilities of ordered and random states are proportional to the number of ways that the state can be achieved.
   a. The toy box has the higher entropy. A file cabinet organized in alphabetical order can be achieved in only one way. Anyone who has ever spent any amount of time with a three year–old knows that there are an infinite number of ways the toys in his toy box can be arranged.
   b. The salt on the road has the higher entropy. Again, a perfectly ordered crystal of salt can be arranged in only one way. However, the crystals of salt found on the road after a winter storm are intermingled with snow and ice and therefore, arranged in many different ways.
   c. The 1 mol of CO_2 gas at STP has the higher entropy because the state of greater volume is more probable. (Remember that 1 mole of gas at STP has a volume of 22.4 L.)

2. a. ΔS is positive; an increase in the volume of a gas at constant temperature leads to a decrease in the pressure, which leads to an increase in the entropy

Appendix C – Self Test Solutions

b. ΔS is positive; more randomness with gaseous particles
c. ΔS is positive; product side of the reaction has more moles of gas
d. ΔS is positive; product side of the reaction has more moles of gas
e. ΔS is negative; moving from more disorder to less disorder by reducing the number of ways the state can be achieved

3. Use the data found in Appendix B of your textbook, and be sure to pay attention to the state of the compound.
   a. Before solving this problem, you may find it useful to write the net ionic equation for the reaction.

$$Na_2CO_3 \,(s) + 2\,H^+ \,(aq) \rightarrow 2\,Na^+ \,(aq) + CO_2 \,(g) + H_2O \,(g)$$

$$\Delta S°_{rxn} = [2 \times S°(Na^+) + S°(CO_2) + S°(H_2O)] - [S°(Na_2CO_3) + 2S°(H^+)]$$

$$\Delta S°_{rxn} = [(2 \times 59.0) + 213.6 + 188.7] - [135.0 + (2 \times 0)]$$

$$\Delta S°_{rxn} = 385.3 \text{ J/K}$$

b. $\Delta S°_{rxn} = [2S°(N_2) + 6S°(H_2O)] - [4S°(NH_3) + 3S°(O_2)]$

$\Delta S°_{rxn} = [(2 \times 191.5) + (6 \times 188.7)] - [(4 \times 192.3) + (3 \times 205.0)]$

$\Delta S°_{rxn} = 131.0$ J/K

4. Using the equation $\Delta S°_{total} = \Delta S°_{rxn} - \dfrac{\Delta H°_{rxn}}{T}$, we can very easily solve for ΔH using the information given.

$$\Delta H°_{rxn} = -T(\Delta S°_{total} - \Delta S°_{rxn}) = -298\left(1.814 \times 10^4 \,\dfrac{J}{mol \cdot K} - 310.8 \,\dfrac{J}{mol \cdot K}\right) = -5.313 \times 10^6 \,\dfrac{J}{mol}$$

$$= -5.313 \times 10^3 \,\dfrac{kJ}{mol}$$

5. Using the equation $\Delta S = R \ln \dfrac{V_{final}}{V_{initial}}$, we can calculate the value of ΔS.

$$\Delta S = 8.314 \,\dfrac{J}{mol \cdot K} \ln \dfrac{30.5 \text{ L}}{22.4 \text{ L}} = 2.57 \,\dfrac{J}{mol \cdot K}$$

6. $\Delta H°_{rxn} = [2\Delta H°_f(PbSO_4) - 2\Delta H°_f(H_2O)] + [\Delta H°_f(Pb) + \Delta H°_f(PbO_2) + 2\Delta H°_f(H_2SO_4)]$

$\Delta H°_{rxn} = [(2 \times -919.9) + (2 \times -285.8)] - [0 + (-277) + (2 \times -814.0)] = -506.4$ kJ

$\Delta S°_{rxn} = [2S°(PbSO_4) + 2S°(H_2O)] - [S°(Pb) + S°(PbO_2) + 2S°(H_2SO_4)]$

$\Delta S°_{rxn} = [(2 \times 148.6) + (2 \times 69.9)] - [64.8 + 68.6 + (2 \times 156.9)] = -10.2$ J/K

Watch your units!

$\Delta G°_{rxn} = \Delta H°_{rxn} - T\Delta S°_{rxn} = (-5.064 \times 10^5 \text{ J}) - (298 \text{ K})(-10.2 \text{ J/K}) = -5.034 \times 10^5 \text{ J} = -503.4 \text{ kJ}$

To determine the temperature at which the reaction becomes nonspontaneous, use

Appendix C – Self Test Solutions

$$T = \frac{\Delta H°}{\Delta S°} = \frac{-5.064 \times 10^5 \text{ J}}{-10.2 \text{ J/K}} = 4.96 \times 10^4 \text{ K}$$

The reaction will become nonspontaneous above this temperature. (The reverse reaction will become spontaneous.)

7. $\Delta G°_{rxn} = [4G°_f(CO_2) + 2G°_f(H_2O)] - [2G°_f(C_2H_2) + 5G°_f(O_2)]$
   $\Delta G°_{rxn} = [(4 \times -394.4) + (2 \times -237.2)] - [(2 \times 209.2) + (5 \times 0)] = -2.470 \times 10^3 \text{ kJ}$

8. $\Delta G = \Delta G° + RT \ln Q$
   $\Delta G° = [2G°_f(SO_3)] - [2G°_f(SO_2) + G°_f(O_2)]$
   $\Delta G° = [(2 \times -371.1)] - [(2 \times -300.2) + 0] = -141.8 \text{ kJ}$

   $$\Delta G = -1.418 \times 10^5 \text{ J} + (8.314 \text{ J/K})(298 \text{ K}) \ln \frac{(3.0)^2}{(3.0)^2 (1.5)} = -1.43 \times 10^5 \text{ J}$$

9. $\Delta G° = -RT \ln K$

   To calculate $\Delta G°$, write the net ionic equation.

   $Na_2CO_3 (s) + 2 H^+ (aq) \rightleftarrows 2 Na^+ (aq) + CO_2 (g) + H_2O (l)$

   $\Delta G°_{rxn} = [2G°_f(Na^+) + G°_f(CO_2) + G°_f(H_2O)] - [G°_f(Na_2CO_3) + G°_f(H^+)]$
   $\Delta G°_{rxn} = [(2 \times -261.9) + (-394.4) + (-237.2)] - [(-1044.5) + (2 \times 0)] = -110.9 \text{ kJ}$

   $-1.109 \times 10^5 \text{ J} = -(8.314 \text{ J/K})(298 \text{ K}) \ln K$
   $\ln K = \dfrac{-1.109 \times 10^5 \text{ J}}{-(8.314 \text{ J/K})(298 \text{ K})} = 45 \qquad K = 2.8 \times 10^{19}$

10. To solve this problem, you need to use the data found in Appendix B in your textbook.

    a. $\Delta G°_{rxn} = [0 + 0] - [(2 \times -409.2)] = 818.4 \text{ kJ/mol}$

    b. To calculate the temperature at which the reaction becomes spontaneous, we need to first calculate the $\Delta S°_{rxn}$ and $\Delta H°_{rxn}$.

    $\Delta S°_{rxn} = [223.0 + (2 \times 64.2)] - [(2 \times 82.6)] = 186.2 \text{ J/mol·K}$

    $\Delta H°_{rxn} = [0 + 0] - [(2 \times -436.7)] = 873.4 \text{ kJ/mol}$

    We can now solve for the temperature (**watch your units**).

$$T = \frac{873.4 \frac{\text{kJ}}{\text{mol}}}{0.1862 \frac{\text{kJ}}{\text{mol} \cdot \text{K}}} = 4.691 \times 10^3 \text{ K}$$

c. $\ln K = \frac{\Delta G^\circ_{rxn}}{-RT} = \frac{818{,}400 \frac{\text{J}}{\text{mol}}}{-\left[\left(8.314 \frac{\text{J}}{\text{mol} \cdot \text{K}}\right)(298\text{ K})\right]} = -330.2 \qquad K = e^{-330.2} = 3.49 \times 10^{-1441}$

Challenge Problem

Using data from Appendix B in your textbook, calculate ΔG° for the reaction.

$\Delta G^\circ = [(2 \text{ mol NO }(g) \times 86.6 \text{ kJ/mol})] = 173.2 \text{ kJ}$

$\Delta G^\circ = -RT\ln K$;

$1.732 \times 10^5 \text{ J} = -\left(8.314 \frac{\text{J}}{\text{mol} \cdot \text{K}}\right)(298\text{ K})\ln K$

$\ln K = -69.87$
$K = 4.5 \times 10^{-31}$

$\Delta H^\circ = [2 \text{ mol NO }(g) \times 90.2 \text{ kJ/mol}] = 180.4 \text{ kJ}$

$\Delta S^\circ = [2 \text{ mol NO }(g) \times 210.7 \text{ J/mol·K}] - [(1 \text{ mol N}_2 (g) \times 191.5 \text{ J/mol·K}) + (1 \text{ mol O}_2 (g) \times 205.0 \text{ J/mol·K})]$

$\Delta S^\circ = 24.9 \text{ J/mol}$

$\Delta G^\circ = 1.804 \times 10^5 \text{ J} - (973\text{ K})\left(24.9 \frac{\text{J}}{\text{K}}\right) = 1.562 \times 10^5 \text{ J}$

$\Delta G^\circ = -RT \ln K$

$1.562 \times 10^5 \text{ J} = -\left(8.314 \frac{\text{J}}{\text{mol} \cdot \text{K}}\right)(973 \text{ K})\ln K$

$\ln K = -19.3$

$K = 4.5 \times 10^{-9}$

To determine the partial pressure of N_2, O_2, and NO at equilibrium, we need to first calculate the initial pressures of N_2 and O_2 using the Ideal Gas Law.

$$P_{N_2} = P_{O_2} = \frac{(5.00 \text{ mol})\left(\frac{0.0821 \text{ L} \cdot \text{atm}}{\text{mol} \cdot \text{K}}\right)(973 \text{ K})}{10.0 \text{ L}} = 39.9 \text{ atm}$$

Appendix C – Self Test Solutions

Principal Reaction	$N_2 (g)$	+	$O_2 (g)$	$\rightleftarrows$	2 NO (g)
Initial Pressure	39.9		37.9		0
Change	$-x$		$-x$		$+2x$
Equilibrium Pressure	$39.9 - x$		$37.0 - x$		$+2x$

Using the equilibrium constant calculated above, we can now solve for the equilibrium pressures.

$$4 \times 10^{-9} = \frac{P_{NO}^2}{P_{N_2} P_{O_2}} = \frac{(2x)^2}{(39.9-x)(39.9-x)} = \frac{(2x)^2}{(39.9-x)^2}$$

Taking the square root both sides gives

$$6 \times 10^{-5} = \frac{2x}{39.9-x} \qquad x = 1 \times 10^{-3} \text{ atm}$$

The equilibrium partial pressures are $P_{N_2} = P_{O_2} = 39.9$ atm and $P_{NO} = 1 \times 10^{-3}$ atm.

Chapter 17

True–False
1. F. In an electrolytic cell, electricity is used to drive a nonspontaneous reaction.
2. T
3. F. Reduction takes place at the cathode.
4. T
5. T
6. T
7. T
8. F. The mechanism described for rusting indicates that the rust is deposited in a spot other than where the pitting occurs.
9. T
10. F. Oxidation takes place at the anode.

Multiple Choice
1. d
2. c
3. b
4. d
5. a
6. b
7. d
8. d
9. d
10. b

Fill-in-the-Blank
1. Electrochemistry
2. half–reactions; electrodes
3. oxidation potential; reduction potential
4. standard hydrogen electrode

5. Tables of standard reduction potentials
6. battery
7. fuel cell
8. galvanizing
9. Hall–Heroult process
10. overvoltage

Matching

Galvanic cell—d
Electrolytic cell—l
Oxidizing agent—f
Reducing agent—m
Cathode—n
Anode—k
Salt bridge—b
Electromotive force—o
Coulomb—h
Standard cell potential—g
Nernst equation—c
Corrosion—a
Electrolysis—j
Electrorefining—i
Electroplating—e

Problems

1. For the reactions to occur in a galvanic cell, they must have a positive $E°_{cell}$. Remember that an oxidizing agent will spontaneously react with a reducing agent that lies below it in the table of standard reduction potentials.

    a. The half–reaction involving MnO_4^- lies above the half–reaction involving NO_3^-; therefore, the oxidizing agent (species reduced; cathode) is the MnO_4^-. The cell could possibly be

    $Pt\ (s)\ |\ NO\ (g)\ |\ NO_3^-\ (aq),\ H^+\ (aq)\ |\ |\ MnO_4^-\ (aq),\ H^+\ (aq),\ Mn^{2+}\ (aq)\ |\ Pt\ (s)$

    b. The half–reaction involving Ag^+ lies above the half–reaction involving Fe^{3+}. Therefore, Ag^+ is the oxidizing agent (species reduced; cathode) and Fe (s) is the reducing agent (species oxidized; anode).

    $Fe\ (s)\ |\ Fe^{3+}\ (aq)\ |\ |\ Ag^+\ (aq)\ |\ Ag\ (s)$

2. We can approach this problem two different ways. We can either use the rules we learned in Chapter 4 on balancing redox reactions, or we can use the table of standard reduction potentials (in which the half–reactions are already balanced) to write the balanced half–reactions. Keep in mind, the reaction involving MnO_4^- is in acid; therefore, H^+, must be involved in the reaction. The two half–reactions are:

    $MnO_4^-\ (aq)\ +\ 8\ H^+\ (aq)\ +\ 5\ e^-\ \rightarrow\ Mn^{2+}\ (aq)\ +\ 4\ H_2O\ (l)$

    $Fe^{2+}\ (aq)\ \rightarrow\ Fe^{3+}\ (aq)\ +\ e^-$

    To write the overall balanced redox reaction, we must multiply the oxidation reaction by five so that electrons lost will equal electrons gained. The overall balanced reaction is

    $MnO_4^-\ (aq)\ +\ 8\ H^+\ (aq)\ +\ 5\ Fe^{2+}\ (aq)\ \rightarrow\ Mn^{2+}\ (aq)\ +\ 4\ H_2O\ (l)\ +\ 5\ Fe^{3+}\ (aq)$

Appendix C – Self Test Solutions

A possible galvanic cell for this reaction would consist of an anode compartment with a Pt (s) electrode immersed in a solution of Fe^{2+}, and a cathode compartment with a graphite electrode immersed in a solution of MnO_4^-. The shorthand notation for this cell is:

$$Pt\,(s)\,|\,Fe^{2+}\,(aq),\,Fe^{3+}\,(aq)\,||\,MnO_4^-\,(aq),\,Mn^{2+}\,(aq),\,H^+\,(aq)\,|\,C\,(s)$$

3. The two half–reactions and their potentials for this cell are

   Anode: $2\,Hg\,(l) + 2\,Cl^-\,(aq) \rightarrow Hg_2Cl_2\,(s) + 2\,e^-$ $E° = -0.28$ V

   Cathode: $Hg_2^{2+}\,(aq) + 2\,e^- \rightarrow 2\,Hg\,(l)$ $E° = 0.80$ V

   $E°_{cell} = 0.80\,V + (-0.28\,V) = 0.52\,V$

4. a. Spontaneous: Cd^{2+}, the oxidizing agent, lies above Ca, the reducing agent, in the table of standard reduction potentials.
   b. Nonspontaneous: Ag, the reducing agent, lies above Ni^{2+}, the oxidizing agent, in the table of standard reduction potentials.
   c. Nonspontaneous: I^-, the reducing agent, lies above SO_4^{2-}, the oxidizing agent, in the table of standard reduction potentials.

5. The half–reactions are:

   Anode: $4\,Fe^{2+}\,(aq) \rightarrow 4\,Fe^{3+}\,(aq) + 4\,e^-$ $E° = -0.77$ V

   Cathode: $O_2\,(g) + 4\,H^+\,(aq) + 4\,e^- \rightarrow 2\,H_2O\,(l)$ $E° = +1.23$ V

   $E°_{cell} = 1.23 + (-0.77) = 0.46$ V

   $\Delta G° = -nFE°_{cell} = -(4)(96,500)(90.46) = -1.8 \times 10^5\,J = -180\,kJ$

   $\log K = \dfrac{nE°_{cell}}{0.0592} = \dfrac{(4)(0.46)}{0.0592} = 31$ $K = 10^{31}$

6. $E°_{cell} = E°_{anode} + E°_{cathode}$ $+1.04\,V = E°_{anode} + (-0.14\,V)$ $E°_{anode} = 1.18$ V

   Remember that the potential at the anode represents the oxidation potential. Therefore, the reduction potential is –1.18 V. From the table of standard reduction potentials we find that Mn (s) has a reduction potential of –1.18 V. The half–reaction at the anode is:

   $Mn\,(s) \rightarrow Mn^{2+}\,(aq) + 2\,e^-$

7. The half–reactions for the cell are

   Anode: $2\,Cl^-\,(aq) \rightarrow Cl_2\,(g) + 2\,e^-$ $E° = -1.36$ V

   Cathode: $MnO_4^-\,(aq) + 8\,H^+\,(aq) + 5\,e^- \rightarrow Mn^{2+}\,(aq) + 4\,H_2O\,(l)$ $E° = +1.51$ V

   The overall reaction for the cell is:

$$10\ Cl^-\ (aq) + 2\ MnO_4^-\ (aq) + 16\ H^+\ (aq) \rightarrow 5\ Cl_2\ (g) + 2\ Mn^{2+}\ (aq) + 8\ H_2O\ (l)$$

$$E°_{cell} = E°_{anode} + E°_{cathode} = (-1.36\ V) + 1.51\ V = 0.15\ V$$

Knowing the overall balanced equation and the $E°_{cell}$, we can now calculate the concentration of $Cl^-\ (aq)$.

$$E_{cell} = E°_{cell} - \frac{0.0592}{n} \log \frac{[Mn^{2+}]^2 P_{Cl_2}^5}{[H^+]^{16}[MnO_4^-]^2[Cl^-]^{10}} \qquad n = 10$$

$$-0.30\ V = 0.15\ V - \frac{0.0592}{10} \log \frac{(0.10)^2 (1)^5}{(1.35 \times 10^{-4})^{16}(0.010)^2[Cl^-]^{10}}$$

$$76.0 = \log \frac{(0.1)^2}{(1.35 \times 10^{-4})^{16}(0.010)^2} - \log[Cl^-]^{10}$$

$$12.1 = -10 \log[Cl^-] \qquad [Cl^-] = 6.2 \times 10^{-2}$$

8. To calculate ΔG, we first need to calculate E_{cell}, using the Nernst equation. The half-reactions are

$$3\ Zn\ (s) \rightarrow 3\ Zn^{2+}\ (aq) + 6\ e^- \qquad E° = +0.76\ V$$

$$2\ Cr^{3+}\ (aq) + 6\ e^- \rightarrow 2\ Cr\ (s) \qquad E° = -0.74\ V$$

$$E°_{cell} = 0.76 + (-0.74) = 0.02\ V$$

$$E_{cell} = 0.02 - \frac{0.0592}{6} \log \frac{(0.035)^3}{(0.050)^2} = 0.04$$

$$\Delta G = -nFE = -(6)(96{,}500)(0.04) = -2 \times 10^4\ J = -20\ kJ$$

9. From the Nernst equation, we have

$$E_{cell} = E°_{cell} - \frac{0.0592}{n} \log \frac{[Ni^{2+}][Ag^+]^2}{[H^+]^4}$$

Substituting the information gives

$$2.06 = 2.48 - \frac{0.0592}{2} \log \frac{(0.01)(0.01)^2}{[H^+]^4}$$

$$14 = \log \frac{(1 \times 10^{-6})}{[H^+]^4} = \log(1 \times 10^{-6}) - \log[H^+]^4$$

$$20 = 4(-\log [H^+]) = 4\ pH$$

Appendix C – Self Test Solutions

pH = 5

10. The overall reaction we are interested in is:

AgBr (s) → Ag$^+$ (aq) + Br$^-$ (aq)

Using the table of standard reduction potentials, we can break this overall reaction into the following two half–reactions:

AgBr (s) + e$^-$ → Ag (s) + Br$^-$ (aq) $E° = 0.07$ V

Ag (s) → Ag$^+$ (aq) + e$^-$ $E° = -0.80$ V

$E°_{cell} = 0.07$ V $+ (-0.80$ V$) = -0.73$ V

-0.73 V $= \dfrac{0.0592}{1} \log K$

$K = 4.67 \times 10^{-13}$

11. Using the equation $\Delta E° = \dfrac{0.0592}{n} \log K$ we find

$\Delta E° = \dfrac{0.0592}{2} \log(2.35 \times 10^{-8}) = -0.226$ V

$\Delta G° = -nFE° = -(2 \text{ mol e}^-)(96,500 \text{ C/mol e}^-)(-0.226 \text{ V}) = 4.36 \times 10^4$ C·V $= 4.36 \times 10^4$ J

12. The reaction occurring at the cathode is

Na$^+$ + e$^-$ → Na (l)

$35 \dfrac{C}{s} \times 6 \text{ h} \times \dfrac{3600 \text{ s}}{h} \times \dfrac{1 \text{ mol e}^-}{96,500 \text{ C}} \times \dfrac{1 \text{ mol Na}}{1 \text{ mol e}^-} \times \dfrac{23.0 \text{ g Na}}{1 \text{ mol Na}} = 180$ g Na

13. To calculate the volume of O$_2$, we first need to calculate the moles of O$_2$ produced and then use the ideal gas law. One of the half–reactions for the electrolysis of water is

2 H$_2$O (l) → O$_2$ (g) + 4 H$^+$ (aq) + 4 e$^-$

1.19×10^3 C $\times \dfrac{1 \text{ mol e}^-}{96,500 \text{ C}} \times \dfrac{1 \text{ mol O}_2}{4 \text{ mol e}^-} = 3.08 \times 10^{-3}$ mol O$_2$

$V = \dfrac{nRT}{P} = \dfrac{(3.08 \times 10^{-3})(0.0821)(298)}{\left(755 \text{ mm Hg} \times \dfrac{1 \text{ atm}}{760 \text{ mm Hg}}\right)} = 0.0759$ L

14. The reaction is Fe (s) + 2 OH$^-$ (aq) → Fe(OH)$_2$ (s) + 2 e$^-$

$$5.00\frac{C}{s} \times 3\,h \times \frac{3600\,s}{1\,h} \times \frac{1\,mol\,e^-}{96{,}500\,C} \times \frac{1\,mol\,Fe(OH)_2}{2\,mol\,e^-} \times \frac{89.8\,g}{1\,mol\,Fe(OH)_2} = 25.1\,g\,Fe(OH)_2$$

15. $\Delta G° = [(6\,mol\,Ag^+ \times 77.1\,kJ/mol) + (3\,mol\,S^{2-} \times 86.0\,kJ/mol) + (2\,mol\,Al^{3+} \times -480.6\,kJ/mol)]$
    $- [3\,mol\,Ag_2S \times -40.7\,kJ/mol] = -118.5\,kJ$

$\Delta G° = -nFE°$

The above reaction has $n = 6$. Keeping in mind that $1\,J = 1\,C \cdot V$, we now have

$$-1.186 \times 10^5\,C \cdot V = -6(96{,}500\,C)E°$$

$$E° = \frac{-1.186 \times 10^5\,C \cdot V}{-6(96{,}500\,C)} = 0.205\,V$$

We can calculate K from either the value of $\Delta G°$ or the value of $E°$.

$\Delta G° = -RT\ln K$

$$-1.186 \times 10^5\,J = -\left(8.314\,\frac{J}{mol \cdot K}\right)(298\,K)\ln K$$

$\ln K = 47.9$

$K = 6.35 \times 10^{20}$

Challenge Problem

It is readily apparent that if we add together the first two chemical equations given, we will obtain the overall equation. We can determine $E°$ for the overall reaction using one of two approaches. One approach is to determine the value of K_{red} for the Pb^{2+}/Pb reduction and multiply this value of K by the K_{sp}. This gives us K_{net} for the overall equation.

$PbC_2O_4\,(s) \rightleftarrows \cancel{Pb^{2+}\,(aq)} + C_2O_4^{2-}\,(aq)$	K_{sp}
$\cancel{Pb^{2+}\,(aq)} + 2\,e^- \rightarrow Pb\,(s)$	K_{red} (calculated from $E°$)
$PbC_2O_4\,(s) + 2\,e^- \rightarrow Pb\,(s) + C_2O_4^{2-}\,(aq)$	$K_{net} = K_{sp} \times K_{red}$

Once we have calculated K_{net}, we can calculate $\Delta G°$ for the overall reaction.

$$E°_{red} = \frac{0.0592}{n}\log K_{red}$$

$n = 2$ for this reaction.

Appendix C – Self Test Solutions

$$\log K_{red} = \frac{2(-0.126)}{0.0592} = -4.26$$

$$K_{red} = 5.50 \times 10^{-5}$$

We can now calculate K_{net} for the overall reaction.

$$K_{net} = (2.8 \times 10^{-13}) \times (5.5 \times 10^{-5}) = 1.5 \times 10^{-17}$$

Using the equation that

$$E°_{net} = \frac{0.0592}{n} \log K_{net}$$

We can calculate $E°_{net}$ for the overall reaction.

$$E°_{net} = \frac{0.0592}{2} \log(1.5 \times 10^{-17}) = -0.50 \text{ V}$$

Chapter 18

True–False
1. F. Hydrogen doesn't completely transfer its valence electron in ordinary chemical reactions. Hydrogen shares its electron with a nonmetallic element to produce a covalent compound.
2. F. Oxygen is the most abundant element on the planet's surface.
3. F. Oxygen is an inexpensive and readily available oxidizing agent.
4. T
5. T
6. F. The three isotopes of hydrogen all have the same electronic configuration, which determines the chemical behavior of an element. The isotopes differ only in the number of neutrons they contain.
7. F. Interstitial hydride compounds are classified as metallic hydrides.
8. T
9. T
10. F. Hydrogen peroxide disproportionates to give water and oxygen.
11. T
12. T

Matching
  Isotope effect – e
  Electrolysis –f
  Binary hydride—b
  Interstitial hydride—a
  Nonstoichiometric compound—g
  Disproportionation reaction—d
  Hydrate—c

Fill–in–the–Blank
1. weak intermolecular forces
2. reaction of a dilute acid with an active metal
3. synthesis gas; CO, and H_2
4. alkali and heavier alkaline earth metals; direct reaction of the elements

Appendix C – Self Test Solutions

5. their composition; the pressure of H_2 gas in the surroundings
6. the noble gases, platinum, or gold
7. elements with intermediate electronegativities.
8. heavier group 1A and 2A metals; heating the metals in an excess of air
9. passing an electric discharge through O_2
10. anhydrous; hydrates; drying agents; water from the air

Problems

1. The balanced equation for the reaction is $Zn\,(s) + 2\,HCl\,(aq) \rightarrow ZnCl_2\,(aq) + H_2\,(g)$.
   To determine the mass of Zn needed for this reaction, we need to calculate the number of moles of hydrogen by using the ideal gas law.

   $$n = \frac{PV}{RT} \qquad n = \frac{\left(740\text{ mm Hg} \times \frac{1\text{ atm}}{760\text{ mm Hg}}\right)(0.750\text{ L})}{\left(0.0821\frac{\text{L}\cdot\text{atm}}{\text{mol}\cdot\text{K}}\right)(295\text{ K})} = 3.02 \times 10^{-2}\text{ mol}$$

   We now can calculate the mass of zinc needed for this reaction.

   $$(3.02 \times 10^{-2}\text{ mol }H_2) \times \frac{1\text{ mol Zn}}{1\text{ mol }H_2} \times \frac{65.4\text{ g Zn}}{1\text{ mol Zn}} = 1.98\text{ g Zn}$$

2. The largest industrial use of hydrogen is the Haber process for the production of NH_3. The balanced chemical equation is $N_2\,(g) + 3\,H_2\,(g) \rightarrow 2\,NH_3\,(g)$

3. a. Ag^+ is reduced to Ag (oxidizing agent) and As^{3-} in AsH_3 is oxidized to As^{5+} (reducing agent) so the two half-reactions are

   $Ag^+ \rightarrow Ag$

   $AsH_3 \rightarrow H_3AsO_4$

   Both half–reactions are balanced for elements other than oxygen and hydrogen. Add H_2O for any oxygens that are needed and H^+ for any hydrogens that are needed.

   $Ag^+ \rightarrow Ag$

   $AsH_3 + 4\,H_2O \rightarrow H_3AsO_4 + 8\,H^+$

   Balance each reaction for charge.

   $Ag^+ + 1\,e^- \rightarrow Ag$

   $AsH_3 + 4\,H_2O \rightarrow H_3AsO_4 + 8\,H^+ + 8\,e^-$

   Make the electron count the same in both reactions.

   $8 \times (Ag^+ + 1\,e^- \rightarrow Ag)$

   $AsH_3 + 4\,H_2O \rightarrow H_3AsO_4 + 8\,H^+ + 8\,e^-$

   Add the two half reactions together, canceling anything that appears on both sides of the equation.

Appendix C – Self Test Solutions

$$8 \, Ag^+ + AsH_3 + 4 \, H_2O \rightarrow 8 \, Ag + H_3AsO_4 + 8 \, H^+$$

b. Al is oxidized (reducing agent) and H^+ is reduced (oxidizing agent), so the two half–reactions are

$$Al \rightarrow Al^{3+}$$

$$2H^+ \rightarrow H_2$$

The equation is balanced, using the method outlined above (see Chapter 4) to give

$$2 \, Al \, (s) + 3 \, H_2SO_4 \, (aq) \rightarrow Al_2(SO_4)_3 \, (aq) + 3 \, H_2 \, (g)$$

c. Sn^{4+} in SnO_2 is reduced (oxidizing agent) and H_2 is oxidized (reducing agent), so the two half–reactions are

$$Sn^{4+} \rightarrow Sn$$

$$H_2 \rightarrow H^+$$

The equation is balanced, using the method outlined above (see Chapter 4) to give

$$SnO_2 + 2 \, H_2 \rightarrow Sn + 2 \, H_2O$$

d. W^{6+} in WO_3 is reduced (oxidizing agent) and H_2 is oxidized (reducing agent), so the two half–reactions are

$$WO_3 \rightarrow W$$

$$H_2 \rightarrow H^+$$

The equation is balanced, using the method outlined above (see Chapter 4) to give

$$WO_3 + 3 \, H_2 \rightarrow W + 3 \, H_2O$$

4. $LiH \, (s) + H_2O \, (l) \rightarrow H_2 \, (g) + LiOH \, (aq)$; reducing agent: H^- in LiH; oxidizing agent: H^+ in H_2O; base: H^-; acid: H_2O

5. Ammonia is synthesized by the Haber process: $N_2 \, (g) + 3 \, H_2 \, (g) \rightarrow 2 \, NH_3 \, (g)$. The large scale industrial method for synthesizing hydrogen is the steam–hydrocarbon re–forming process. The primary source of hydrocarbons is natural gas. The chemical equations for the steam–hydrocarbon re–forming process are

$H_2O + \text{hydrocarbon} \rightarrow CO + H_2$
$CO + H_2O \rightarrow CO_2 + H_2$

Since the primary source for synthesizing hydrogen is fuel oil, it follows that the price of ammonia depends on the price of fuel oil.

6. a. ionic   b. covalent   c. covalent   d. metallic   e. metallic

7. The balanced equation for the reaction is $2 \, KClO_3 \, (s) \rightarrow 3 \, O_2 \, (g) + 2 \, KCl \, (s)$. To determine the mass of $KClO_3$ needed for this reaction, we need to calculate the number of moles of oxygen, using the ideal

Appendix C – Self Test Solutions

gas law.

$$n = \frac{\left(758 \text{ mm Hg} \times \dfrac{1 \text{ atm}}{760 \text{ mm Hg}}\right) \times 0.500 \text{ L}}{\left(0.0821 \dfrac{\text{L} \cdot \text{atm}}{\text{mol} \cdot \text{K}}\right)(295 \text{ K})} = 2.06 \times 10^{-2} \text{ mol O}_2$$

We can now calculate the mass of $KClO_3$ needed for this reaction.

$$2.06 \times 10^{-2} \text{ mol O}_2 \times \frac{2 \text{ mol KClO}_3}{3 \text{ mol O}_2} \times \frac{122.6 \text{ g KClO}_3}{1 \text{ mol KClO}_3} = 1.68 \text{ g KClO}_3$$

8. a. Vanadium is oxidized from V^{4+} to V^{5+} (reducing agent = VO^{2+}), and chromium is reduced from Cr^{6+} to Cr^{3+} (oxidizing agent = $Cr_2O_7^{2-}$). The half–reactions are

   $VO^{2+} \rightarrow VO_2^+$

   $Cr_2O_7^{2-} \rightarrow Cr^{3+}$

   Using the half–reaction method (see Problem 3 and Chapter 4), the balanced equation is

   $6 \, VO^{2+} + 2 \, H^+ + Cr_2O_7^{2-} \rightarrow 6 \, VO_2^+ + 2 \, Cr^{3+} + H_2O$

   b. Magnesium is oxidized to Mg^{2+} (reducing agent = Mg), and oxygen is reduced to O^{2-} (oxidizing agent = O_2). The half–reactions are

   $Mg \rightarrow Mg(OH)_2 \, (s)$

   $O_2 + H_2O \rightarrow 4 \, OH^-$

   Using the half–reaction method (see Problem 3 and Chapter 4), the balanced equation is

   $O_2 + 2 \, H_2O + 2 \, Mg \rightarrow 2 \, Mg(OH)_2 \, (s)$

   c. Copper is oxidized to Cu^{2+} (reducing agent = Cu), and oxygen is reduced to O^{2-} (oxidizing agent = O_2). The half–reactions are

   $Cu \rightarrow Cu^{2+}$

   $O_2 + 4 \, H^+ \rightarrow H_2O$

   Using the half–reaction method (see Problem 3 and Chapter 4), the balanced equation is

   $O_2 + 4 \, H^+ + 2 \, Cu \rightarrow 2 \, H_2O + 2 \, Cu^{2+}$

9. a. basic   b. acidic   c. amphoteric   d. acidic

10. Some of these equations can be balanced by inspection. Others are redox reactions, which need to be balanced using either the half–reaction method (see Problems 3 and 8) or the oxidation number method (see Chapter 4).

    a. $CaH_2 \, (s) + 2 \, H_2O \rightarrow 2 \, H_2 \, (g) + Ca(OH)_2 \, (s)$

Appendix C – Self Test Solutions

b. $2\,KH\,(s) + O_2\,(g) \rightarrow H_2O\,(l) + K_2O\,(s)$

c. $4\,Li\,(s) + O_2\,(g) \rightarrow 2\,Li_2O\,(s)$

d. $P_4\,(s) + 5\,O_2\,(g) \rightarrow P_4O_{10}\,(s)$

e. $Na_2O\,(s) + H_2O\,(l) \rightarrow 2\,NaOH\,(aq)$

f. $BaO_2\,(s) + H_2SO_4\,(aq) \rightarrow BaSO_4\,(s) + H_2O_2\,(aq)$

g. $N_2O_5\,(s) + H_2O\,(l) \rightarrow 2\,HNO_3\,(aq)$

h. $2\,CsO_2\,(s) + H_2O\,(l) \rightarrow O_2\,(g) + 2\,Cs^+\,(aq) + HO_2^-\,(aq) + OH^-\,(aq)$

i. $5\,H_2O_2\,(aq) + 2\,MnO_4^-\,(aq) + 6\,H^+\,(aq) \rightarrow 5\,O_2\,(g) + 2\,Mn^{2+}\,(aq) + 8\,H_2O\,(l)$

j. $O_3\,(g) + 2\,I^-\,(aq) + H_2O\,(l) \rightarrow O_2\,(g) + I_2\,(s) + 2\,OH^-\,(aq)$

11. The balanced equation for this reaction is $CaH_2\,(s) + 2\,H_2O\,(l) \rightarrow 2\,H_2\,(g) + Ca(OH)_2\,(aq)$. We first need to calculate the number of moles and grams of H_2 from the number of moles of CaH_2.

$$0.2348\text{ g CaH}_2 \times \frac{1\text{ mol CaH}_2}{42.1\text{ g CaH}_2} = 5.577 \times 10^{-3}\text{ mol CaH}_2$$

$$5.577 \times 10^{-3}\text{ mol CaH}_2 \times \frac{2\text{ mol H}_2}{1\text{ mol CaH}_2} = 1.115 \times 10^{-2}\text{ mol H}_2$$

$$1.115 \times 10^{-2}\text{ mol H}_2 \times \frac{2.00\text{ g H}_2}{1\text{ mol H}_2} = 2.230 \times 10^{-2}\text{ g H}_2$$

Knowing moles of H_2 gas, we can now use the ideal gas law to calculate the volume of the gas.

$$V = \frac{(1.115 \times 10^{-2}\text{ mol H}_2)\left(0.0821\,\frac{\text{L}\cdot\text{atm}}{\text{mol}\cdot\text{K}}\right)(273\text{ K})}{1\text{ atm}} = 0.2499\text{ L}$$

We now know both the grams and volume of H_2 produced and can calculate the density.

$$\frac{2.230 \times 10^{-2}\text{ g H}_2}{0.2499\text{ L H}_2} = 0.08924\text{ g/L}$$

12. The balanced equation for this reaction is

$$5\,H_2O_2\,(aq) + 2\,MnO_4^{2-} + 6\,H^+\,(aq) \rightarrow 5\,O_2\,(g) + 2\,Mn^{2+}\,(aq) + 8\,H_2O\,(l)$$

To calculate the volume of $KMnO_4$ needed to produce 3.000 g of O_2, we need to first calculate the number of moles of $KMnO_4$ needed to produce 3.000 g of O_2. This can be accomplished by calculating the number of moles of O_2 in 3.000 g of O_2.

Appendix C – Self Test Solutions

$$3.000 \text{ g O}_2 \times \frac{1 \text{ mol O}_2}{32 \text{ g O}_2} = 9.375 \times 10^{-2} \text{ mol O}_2$$

$$9.375 \times 10^{-2} \text{ mol O}_2 \times \frac{2 \text{ mol MnO}_4^-}{5 \text{ O}_2} \times \frac{1 \text{ mol KMnO}_4}{1 \text{ mol MnO}_4^-} = 3.750 \times 10^{-2} \text{ mol KMnO}_4$$

$$\left(3.750 \times 10^{-2} \text{ mol KMnO}_4\right) \times \frac{1 \text{ L}}{0.7500 \text{ mol KMnO}_4} = 5.000 \times 10^{-2} \text{ L KMnO}_4 \text{ or } 50.000 \text{ mL}$$

To determine the density of O_2, we need to calculate the volume of O_2 using the ideal gas law.

$$V = \frac{\left(9.375 \times 10^{-2} \text{ mol O}_2\right) \times \left(0.0821 \frac{\text{L} \cdot \text{atm}}{\text{mol} \cdot \text{K}}\right)(273 \text{ K})}{1 \text{ atm}} = 2.101 \text{ L}$$

$$\frac{3.000 \text{ g O}_2}{2.101 \text{ L O}_2} = 1.428 \text{ g/L}$$

13. a. peroxide   b. superoxide   c. oxide   d. peroxide   e. oxide

14. The difference between the mass of the hydrate and the anhydrous compound is the mass of water.

    2.879 g $MgSO_4 \cdot XH_2O$ – 1.406 g $MgSO_4$ = 1.473 g H_2O

    We can now calculate the number of moles of water from the mass of water.

    $$1.473 \text{ g H}_2\text{O} \times \frac{1 \text{ mol H}_2\text{O}}{18.0 \text{ g H}_2\text{O}} = 8.183 \times 10^{-2} \text{ mol H}_2\text{O}$$

    We now must determine the number of moles of H_2O relative to the number of moles of $MgSO_4$. (But first, we have to calculate the number of moles of $MgSO_4$.)

    $$1.406 \text{ g MgSO}_4 \times \frac{1 \text{ mol MgSO}_4}{120.3 \text{ g MgSO}_4} = 1.169 \times 10^{-2} \text{ mol MgSO}_4$$

    $$\frac{8.183 \times 10^{-2} \text{ mol H}_2\text{O}}{1.169 \times 10^{-2} \text{ mol MgSO}_4} = 7$$

    From the above equation, we learn that we have a ratio of 7 mol H_2O to 1 mol $MgSO_4$. Therefore, the formula of the hydrate is $MgSO_4 \cdot 7 H_2O$.

15. Again, the difference between the mass of the hydrate and anhydrous compound is the mass of water. This problem is worked in exactly the same manner as the previous problem. Following this procedure gives rise to the formula $Na_2CO_3 \cdot 10 H_2O$.

Challenge Problem

Appendix C – Self Test Solutions

a. $I^- \rightarrow I_2$, iodide is being oxidized. Therefore, H_2O_2 is being reduced and is the oxidizing agent. From the description given of H_2O_2, we know that the peroxide ion is changed to the oxide ion when H_2O_2 is reduced. In this case, $H_2O_2 \rightarrow OH^-$. Our unbalanced net ionic equation is

$$H_2O_2\,(aq) + 2\,I^-\,(aq) \rightarrow OH^-\,(aq) + I_2\,(s)$$

This chemical equation can easily be balanced by inspection

$$H_2O_2\,(aq) + 2\,I^-\,(aq) \rightarrow 2\,OH^-\,(aq) + I_2\,(s)$$

b. If $MnO_4^- \rightarrow Mn^{2+}$, then manganese is reduced. Therefore, H_2O_2 is oxidized and is the reducing agent. When H_2O_2 undergoes oxidation, oxygen gas is produced. The unbalanced net ionic equation is

$$H_2O_2\,(aq) + MnO_4^-\,(aq) \rightarrow O_2\,(g) + Mn^{2+}\,(aq)$$

Using the half–reaction method to balance this equation, we obtain

$$5\,H_2O_2\,(aq) \rightarrow 5\,O_2\,(g) + 10\,H^+\,(aq) + 10\,e^-$$

$$10\,e^- + 2\,MnO_4^-\,(aq) + 16\,H^+\,(aq) \rightarrow 2\,Mn^{2+}\,(aq) + 8\,H_2O\,(l)$$

Adding these two equations together gives:

$$5\,H_2O_2\,(aq) + 2\,MnO_4^-\,(aq) + 6\,H^+\,(aq) \rightarrow 5\,O_2\,(g) + 2\,Mn^{2+}\,(aq) + 8\,H_2O\,(l)$$

Chapter 19

Fill–in–the–Blank
1. valence d electrons
2. molecular
3. three–center, two–electron bonds
4. nitric acid
5. contact process
6. diamond, graphite, fullerene
7. SiO_4 tetrahedra
8. carbides
9. Haber process
10. white phosphorus and red phosphorus

General Questions
1. a) Ba   b) Ga   c) Rb

2. a) Na   b) Sn   c) Bi

3. Ionization energy— decreases down a group; atomic radius—increases down a group; electronegativity—decreases down a group; basicity of oxides—increases down a group

4. The diamond structure for carbon and silicon involves the use of sp^3 orbitals and a tetrahedral array of σ bonds. Graphite, on the other hand, involves the use of sp^2 orbitals to form σ bonds along with a p orbital perpendicular to the plane, which forms π bonds. It is the small size of carbon that allows for the

Appendix C – Self Test Solutions

overlap of p orbitals to form π bonds. The $3p$ orbitals in silicon are more diffuse, and therefore are unable to form π bonds. As a consequence, silicon cannot form an allotrope with the graphite structure.

5. It is the much higher electronegativity and much smaller atomic radius of boron that distinguishes the chemistry of this element from the other elements in group 3A.

6. The boron atom in boron halides does not have a complete octet surrounding it. Therefore, it is able to accept a pair of electrons from a Lewis base.

7. ns^2np^1. Stable oxidation states — Al, Ga, and In: +3; Tl: +1.

8. Each boron in diborane uses an sp^3 hybrid orbital to overlap with a terminal hydrogen $1s$ orbital. Each bridging H atom is joined to both boron atoms through a three–center two–electron bond in which the two electrons in the B–H–B bridge are spread out over three atoms.

9. ns^2np^2. Common oxidation states for group 4A — +4; Sn and Pb: +2.

10. The carbon sheets in graphite are held together by London forces that can easily slide over each other, giving it a slippery feel and allowing it to be used as a lubricant.

11. Beverages and fire extinguishers.

12. Silicon is first converted to $SiCl_4$: $Si\,(s) + 2\,Cl_2\,(g) \rightarrow SiCl_4\,(g)$. The $SiCl_4$ is removed by fractional distillation and then converted back to silicon by reduction with hydrogen: $SiCl_4\,(g) + 2\,H_2\,(g) \rightarrow Si\,(s) + 4\,HCl\,(g)$. Further purification is accomplished with zone refining.

13. a. $Si_6O_{18}^{12-}$; number of shared oxygens per silicon is 2; cyclic anion
    b. $Si_4O_{11}^{6-}$; number of shared oxygens per silicon is 2.5; double–stranded chain anion
    c. $Si_4O_{10}^{4-}$; number of shared oxygens per silicon is 3; infinitely extended two–dimensional layer anion
    d. SiO_2; number of shared oxygens per silicon is 4; infinitely extended three–dimensional covalent network
    e. $AlSi_3O_8^-$; number of shared oxygens per silicon is 4; extended three–dimensional anion

14. N and P oxides—acidic; As and Sb oxides—amphoteric; Bi_2O_3—basic

15. $-3 \rightarrow +5$

16. ox. state = –3, NH_3; ox. state = –2, N_2H_4; ox. state = –1, NH_2OH; ox. State = 0, N_2; ox. state = +1, N_2O; ox. state = +2, NO; ox. state = +3, HNO_2; ox. state = +4, NO_2; ox. state = +5, HNO_3

17. $2\,H_3PO_4 \rightarrow H_4P_2O_7 + H_2O$ (diphosphoric acid)
    $H_4P_2O_7 + H_3PO_4 \rightarrow H_5P_3O_{10} + H_2O$ (triphosphoric acid)

18. O: –2; S, Se, Te: –2, +4, +6

19. $HClO_4$ – perchloric acid; $HClO_3$ – chloric acid; $HClO_2$ – chlorous acid; HClO – hypochlorous
    $HBrO_4$ – perbromic acid; $HBrO_3$ – bromic acid; $HBrO_2$ – bromous acid; HBrO – hypobromous
    HIO_4 – periodic acid; HIO_3 – iodic acid; HIO_2 – iodous acid; HIO – hypoiodous acid

20. Washing soda:
    $Ca^{2+}\,(aq) + CO_3^{2-}\,(aq) \rightarrow CaCO_3\,(s)$

Appendix C – Self Test Solutions

$$CO_3^{2-} (aq) + H_2O (l) \rightleftarrows HCO_3^- (aq) + OH^- (aq)$$

Baking soda:
$$NaHCO_3 (s) + H^+ (aq) \rightarrow Na^+ (aq) + CO_2 (g) + H_2O (l)$$

21. $2 Ag(CN)_2^- (aq) + Zn (s) \rightarrow 2 Ag (s) + Zn(CN)_4^{2-} (aq)$

22. $Cu (s) + 2 NO_3^- (aq) + 4 H^+ (aq) \rightarrow 2 NO_2 (g) + 2 H_2O (l) + Cu^{2+} (aq)$

23. a) air oxidation of NH_3 to nitric oxide: $4 NH_3 (g) + 5 O_2 (g) \rightarrow 4 NO (g) + 6 H_2O (g)$
    b) oxidation of nitric oxide to nitrogen dioxide: $2 NO (g) + O_2 (g) \rightarrow 2 NO_2 (g)$
    c) disproportionation of NO_2 in water: $3 NO_2 (g) + H_2O (l) \rightarrow 2 HNO_3 (aq) + NO (g)$

24. a) S burns in air to give SO_2: $S (s) + O_2 (g) \rightarrow SO_2 (g)$
    b) SO_2 is oxidized to SO_3: $2 SO_2 (g) + O_2 (g) \rightarrow 2 SO_3 (g)$
    c) SO_3 reacts with water to H_2SO_4: $SO_3 (g) + H_2O$ (in conc. H_2SO_4) $\rightarrow H_2SO_4 (l)$

25. $I_2 (s) + H_2O (l) \rightleftarrows HOI (aq) + H^+ (aq) + X^- (aq)$

Challenge Problem

$$XeF_2 (aq) + 2 H^+ (aq) + 2 e^- \rightarrow Xe (g) + 2 HF (aq) \qquad E^\circ = 2.32 \text{ V}$$

To determine if the decomposition of XeF_2 in water to produce O_2 (g) is spontaneous, we need to find the appropriate half–reaction and E°. The most appropriate half–reaction in Appendix D in your textbook is

$$O_2 (g) + 2 H_2O (l) + 4 e^- \rightarrow 4 OH^- (aq) \qquad E^\circ = 0.40 \text{ V}$$

This half–reaction is considered the most appropriate since it is the only one that contains only water and some form of oxygen.

Let's manipulate the two half–reactions to obtain an overall reaction. First we need to reverse the second reaction so that the O_2 (g) is a product.

$$4 OH^- (aq) \rightarrow O_2 (g) + 2 H_2O (l) + 4 e^- \qquad E^\circ = -0.40 \text{ V}$$

$$XeF_2 (aq) + 2 H^+ (aq) + 2 e^- \rightarrow Xe (g) + 2 HF (aq) \qquad E^\circ = 2.32 \text{ V}$$

Notice that the number of electrons lost does not equal the number of electrons gained. We need to multiply the bottom equation by 2.

$$4 OH^- (aq) \rightarrow O_2 (g) + 2 H_2O (l) + 4 e^- \qquad E^\circ = -0.40 \text{ V}$$

$$2 XeF_2 (aq) + 4 H^+ (aq) + 4 e^- \rightarrow 2 Xe (g) + 4 HF (aq) \qquad E^\circ = 2.32 \text{ V}$$

$$4 OH^- (aq) + 2 XeF_2 (aq) + 4 H^+ (aq) \rightarrow 2 H_2O (l) + 2 Xe (g) + 4 HF (aq) + O_2 (g)$$

Is this the reaction we are seeking? Actually, it is. Remember that anytime you have both OH^- and H^+ on the same side of the reaction, you can combine them to form water. Remember that the $K_n = 1 \times 10^{14}$ for the reaction $OH^- (aq) + H^+ (aq) \rightarrow H_2O (l)$

Appendix C – Self Test Solutions

The overall reaction becomes

4 H$_2$O (l) + 2 XeF$_2$ (aq) → 2 H$_2$O (l) + 2 Xe (g) + 4 HF (aq) + O$_2$ (g)

Canceling the 2 H$_2$O's on the product side leads to

2 H$_2$O (l) + 2 XeF$_2$ (aq) → 2 Xe (g) + 4 HF (aq) + O$_2$ (g)

Now that we know we used the correct half–reaction, let's calculate $E°$ for the overall reaction.

$E°$ = 2.32 V + (–0.40 V) = +1.92 V

The reaction is spontaneous.

Chapter 20

Fill–in–the–Blank
1. chiral
2. *cis*
3. more readily
4. increase; *d* electrons
5. decrease; pair up
6. reducing
7. increasing polarity of the O–H bond
8. powerful oxidizing agent; acidic
9. +2 (ferrous) and +3 (ferric)
10. disproportionation
11. Lewis acid–base; Lewis base; Lewis acid
12. 4 and 6
13. Polydentate
14. metal chelate
15. coordinate covalent
16. Isomers
17. d^8
18. achiral
19. plane–polarized light
20. Greek prefix
21. optical isomers
22. absorbance spectrum
23. paramagnetic
24. valence bond theory
25. high–spin complex
26. Crystal field theory
27. large
28. high; the small value of Δ
29. low
30. electronic transitions

General Questions

1. Mn: [Ar] $4s^2\ 3d^5$ Pt: [Xe] $6s^2\ 4f^{14}\ 5d^8$ Co^{3+}: [Ar] $3d^6$ V^{2+}: [Ar] $3d^3$

2. The sharing of *d*, as well as *s*, electrons gives rise to stronger metallic bonding. As a consequence, transition metals are harder, have higher melting and boiling points, and are more dense than the group

Appendix C – Self Test Solutions

1A and 2A metals. The melting points increase as the number of unpaired d electrons available for metallic bonding increases and then decrease as the d electrons pair up and become less available for bonding.

3. As the $4f$ subshell is filled, there is an increase in the effective nuclear charge. However, the size decrease, due to a larger Z_{eff}, is almost exactly compensated for by the expected size increase that is due to an added quantum shell of electrons in the third-transition series. As a consequence, the atoms of the third-transition series have atomic radii (or volumes) very similar to the second-transition series. The increase in atomic mass of the third-transition series coupled with the similarity in atomic volume for the second- and third-transition series gives rise to unusually high densities for the atoms in the third-transition series.

4. The ease of oxidation of the metal decreases as the ionization energies increase across the first-transition series. Therefore, the strength of the reducing agent decreases across the first-transition series and Mn is a better reducing agent than Fe.

5. Ions that have transition metals in a high oxidation state are good oxidizing agents.

6. Os^{6+} is more stable because the stability of the higher oxidation states increases down a periodic group.

7. In acidic solution, $Cr_2O_7^{2-}$ undergoes the following half-reaction:

$$Cr_2O_7^{2-}\ (aq) + H^+\ (aq) \rightarrow Cr^{3+}\ (aq)$$

The other half-reaction is:

$$Pb\ (s) + HSO_4^-\ (aq) \rightarrow PbSO_4\ (s)$$

Using the rules to balance a redox equation, we end up with the following balanced half-reactions:

$$Cr_2O_7^{2-}\ (aq) + 14\ H^+\ (aq) + 6\ e^- \rightarrow 2\ Cr^{3+}\ (aq) + 7\ H_2O\ (l)$$

$$Pb\ (s) + HSO_4^-\ (aq) \rightarrow PbSO_4\ (s) + H^+\ (aq) + 2\ e^-$$

To get the overall balanced net ionic equation, we need to multiply the bottom equation by 3 and cancel out what is common. This gives

$$Cr_2O_7^{2-}\ (aq) + 11\ H^+\ (aq) + 3\ Pb\ (s) + 3\ HSO_4^-\ (aq) \rightarrow 2\ Cr^{3+}\ (aq) + 7\ H_2O\ (l) + 3\ PbSO_4\ (s)$$

8. $4\ Fe^{2+}\ (aq) + O_2\ (g) + 4\ H^+\ (aq) \rightarrow 4\ Fe^{3+}\ (aq) + 2\ H_2O\ (l)$

$Fe^{3+}\ (aq) + 3\ OH^-\ (aq) \rightarrow Fe(OH)_3\ (s)$

9. a. coordination compound—a compound in which a metal ion is attached to a group of surrounding molecules or ions by coordinate covalent bonds
   b. ligand—the molecule or ions that surround the central metal ion in a complex
   c. donor atom—the atoms that are attached directly to the metal ion
   d. coordination number—the number of ligand donor atoms that surround a central metal ion in a complex
   e. chelating agent—a polydentate ligand in which electron pairs on more than one donor atom can bond to the metal ion
   f. linkage isomers—isomers that result from a ligand's ability to bond to a metal through either of two different donor atoms
   g. ionization isomers—isomers that differ in the ion that is bonded to the metal ion

Appendix C – Self Test Solutions

    h. diastereoisomers—isomers which have different relative orientations of their metal–ligand bonds
    i. enantiomers—molecules or ions that are nonidentical mirror images of one another
    j. racemic mixture—a 50:50 mixture of the (+) and (–) isomers which produces no net optical rotation

10. a. Ligands: NH_3 (donor atom—N); H_2O (donor atom—O); complex ion charge = +3.
       Coordination number is 6; oxidation state of metal is +3; has the general formula MA_2B_4; therefore can be geometrical isomers.
   b. Ligands: NH_3 (donor atom—N); H_2O (donor atom—O).
       Coordination number is 6; oxidation state of metal is +2; complex ion charge = +2.
   c. Ligands: ethylenediamine (donor atoms—N), Cl^-, SCN^- (donor atom either N or S).
       Coordination number is 6 (en is a bidentate ligand); oxidation state of metal is +3; complex ion chare = +1. Because SCN^- can bond through either the N or S, this compound can display linkage isomerism. The en ligand can spiral either to the right or to the left, indicating the compound can exist as enantiomers. It is also possible for this complex to have ionization isomers with the NO_2^- changing places with either the Cl^- or SCN^- ligands. If the NO_2^- isomer changes places with either the Cl^- or SCN^- ligands, the NO_2^- can bond either through the N or one of the O's providing another example of linkage iosmerism. It can also have *cis* and *trans* diastereoisomers.
   d. Ligands: CN^- (donor atom—C)
       Coordination number is 6; oxidation state of metal is +2; Charge on complex ion is –4
   e. Ligands: en (donor atoms—N)
       Coordination number is 6 (en is a bidentate ligand); oxidation state of metal is +3; complex ion charge = +3. The en ligand can spiral either to the right or to the left, indicating the compound can exist as enantiomers.

11. a. tetraaminediaquacobalt(III) bromide
   b. pentaammineaquacobalt(II) chloride (linkage through O) depending on the linkage isomer.
   c. chlorobis(ethylenediamine)thiocyanatocobalt(III) nitrite or
      chlorobis(ethylenediamine)isothiocyanatocobalt(III) nitrite depending on the linkage isomer.
   d. sodium hexacyanoferrate(II)
   e. tris(ethylenediamine)cobalt(III) chloride

12. a. $[CrCl_2(H_2O)_4]Cl$
   b. $[Ru(NH_3)_5(N_2)]Cl_2$
   c. $Na_2[PdCl_6]$
   d. $Na_4[Fe(CN)_5I] \cdot 2\,H_2O$
   e. $[Cu(NH_3)_4]SO_4$

13. a. Since $[FeCl_4]^-$ is tetrahedral, we know that the hybridization on the metal ion is sp^3. The Fe^{3+} ion has the electron configuration [Ar] $3d^5$. The orbital diagram for this ion is

    [Ar] ↑ ↑ ↑ ↑ ↑
             3d   4s   4p

To accept a share in the 4 electron pairs provided by the chloride ion, Fe^{3+} must use the empty $4s$ and $4p$ orbitals to form hybrid orbitals. The orbital diagram for the complex is

    [Ar] ↑ ↑ ↑ ↑ ↑   ↑↓ ↑↓ ↑↓ ↑↓
            3d sp^3

This complex is high spin since it gives the maximum number of unpaired electrons.

Appendix C – Self Test Solutions

b. $[Ni(CN)_4]^{2-}$ is square planar; therefore, the hybrid orbitals being used by the Ni^{2+} ion are dsp^2. The Ni^{2+} ion has the electron configuration [Ar] $3d^8$. The orbital diagram for this ion is

[Ar] ↑↓ ↑↓ ↑↓ ↑ ↑   __ __ __ __
     __ __ __ __ __
        3d   4s   4p

For Ni^{2+} to use dsp^2 hybrid orbitals, the two unpaired electrons must first pair up. Ni^{2+} can then accept a share in the 4 electron pairs provided by the CN^- ion. The orbital diagram for the complex is

[Ar] ↑↓ ↑↓ ↑↓ ↑↓   ↑↓ ↑↓ ↑↓ ↑↓   __
     __ __ __ __ __   __ __ __ __
        3d   dsp²   4p

The complex is low spin since there are no unpaired electrons.

14. From the spectrochemical series, we know that F^- is a weak–field ligand and that CN^- is a strong–field ligand. Therefore, the Δ for $[CoF_6]^{3-}$ is small, whereas the Δ for $[Fe(CN)_6]^{3-}$ is large. The Co^{3+} ion has the electron configuration [Ar] $3d^6$. These six electrons will first fill all of the d orbitals before pairing up, since Δ < P. Therefore, the number of unpaired electrons equals four. The Fe^{3+} ion has the electron configuration [Ar] $3d^5$. Since Δ > P, the electrons will pair up in the lower–energy d orbitals in the complex, leaving only one unpaired electron.

For $[CoF_6]^{3-}$

    ↑   ↑
   __   __
   d_{z^2} $d_{x^2-y^2}$

        ↑
        Δ
        ↓

  ↑↓   ↑   ↑
  __   __   __
  d_{xy} d_{xz} d_{yz}

For $[Fe(CN)_6]^{3-}$

   __   __
   d_{z^2} $d_{x^2-y^2}$

        ↑
        Δ
        ↓

  ↑↓   ↑↓   ↑
  __   __   __
  d_{xy} d_{xz} d_{yz}

15. The crystal field splitting in tetrahedral complexes is half of the crystal field splitting in octahedral complexes because none of the *d* orbitals points directly at the ligands and there are only four ligands, as opposed to six in an octahedral complex.

Challenge Problem

The question centers around the bonding of the 6 H$_2$O's and the 3 chlorine ions. Are the chlorine ions directly bonded to the metal, are the water molecules directly bonded to the metal, or is it a combination of both? We know that Cr^{3+} complexes usually have a coordination number of 6 and form octahedral complexes. Therefore, the possible structural formulas are [Cr(H$_2$O)$_4$Cl$_2$]Cl · 2H$_2$O) (*cis* or *trans*), [Cr(H$_2$O)$_5$Cl]Cl$_2$·H$_2$O, and Cr(H$_2$O)$_6$]Cl$_3$. The change in mass, due to the loss of water and the amount of silver chloride produced, provide the information needed to determine the structural formulas of each isomer. The molar mass for all three isomers is 266.4 g/mol. For the dehydration reaction, any loss in mass is due to a loss of water, hence the term dehydration. This loss of water will occur only if the water is not part of the complex ion. Isomer *C* has no change in mass; therefore, we may assume that the six water molecules are bonded to the metal. If all six water molecules are bonded to the metal, then three moles of AgCl for every one mole of the complex should precipitate upon reaction with silver nitrate. Therefore, for isomer *C*

$$\text{mass AgCl} = (0.100 \text{ L}) \times \frac{0.100 \text{ mol } C}{1 \text{ L}} \times \frac{3 \text{ mol AgCl}}{1 \text{ mol } C} \times \frac{143.4 \text{ g AgCl}}{1 \text{ mol AgCl}} = 4.30 \text{ g AgCl}$$

The data support that isomer *C* has the formula [Cr(H$_2$O)$_6$]Cl$_3$.

For isomers *A* and *B*, the loss of mass on dehydration equals 0.036 g. The number of moles of either isomer is

$$\text{mol} = 0.248 \text{ g} \times \frac{1 \text{ mol}}{266.4 \text{ g}} = 9.31 \times 10^{-4} \text{ mol}$$

The isomers will have either one or two water molecules not bonded to the complex. We now can calculate the mass of water lost depending upon whether one or two moles of water are lost.

$$\text{g H}_2\text{O} = 9.31 \times 10^{-4} \text{ mol isomer} \times \frac{1 \text{ mol H}_2\text{O}}{1 \text{ mol isomer}} \times \frac{18.0 \text{ g H}_2\text{O}}{1 \text{ mol H}_2\text{O}} = 0.0168 \text{ g H}_2\text{O}$$

$$\text{g H}_2\text{O} = 9.31 \times 10^{-4} \text{ mol isomer} \times \frac{2 \text{ mol H}_2\text{O}}{1 \text{ mol isomer}} \times \frac{18.0 \text{ g H}_2\text{O}}{1 \text{ mol H}_2\text{O}} = 0.0335 \text{ g H}_2\text{O}$$

Isomer *A* lost 0.0360 g on dehydration and isomer *B* lost 0.0180 g. It appears that isomer *A* has the formula [Cr(H$_2$O)$_4$Cl$_2$]Cl·2 H$_2$O and isomer *B* has the formula [Cr(H$_2$O)$_5$Cl]Cl$_2$· H$_2$O. However, we can verify this conclusion by calculating the expected amount of silver chloride that should precipitate.

One mol of AgCl will precipitate for every one mol of isomer *A*. 2 mol of AgCl will precipitate for every 1 mol of isomer *B*. Knowing this, we can calculate the expected amount of silver chloride that will precipitate for each isomer, given the concentrations of isomers *A* and *B*.

Appendix C – Self Test Solutions

$$\text{g AgCl} = 0.100 \text{ L} \times \frac{0.100 \text{ mol } A}{1 \text{ L}} \times \frac{1 \text{ mol AgCl}}{1 \text{ mol } A} \times \frac{143.4 \text{ g AgCl}}{1 \text{ mol AgCl}} = 1.43 \text{ g AgCl}$$

$$\text{g AgCl} = 0.100 \text{ L} \times \frac{0.100 \text{ mol } B}{1 \text{ L}} \times \frac{2 \text{ mol AgCl}}{1 \text{ mol } B} \times \frac{143.4 \text{ g AgCl}}{1 \text{ mol AgCl}} = 2.87 \text{ g AgCl}$$

The calculated and measured amounts of AgCl agrees for both isomers A and B. Therefore, the structural formulas for A and B are correct.

Chapter 21

True–False
1. F. The early transition metals occur in nature as oxides because the less electronegative metals form compounds by losing electrons to highly electronegative nonmetals.
2. F. The *s*–block metal oxides are strongly basic and too reactive to exist with acidic oxides. The *s*–block metals occur as carbonates, silicates, and chlorides (for Na and K).
3. T
4. F. Insulators have very large band gaps, which is why they are unable to conduct an electrical current.
5. F. Because the electrons in a metal extend in all directions, no localized bonds are broken when the metal is hammered and therefore, it doesn't break.
6. T
7. F. The valence band contains the lower–energy, bonding MOs.
8. T
9. F. Most ceramics are electrical insulators.
10. T

Matching
Minerals—i
Ore—k
Metallurgy—a
Bayer process—l
Semiconductor—h
Doping—j
Ceramics—c
Superconductors—b
Sintering—f
Sol–gel method—g
Composite—e
Whiskers—d

Fill–in–the–Blank
1. location of the metal in the periodic table
2. heating the mineral in air; are more easily reduced
3. concentration of the ore and chemical treatment prior to reduction, reduction of the mineral to the free metal, refining or purification of the metal
4. no chemical reducing agent is strong enough to reduce the compounds.
5. $Fe_2O_3 (s) + 3 CO (g) \rightarrow 2 Fe (l) + 3 CO_2 (g)$
6. metal cations; delocalized electrons; move about the crystal
7. ability to carry kinetic energy from one part of the crystal to another
8. the band is partially filled
9. completely filled bands
10. partially filled bands
11. band gap

12. valence band; large; conduction band
13. band gap
14. doped; positive holes
15. they are three-dimensional superconductors
16. the highly directional covalent bonds prevent the planes of atoms from sliding over one another
17. Fibers and whiskers

Challenge Problem

The feasibility of using cyanidation to remove silver from argentite can be determined from the ΔG of the reaction. If ΔG is negative, then this process may be a possible means of obtaining silver from the ore. We can determine ΔG for the process from the K_{net} for the reaction. The two equations involved are

$$Ag_2S\,(s) \rightleftarrows 2\,Ag^+\,(aq) + S^{2-}\,(aq) \qquad K_{sp} = 6 \times 10^{-51}$$

$$2\,Ag^+\,(aq) + 4\,CN^-\,(aq) \rightleftarrows 2\,[Ag(CN)_2]^- \qquad K_f^2 = 1 \times 10^{42}$$

(We square the value of K_f because the stoichiometry of the chemical equation is two times that of the chemical equation for the formation of $[Ag(CN)_2]^-$.)

The overall reaction and K_{net} are:

$$Ag_2S\,(s) + 4\,CN^-\,(aq) \rightleftarrows 2\,[Ag(CN)_2]^-\,(aq) + S^{2-}\,(aq) \qquad K_{net} = 6 \times 10^{-9}$$

We can now calculate ΔG assuming that the temperature is at 25°C which is the temperature for the reported equilibria constants.

$$\Delta G = -RT(\ln K)$$

$$\Delta G = -(8.314\,J/mol \cdot K)(298K)\ln(6 \times 10^{-9}) = 4.69 \times 10^4\,J/mol = 46.9\,kJ/mol$$

Both the small value of K_{net} and the positive value of ΔG indicate that the cyanidation process for the removal of silver from argentite is not feasible.

We will use the same approach to determine if it is possible to use cyanidation to remove silver from horn silver.

$$AgCl\,(s) \rightleftarrows Ag^+\,(aq) + Cl^-\,(aq) \qquad K_{sp} = 1.8 \times 10^{-10}$$

$$Ag^+\,(aq) + 2\,CN^-\,(aq) \rightleftarrows [Ag(CN)_2]^-\,(aq) + Cl^-\,(aq) \qquad K_f = 1 \times 10^{21}$$

The overall chemical equation is

$$AgCl\,(s) + 2\,CN^-\,(aq) \rightleftarrows [Ag(CN)_2]^-\,(aq) + Cl^-\,(aq) \qquad K_{net} = 1.8 \times 10^{11}$$

$$\Delta G = -(8.314\,J/mol \cdot K)(298K)\ln(1.8 \times 10^{11}) = -6.42 \times 10^4\,J/mol = -64.2\,kJ/mol$$

Both the large value of K_{net} and the negative value of ΔG indicate that cyanidation is a process that can be used to extract silver from horn silver.

Appendix C – Self Test Solutions

Chapter 22

1. Nuclear reactions cause a change in an atom's nucleus; chemical reactions cause a change in the distribution of the outer shell electrons around an atom. Different nuclides of an element have essentially the same behavior in chemical reactions but have different behavior in nuclear reactions. Changes in temperature or pressures or the addition of a catalyst do not affect the rate of a nuclear reaction. The nuclear reaction of an atom is essentially the same whether it is combined with other elements or not. The energy changes accompanying nuclear reactions are far greater than those accompanying chemical reactions.

2. a. $^{81}_{36}Kr + ^{0}_{-1}e \rightarrow ^{81}_{35}Br$

   b. $^{104}_{47}Ag \rightarrow ^{0}_{1}e + ^{104}_{46}Pd$

   c. $^{73}_{31}Ga \rightarrow ^{0}_{-1}e + ^{73}_{32}Ge$

   d. $^{104}_{48}Cd \rightarrow ^{104}_{47}Ag + ^{0}_{1}e$

3. a. $^{11}_{5}B \rightarrow ^{4}_{2}He + ^{7}_{3}Li$

   b. $^{121}_{51}Sb \rightarrow ^{0}_{-1}e + ^{121}_{52}Te$

   c. $^{70}_{35}Br \rightarrow ^{1}_{0}n + ^{69}_{35}Br$

   d. $^{41}_{19}K \rightarrow ^{1}_{1}p + ^{40}_{18}Ar$

4. The ratio of the decay rate at any time t to the decay rate at time $t = 0$ is the same as the ratio of N and N_0.

$$\frac{\text{Decay rate at time } t}{\text{Decay rate at time } t = 0} = \frac{kN}{kN_0} = \frac{N}{N_0}$$

$$\ln\left(\frac{N}{N_0}\right) = -(0.693)\left(\frac{t}{t_{1/2}}\right)$$

Let $N = 3.5$ d/min/g and $N_0 = 15.3$ d/min/g. Solving for t

$$\ln\left(\frac{3.5}{15.3}\right) = -0.693\left(\frac{t}{5715 \text{ y}}\right) \qquad t = 12{,}000 \text{ y}$$

5. $0.693 = kt_{1/2}$ $\qquad 0.693 = k(120.0 \text{ d}) \qquad k = 5.78 \times 10^{-3} \text{ d}^{-1}$

6. $\ln\left(\dfrac{N}{N_0}\right) = -(0.693)\left(\dfrac{t}{t_{1/2}}\right) \qquad \ln\left(\dfrac{N}{0.500}\right) = -0.693\left(\dfrac{5.00}{28.1}\right) \qquad \ln\left(\dfrac{N}{0.500}\right) = -0.123$

Taking the antilog of both sides gives:

Appendix C – Self Test Solutions

$$\frac{N}{0.500} = 0.884 \qquad N = 0.442 \text{ mg}$$

7. This particular element will not be stable because the neutron/proton ratio is less than 1. This element can undergo positron emission or electron capture to increase the neutron/proton ratio and gain stability.

8. Total mass of the nucleons (50 n + 42 p)

   Mass of 50 neutrons = (50) (1.008 66 amu) = 50.443 00 amu
   Mass of 42 protons = (42) (1.007 28 amu) = 42.305 76 amu
   _____
   Mass of 50 neutrons and 42 protons   = 92.738 76 amu

   Mass of nucleus = Atomic mass − Mass of 42 e⁻ = 91.906 91 − 0.02304 = 91.883 87

   Mass defect = mass of nucleons − mass of nucleus
              = 92.738 76 amu − 91.833 87 amu
              = 0.854 89 amu

   $$0.85489 \text{ amu} \times \left(1.660\,54 \times 10^{-24} \frac{g}{amu}\right)\left(6.022 \times 10^{23} \text{ mol}^{-1}\right) = 0.85489 \frac{g}{mol}$$

   $$\Delta E = \Delta mc^2 = \left(0.8549 \frac{g}{mol}\right)\left(10^{-3} \frac{kg}{g}\right)\left(3.00 \times 10^8 \frac{m}{s}\right)^2 = 7.69 \times 10^{13} \text{ J/mol} = 7.69 \times 10^{10} \text{ kJ/mol}$$

9. Mass change = (mass of reactants) − (mass of products)

   $$= [1.008\,66 + 235.0439] - [136.9254 + 96.9111 + (2 \times 1.008\,66)]$$
   $$= 0.118\,74 \text{ amu}$$

   $$\Delta E = \Delta mc^2 = (0.118\,74 \text{ amu})\left(1.6605 \times 10^{-27} \frac{kg}{amu}\right)\left(6.022 \times 10^{23} \text{ mol}^{-1}\right)\left(3.00 \times 10^8 \frac{m}{s}\right)^2$$

   $$\Delta E = 1.07 \times 10^{13} \text{ kg} \cdot m^2 / (s^2 \cdot mol) = 1.07 \times 10^{13} \text{ J/mol}$$

10. a. $^{242}_{96}\text{Cm} + ^{4}_{2}\text{He} \rightarrow ^{245}_{98}\text{Cf} + ^{1}_{0}\text{n}$

    b. $^{14}_{7}\text{N} + ^{4}_{2}\text{He} \rightarrow ^{17}_{8}\text{O} + ^{1}_{1}\text{p}$

11. $\ln\left(\dfrac{2.2 \text{ d/min/g}}{15.3 \text{ d/min/g}}\right) = -0.693\left(\dfrac{t}{5715 \text{ y}}\right) \qquad t = 16{,}000 \text{ y}$

Challenge Problem

We begin by calculating the number of disintegrations per mole of ^{51}Cr in $K_2{}^{51}Cr_2O_7$ and the number of disintegrations per mole ^{14}C in $H_2{}^{14}C_2O_4$.

Appendix C – Self Test Solutions

$$843 \frac{\text{cpm}}{\text{g K}_2{}^{51}\text{Cr}_2\text{O}_7} \times \frac{292.2 \text{ g K}_2{}^{51}\text{Cr}_2\text{O}_7}{1 \text{ mol K}_2{}^{51}\text{Cr}_2\text{O}_7} \times \frac{1 \text{ mo K}_2{}^{51}\text{Cr}_2\text{O}_7}{2 \text{ mol }{}^{51}\text{Cr}} = 1.23 \times 10^5 \frac{\text{cpm}}{\text{mol }{}^{51}\text{Cr}}$$

$$345 \frac{\text{cpm}}{\text{g H}_2{}^{14}\text{C}_2\text{O}_4} \times \frac{94.0 \text{ g H}_2{}^{14}\text{C}_2\text{O}_4}{1 \text{ mol H}_2{}^{14}\text{C}_2\text{O}_4} \times \frac{1 \text{ mol H}_2{}^{14}\text{C}_2\text{O}_4}{2 \text{ mol }{}^{14}\text{C}} = 1.62 \times 10^4 \text{ } \gamma \text{ counts}$$

We know that the γ count from the resulting product is due to the ^{51}Cr and that the β count is from the ^{14}C. Therefore, we can calculate the number of moles of ^{51}Cr and ^{14}C present.

$$165 \text{ cpm} \times \frac{1 \text{ mol }{}^{51}\text{Cr}}{1.23 \times 10^5 \text{ cpm}} = 1.34 \times 10^{-3} \text{ mol }{}^{51}\text{Cr}$$

$$83 \text{ cpm} \times \frac{1 \text{ mol }{}^{14}\text{C}}{1.62 \times 10^4 \text{ cpm}} = 5.12 \times 10^{-3} \text{ mol }{}^{14}\text{C}$$

The ^{14}C:^{51}Cr = (5.12×10^{-3}):(1.34×10^{-3}) = 3.8 ≈ 4

Therefore, four oxalate ions are bonded to the chromium ion.

Chapter 23

True–False
1. F. Organic molecules have polar covalent bonds when carbon bonds to an element on the right or left side of the periodic table leading to a difference in electronegativity of 0.40 or greater.
2. T
3. T
4. F. Different isomers are different chemical compounds with different physical properties.
5. T
6. F. Alkanes have low reactivity.
7. T
8. F. Amines are organic derivatives of NH_3.
9. T
10. T

Fill–in–the–Blank
1. tetravalent
2. isomers
3. conformers
4. cycloalkanes
5. functional group
6. benzene
7. Ethers; inert
8. amines
9. alcohol; carboxylic acid
10. diacids; dialcohols

General Questions
1. a. 3,4–dimethylhexane; b. 2,4–dimethylpentane; c. 2–ethyl–4–methyl–1–pentene;
   d. 4–methylcyclopentene; e. 1–bromo–2–pentyne; f. bromobenzene; g. *meta*–bromochlorobenzene;
   h. 1,3–dichloro–2–methylcyclopentane; i. 2,3–dimethylbutane;
   j. bromocyclopentane; k. 1,1,3–trimethylcyclohexane; l. 2,3–dibromo–2–butene

2. a. ethyl b. isopropyl c. *tert*-butyl

3.

a. 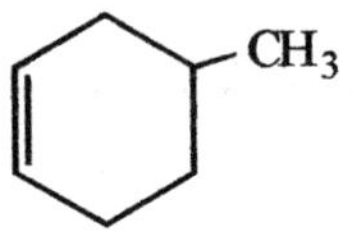 [structure: cyclohexene with CH₃ substituent]

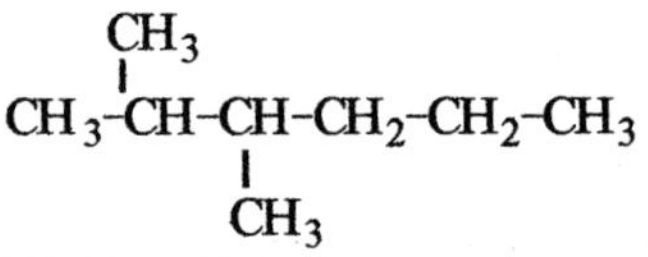

b. CH₃CH₂−C≡CCH₃

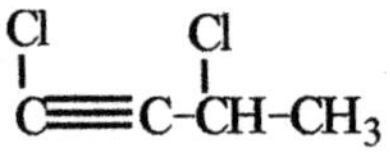

c. 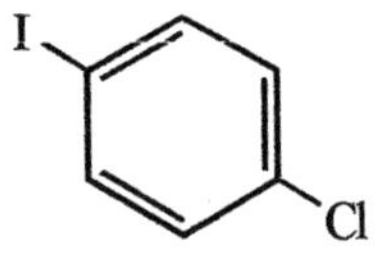 [structure: benzene ring with I (para) and Cl]

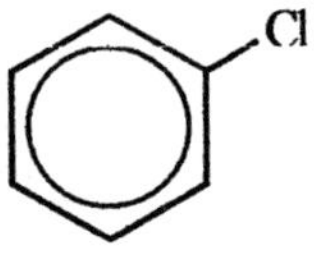

d. 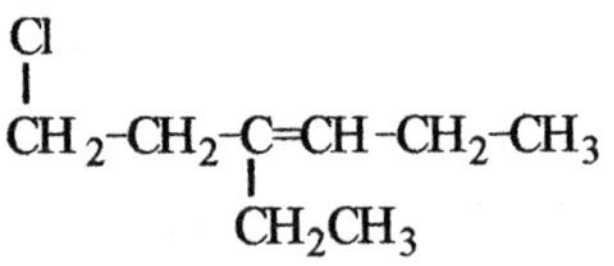
$$\text{CH}_2\text{-CH}_2\text{-}\underset{\underset{\text{CH}_2\text{CH}_3}{|}}{\overset{\overset{\text{Cl}}{|}}{\text{C}}}\text{=CH-CH}_2\text{-CH}_3$$

e.
$$\text{CH}_3\text{-}\underset{\underset{\text{CH}_3}{|}}{\overset{\overset{\text{CH}_3}{|}}{\text{CH}}}\text{-CH-CH}_2\text{-CH}_2\text{-CH}_3$$

f.
$$\overset{\overset{\text{Cl}}{|}}{\text{C}}\text{=C-}\overset{\overset{\text{Cl}}{|}}{\text{CH}}\text{-CH}_3$$

g. [structure: benzene ring with Cl]

4. CH₃CH₃ + Br₂ → CH₃CH₂Br

5. a. 1,2-dibromopropane; b. propane; c. cyclohexanol

6. a. *cis* b. *trans*

7. a. benzene + H₂SO₄ + SO₃; b. benzene + HNO₃ + H₂SO₄

8. a. carboxylic acid: pentanoic acid; b. ketone: 3-pentanone
   c. aldehyde: hexanal d. ether: ethylpropylether
   e. ester: propylbutanoate f. alcohol: 3-pentanol
   g. amine: isopropylamine h. amide: *N,N*-dimethylacetamide

9. a. propanol + butanoic acid; b. 2-methylbutyl-3-methylbutanoate; c. *N*-ethylpropanamide

10.

a. 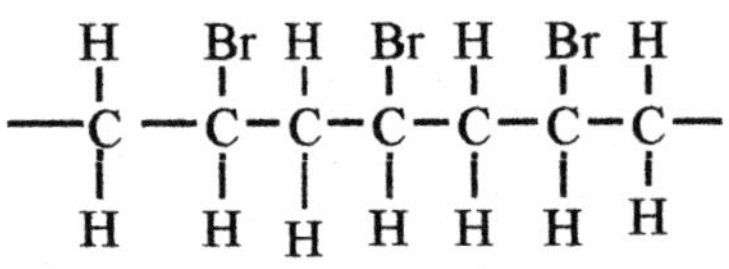 [polymer structure: −C(H)(H)−C(Br)(H)−C(H)(H)−C(Br)(H)−C(H)(H)−C(Br)(H)−C(H)(H)−]

Appendix C – Self Test Solutions

b. Dacron —

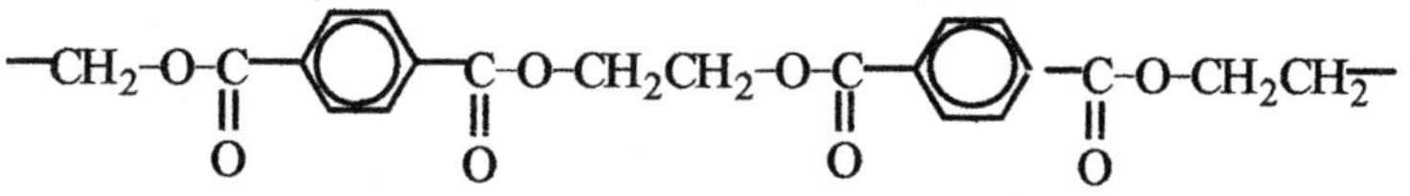

$$-CH_2-O-\underset{\underset{O}{\|}}{C}-\underset{}{\bigcirc}-\underset{\underset{O}{\|}}{C}-O-CH_2CH_2-O-\underset{\underset{O}{\|}}{C}-\underset{}{\bigcirc}-\underset{\underset{O}{\|}}{C}-O-CH_2CH_2-$$

Challenge Problem

Grams of C from 0.266 g CO_2 = 0.0725 g C = 6.05×10^{-3} mol C

Grams of H from 0.0544 g H_2O = 6.04×10^{-3} g H = 6.04×10^{-3} mol H

Grams of O = 0.175 g acid – 0.0725 g C – (6.04×10^{-3} g H) = 0.0965 g O

0.0965 g O = 6.03×10^{-3} mol O

Empirical formula = CHO

The molecular molar mass is determined from the moles of NaOH used in the titration and the fact that since maleic acid is a diprotic acid, two moles of NaOH will react for every mole of maleic acid.

$$\text{mol maleic acid} = 0.02780\,L \times \frac{0.150\,\text{mol NaOH}}{1\,L} \times \frac{1\,\text{mol maleic acid}}{2\,\text{mol NaOH}} = 2.09 \times 10^{-3}\,\text{mol maleic acid}$$

$$\text{molar mass} = \frac{0.242\,\text{g maleic acid}}{2.09 \times 10^{-3}\,\text{mol maleic acid}} = 116.1\,\text{g/mol}$$

The ratio of the molecular molar mass to the empirical molar mass is 4. Therefore, the molecular formula is $C_4H_4O_4$.

Chapter 24

Matching

    Metabolism—i
    Protein—k
    Dipeptide—l
    Residue—n
    Fibrous proteins—c
    Globular proteins—h
    Enzyme—b
    Cofactor—j
    Carbohydrates—a
    Lipids—m
    Nucleoside—e
    Chromosomes—f
    Replication—g
    Translation—d

Appendix C – Self Test Solutions

Fill–in–the Blank
1. 2–carbon acetyl groups; coenzyme A
2. ATP $\rightleftarrows$ ADP
3. the basic amino group, –NH$_2$, and the acidic group (–OOH)
4. side chain attached to the α–carbon
5. carboxylic acid group
6. N–terminal amino acid; C–terminal residue
7. hydrogen bonds; N–H group; C=O
8. turnover number
9. an aldehyde carbonyl; ketone carbonyl
10. Cellulose
11. solubility; chemical properties
12. deoxyribonucleic acid; ribonucleic acid
13. adenine, guanine, cytosine, thymine
14. uracil; thymine
15. ribonucleotides; uracil

Problems
1. a. tryptophan; b. glutamic acid; c. histidine; d. arginine

2.

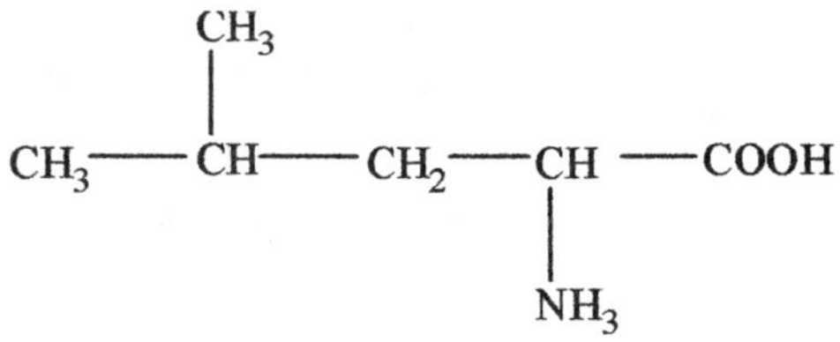

$$CH_3-\underset{\underset{CH_3}{|}}{CH}-CH_2-\underset{\underset{NH_3}{|}}{CH}-COOH \qquad CH_3-CH_2-\underset{\underset{CH_3}{|}}{CH}-\underset{\underset{NH_3}{|}}{CH}-COOH$$

Leucine 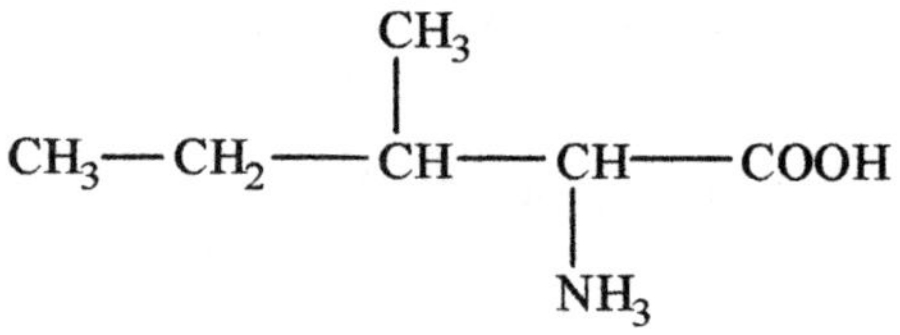   Isoleucine

3. Asp–Pro–Gln–His

4. a. outside, hydrophilic; b. inside, hydrophobic; c. outside, hydrophilic d. outside, hydrophilic

5. T–C–A–T–T–A

6. U–C–A–U–U–A

7.

[Structure: A dinucleotide fragment of DNA showing two deoxyribose-phosphate units linked together. The top nucleotide contains Adenine attached to deoxyribose with a phosphate group (with O⁻ charges). The bottom nucleotide contains Guanine attached to deoxyribose with a phosphate group, terminating in an OH group. Labels: Adenine, Guanine, Deoxyribose.]

Challenge Problem

We know that an increase in temperature is directly proportional to an increase in kinetic energy. This increase in kinetic energy increases the total number of collisions that occur, thereby increasing the total number of effective collisions that occur between the enzyme and the substrate. This increase in effective collisions leads to an increase in the rate. However, eventually the increase in kinetic energy will be so great that the intermolecular forces that hold the tertiary structure of the enzyme together are disrupted. The active site on the enzyme is no longer available, and the enzyme activity decreases.

 **Inquiry Based Problems**

Chapter 1

Density is defined as g/mL or g/cm^3. Therefore, you need to determine the mass and the volume of the cylinder.

The volume of a cylinder is determined using the formula $V = \pi r^2 h$, where r = the radius of the cylinder and h = the height of the cylinder.

Mass can be determined with the balance supplied. The volume of the metal cylinder can be determined by determining the amount of water displaced in the graduate cylinder or by determining the diameter (which is twice the radius) and the height of the metal cylinder.

Chemists always want to collect precise data. Precision is defined as how well a number of independent measurements agree with one another. Given that two points define a line, it is always a good idea to collect at least three data points.

Depending upon the technique of the person taking the data, the more accurate approach is the use of the caliper to determine the volume of the cylinder given that the measurement is carried out to more significant figures.

Procedure 1:

1. Using the balance, determine the mass of the cylinder.

2. Place 50 mL of water in the graduated cylinder. Place the metal cylinder in the graduated cylinder and record the new volume. Determine the volume of water displacement.

3. Calculate the density from the mass of the cylinder and the volume of water displacement.

4. Repeat two more times.

5. Determine the average density of the metal cylinder and compare the answer to the values in the table provided. Determine the identity of the metal cylinder.

Procedure 2:

1. Using the balance, determine the mass of the cylinder.

2. Using the caliper, measure the diameter and the height of the cylinder to 0.0001 cm (1 μm).

3. Using the mathematical formula, determine the volume of the metal cylinder.

4. Calculate the density of the metal cylinder.

5. Repeat two more times.

Appendix D – Inquiry Based Problems Solutions

6. Determine the average density of the metal cylinder and compare the answer to the values in the table provided. Determine the identity of the metal cylinder.

Using Procedure 1 to determine the density:

$$D = \frac{28.73 \text{ g}}{3.4 \text{ mL}} = 8.4 \frac{\text{g}}{\text{mL}}$$

$$D = \frac{28.79 \text{ g}}{3.5 \text{ mL}} = 8.2 \frac{\text{g}}{\text{mL}}$$

$$D = \frac{28.85 \text{ g}}{3.5 \text{ mL}} = 8.2 \frac{\text{g}}{\text{mL}}$$

$$\text{Avg. D} = \frac{8.4 \frac{\text{g}}{\text{mL}} + 8.2 \frac{\text{g}}{\text{mL}} + 8.2 \frac{\text{g}}{\text{mL}}}{3} = 8.3 \frac{\text{g}}{\text{mL}}$$

Based on the average density, the metal cylinder could be either Yellow Brass (high brass) with a reported density of 8.47 g/cm³ (g/mL) or Beryllium copper 25 with a reported density of 8.23 g/cm³ (g/mL). (At this point, you may be able to identify the cylinder based on the actual color. Does it look like bronze or copper?)

Using Procedure 2 to determine the density:

$$r = \frac{0.6293 \text{ cm}}{2} = 0.3147 \text{ cm}; \qquad V = \pi r^2 h = \pi (0.3147 \text{ cm})^2 (11.1742 \text{ cm}) = 3.4766 \text{ cm}^3$$

$$D = \frac{28.73 \text{ g}}{3.4766 \text{ cm}^3} = 8.2638 \frac{\text{g}}{\text{cm}^3}$$

$$r = \frac{0.6302 \text{ cm}}{2} = 0.3151 \text{ cm}; \qquad V = \pi r^2 h = \pi (0.3151 \text{ cm})^2 (11.2437 \text{ cm}) = 3.5072 \text{ cm}^3$$

$$D = \frac{28.79 \text{ g}}{3.5072 \text{ cm}^3} = 8.2088 \frac{\text{g}}{\text{cm}^3}$$

$$r = \frac{0.6298 \text{ cm}}{2} = 0.3149 \text{ cm}; \qquad V = \pi r^2 h = \pi (0.3149 \text{ cm})^2 (11.1765 \text{ cm}) = 3.4818 \text{ cm}^3$$

$$D = \frac{28.85 \text{ g}}{3.4818 \text{ cm}^3} = 8.2859 \frac{\text{g}}{\text{cm}^3}$$

Appendix D – Inquiry Based Problems Solutions

$$\text{Avg. D} = \frac{8.2638 \frac{g}{cm^3} + 8.2088 \frac{g}{cm^3} + 8.2859 \frac{g}{cm^3}}{3} = 8.2528 \frac{g}{cm^3}$$

Based on the average density of the metal cylinder and the reported density values, it is highly likely that the metal cylinder is Beryllium copper 25 with a reported density value of 8.23 g/cm^3.

Chapter 3

The number of moles of a compound is determined either from knowing the mass of the compound and the formula mass

$$\text{mol compound} = \text{g compound} \times \frac{\text{mol compound}}{\text{g compound}}$$

or from knowing the molarity of a solution and the volume of the solution.

$$\text{mol compound} = \text{L solution} \times \frac{\text{mol compound}}{\text{L solution}}$$

In this particular case, AgCl is a solid. Knowing the mass of AgCl, we can determine the number of moles of AgCl.

Once we know the mol of AgCl, we can then determine the number of moles of Cl⁻ present based on the chemical equation.

$$\text{mol Cl}^- = \text{mol AgCl} \times \frac{1 \text{ mol Cl}^-}{1 \text{ mol AgCl}}$$

Once we know the mol Cl⁻, we can calculate the grams of Cl⁻ present in the unknown salt.

$$\text{g Cl}^- = \text{mol Cl}^- \times \frac{35.45 \text{ g Cl}^-}{1 \text{ mol Cl}^-}$$

Working backwards, we know that we can determine the mass of Cl⁻ if we know the moles of Cl⁻. We can find the moles of Cl⁻ if we know the moles of AgCl. We can find the moles of AgCl if we know the mass of AgCl. To find the mass of AgCl, we need to **make** AgCl. We know, from the chemical equation provided, that if we react Cl⁻ (the unknown salt) with AgNO$_3$, we produce AgCl. We now have everything we need to write a simple laboratory procedure.

A sample of the unknown salt is reacted with a solution of AgNO$_3$. The precipitate produced is filtered, dried, and weighed. From the weight of the precipitate, we can determine the moles of AgCl, moles of Cl⁻ and the mass of Cl⁻ in the salt. (The actual procedure for this experiment requires a few other steps, such as adding HNO$_3$ to produce an acidic solution and gently heating the reaction mixture to assure complete dissolution. However, your laboratory instructor just wanted to see if you could apply stoichiometric calculations in the laboratory. This simple procedure proves that you can.)

Appendix D – Inquiry Based Problems Solutions

Chapter 4

The limiting reactant is the reactant present in the least molar amount, and the excess reactant is the reactant that is present in more than the amount that is needed according to the stoichiometry. Therefore, the excess reactant is not completely used up and should be present after the reaction is complete.

Unbalanced molecular equation:

$Na_2SO_4 \,(aq) + BaCl_2 \,(aq) \rightarrow NaCl \,(aq) + BaSO_4 \,(s)$

Balanced molecular equation:

$Na_2SO_4 \,(aq) + BaCl_2 \,(aq) \rightarrow 2\,NaCl \,(aq) + BaSO_4 \,(s)$

Ionic equation:

$2\,Na^+ \,(aq) + SO_4^{2-} \,(aq) + Ba^{2+} \,(aq) + 2\,Cl^- \,(aq) \rightarrow 2\,Na^+ \,(aq) + 2\,Cl^- \,(aq) + BaSO_4 \,(s)$

Net ionic equation:

$Ba^{2+} \,(aq) + SO_4^{2-} \,(aq) \rightarrow BaSO_4 \,(s)$

From the net ionic equation, we know that the excess reactant (either the Ba^{2+} or SO_4^{2-}) will be in the reaction solution.

We know, based on the chemical equation and solubility guidelines, that Na_2SO_4 and $BaCl_2$ will dissolve when placed in water. We also know that the reaction of Na_2SO_4 and $BaCl_2$ will produce a solid and that the excess reactant will be in the reaction solution. If we separate the solid from the reaction solution, we can determine the mass of the solid. We can test the reaction solution to see which reactant is left over.

We can take some of the reaction solution and place it into two different beakers. We then would add SO_4^{2-} to one beaker and add the Ba^{2+} to another beaker. A precipitate should form in one of the beakers. If the precipitate forms when the SO_4^{2-} is added, then Ba^{2+} is present in the reaction solution and is the excess reactant. If the precipitate forms when Ba^{2+} is added, then SO_4^{2-} is present in the reaction solution and is the excess reactant.

Experimental Procedure:

1. Weigh out 1.25 g of the solid mixture and place in the 400 mL beaker. Dissolve the solid in 250 mL of water and add 1 mL of conc HCl (aq). Stir to dissolve.

2. Cover the beaker with a watch glass (to avoid evaporation) and heat over a low flame. Make sure the mixture stays between 80° C and 90° C.

3. Once the precipitate has settled, decant (carefully remove the reaction solution without disturbing the precipitate) 25 mL of the reaction solution (supernatant) into the 50 mL beaker and label Ba^{2+} test. Repeat, using a different 50 mL beaker, and label SO_4^{2-} test.

4. Vacuum filter the $BaSO_4$ precipitate and let dry.

5. Add a few drops of the SO_4^{2-} solution to the beaker labeled Ba^{2+} test. Record your observations.

6. Add a few drops of the Ba^{2+} solution to the beaker labeled SO_4^{2-} test. Record your observations.

Laboratory data:

Mass of BaSO$_4$ precipitate = 0.478 g
Addition of SO$_4^{2-}$ to reaction mixture – no precipitate.
Addition of Ba^{2+} to reaction mixture – precipitate forms.

$$\text{mol BaSO}_4 = 0.478 \text{ g BaSO}_4 \times \frac{1 \text{ mol BaSO}_4}{233.33 \text{ g BaSO}_4} = 2.05 \times 10^{-3} \text{ mol BaSO}_4$$

Excess reactant = SO$_4^{2-}$ given the precipitate that formed upon the addition of Ba^{2+}. Therefore the limiting reactant is Ba^{2+}.

Ba^{2+} is present in the form of BaCl$_2 \cdot$2 H$_2$O. The amount of BaCl$_2 \cdot$2 H$_2$O present in the solid mixture is:

$$\text{g BaCl}_2 \bullet 2\text{H}_2\text{O} = 2.05 \times 10^{-3} \text{ mol BaSO}_4 \times \frac{1 \text{ mol BaCl}_2 \bullet 2\text{H}_2\text{O}}{1 \text{ mol BaSO}_4} \times \frac{244.23 \text{ g BaCl}_2 \bullet 2\text{H}_2\text{O}}{1 \text{ mol BaCl}_2 \bullet 2\text{H}_2\text{O}} = 0.501 \text{ g BaCl}_2 \bullet 2\text{H}_2\text{O}$$

mass of solid mixture = mass of limiting reactant + mass of excess reactant

1.25 g = 0.501 g + mass (Na$_2$SO$_4$)

Mass Na$_2$SO$_4$ = 1.25 g – 0.501 g = 0.75 g.

$$\% \text{ Na}_2\text{SO}_4 = \frac{0.75 \text{ g Na}_2\text{SO}_4}{1.25 \text{ g solid mixture}} \times 100 = 60\%$$

$$\% \text{ BaCl}_2 \bullet 2\text{H}_2\text{O} = \frac{0.501 \text{ g BaCl}_2 \bullet 2\text{H}_2\text{O}}{1.25 \text{ g solid mixture}} \times 100 = 40\%$$

Chapter 6

Step 1:

A good resource for aluminum metal in the General Chemistry laboratory, as well as for other uses, are the aluminum soda cans which proliferate on college and university campuses.

Knowing that the aluminum metal in this reaction goes from its elemental state to Al^{3+} in the form of KAl(OH)$_4$, this reaction must be a redox reactions.

The aqueous solution of KOH contains KOH and water. If 20% of the solution is KOH, then 80% of the solution is water. Therefore, water is also present and may be part of the chemical reaction.

Unbalanced molecular equation:
Al (s) + KOH (aq) + H$_2$O → KAl(OH)$_4$ (aq) + H$_2$ (g)

Ionic equation:
Al (s) + K$^+$ (aq) + OH$^-$ (aq) + H$_2$O → K$^+$ (aq) + Al^{3+} (aq) + 4 OH$^-$ (aq) + H$_2$ (g)

Spectator ions: K$^+$, OH$^-$

Net ionic equation:

Appendix D – Inquiry Based Problems Solutions

$Al\ (s) + H_2O \rightarrow Al^{3+}\ (aq)\ H_2\ (g)$

Half-reactions:
$Al\ (s) \rightarrow Al^{3+}\ (aq)$

$H_2O \rightarrow H_2\ (g)$

Each half-reaction is balanced for atoms other than H and O.
$Al\ (s) \rightarrow Al^{3+}\ (aq)$

$H_2O \rightarrow H_2\ (g)$

Add H_2O for any oxygens that are needed and H^+ for any hydrogens that are needed.
$Al\ (s) \rightarrow Al^{3+}\ (aq)$

$2H^+\ (aq) + H_2O \rightarrow H_2\ (g) + H_2O$

Balance each reaction for charge.
$Al\ (s) \rightarrow Al^{3+}\ (aq) + 3\ e^-$

$2\ e^- + 2\ H^+\ (aq) + H_2O \rightarrow H_2\ (g) + H_2O$

Make the electron count the same in both reactions. Multiply the first reaction by 2 and the second reaction by 3.
$2\ Al\ (s) \rightarrow 2\ Al^{3+}\ (aq) + 6\ e^-\ (aq)$

$6\ e^- + 6\ H^+\ (aq) + 3\ H_2O \rightarrow 3\ H_2\ (g) + 3\ H_2O$

Add the two half-reactions together, canceling anything that appears on both sides of the equation.

$2\ Al\ (s) + 6\ H^+\ (aq) \rightarrow 2\ Al^{3+}\ (aq) + 3\ H_2\ (g)$

The solution is basic, given the presence of KOH. Therefore, we need to make the chemical equation basic by adding OH^- for any H^+ present.
$2\ Al\ (s) + 6\ H^+\ (aq) + 6\ OH^-\ (aq) \rightarrow 2\ Al^{3+}\ (aq) + 3\ H_2\ (g)$

Combine H^+ and OH^- to form water
$2\ Al\ (s) + 6\ H_2O \rightarrow 2\ Al^{3+}\ (aq) + 3\ H_2\ (g) + 6\ OH^-\ (aq)$

We now go back to the original molecular equation and balance the spectator ions:
$2\ Al\ (s) + 2\ KOH\ (aq) + 6\ H_2O \rightarrow 2\ KAl(OH)_4\ (aq) + 3\ H_2\ (g)$

Step 2:

The reactants in this reaction are H_2SO_4 and $KAl(OH)_4$. We know that the insoluble salt formed is $Al(OH)_3$. From the Law of Conservation of Mass, we know that the potassium and sulfate ion are left in solution. Therefore, the soluble salt must be K_2SO_4. (We also know from the solubility guidelines that K_2SO_4 is soluble.)

Unbalanced molecular equation: [HINT: It may not be readily apparent that water is one of the products in this reaction]
$H_2SO_4 + KAl(OH)_4\ (aq) \rightarrow K_2SO_4\ (aq) + Al(OH)_3\ (s) + H_2O$

Balancing for atoms other than H and O gives
$H_2SO_4 + 2\ KAl(OH)_4 \rightarrow K_2SO_4\ (aq) + 2\ Al(OH)_3\ (s) + H_2O$

Balance for hydrogen:
$H_2SO_4 + 2\ KAl(OH)_4 \rightarrow K_2SO_4\ (aq) + 2\ Al(OH)_3\ (s) + 2\ H_2O$

Step 3:
Reactants present in step 3 include $Al(OH)_3$, K_2SO_4, and H_2SO_4.

The conversion of $Al(OH)_3$ to $Al_2(SO_4)_3$ does not involve the presence of K_2SO_4. Therefore, the unbalanced chemical reaction is: [HINT: water is also a product.]

$Al(OH)_3\ (s) + H_2SO_4 \rightarrow Al_2(SO_4)_3\ (aq) + H_2O$

$2\ Al(OH)_3\ (s) + 3\ H_2SO_4 \rightarrow Al_2(SO_4)_3\ (aq) + 6\ H_2O$

In the final step of the procedure, the K_2SO_4 must be one of the reactants given that the formula for the alum, $KAl(SO_4)_2 \cdot 12\ H_2O$ contains potassium. The water produced in the reaction above also is included as a reactant. The unbalanced chemical equation is:

$K_2SO_4\ (aq) + Al_2(SO_4)_3\ (aq) + H_2O \rightarrow KAl(SO_4)_2 \cdot 12\ H_2O\ (s)$

This unbalanced chemical equation can be balanced by inspection.

$K_2SO_4\ (aq) + Al_2(SO_4)_3\ (aq) + 12\ H_2O \rightarrow 2\ KAl(SO_4)_2 \cdot 12\ H_2O\ (s)$

Chapter 8

We know that heat lost must equal heat gained. If we want to determine the heat lost by one quantity of water and the heat gained by the other quantity of water and the calorimeter, then one quantity of water must be at a higher temperature than the other quantity of water.

By collecting the change in temperature over time, we can determine the heat lost and the heat gained once the change in temperature becomes constant:

$q = \Delta T \times C$

The mathematical equation which allows you to calculate the heat capacity of the calorimeter is:

$q_{\text{lost by water}} = q_{\text{gained by water}} + q_{\text{gained by calorimeter}}$

If T_1 = the temperature of the warmer water, T_2 = the temperature of the cooler water and T_f is the final temperature, then

$\Delta T_{\text{warmer water}} = T_1 - T_f;$ $\quad\quad \Delta T_{\text{cooler water}} = T_f - T_2;$ $\quad\quad \Delta T_{\text{calorimeter}} = T_f - T_2$

Substituting the ΔT expressions into the heat loss equation gives:

$(T_1 - T_f)C_{\text{water}} = (T_f - T_2)C_{\text{water}} + (T_f - T_2)C_{\text{calorimeter}}$

Appendix D – Inquiry Based Problems Solutions

$$(T_1 - T_f)C_{water} - (T_f - T_2)C_{water} = (T_f - T_2)C_{calorimeter}$$

Knowing, the heat capacity of water (4.18J/K-g), T_1, T_2, and T_f, we can determine the heat capacity of the calorimeter.

Chapter 9

To determine the molar mass of any compound, you need to know both the mass of the compound and the number of moles contained in that mass. We can determine the number of moles of the compound from the ideal gas law:

$PV = nRT$.

However, we need to know the volume of the vapor if we are going to use the ideal gas law to determine the number of moles. Therefore, we need to vaporize the volatile liquid in an Erlenmeyer flask of known volume. Knowing the temperature needed for vaporization, as well as the barometric pressure in the room, we can calculate the number of moles of gas.

The mass of the compound is determined by weighing the flask before the volatile liquid is added and weighing the cooled flask after vaporization. The difference in mass is the mass of the vapor. Knowing mass and number of moles, we can calculate the molar mass.

To determine the density of the gas, we would need to determine the volume at STP using the equation:

$$\frac{P_{measured} V_{measured}}{T_{measured}} = \frac{1\,\text{atm} V_{STP}}{273\,\text{K}}$$

Chapter 11

Given that the experimental design involves the use of an ice-water bath, it is obvious we will determine the freezing point depression of the unknown. The equation for freezing point depression is:

$$\Delta T_f = K_f \cdot m \quad \text{where } m = \frac{\text{mol solute}}{\text{kg solvent}}$$

Given that $\text{molar mass} = \dfrac{\text{g solute}}{\text{mol solute}}$ or $\text{mol solute} = \dfrac{\text{g solute}}{\text{molar mass}}$; we can re-write the equation for freezing point depression as

$$\Delta T_f = K_f \cdot \left(\frac{\text{g solute} / \text{molar mass}}{\text{kg solvent}} \right)$$

$$\frac{\Delta T_f \cdot \text{kg solvent}}{K_f} = \frac{\text{g solute}}{\text{molar mass}}$$

$$\text{molar mass} = \frac{\text{g solute} \cdot K_f}{\Delta T_f \cdot \text{kg solvent}}$$

From this re-arranged equation, we now know that we need to have the mass of the unknown liquid and cyclohexane as well as ΔT_f. (Remember ΔT_f = cyclohexane freezing point – solution freezing point.

To carry out this experiment, we need to measure the freezing point of cyclohexane and the freezing point of the solution. From the phase change diagrams in chapter 10. we know that freezing we reach the freezing point when we have an equilibrium between liquid and solid (liquid ⇌ solid). During this equilibrium, the temperature remains constant. Once all of the liquid has converted to solid, the temperature begins to drop off. Therefore, we should take temperature measurements at timed intervals (once a minute, for example) through an initial drop in temperature to a constant reading of temperature to another drop in temperature. A plot of the cooling curve should look like:

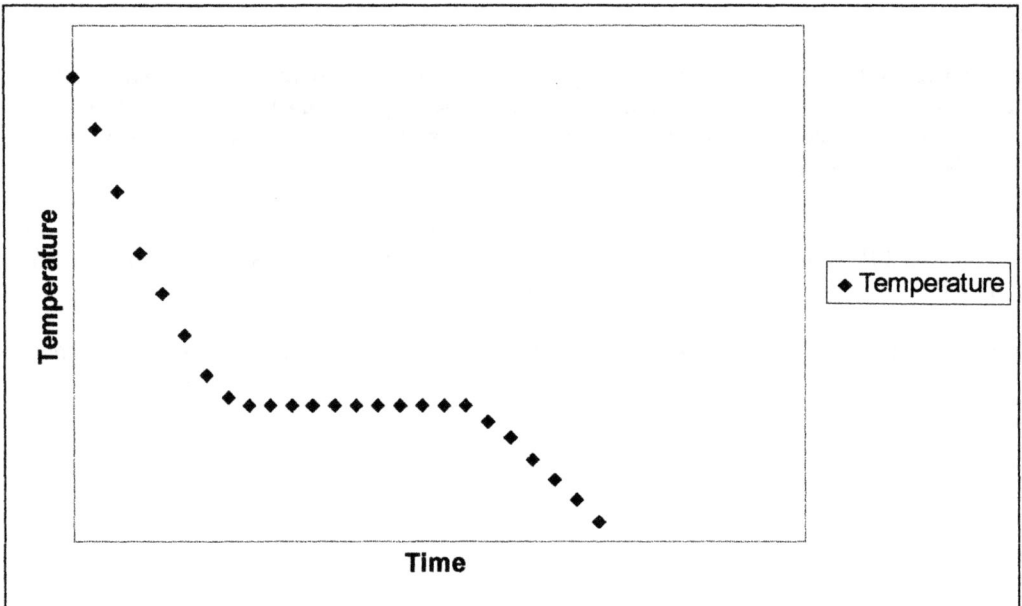

A brief outline of the experimental procedure:
1. Weigh the clean, dry empty test tube in a beaker.
2. Add cyclohexane and reweigh the test tube and the beaker. (The amount of cyclohexane will depend upon the size of the test tube. You want to add enough so that you can easily submerge the thermometer without touching the bottom of the test tube.)
3. Clamp the test tube to a ring stand and place in the ice-water bath.
4. Place the stopper with the thermometer and stirrer on the test tube.
5. With constant stirring, record the temperature at 60 s intervals.
7. Plot temperature vs. time to determine the freezing point of cyclohexane.
8. Repeat the experiment for the solution keeping in mind that the amount of unknown should be much less than the cyclohexane (amount of unknown should be 10% of cyclohexane. Remember that the solute is always present in the smaller amount.)

Chapter 12

The reaction mechanism we are considering is:

$IO_3^- (aq) + 3\ SO_3^{2-} (aq) \rightarrow I^- (aq) + 3\ SO_4^{2-} (aq)$

$5\ I^- (aq) + 6\ H^+ (aq) + IO_3^- (aq) \rightarrow 3\ H_2O + 3\ I_2 (s)$

Appendix D – Inquiry Based Problems Solutions

$$3\,I_2(s) + 3\,SO_3^{2-}(aq) + 3\,H_2O \rightarrow 6\,I^-(aq) + 3\,SO_4^{2-}(aq) + 6\,H^+(aq)$$

$$\overline{2\,IO_3^-(aq) + 6\,SO_3^{2-}(aq) \rightarrow 2\,I^-(aq) + 6\,SO_4^{2-}(aq)}$$

The overall rate law for this reaction is:

$$\text{Rate} = k[IO_3^-]^m[SO_3^{2-}]^n$$

To determine the impact of each individual reactant on the rate, we need to hold one reactant constant at a fairly high concentration, while varying the concentration of the other reactant.

Notice that I_2 is produced in the second step and consumed in the third step. However, the amount of I_2 consumed in the third step is dependent upon the amount of SO_3^{2-} available to react with the I_2. Any unreacted I_2 will react with the starch to form the deep blue color from the I_2•complex. Therefore, the availability of unreacted I_2 in the third step is dependent upon the consumption of SO_3^{2-}. Based on the overall chemical equation, once the SO_3^{2-} is consumed the reaction is completed. Therefore, we can monitor the reaction by timing the amount of time it takes for the deep blue color of the I_2•complex to appear. The rate will then be the formation of I_2•complex/s. The unit for the rate is s^{-1}.

Based on the discussion in the chapter on determining experimental rate laws, we can use the following procedure.

1. Create a table recording the volume of each solution used to carry out the reaction as well as a space for recording the time it takes for the deep blue color to appear. This table should provide enough space for entry of several experiments. An example in which the $[H_2SO_3]$ is held constant is shown below:

Test Reaction	0.1 M HIO₃	Starch	0.05 M H₂SO₃	Volume of Water	Total Volume	Time
1	40 mL	5 mL	25 mL	30 mL	100 mL	
2	20 mL	5 mL	25 mL	50 mL	100 mL	
3	10 mL	5 mL	25 mL	60 mL	100 mL	
4	5 mL	5 mL	25 mL	65 mL	100 mL	
5	2.5 mL	5 mL	25 mL	67.5 mL	100 mL	

NOTE: Keeping the $[H_2SO_3]$ constant requires that we add enough water so that the total volume for each experiment is the same.

2. The above experiment should be repeated, now holding the HIO_3 concentration constant and changing the concentration of H_2SO_3. Be sure to maintain a constant temperature for each experiment and to record this temperature.

3. Once the rate is recorded for each experiment, you can determine the value of the exponents m and n, using the method described in section 12.3. It may be that the data you have collected does not lend itself to a simple interpretation of the impact of changing concentrations on the rate, *i.e.* if you double the concentration, the rate doubles. The exponents can also be determined graphically. Consider the following:

When you hold the H_2SO_3 constant, the rate law for the reaction becomes:

rate = $k[HIO_3]^m$

If we take the logarithm of both sides, we have:

log rate = log $k[HIO_3]^m$

Based on the rules of logarithms, this equation becomes:

log rate = log k + log $[HIO_3]^m$ = log k + m· log $[HIO_3]$

This equation takes the form of an equation for a straight line where y = log rate, the intercept = log k, the slope = m, and x = log$[HIO_3]$.

Once you have determined the value of both exponents, m and n, you may then go back and substitute these values along with one set of experimental conditions into the overall rate law to determine the value of k for the overall rate law.

5. Once you have determined the rate law, you need to repeat the reaction at a higher temperature. To determine the activation energy, use the equation:

$$\ln\left(\frac{k_2}{k_1}\right) = \left(\frac{-E_a}{R}\right)\left(\frac{1}{T_2} - \frac{1}{T_1}\right)$$

Chapter 13

1. From the density of the 6 M HCl solution, we can determine the mass of 5.00 mL of 6 M HCl.

$$5.00 \text{ mL} \times \frac{1.11 \text{ g}}{\text{mL}} = 5.55 \text{ g solution}$$

2. The mass of water in this solution can be calculated based upon the number of moles of NaOH needed for the titration.

$$0.0289 \text{ L NaOH} \times \frac{1 \text{ mol NaOH}}{1 \text{ L}} = 0.0289 \text{ mol NaOH}$$

The mole:mole ratio for the titration is 1:1; therefore, 0.0298 mol HCl is present in 5.00 mL of 6 M HCl.

Knowing the number of moles of HCl, we can calculate the number of grams of HCl in the 6 M solution.

$$0.0289 \text{ mol HCl} \times \frac{36.5 \text{ g HCl}}{1 \text{ mol HCl}} = 1.05 \text{ g HCl}$$

Given that grams of solution = grams of HCl + grams of water, we can calculate the grams of water.

5.55 g soln. = 1.05 g HCl + g water;   g water = 5.55 g soln. − 1.05 g HCl = 4.50 g water

3. The mass of ethyl acetate is calculated in the same way we calculated the mass of the 6 M HCl solution, using the density of ethyl acetate.

$$2.00 \text{ mL} \times \frac{0.893 \text{ g}}{1 \text{ mL}} = 1.79 \text{ g ethyl acetate}$$

Appendix D – Inquiry Based Problems Solutions

Water is present in both the 6 M HCl solution used to catalyze the reaction and the water added to the solution. Therefore, the total mass of water is calculated from the mass of 3.0 mL of water (3.0 g water assuming the density of the water is 1.0 g/mL) and the mass of water in the 6 M HCl.

total mass H_2O = 4.50 g (from 6 M HCl) + 3.00 g (from 3 mL H_2O) = 7.50 g H_2O.

Knowing these masses allows us to calculate the initial number of moles present in the reaction mixture.

$$7.50 \text{ g } H_2O \times \frac{1 \text{ mol } H_2O}{18.0 \text{ g } H_2O} = 0.417 \text{ mol } H_2O$$

$$1.79 \text{ g ethyl acetate} \times \frac{1 \text{ mole ethyl acetate}}{88.0 \text{ g ethyl acetate}} = 0.0203 \text{ mol ethyl acetate}$$

4. A total of 38.70 mL of 1.0 M NaOH is needed for the titration of solution 2. However, this solution contains 5 mL of 6 M HCl, and we know that 28.90 mL of 1.0 M NaOH is needed to titrate the 6 M HCl. Therefore, the amount of NaOH needed to titrate the acetic acid produced is:

38.70 mL – 28.90 mL = 9.80 mL

We can calculate the moles of acetic acid produced knowing the volume of 1.0 M NaOH needed to titrate the acetic acid.

$$0.0098 \text{ L NaOH} \times \frac{1 \text{ mol NaOH}}{1 \text{ L NaOH}} \times \frac{1 \text{ mol acetic acid}}{1 \text{ mol NaOH}} = 0.0098 \text{ mol acetic acid}$$

This represents the number of moles of acetic acid at equilibrium.

5. Knowing the number of moles of acetic acid at equilibrium, we can determine the number of moles of ethyl alcohol at equilibrium given that the products exist in a 1:1 mole ratio. Therefore, the number of moles of ethyl alcohol at equilibrium equals 0.0169.

6. We now can calculate the equilibrium constant for this reaction.

	Ethyl Acetate	+	Water	⇌	Ethyl Alcohol	+	Acetic Acid
Initial moles	0.0203		0.417		0		0
Change (moles)	– 0.00980		– 0.00980		+ 0.00980		+ 0.00980
Equilibrium (moles)	0.0.105		0.407		0.00980		0.00980

$$K_c = \frac{(0.00980)(0.00980)}{(0.0105)(0.407)} = 0.0225$$

Chapter 14 Solution

1. Anions, which are the conjugate base of a weak acid, undergo reaction with water (hydrolysis) to form the acid and the OH^-.

$A^- (aq) + H_2O \rightleftarrows HA (aq) + OH^- (aq)$

Appendix D – Inquiry Based Problems Solutions

$$K_b = \frac{[HA][OH^-]}{[A^-]}$$

2. Given that the reaction produces OH^-, the pH of the solution should be > 7.0.
3. The pH can be determined by a variety of methods. You can use pH paper, which gives a very broad range of pH values, acid-base indicators, which give a more narrow range of pH values, or a pH meter, which gives a precise reading of pH. If you use acid-base indicators, you may need to test several solutions of the salt before getting a positive result. The best method to use, given that you need to identify the identity of the salt, is the pH meter, which gives a very precise reading of the pH of the solution.

4. Once you know the pH of the solution, you can determine the $[OH^-]_e$. $[OH^-]_e = [HA]_e$. $[A^-]_e = [A^-]_i - [OH^-]_e$. Armed with this information, you can calculate the K_b for the anion. Once you know the K_b of the anion, you can use a table of K_b values to determine the identity of the anion. If no such table is available, you can calculate the K_a of the conjugate acid. The anion of that conjugate acid is the base you investigated in the hydrolysis reaction.